Current Studies in Archaeoastronomy

Current Studies in Archaeoastronomy

Conversations Across Time and Space

Selected Papers from the Fifth Oxford International Conference at Santa Fe, 1996

Edited by

John W. Fountain

and

Rolf M. Sinclair

Carolina Academic Press

Durham, North Carolina

ISBN 0-89089-771-9
LCCN 2004118386

Carolina Academic Press
700 Kent Street
Durham, North Carolina 27701
Telephone (919) 489-7486
Fax (919) 493-5668
E-mail: cap@cap-press.com
www.cap-press.com

Printed in the United States of America

Contents

Introduction

John W. Fountain
Tucson, Arizona

Rolf M. Sinclair
Chevy Chase, Maryland

The "Oxford" Meetings

It is a pleasure to introduce this volume of papers presented at the 1996 meeting on "Cultural Aspects of Astronomy", held at St. John's College in Santa Fe, New Mexico, USA, August 3–9, 1996. This was the fifth in a series of "Oxford Conferences on Archaeoastronomy" and the first held in the US Southwest, and got the working name "Oxford V".

The Oxford Conferences are a series of triennial meetings that have focused on the role that astronomical phenomena have played in human societies, ranging from the applied (such as the basis for calendrics and orientations) to the ceremonial (the significance given the "ritual landscape" of the sky). These conferences serve as a meeting place for those working in anthropology, archaeology, history and prehistory, archaeoastronomy, and other studies of human cultures, who share a common interest in the importance of astronomical phenomena (or "skywatching") to traditional societies of the past and present. The emphasis of these meetings is on the reaction of human societies to these astronomical phenomena as part of the worldview of each society. The astronomical phenomena are universal, while the reactions to them are part of (and unique to) each culture

The first of these meetings (called in retrospect "Oxford I") was held at Oxford University (England) in 1981, and was the first major international conference on archaeoastronomy. Further ones have converged on a three-year interval—Oxford II in Mexico (1986), Oxford III in Scotland (1990), Oxford IV in Bulgaria (1993), and then Oxford V in Santa Fe. We note that Oxford VI (1999, Canary Islands) and Oxford VII (2004, Flagstaff) have in turn been successes, and we hope that the time and place of Oxford VIII will be announced soon.

This series of "Oxford" meetings and the publications that have come out of them show clearly the powerful effect that simply getting like-minded people together for a week to compare notes and ideas can have on a field of study. Prior to Oxford I there had been only regional and specialized meetings on various aspects of past astronomical practices. Oxford I brought together this past work as well as newcomers to the field, and helped define a more general study of archaeoastronomy. The later meetings built on this beginning and explored the variety of research in this area.

Through the Oxford meetings, many from a variety of disciplines have come to realize the importance of astronomical phenomena in the development and even the daily life of peoples throughout the world. This has led to a broader knowledge of past cultures, and some understanding of the motivations that led to the traditions, constructions and artifacts by which we now know them.

The Nature of Oxford V

Oxford V was the first meeting of this series to be announced on the World Wide Web at <http://www.lpl.arizona.edu/oca/oxv.htm>. Also, it was the first to be publicized widely in professional circles in archaeology and anthropology. The result was a larger and more varied attendance, and a broader range of topics discussed, than had been the case at the earlier meetings. Astronomers made up a third of the attendees, and archeologists and anthropologists another third. The remaining third were distributed among a number of disciplines (Gumerman and Warburton, this volume, Table I.). There was unfortunately only one person from religious studies—this is an area that future meetings could contact profitably. Another result was that Oxford V avoided becoming a regional meeting—papers dealing with the US Southwest were in the minority, and most presentations dealt with other parts of the World. The Oxford meetings are now established internationally through a wide range of disciplines, and this continued at Oxford VI and Oxford VII.

As usual, the attendees define a meeting. Get enough good people there and it works. By publicizing the meeting widely and accepting the great majority of applicants, we let Oxford V define itself (and archaeoastronomy) through the interests of the attendees rather than set up *a priori* guidelines. This allows the subject areas of the meetings to evolve with time. The attendees are listed in an appendix, as are the meeting program and the list of poster papers presented. We note that about half the presentations at the meeting were in the form of posters. This allowed for fewer and longer talks, and more time for discussion. Each lot of posters was on display for three days. Specific times each day were set aside for the posters, with no competition from talks or other events, so attendees could all view them. The poster style of presentation worked well at Oxford V, and allowed more time for discussion of each than would have been possible had it been given as a talk The papers in this volume are based equally on talks and posters presented at Oxford V.

Santa Fe was the clear choice for the site of Oxford V in the United States. It is a fascinating town ("The City Different") ringed by mountains in the lovely high plateau country of New Mexico. It is easily accessible, yet avoids the unpleasant sprawl and summer heat of the major cities of the Southwest. It brings together several threads of history—the Spanish and then Mexican rule of the area, starting in the 16th century, followed by the 19th century acquisition by the United States—while around it is the living Indian Pueblo world which derives from the complex prehistoric Pueblo cultures that are proving to have had a rich astronomical tradition. The surrounding area bears witness to many millennia of human activity.

The meeting was held on the campus of St. John's College, with magnificent views down to Santa Fe and across the Rio Grande to Los Alamos and the Jemez Mountains. The main lecture hall held the conference comfortably, and opened onto the poster area where a bar was set up each evening at "poster time". The attendees stayed in the dormitories and ate together in the main dining room, letting us all enjoy a collegial atmosphere for the meeting. One day was spent on a visit to the world-famous silver and pottery collections at the School of American Research in Santa Fe, and on a trip to the Comanche Gap petroglyph field near Lamy. After the meeting, a number of attendees went on a three-day guided tour of the archaeological riches of Chaco Canyon (the Chaco Culture National Historic Park) (<http://www.nps.gov/chcu/>).

The petroglyph on the cover of this volume was suggested by Polly Schaafsma as the identifying symbol of Oxford V. It was used on the poster and on the Web Site <http://www.lpl.arizona.edu/oca/pet.htm> that announced the meeting, and on the lovely commemorative T-shirt that Abbie Morris designed for us (as seen in the Oxford V group picture at <http://www.lpl.arizona.edu/oca/group400.htm>). This glyph is located in the Petroglyph National Monument in New Mexico <http://www.nps.gov/petr/>, across the Rio Grande from Albuquerque. It is one of several in that area that Polly interprets as Venus glyphs (Schaafsma, this volume).

Acknowledgments

The success of Oxford V was due to the work of many people. We would first like to single out Ray and Ruby White and Suzy Chippindale for praise for actually making the meeting happen day by day. Mike Zeilik made Oxford V known internationally through the World Wide Web and the posters. In addition to them, the others on the Organizing Committee—George Gumerman, Ed Krupp, Bill Longacre, Steve McCluskey, Peter Pesic, and Ray Williamson—played important and sometimes vital roles in the planning of the meeting. We are grateful to Joe Shaffer and Irene Sadler and the staff of St. John's College for being our hosts and for making our stay at the College so pleasant.

Doug Schwartz and Duane Anderson of the School of American Research are to be thanked for inviting us to visit their world-famous collections. Henry Singleton graciously allowed us to visit the Comanche Gap site on the San Cristóbal Ranch. J. D. Arnold of Rocky Mountain Tours, Santa Fe, did a splendid job of supplying transportation and very comfortable life support for the trip to Chaco Canyon. Dave Phillips and Steve Lekson served ably as our guides in Chaco, and made our trip both instructive and enjoyable.

And a final thanks to all who came to Oxford V and made the meeting worth attending.

References

Gumerman, George and Miranda Warburton, this volume.
Schaafsma, Polly, this volume.

Section 1

Worldwide Archaeoastronomy

1

The Nature of Archaeoastronomy: Archaeoastronomy and the Oxford Conferences

Rolf M. Sinclair
Chevy Chase, MD

We discuss the history of archaeoastronomy and show how it evolved to its present form. We suggest that archaeoastronomy cannot be viewed as a discipline in its own right, but is better considered as a part of archaeology and anthropology. We then look at the tools and methods of archaeoastronomy and find that there are a number of simplifying assumptions made in practice that can bias the results if the investigations are not done carefully. We distinguish between "skywatching"—the observation of naked-eye phenomena—and astronomy/astrophysics as a part of the physical sciences. We then discuss briefly the papers brought together in this volume.

* * *

A History and Definition of Archaeoastronomy

Since we have used the term archaeoastronomy to define this meeting, I should explore the various uses of the term and find one appropriate to this volume. Archaeoastronomy grew out of the realization of the importance of astronomical phenomena throughout prehistory and history (see Krupp 1992 for a summary). Indeed, Emerson observed "The Sky is the daily bread of the eyes." (Emerson 1843) One way to appreciate the effect of astronomical phenomena on human culture is to imagine that mankind had lived on a perpetually cloudy planet, with nothing other than diffuse light by day and darkness at night. Let the clouds suddenly part, and imagine the effect on human beings of their first glimpse of the heavens. This is a measure of the importance of astronomical phenomena to mankind. We cannot study any aspect of mankind without factoring the reaction to these phenomena into the cultural equation. [This idea is not new (Sinclair 1995), and goes back at least two millennia (Lucretius 1994).]

The conventional disciplines of archaeology, anthropology, and history had in general at first not given this aspect of human life the attention that it was due, and tended to disregard it or relegate it to an aside. It was up to astronomers initially to map out and develop this hitherto unexplored portion of human experience. Thus the first definition, historically, was that archaeoastronomy was that which interested astronomers, and was the study of what they found to be evidence of astronomically-driven practices in past cultures.

What then is archaeoastronomy today? By the accident of history, archaeoastronomy was started by people outside of archaeology/anthropology and has been treated separately ever since to a certain extent. Although it has somewhat of a life of its own and is a convenient term to use, it is not really a separate discipline. Archaeoastronomy now can perhaps best be considered as a historically distinct sub-discipline of archaeology (and to some extent of anthropology) rather than a discipline in its own right. It exists because archaeologists and anthropologists have rarely included in their studies or their work the cultural impacts of astronomical phenomena.

Archaeoastronomy has brought an important point of view to archaeology and anthropology, namely that for all the peoples that we know about, the sky above is as real and as important as the Earth around. Archaeoastronomy can be defined as the study of what peoples throughout history and prehistory have made of the phenomena in the sky, how they have used these phenomena, and what role they played in their cultures. The social sciences have started to include this as part of their wider studies of humanity. Archaeoastronomy is still a convenient term to use to describe this aspect of the social sciences, with no implication that we are describing a separate discipline.

The specific tools of archaeoastronomy are positional astronomy and surveying, but it also relies on the techniques of archaeology and anthropology. Its subject matter is not astronomy in the sense of the physical sciences, but is limited to a much simpler and minor set of naked-eye observations. Its contribution has been to add the astronomical point of view to archaeology and anthropology, and thus broaden the approach of those disciplines. It contributes to the more general findings of archaeology and anthropology, and must be consistent with those findings.

Archaeoastronomy has settled into its role as a sub-discipline with its own name and customs, but it cannot be considered independently of archaeology and anthropology. There are other such sub-disciplines that represent the intersection of the physical sciences with the social sciences. For example, a part of nuclear physics—isotope dating and sourcing—has similarly broadened archaeology, and has a bit of a life of its own without necessarily being considered a discipline in its own right. (I don't think anyone has proposed "Archaeonucleonics" as a separate discipline).

The study of the techniques of ancient astronomical practices *per se* is a study in its own right, but the goal of much of the work reported at the Oxford conferences has been to attempt to place these practices in their cultural contexts. One study we would categorize somewhat separately from archaeoastronomy is that of the path that led to the present-day science of astronomy—the "history of astronomy" as a part of the history of science.

The *Modus Operandi* of Archaeoastronomy

Archaeoastronomy deals with astronomical phenomena that are universals. All peoples everywhere see them, suitably altered by latitude. True, a particular solar eclipse will be seen as total only along one narrow track, but wait long enough and any place on Earth will see one every three centuries or so on the average. How people react to these phenomena is strongly culturally dependent.

If the only record of a culture is archaeological, then the existence of astronomical practices must perforce be teased out from that record. It is of course first necessary to demonstrate or hypothesize (or assume) just what astronomical practices would be preserved in the archaeological record, and how they would be indicated. Archaeoastronomy has made some important simplifying assumptions that are often taken for granted. It has concentrated on what we consider the most basic phenomena—first, the extremes and midpoints of the annual solar cycle and the annual motion of stars, and later on the extremes of the more complicated motions of the Moon (and occasionally the planets). Second, it has focused on the rising and setting points of stars or at the solar and lunar extrema, and assumed that these points on the horizon would be indicated by the alignment of structures, uncovered by archaeology, as the principle way of showing calendric or ritual interest in these events. A modification of this is the idea of calendars based on the movement of sunrise along the horizon ("horizon calendars"), which would leave no physical record of use but instead is based on the assumed use of features on the horizon. The strength of these assumptions is that they focus on a few quite evident events that are arguably of universal interest, while there is an infinity of other events we could imagine that would be of only local interest (such as the azimuth of the sun at mid-morning on a particular leader's accession day). The weakness is that many archaeoastronomical studies have strong assumptions built in, and can easily miss other ways of indicating interest in astronomical phenomena. This has led in many cases to a virtually automatic search for solar- and lunar-aligned constructions at many sites, or a scan of the horizon for possible calendrically useful features. However, recent work has let some observations be recognized more on their own merit. Studies of cardinally oriented structures and of the illumination of rock art have shown how some peoples could have used more generally the position of the sun during the day, not just at rising and setting.

We find the beginning of formal archaeoastronomy (or astroarchaeology as it was first termed) in Great Britain. There had been both popular and informed speculation for centuries concerning astronomical alignments at Stonehenge (Ruggles 1999). The first hypothesis was the simplest possible one—that the structures had a clear alignment to the rising point of the sun on the horizon at summer solstice. The first serious quantitative study of this aspect of Stonehenge seems to have been the work of Sir W. M. Flinders Petrie (1880), who made the first survey of such accuracy that his measurements were not the limiting factor in his work. [He was also perhaps the first to argue for a well-defined unit of measure to have been used in the construction of such monuments (Petrie 1877).] These ideas of solar oriented constructions were picked up on and generalized by Sir Norman Lockyer. Lockyer is best known as one of the founders of modern astrophysics, for the discovery of the element helium in solar spectra, and for founding and guiding the journal *Nature*, but it was his studies of the astronomical purposes of such archaeological sites as the Pyramids and Stonehenge that became role models in archaeoastronomy (Lockyer 1894, 1909a). Indeed these studies, together with his "Surveying for Archæologists " (Lockyer 1909b), laid down the basic principles of that branch of archaeoastronomy involved with astronomical alignments to rising and setting points. While Petrie focused on solar rise alignments, Lockyer looked further afield and added stellar alignments as a way to date the construction of monuments. Lockyer assumed that the rising points of various bright stars had been marked by alignments. Since these stellar rising points would be shifted noticeably over centuries and millennia by precession (while solar rising points are shifted only imperceptibly over such times), deviations from the present rising points would allow the construction of the alignments to be dated. Although the stellar argument has proven to be unproductive, the methods Lockyer pioneered are still the basis of much of current archaeoastronomy. If we are compelled to seek paternity for our field, we could easily call Sir Norman the "father of archaeoastronomy". [Although much of Lockyer's work has been called more "adventurous than critical", one biographer says, "His effect on the course of science...is undoubtedly greater than the fate of his particular speculations would suggest" (Dingle, 1973).]

There have been only two main additions to the "standard" methodology of archaeoastronomy since Lockyer. The first concerned the Moon. This was the inclusion into the observational reckoning of the complex apparent path of the Moon on a par with that of the Sun by Capt. Boyle T. Somerville, R. N. (a naval surveyor) in his study of the Callanish site (Scotland) (Somerville 1912). Capt. Somerville deduced that one of the Callanish alignments was to the rising point of the moon at the major lunastice (or "standstill"). This idea was expanded on by Alexander Thom (a civil engineer) (Thom and Thom 1978), Gerald Hawkins (an astronomer) (Hawkins 1965), and Sir Fred Hoyle (another astronomer) (Hoyle 1977), who demonstrated that it was possible to ask quite detailed astronomical and engineering questions of hitherto mute stones. Even though their work has been open to criticism and revision, they deserve credit for the initial impetus of what we could term "modern archaeoastronomy"—the subject of the Oxford meetings. It was no coincidence that Oxford I was held in England, that many of its talks were on British megaliths, and a tour of Stonehenge was the high point. The work of Thom, Hawkins, and Hoyle has spawned a worldwide search for "standstill alignments" (i. e., alignments associated with moonrises/sets at the lunastices) to match the earlier interest in solstice alignments.

The other new addition to the repertoire of archaeoastronomy was the observation that patterns of light and shadow fall on isolated rock paintings and carvings at various times of day (not necessarily at sunrise and sunset) in such a manner as to indicate the extrema and midpoints of the annual solar motion and possibly finer divisions of the year (and in a few cases possibly the lunastices). The inference is, of course, that those who fashioned the rock art deliberately sought the effect. The first report of this phenomenon seems to have been made by Ken Hedges (an anthropologist) in 1976 of a site in Baja California (Hedges 1977), while the main impetus for such studies was the recognition by Anna Sofaer (an artist) of the Fajada Butte (Chaco Canyon) "Sun Dagger" site (Sofaer *et al.* 1979). In fact "sun dagger" has now become a generic term for one type of moving light pattern that interacts with rock art, and such patterns are sought far and wide.

Further work has shown that such seasonal marking of petroglyphs was actually common in a large region of the prehistoric Southwest of the US (Fountain and Preston, this volume). This work has shown that in addition to all the previous questions to ask of rock art, others that can be meaningfully asked concern how light and shadows fall on it at certain times of day and year. The interaction of rock art with sunlight (and possibly moonlight) begins to supply one motivation for the observed placement of rock art, and indeed for the very production of the art in the first place. For example, little rock art in this large region is placed so that sunlight does not fall on it, so to begin with the distribution of rock art is far from uniform in azimuth (Fountain 1999). [These studies are distinct from those of similar phenomena in architectural settings, in which the orientation of some element(s) to an astronomically significant azimuth can result in a special illumination of a carving or other feature, such as at Newgrange (Ireland) (Ruggles 1999, pp. 12–19) and at Maes Howe (Scotland) (Ruggles 1999, p. 129).]

Cardinal orientations of constructions or alignments can only be done astronomically, and their existence where verified would show most likely the use of the symmetry of the sun's apparent motion during the day to determine the cardinal directions. The location of equinox sunrise/sunset could also be used, of course, but this would probably be less accurate. It is possible (but in my opinion unlikely) that the meridian could have been established by stellar observations. As a generalization of locating the cardinal directions, evidence for repeated use at separate sites of certain non-cardinal bearings that are not clearly related to the extrema of solar or lunar risings/settings are most likely offsets from astronomically-determined azimuths.

It is interesting to note that these new ideas in archaeoastronomy were first made by people from other disciplines—the importance of the lunastices noted by a person trained in celestial navigation, and of light patterns on petroglyphs by an artist. Archaeology and prehistory are the richer for these ideas brought in via archaeoastronomy.

Archaeoastronomy has thus largely come to mean in practice the search in the archaeological record for the alignment of constructions to astronomically significant azimuths, or the search for features (such as tombs, altars, or carvings) so placed that light shines on them only at certain times and in certain patterns, or the inspection of the horizon for features that coincide with certain sunrises. Obviously the field is richer and the task easier, and more things can be considered, if one has in addition present or past ethnographic records so that there is a broader base of knowledge. Then the study is better called ethnoastronomy.

However, in the absence of much (or even any) other knowledge of a site (or even of its relevant culture), research in archaeoastronomy must rely principally on astronomical evidence alone, and we are forced to accept the importance of this evidence in isolation. This has led to the concern that some astronomically-related alignments or exposures were never intended (or even realized to exist) by the people who made the sites in the first place but are really in the eye of the contemporary beholder, and that some reports are examples of a "Matthew Principle" at work: "Seek and ye shall find" (Matthew, n. d.).

[I will not include the (mostly British) reports of "ley lines" and other long-range relations of a mixture of sites as examples of astronomical alignments. These claims so far have raised more questions than answers.]

The astronomical phenomena that people knew until the invention of the telescope were limited to what could be seen by watching the sky with the naked eye. [The impact of instrumentation on the development of astronomy is summarized in Harwit (1981).] I speak here deliberately of "reactions to (naked eye) astronomical phenomena" or simply "skywatching" rather than "early astronomies". The practices that are uncovered through archaeoastronomy are not what is now called astronomy. The term astronomy (which is now essentially synonymous with astrophysics) is reserved for that branch of the physical sciences that deals with the nature of celestial bodies. We have every indication that the physical sciences as a whole are self-consistent and describe the entire Universe that we can observe. Gravity is universal, fundamental particles and their interactions are the same throughout the Universe, and the chemistry of stars can be described in terms of the same chemistry that obtains on Earth. Our instruments show us that the Earth and the sky are made of the same stuff and follow the same behavior. This cannot be demonstrated by naked eye observations alone, which are limited to a narrow band of optical frequencies of an astronomically trivial collection of bodies in the Solar System and some nearby stars. To the naked eye the heavens are different from Earth. True, the Milky Way does inform us of our galaxy, but the unaided eye doesn't allow its nature to be determined. The darkness of the night sky is one critical cosmological observation that can be made with the naked eye, but the importance of this observation was first realized only in 1576 by Thomas Digges (Harrison 1987). In short, the nature of the Universe cannot be inferred from what we see with the naked eye from the surface of Earth. Naked eye observations are in general lost in the trivia of contemporary astronomy.

The phenomena that are visible to the unaided eye have been much the same for millennia. The Sun, Moon, planets, and stars cycle endlessly in the sky, defining the day, the month, and the year for all peoples. They serve as time givers, both defining the passage of time and giving the means to measure it. The symmetry of their motions defines the cardinal and solstice directions. Their repetition was the basis for orientation and navigation. The seemingly complicated apparent motions of the Moon and the planets proved a surmountable challenge to some cultures. [I have described the pattern of these phenomena in more detail elsewhere (Sinclair, 1995).]

But sometimes this regular pattern of the celestial clockwork goes awry. Occasional spectacular events such as solar eclipses, comets, and supernovæ disturbed the regularity and set off cultural shock waves (Masse, this volume). Although virtually all meteors are harmless scratches of light across the night sky, there is evidence that Earth has had periods of enhanced meteoritic bombardment that had strong climatic and cultural effects (Peiser, this volume). At times the skies were to be watched with foreboding and alarm, and comets and meteors came to be feared.

Skywatching was certainly a precursor to contemporary positional astronomy, and laid a foundation of interest in the nature of celestial bodies and the reasons for their apparent motions. [One recent study (Gutzwiller 1998) shows that key features of the Moon's motion were known 3,000 years ago, even though the underlying physical principles of planetary motion were not found until the 17th century.] These naked eye observations served a number of purposes. Some were of an applied sort: the use of astronomical phenomena as timekeepers and calendar-definers, or for orientation and navigation, with no understanding of the physical phenomena themselves. Most however concerned themselves with the models developed to explain the phenomena. These were of a religious nature that tended to set the sky apart from the Earth, and most were dead ends off the trail that led to contemporary astronomy. Out of this came what we now term astrology: a belief that things seen in the sky affected mankind in mysterious ways, at odds with what we know of the physical sciences, and that these effects could be understood and predicted by watching the motion of the sky. The ideas of astrology persist to the present day as a fossilized religion, although with accretions first in the Renaissance and again in this century.

The corpus and tradition of observations of the sky through millennia did serve as a basis that made possible the development of a science of positional astronomy as part of a larger concept of the physical sciences. The roots lay in many cultures, and the threads came together in the late 16th century—after which there was no turning back.

Archaeoastronomy has its fringe elements, as does any other part of science. There is something about "ancient astronomies" that attracts more than the sober attendees at the Oxford meetings. A look at the World Wide Web or the books for sale in "New Age" stores show that there is no limit to what people can do with remarkably little data. The usual pitch is to ascribe wide-ranging knowledge of science, particularly astronomy, to people some millennia ago, as part of ancient (or lost) cultures. [Egypt has been a favorite place to locate this knowledge ever since pyramidology got its start in the 19th century by Charles Piazzi Smyth (also an astronomer).] At its most dense, this work gives rise to long convoluted mock-scholarly studies that defy analysis. Some recent "archaeoastronomical" work is even done on Mars, copying the above assumptions about the importance of rising/setting points and using verbatim the methods laid down by Lockyer, to demonstrate supposed links to (or from) Egypt. (An example of this work is described in Hoagland 1987.) The hard core of all this work relies on conspiracy theories and massive (if unexplained) "government (or academic) cover-ups" for its defense against review or criticism. Some of this work is pure invention; some of it selects subsets of known data with great care; some just muddles along with no particular logic. There is caution to be learned from the excesses of such works and their ability to find archaeoastronomical clues buried in stray elements of natural sites or of history. The Oxford meetings have been liberal in letting some of these voices be heard, believing that exposure and discussion is the best way to sift new ideas. (There are, after all, no embarrassing questions—only embarrassing answers.) This must be balanced against the desire of some to cite a presentation at an Oxford meeting as "authentication" of odd ideas.

The Oxford V Papers

Having discussed what we mean by archaeoastronomy, we can now use that term to discuss the papers in this volume. We group the papers here by subject matter and geographical region. This allows for easy intercomparison of similar papers.

Worldwide Archaeoastronomy

The first section of this volume consists of six other papers that treat general topics that apply to archaeoastronomy worldwide. By looking at general questions in sociology or astronomy they provide parts of a framework on which the other papers can rest. Gumerman and Warburton make the important point that anthropological archaeology and archaeoastronomy have failed to use each other's results, and offers suggestions to bring the two together. Peiser explores the potentially significant idea that Earth goes through periods of "coherent catastrophism" caused by the breakup of large comets in the inner Solar System, and that this enhanced bombardment would be reflected in both world-wide archaeoastronomical and historical studies of those times. Schaefer discusses the nature of astronomical "omens", and shows that they happen often enough to serve as fresh input for even the most prolific soothsayer. Masse applies the reasoning of the previous two papers to solar eclipses in the prehistoric US Southwest, and argues that they left definite cultural marks. McCluskey argues for a broader definition of astronomy than I have used above, speaking of "astronomies" (which might alternatively be termed "folk astronomies"). Finally, Poss compares Western European art with Southwest rock art and argues for a common method of analysis.

All six of these papers are excellent examples of interdisciplinary studies, relating archaeoastronomy to other disciplines and showing the power of a common approach.

Archaeoastronomy in the United States (Principally in the US Southwest)

Eleven papers describe various sites in the Southwest, and two describe sites in the Great Plains and the upper Midwest. This distribution reflects the activity in archaeoastronomy in the US, and is probably due to the density of relevant sites (and possibly their preservation in the arid and largely empty Southwest). This apparent emphasis on astronomical phenomena among the prehistoric peoples of the Southwest is perhaps a reflection of their proximity to the astronomically hyperactive cultures of Mesoamerica (Schaafsma, this volume). (The prehistoric US Southwest is better described as the "Mesoamerican Upper Northwest" to reflect the cultural radiance of Mesoamerica, since the present-day political boundaries are irrelevant to prehistory.)

The first two papers (Fountain and Preston) argue for an important discovery—that a number of petroglyphs distributed over a wide region of the Southwest were so placed as to "mark" certain common dates by their interaction with sunlight. This is an unusual and important cultural factor found among the diverse peoples of this region, discovered through archaeoastronomical studies. Rowe *et al.* describe the dating of a particular rock painting and argue that it is not associated with the 1054 supernova, pointing out the hazards of the assignment of rock art to particular events. The next six papers (Bates, Bostwick, Chamberlain, Hoskinson, Olowin, and Schaafsma) describe particular sites in the Southwest (mostly in the Anasazi world), and the next two (Ambruster and Chamberlain) describe astronomical practices among the Navajo of the Southwest. Hollabaugh takes us into the astronomical patterns of the Lakota of the Great Plains. Finally, Stekel shows how it is possible to use century-old surveys of since-disappeared sites in Wisconsin to reconstruct their properties. What is interesting in this collection of papers is the variety of the sites, and the importance that astronomical phenomena had for these diverse people.

Central and South America

The Precolumbian cultures of Central and South America were unique in the strength of their reactions to astronomical phenomena. This is most likely due to their historical isolation that allowed this aberration to become dominant without having to be tempered by outside ideas (Paz 1997). The six papers in this group show some of the varied uses of astronomical phenomena by these cultures. Macri and Robbins describe Mesoamerican customs that lasted for two millennia, and Milbrath shows some of the astronomical richness of Maya iconography. Pitluga describes the enigmatic, large-scale constructions at Nazca, while Westerman and Zuidema discuss new findings in the Inca Empire.

Europe, Asia Minor, and Egypt

Oxford I had thirteen papers from the British Isles and only two from elsewhere in Europe. Oxford V reversed this trend: the nine papers in this section come from continental Europe (with one from the Canary Islands), Asia Minor, and Egypt. These papers offer glimpses into quite varied astronomical practices. Blomberg, Gardner, Krupp, Locher, and Simonia provide welcome unions with the classical archaeology of the Mediterranean region. Esteban describes some astronomical customs of the original proto-Berber inhabitants of the Canary Islands, which may allow comparison with similar studies in North Africa. Frank uses very different methods to find the astronomical practices of the early inhabitants of Iberia. Henriksson discusses early Sweden, and Nedialkov 17th century Bulgarian art; both find evidence for the portrayal of comets.

Far East and Australia

There were only three papers from this part of the World. Pankenier describes a correspondence formed in China between the order of the heavens and that of Earth. Chiu shows that the site of present-day Canton may have been chosen in the 3rd century BC for its location on the Tropic of Cancer. Cairns describes newly discovered sky maps of the Australian aboriginal population. We hope that future Oxford meetings will have more participation from such interesting and diverse areas.

Education in Archaeoastronomy

The final three papers describe quite different ways of bringing archaeoastronomy into education. Nelson describes an upper-level undergraduate interdisciplinary course that combines anthropology, astronomy, and history. Penprase describes an Internet-based resource center in archaeoastronomy that can serve its clientele via the World Wide Web. Sumners is developing a planetarium program to teach archaeoastronomy to the general public and selected school group.

References

Dingle, Herbert in *Dictionary of scientific biography* Vol. VIII, C. G. Gillispie, Ed., pp. 440–43. New York City: Charles Scribner's Sons, 1973.

Emerson, Ralph Waldo. "Journal", Vol. 6 (1911), entry for 25 May 1843.

Flinders Petrie, Sir W. M.: *Inductive metrology: or the recovery of ancient measures from the monuments.* London: H. Saunders, 1877.

Flinders Petrie, Sir W. M.: *Stonehenge: plans, description, and theories.* London: Edward Stanford, 1880.

Fountain, John, this volume.

Fountain, John, private communication, 1999.

Gumerman, George and Miranda Warburton, this volume.

Gutzwiller, Martin C.: *Moon-Earth-Sun: the oldest three-body problem.* Revs. Modern Phys. 70, pp. 589–640 (1998).

Harrison, Edward Robert: *Darkness at night: a riddle of the universe.* Harvard University Press, 1987.

Harwit, Martin: *Cosmic discovery.* New York: Basic Books, 1981.

Hawkins, Gerald: *Stonehenge decoded.* New York: Doubleday, 1965.

Hedges, K., "Rock Art in the Piñon Forest of Northern Baja California", in *American Indian Rock Art Vol. III*, pp. 1–8. Edited by A. J. Buck, Frank Buck, and John Cawley (American Rock Art Research Association, Whittier CA, 1977).

Hoagland, Richard C.: *The monuments of Mars.* Berkeley: North Atlantic Books, 1987.

Hoyle, Sir Fred: *On Stonehenge.* W. H. Freeman, 1977.

Krupp, E. C.: *Beyond the blue horizon: myths and legends of the Sun, Moon, stars, and planets.* Oxford U. Press, 1992.

Lockyer, Sir Norman: *The dawn of astronomy.* London: Macmillan, 1894.

Lockyer, Sir Norman: *Stonehenge and other British stone monuments astronomically considered.* London: Macmillan, 1909a.

Lockyer, Sir Norman: *Surveying for archæologists.* London: Macmillan, 1909b.

Lucretius: *On the nature of the Universe*, Book II, line 1026 ff. (ca. 60 BC). London: Penguin Classics, 1994.

Masse, W. Bruce, this volume.

Matthew (Saint): *Gospel according to St. Matthew* (no date) 7:7–8: "...seek and ye shall find...he that seeketh findeth..."

Paz, Octavio: *In light of India.* Harcourt Brace, 1997, pp. 90–92, 97–99. (Trans. by E. Weinberger of *Vislumbres de la India.* Barcelona: Editorial Seix Barral, 1995).

Peiser, Benny, this volume.

Preston, Robert, this volume.

Ruggles, Clive: *Astronomy in prehistoric Britain and Ireland.* New Haven: Yale University Press, 1999 (In particular see "Introduction".)

Schaafsma, Polly, this volume.

Sinclair, Rolf: *Introduction to the INSAP Meeting.* Vistas in Astronomy 39, Part 4 (1995), pp. 381–94 [http://ethel.as.arizona.edu/~white/rmsintro.htm].

Sofaer, A., V. Zinser, and R. M. Sinclair: Science 206 (1979), pp. 283–91.
Somerville, Capt. (later Vice-Adm.) Boyle T., R. N.: J. Royal Anthrop. Institute Gr. Brit. & Ireland 42 (1912), pp. 23–52; J. Brit. Astron. Assn. 23 (1912), pp. 83–96.
Thom, A. and A. S. Thom: *Megalithic remains in Britain and Brittany.* Oxford: Clarendon, 1978.

2

The Universe in a Cultural Context: An Essay

George J. Gumerman
Santa Fe Institute
Santa Fe, NM

Miranda Warburton
Navajo Nation Archaeology Department
Flagstaff, AZ

Researchers who have made a study of past cultures' interest in and conception of the heavens have often treated the topic narrowly, rather than in the context of the entire culture. Archaeoastronomers, coming from many different disciplines, have explored the subject by trying to identify archaeological features such as architectural alignments and rock art in relationship to celestial events and processes. Anthropological archaeologists on the other hand, whose goal it is to understand the evolutionary trajectory of a past culture, often completely ignore the entire concept of cosmology and the role of celestial phenomena. This essay explores the reasons for the failure of anthropological archaeologists and archaeoastronomers to effectively utilize the results of each other's research, and offers suggestions for the integration of archaeoastronomy into more systemic cultural studies.

* * *

Introduction

It may seem like the height of hubris that we as anthropologists and not astronomers or physicists entitle our essay *The Universe in a Cultural Context.* Our basic premise, however, is that to truly comprehend a culture we must have some sense of its cosmology—the group's conception of themselves in relation to the heavens. Throughout the world, human social behavior is inextricably (and perhaps inexplicably) intertwined with its world view. We propose that the following questions are among the most appropriate if a researcher is to use an anthropological perspective for understanding behavior: What is a given culture's view of its place in the universe? How does that perception affect the character of their culture? And, how does the character of the culture affect their cosmology?

In order to make our points, we briefly review the nature of anthropological archaeology, and apologize to those of you for whom this may be "old hat". Because the study of archaeoastronomy is peopled by many who have their disciplinary home in any of a great number of academic departments, we provide a very brief history of the recent development of the discipline within which archaeoastronomy is attempting to carve a niche.

This commentary is not the result of research, nor of any specific observations about the relationship between culture or cultures and the heavens. Rather, we offer some general observations about the state of the study of archaeoastronomy within anthropology, and offer some suggestions that may help to better integrate archaeoastronomical studies into the most encompassing context, that of an entire culture. The reader will note that most of our illustrative examples are from the American Southwest. This is not because there are exceptionally good examples in this area, but simply because we know the literature better.

Culture as a System

It is more than a little presumptuous of us to offer these comments because neither of us has ever done any research involving archaeoastronomy, and in the past might have been considered "non-believers." Over the last few years, however, we have realized that in order to understand the nature of a prehistoric culture it is important to view it in a truly systemic or holistic context—a view that treats the economic, technological, sociological, and ideological (including cosmology) realms as being in various degrees of harmony with one another. As anthropological archaeologists, we operate under the fundamental principle that culture, including cosmology, is writ large in the material record left behind by former inhabitants. This is, of course, an idealistic and unrealistic conception: none of us has ever been able to reconstruct even a small percentage of a past culture, but operationalizing the concept of culture as a system has brought us closer to the goal, and we continue to refine our means of accomplishing this.

The concept of culture as a system was the reigning paradigm in anthropological archaeology from the late 1960s to the mid 1980s. Conceptually, because culture can be viewed as a highly integrated system, many archaeologists believe that an understanding of one relatively easily derived aspect of a prehistoric society (usually technology or economy), allows inferences to be made about those components less amenable to archaeological reconstruction, such as religion.

One of the most striking examples of this is the difference in religious practice between Native Americans on the Plains and in the Pueblo Southwest. Most people are familiar with the stories of adolescent male Plains Indians searching for a personal guardian spirit during a vision quest. And many know about Pueblo katsina dances involving whole villages, the purpose of which is largely to bring rain for the crops. These divergent forms of religious behavior are predictable by knowing the nature of the two economies. For the hunting and gathering economy of the Great Plains where success is often dependent on the individual hunter's skill, the ide-

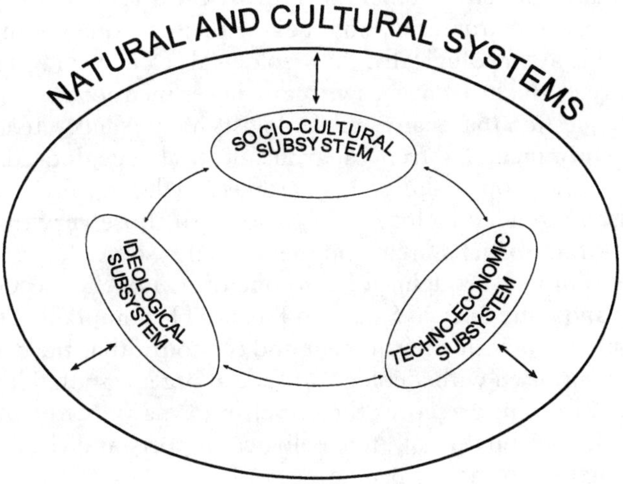

Figure 1. A culture depicted as a system with three sub-systems that interact with one another, with the natural environment, and with other cultural systems.

ological system focuses on the individual. For the farming Pueblos, rainfall benefits the entire village and therefore it is not surprising that the ideological sub-system reflects that group focus.

There has been, however, increasing criticism of the concept of culture as a system because not all cultures are tightly integrated (Trigger 1984). In other words, some components of the system may have various relationships of different intensity with other components of the same system. For example, Christians in the U.S. are often exhorted to extend their Christian fellowship beyond the one hour spent at church on Sunday, suggesting perhaps that religion is less integrated into Western culture than in some other cultures. This observation notwithstanding, herein we follow the archaeological convention of viewing culture as a system.

Figure 1 illustrates the interrelationship of the various components of any culture which conventionally are grouped under three headings. The ideological realm comprises a culture's system of beliefs and values, including religion and cosmology. The socio-cultural realm comprises a network of status and roles, kinship systems, and ways in which a culture's individuals relate to one another. The techno-economic realm comprises a culture's technology and that part of the economy that is used to wrest a living from the environment. These components make up the system and interact not only with the natural environment but with other cultural systems.

Most successful archaeological studies that endeavor to reconstruct a past culture, focus on the economy, technology, and the peoples' relationship to the environment—either because these may be the easiest parts of the culture to reconstruct, or because the investigator believes the economy and peoples' relationship to the environment were the primary factors that determined the character of a particular culture. The economy is relatively easy to reconstruct because we can excavate hoes, spear points, plant remains, and animal bones, and even determine diet from the chemical analysis of human skeletal remains. Additionally, tree ring analysis, geomorphology, macrobotanical studies, and other analytical techniques provide independent and accurate environmental records. Until very recently, however, the sociological sphere of culture, i.e., how people relate to one another, was considered less knowable archaeologically, the ideological realm even less so, and the culture's perspective of the heavens was virtually never mentioned.

There is no question that many archaeologists have exaggerated the importance of the role of environment and technology in culture change. Indeed, in many cases, we treated prehistoric peoples almost as a barometer reflecting environmental change. In other words, we tended to lose sight of some of those very things that made people human—their belief system and their social system.

This realization was brought home to one of us by Alfonso Ortiz, a cultural anthropologist and a member of San Juan Pueblo. He complained that archaeologists are too concerned with environment and economy, that the decisions of most Pueblo people are infused with ideology and social organization. He, of course, was right. His observation underscores the criticism of many Native Americans concerning the unidimensionality of archaeological reports and their general lack of relevance and interest for native peoples.

Furthermore, with our focus on economy and technology, we had also lost sight of one of the basic principles of systems research—that all the subsystems interact to form more than the sum of the parts. More crucial than the primacy of one subsystem over another, is the interaction of the sub-systems with one another and with the environment.

One of the most colorful and compelling examples of the integration of culture occurs in the Mexican town of Magdalena de Kino. A community of some 40,000 people, it lies about 60 miles south of the Arizona-Mexican border, in the state of Sonora. On the fourth of October each year, thousands of people from Arizona and northern Mexico gather to honor St. Francis Xavier, but it is somewhat confusing because this date is actually the saint day of a totally different saint, St. Francis of Assisi. It is further complicated by the fact that many of the pilgrims are really there to honor Father Kino who first missionized much of what is now northwestern Mexico and southern Arizona and who is buried in Magdalena. In any case, the roads are filled with pilgrims—Yaqui Indians, Mayos, Tohono O'odam from Arizona, and Mexicans—many of whom have walked for a week or more, some on their knees. They stand in long lines to kiss the reclining statue of St. Francis Xavier in the church.

Once that obligation is over, however, and often before, the fun begins. The town's huge plaza is filled with double and even triple lines of stalls with vendors selling toys, souvenirs, religious items, and folk medicines. There are outdoor temporary restaurants and bars, each with a special clientele—one bar, for example, is where the Yaquis drink. There is a carnival and even a large section of town for prostitutes who come in from all over northern and central Mexico. Sellers of cloth and plastic housewares hawk their goods from the back of vans using loudspeakers which compete with the lively, but noisy mariachi bands, marimbas from distant Chiapas, and harps from Vera Cruz. There are even Yaqui deer dancers, the origin of which is in pre-Spanish contact ritual.

So what is this event? Is it a religious experience of Spanish content, a prehispanic celebration, a mixture of both, a social event for the release of sexual tension, an opportunity to meet a potential spouse, an occasion to increase family solidarity, an opportunity to drink with your buddies, an economic opportunity, a chance to sell your wares, pick pockets, or make a deal on religious medals? The answer is, of course, it is all those things and more. That is how culture as a system works.

Archaeology and the Problem of Ideology

In recent years some archaeologists have tried to redress the relative neglect of ideology in understanding past human behavior. One of the major advances in this effort was a School of American Research advanced seminar devoted to ideology and Pre-Columbian civilizations. The published proceedings (Demarest and Conrad 1992) include papers by eleven scholars, and while the effort to incorporate ideology into an understanding of New World civilizations should be applauded, as Rolf Sinclair (personal communication) has pointed out, astonishingly no terms associated with the heavens appear in the index—and this in a book devoted to the Andean and Mesoamerican civilizations whose knowledge of the heavens was without parallel in the world at that time.

It is clear that all cultures have a concept of the universe which includes the heavens and further, which is an integral part of any culture's world view. Western societies within the paradigms of the "big bang" theory and various religious constructs are no exception, of course. Despite this knowledge on the part of anthropological archaeologists, as we have noted, those aspects of a culture's world view that include the celestial have been largely ignored. Ignored, that is, by most, except for a growing band of archaeoastronomers. Herein lies the conundrum. While the broader scholarly discipline of anthropological archaeology has generally not dismissed, but more commonly ignored, the celestial realm of the group they are studying, many have also ignored the results of the research of archaeoastronomers. This latter neglect is, in part the result of archaeoastronomy being a specialty, requiring particular kinds of knowledge not normally found in anthropology departments. In addition, the pursuit of a prehistoric group's knowledge of astronomical phenomenon does not typically require excavation or large teams of investigators.

Archaeoastronomical study is often considered by non-archaeologists as a discrete and bounded sphere of study. This view, conceptually divorced from the idea of culture as a system, has encouraged researchers from many fields of study to make archaeoastronomy a second area of interest. On the other hand, a few archaeoastronomers have felt, usually unrealistically we believe, that a culture's cosmology equaled the sum of a culture, that is, that the entire population danced to the music of the spheres. We recall a conversation with a researcher who has done some very important work in Chaco Canyon archaeoastronomy. After we admitted that in the past archaeologists had placed too much emphasis on ecology when trying to explain the so-called "Chaco phenomenon," we asserted that the archaeoastronomer had erred in the opposite direction by explaining Chaco solely in terms of cosmology. The archaeoastronomer responded that it was necessary to exaggerate to counteract the prejudice of anthropologically trained archaeologists. We believe the time has now come for both camps to take more reasoned stances.

Do not misunderstand us. We are not criticizing these specialized interests in archaeoastronomy. Archaeology has a long tradition of valuable contributions made by avocational or part time students of the past. And archaeology would be a moribund science if we did not have the contributions of specialists ranging from radiocarbon determinations to models borrowed from theoretical evolutionary biologists. After all, it is no accident that the subtitle of this conference is *An Intersection of Disciplines*. Table 1 illustrates the diversity of disciplines and academic backgrounds of individuals interested in archaeoastronomy, and in attendance at the Oxford V Conference.

Table 1. Area of Interest or Academic Department as Listed by Those Registered at the Oxford V Conference on Archaeoastronomy

Dept/Interest	N	%
Astronomy	42	35.3
Anthropology	17	14.3
Archaeology	17	14.3
History	7	5.9
Architecture	4	3.4
Art	4	3.4
Physics	4	3.4
Archaeoastronomy	2	1.7
Mathematics	2	1.7
Sociology	2	1.7
Biology	1	0.8
Chemistry	1	0.8
Engineering	1	0.8
Humanities	1	0.8
Iberian Studies	1	0.8
Language	1	0.8
Medicine	1	0.8
Near Eastern	1	0.8
NPS Ranger	1	0.8
Religion	1	0.8
Students-Undeclared	2	1.7
Undeclared	6	5.0
TOTAL	119	100.0

Integrating Archaeoastronomy and Archaeology — A Parallel Case

The problem is as follows: although all cultures, both past and present, consider astronomical phenomena an integral part of their world view, most archaeologists reconstructing the past ignore this critical aspect despite the fact that it is a window into that culture's perspective of their cultural universe. On the other hand, those studies of a culture's perspective on astronomy tend to treat the subject as independent of the rest of the culture, or, less commonly, to make astronomical considerations virtually the entire focus of those prehistoric groups they consider.

In order to better understand the problem of the relationship between archaeoastronomy and anthropological archaeology, it is instructive to examine the changing relationship between Mayan epigraphic studies and Mayan archaeology. There are a number of interesting parallels between the two fields of study, and perhaps the Maya example offers some insights into possible solutions for the integration of astronomical studies into cultural reconstructions. The historical development of Mayan epigraphic studies, its early denigration by many Mayanists, its gradual acceptance, and finally integration into anthropological studies is similar to archaeoastronomy and anthropological archaeology with one major difference—archaeoastronomy has not as yet been accepted and therefore not integrated into the main stream of archaeology.

Also, unlike archaeoastronomy in North America, Mayan epigraphic studies had an early start in Mesoamerican archaeology, largely because the glyphs were so obvious and were clearly some form of recording system. As early as 1841, with the publication of Stephens and Catherwood's *Incidents of Travels in Central America, Chiapas and Yucatan*, the general public became aware of the ruins and writings of the Maya. These discoveries caused a firestorm of interest in the Euro-American world.

Since the middle of the nineteenth century many scholars have added to our knowledge of the Maya and their writing system. But as the efforts at decipherment continued, the subject of the glyphs was either hotly debated or ignored, much as the astronomical symbols discovered relatively recently were in the Southwestern U. S. The best Southwestern example is that of the Sun Dagger petroglyph, a solar marker in Chaco Canyon. The majority view was that the Mayan glyphs recorded calendrical data only, although a few scholars maintained that they were also a record of events and social ranking.

The most influential Maya scholar of the 20th century, J. Eric S. Thompson, convinced that Mayan glyphs were concerned only with the passage of time, wrote so persuasively that his voice overrode those of other scholars who disagreed with him. Archaeoastronomers can understand this rejection of ideas at odds with the reigning views of the times. Gradually, however, the accumulated work with its carefully constructed logic of the great scholar Tatiana Proskouriakoff (Schele and Friedel 1990:47) finally convinced the field that the contents of the inscriptions did indeed concern the affairs of the elites.

Knowing that the contents of the inscriptions concerned more than calendrical data did not help the historical epigraphers understand how the Maya spelled their words. A major advance in decipherment came from a Russian, Yuri Knorozov, who in 1952, proposed that Maya writing was "a mixed system composed of full word signs combined with signs representing the sounds of syllables" (Schele and Freidel 1990:48). None of the renowned Mayanists accepted Knorozov's thesis of phoneticism, and only a few scholars were willing to consider that Knorozov might be right.

It was not until December of 1973 that the big breakthrough was made at small conference held in the village near the ruins of Palenque. At that conference, a new generation of epigraphers was initiated into the mysteries of glyphic decipherment. Blending Knorozov's phoneticism with Proskouriakoff's 'historical approach' a group of epigraphers in the next few years developed a highly successful collaborative approach and forged the last link—the axiom that the writing reflected spoken language and thus had word order which could be used to determine the function of glyphs (Schele and Freidel 1990:49).

While archaeologists came to appreciate the rich textual potential of glyphs, they were at a loss as how to incorporate the epigraphic findings into their general picture of Mayan society. Hammond and Willy, for example, in their major compendium on Mayan society, *Maya Archaeology and Ethnohistory* published in 1979, include epigraphic studies in all the major sections of their work, but an examination of the papers shows a failure to integrate the two approaches.

This divergence reflects the fact that the questions asked by the archaeologists and epigraphers lacked sufficient overlap to combine the two perspectives. An example is an article entitled "The Representation of Underworld Processions in Maya Vase Painting: An Iconographic Study." The author (Quirarte in Hammond and Willey 1979) examined a number of Mayan vases related to death and the underworld, and discusses the procession scenes depicted on them. The discussion concentrated on the individuals in the processions and their identification in order to determine the significance and meaning of the themes portrayed. This certainly adds to our understanding of Mayan society, but as archaeologists conducting an analysis of Mayan culture as a system, or a settlement pattern study, or an economic study, we ask how do these data contribute and how can our studies benefit from them? It may well be that the way archaeologists have defined or identified archaeological data has limited the range of questions we ask and thus are self-limiting on what we include as data.

We believe that archaeoastronomical data suffer from the same syndrome as Mayan epigraphic studies. The questions asked and issues addressed by the archaeoastronomers tend to fall outside the self-defined parameters of what archaeologists consider to be significant and relevant. As with Mayan studies, the results of archaeoastronomy research have evolved largely independently from the broader context of archaeological investigations. Ways must be found to integrate the two spheres of inquiry so that they may complement one another, much like Mayan archaeological research and epigraphic studies have come to complement each other to produce a more complete understanding of Mayan society.

Keith Kintigh, an archaeologist, made an interesting comment on this problem in a short note, in "The Archaeoastronomy and Ethnohistory Newsletter"(1992). Kintigh acknowledged that most archaeoastronomical research had been ignored by archaeologists, and attributed this largely to the fact that most of the efforts of archaeoastronomers were outside the purview of mainstream archaeology, and instead fell into the category of "gosh, that's neat that a building is oriented to some celestial event or other." This was, he asserted, simple data collecting, about on the same level as stamp collecting.

But Kintigh was wrong on several counts. First, it is important to understand that in many parts of the world, such as in the American Southwest, the evidence for prehistoric astronomical recording is not as obvious as it is in the complex cultures of Mesoamerica. It is necessary to first document the existence of such data, before they can be interpreted and integrated into more general anthropological questions. After all, archaeology had to go through its period of data collection and lower level interpretation before it could address human behavior. Second, it is not so much the archaeoastronomers who have to shape their questions to fit the research framework of anthropological archaeologists, but the latter who must strive to incorporate the insights and knowledge derived by archaeoastronomers into the context of the culture they are trying to understand. After all, as we have said, there is probably no culture that does not include the heavens as a major part of their worldview, and to truly understand a culture it is necessary to comprehend some aspects of their cosmology.

Again the Mayanists provide a good example of the potential of expanding anthropological inquiry to include celestial data. As the results of epigraphic research have been published, more and more archaeologists have realized that the Maya inscriptions offer an important source of data about how the Maya thought about themselves. They are merging epigraphic and iconographic studies with archaeological projects designed to understand the nature of Mayan society. The classic work which combines the efforts of an archaeologist and an epigrapher is *A Forest of Kings: The Untold Story of the Ancient Maya* by Linda Schele and David Freidel published in 1990.

In sum, there is a history of over 150 years of epigraphic and iconographic research conducted on the peripheries of Mayan archaeological research. It was not, however, until the 1980s and 1990s that archaeologists began to accept the contributions of epigraphers.

The problem, as we see it, centers on what questions are asked and how they are framed. While archaeologists now see that the glyphs from a site may provide a framework from within which we can structure archaeological inquiry, prior to their decipherment it was thought the glyphs might only provide chronological information or perhaps the names of rulers. This kind of information contributes little to economic theory, settlement patterns, demography, culture ecology, or other traditional topics of interest in anthropological archaeology. So archaeologists are now considering how to reframe questions to utilize these data, thereby providing a greater contribution to archaeological knowledge.

We might say that archaeoastronomers too have been laboring in the far reaches of mainstream anthropological archaeology, and that archaeologists have been remiss in not considering more carefully the accumulating body of astronomical knowledge and using that knowledge to restructure aspects of their inquiry into the nature of prehistoric society.

Suggestions and Conclusions

This then is the challenge for those archaeologists who attempt to understand culture in a systemic or holistic fashion as well as for archaeoastronomers—how can we work together to complement one another's work?

First, anthropological archaeologists must be made aware of the potential contributions of archaeoastronomers. This paper does not enumerate these because other papers in this volume do so very effectively, and anyone reading this volume is not likely to need convincing. It is those who do not read this literature or attend these kinds of conferences that need some persuading. But on a more positive note, the seeds have been sown, and archaeological theory is increasingly susceptible to the incorporation of astronomical phenomenon because of the trend for the more inclusive study of ideological aspects of culture.

Another hopeful trend is that as archaeologists attempt to understand not only selected aspects of a culture, but as much of the culture as possible, the lone scholar has a less important role in archaeology. Understanding culture in a systemic way is an enormous task and requires teams of researchers (Gumerman 1989). The accumulating results of archaeoastronomical research suggest that in the study of many cultures the inclusion of an archaeoastronomer on the team would enhance understanding of that culture. This will require a willingness to interact and compromise on the parts of both archaeologists and archaeoastronomers cooperating on collaborative efforts.

There is also an untapped potential for understanding the relationship between people in the past and their concern with astronomical phenomena—that is, an ethnographic link between living peoples and their ancestors. In the American Southwest alone, there are numerous ethnic groups that have maintained their oral histories explaining the way the universe works based on cosmological principles. We should explore this potential.

Another final optimistic note: in contrast with our Maya example, the study of North American prehistory did not evolve contemporaneously with the archaeoastronomical studies, like the co-evolution of epigraphy and Mayan archaeology. Only recently, in many cases, have archaeologists been willing to accept the cosmological significance of archaeological remains. Much of the efforts of archaeoastronomers are still in the "Gee whiz—look at this," stage. It is necessary, of course, that data first be observed before they can be explained, but given the relatively late start of archaeoastronomy in North America, it seems obvious that the field is progressing from description to explanation, and therefore has a greater likelihood of being incorporated into a more comprehensive understanding of past cultures.

Acknowledgments

We would like to thank Rolf Sinclair and the Oxford V Conference organizers for inviting our participation in this symposium and for stimulating our thoughts regarding the relationship of archaeology and archaeoastronomy. Robert Euler and David Ortiz provided helpful editorial suggestions. The authors, however, accept full responsibility for any oversights or omissions

References

Demarest, A. A. and G.W. Conrad (editors) 1992, Ideology and Precolumbian Civilizations (Santa Fe: School of American Research Press).

Gumerman, G. J. 1989, "Understanding the Hohokam", in Exploring the Hohokam: Prehistoric Desert Dwellers of the Southwest, Albuquerque: University of New Mexico Press.

Hammond, N. and G. Willey (editors) 1979, Maya Archaeology and Ethnohistory (Austin: University of Texas Press).

Kintigh, K. W. 1992, "I Wasn't Going to Say Anything, But Since You Asked: Archaeoastronomy and Archaeology", in Archaeoastronomy and Ethnoastronomy News, September Equinox, pp.1–2.

Schele, L. and D. Freidel 1990, A Forest of Kings: The Untold Story of the Ancient Maya (New York: William Morrow and Company, Inc.).

Trigger, B. G. 1984, "Archaeology at the Cross Roads: What's New?", in Annual Review of Anthropology, 13:4. Eds. D. Siegel, A. Beals, S. Tyler, Palo Alto: Annual Reviews, Inc.

3

Cultural Aspects of Neo-Catastrophism: Implications for Archaeoastronomy

Benny J. Peiser
Liverpool John Moores University
School of Human Sciences
United Kingdom

During the last 20 years, some of Britain's leading astronomers (who now loosely form the 'British School of Neo-Catastrophism') have developed the 'giant comet' theory and claim that a more 'active' and threatening sky might have caused the collapse of prehistoric civilizations as well as major cultural punctuations, changes of ancient belief systems and religious rituals. Can the astronomical evidence brought forward by these scholars be substantiated by historical and archaeological evidence?

* * *

Introduction

Since the early 1980s, when Luis Alvarez and his colleagues linked the extinction of the dinosaurs to a cosmic impact catastrophe, research in the field of astronomical neo-catastrophism has quickened its pace. During the course of the last decade, most scientists have gradually accepted the idea of global disasters caused by extra-terrestrial bodies. This acceptance depended on the assumption that cosmic induced cataclysms were restricted to primordial times, millions of years before the origins of the human ace.

Yet an increasing number of astronomers claim that a series of cosmic disasters on a much smaller yet global scale punctuated the Earth in more recent, i.e. prehistoric times (Clube & Napier 1982, 1990; Hoyle 1993; Asher, Clube, Napier & Steel 1994, Bailey 1995; Verschuur 1996, Peiser 1997a).

These authors envisage trains of cometary debris which repeatedly encounter the Earth, the largest and most dangerous representatives of which may be between one and several hundred metres in diameter. Such bodies, depending on their cometary constitution and hence their cohesive strength, can have catastrophic effects on the ecological system in a variety of ways. According to recent research findings, the encounters between cometary debris and the Earth are capable of producing (but do not always do so) three types of natural disasters: I) low or high level multi megaton explosions of fireballs which destroy natural and cultural features on the surface of the earth by means of floods, blasts and seismic damage and which can destroy regions up to the size of a large country; ii) massive high level influx of cosmic dust high above the stratosphere which causes a dramatic drop of global temperature. Such cosmically induced climatical disasters can last for decades and consequently lead to the suspension of agriculture; iii) massive high level influx of cosmic chemicals (associated with dust) with as yet incalculable biochemical potentials which may be harmful to DNA and can trigger evolutionary mutation.

Coherent Catastrophism

Dozens of astronomers around the globe are now involved in projects which focus on past, present and future impact catastrophes (Gehrels 1994; Asher, Clube, Napier & Steel 1994; Bailey, Clube, Hahn, Napier & Valsecci 1994; Bailey 1995; Steel 1995; Gribbin & Gribbin 1996; Bailey, Palmer & Peiser 1998).

Research in many other scientific fields has already been affected by the emergence of neo-catastrophism. Over the last few years, scholars have started to investigate the implications of catastrophic events on societal evolution (Heinsohn 1990, 1992; Boulding 1992), anthropology (Cachel 1992; Peiser 1994b; Peiser 1995; 1996a), human social behaviour (Mazur 1992) and the development of religion (Wolfe 1994a, 1994b; Peiser 1996b).

Above all, neo-catastrophist astronomers have argued that both the emergence and the collapse of human cultures, e.g. the Neolithic Revolution (Lewis 1996), the origin of metallurgy (Hoyle 1993), the onset and collapse of the Bronze Age civilisations (Bailey, Palmer & Peiser 1998), the construction of the Egyptian pyramids (Hoyle 1993; Steel 1995), the rise of Monotheism (Hoyle 1993), and even the collapse of the Roman Empire (Clube & Napier 1990; Baillie 1994, 1995), etc., might be associated with episodes of increased meteoric activity and multiple impacts. Such episodes punctuating the evolution of human cultures are looked upon as a primary agency determining the rise and fall of ancient civilisations (Clube 1994a, 1994b, Asher & Clube 1993).

The new findings of cometary astronomy and archaeological research, therefore, point to a new scientific hypothesis which deserves careful scrutiny by archaeoastronomers specifically that the multiple destructions of the Bronze Age cultures, documented by unambiguous destruction layers, might be related to cosmically induced catastrophes.

The Archaeological Evidence

During the last two decades a major paradigm shift towards neo-catastrophism has also occurred in the field of archaeology. A growing number of archaeologists and climatologists have documented strong evidence for seismic, climatical and ecological disasters at the end of the Early, Middle and Late Bronze Age in many parts of Greece (Dalfes, Kukla & Weiss 1997; Bailey, Palmer & Peiser 1998), the Near East (Weiss, et al. 1993; Weiss 1996; Egypt (Baillie 1995), China (Pang & Srivastava 1988; Pang, Keston & Srivastava 1989; Baillie 1995) and India (Raikes 1964; Raikes 1979; Raikes & Dales 1977; Tainter 1988).

Ever since Claude Schaeffer published his book *Stratigraphie Comparée et Chronologie de l'Asie Occidentale* (Schaeffer 1948), there has been continuous scientific debate about the nature and extent of the multiple destructions of Bronze Age civilisations. Schaeffer claimed that the repeated collapses of Bronze Age cultures were not caused by the action of man but instead by seismic activity. During the last decade, eminent archaeologists have substantiated his claim and have linked destruction layers in Aegean and Near Eastern sites with natural disasters rather than with military conquests (Kilian 1980; Gammon 1980; Weiss 1996).

Schaeffer analysed and compared the destruction layers of more than 40 archaeological sites in the Near and Middle East. He detected that all of these sites had been totally destroyed at least four times and that these destructions occurred simultaneously. Four major destruction levels were clearly detectable throughout Anatolia, the Levant and the Middle East: Two destructions dated from the end of the Early Bronze Age (EBII & EBIII), the other two destruction layers were detected at the end of the Middle Bronze Age (MBII) and the Late Bronze Age (LBIII) (Schaeffer 1948).

Now, almost fifty years after the launch of Schaeffer's study, a new publication presents unambiguous evidence for natural disasters and climate change during the Early Bronze Age (Dalfes, Kukla & Weiss 1997). The book presents the findings of more than 40 researchers and provides a review of relevant information about the Early Bronze Age collapse which brought down the first high civilisations of mankind. Most contributors suggest that a major climate change around 2200 ± 200 BC caused agricultural disasters and prolonged droughts which consequently led to a breakdown of social order in population centres in Asia, Africa and Eastern Europe. The exact physical causes, however, for this ecological catastrophe which coincided with simultaneous fire, earthquake and flood disasters in the same geographical areas, are still unclassified and require further scrutiny.

Endogenous causes such as volcanic eruptions and earthquakes appear to be out of the question for such widespread destructions. Extra-terrestrial catastrophes, on the other hand, can lead to a variety of ecological effects and physical damage. One should also bear in mind that only eighty years after the cosmic catastrophe over the Tunguska river, it is quite difficult to detect any hard evidence for large scale destructions even at the epicentre of the explosion.

Yet the likelihood of large-scale atmospheric impact events makes it much more difficult to detect unambiguous evidence for the occurrence of celestial disasters in the historical period. Not only is the smoking gun (i.e. the cosmic agent or a major impact crater) lacking; moreover, the body itself (corresponding destruction layers of regional, continental or global extent) is still missing. Consequently, scholars doubt whether a continental or even global assault during the last 10,000 years ever took place.

Should the 'giant comet' hypothesis by Clube & Napier (1982; 1990) be substantiated, repetitive punctuations and multiple 'Super-Tunguska' showers of fireballs might have interrupted, strongly influenced, damaged or utterly destroyed some of the main Bronze Age civilisations. These multiple impacts might have affected the climate, civilisations and belief systems on a global scale; the particular characteristics and the degree of damage afflicted upon the eco-system, however, would considerably differ from continent to continent, from region to region (Toon, Zahnle, Turco & Covey 1994).

The Actualistic Interpretation of Mythology

Can we really link historical impact catastrophes with mythological catastrophe legends? According to the generally accepted view among scholars, myths and legends about astral objects and sky gods and their battles represent astronomical movements which occur in a periodic manner and are observable today—such as the movement of the sun, the moon and the planets, eclipses, constellations etc. If we accept this actualistic paradigm, myths and legends and their playful re-enactment in ancient rituals incorporated the function to enhance the vitality of society, the fertility of the crops or the stability of the sun by way of "sympathetic magic." According to this view, ancient people modelled their myths basically according to the same sky that we can see today. Most myths and rituals have been interpreted in similar manner by archaeoastronomers and anthropologists. Yet the classical scholar Lewis Farnell stressed as early as 1919 that normal events and regular occurrences in nature would hardy have been incorporated in myths: "What is normal in nature and society rarely excites the myth-making imagination, which is more likely to be kindled by the abnormal, some startling catastrophe, some terrible violation of the social code" (Farnell 1919). If this observation is true—and the majority of chaos and catastrophe myths seem to confirm Farnell's statement—then the question arises whether it was not in fact extraordinary celestial events which led to the creation of chaos legends and catastrophe rituals.

Re-Assessing Catastrophe Myths and Rituals

Only 150 years ago, most of Europe's leading scholars were convinced that the ancient catastrophe and flood traditions originated from natural upheavals which had occurred during man's early history. Both historians and anthropologists were able to point to a rich variety of catastrophe legends found throughout the world.

For more than 2500 years, these ancient traditions were generally viewed as historical. Not only did most of the famous scholars of antiquity write about these disastrous turning points; during the 18th and 19th century the newly established science of geology gave rise to findings and strata which clearly appeared to testify to natural cataclysms. By consequence, scientists of Europe's enlightenment such as Vico, Boulanger, Whiston, Halley, Cuvier, and others (Schechner Genuth 1997) related the origins of ancient customs and religions to these past catastrophes. But when biblical fundamentalists started to use this catastrophic reading of natural history to accentuate their apocalyptic views and policies, a general opposition to catastrophe theory emerged.

Strictly speaking, however, it has to be said that most of the so-called 'biblical catastrophists' did not (and do not) seriously connect the geological timescale with the apparent historical timescale: rather they continued to identify "creation" and its aftermath with dramatic and/or catastrophic events, albeit of a supernatural nature, occuring within a timescale of some 6000 years. Yet, with the emergence of 19th century evolutionary gradualism, many fields of science directed their opposition to the apocalyptic and 'revolutionary' interpretation of natural history. Since scientists correctly associated apocalyptic radicalism with catastrophism, they threw the 'baby' out with the bathwater and simply banned catastrophe theories from universities, research and scientific journals. Over and over again, historians and anthropologists rejected any historical reading of the ancient catastrophe legends by reference to the works of Darwin and Lyell.

Since then, cultural anthropology and archaeoastronomy has tended to view the ancient chaos traditions as mere reflections of "mythical" thinking. As a result, archaeoastronomy, which focuses on the relationship between ancient astronomy, culture and religious ceremonies, has to a large extent ignored these traditions.

Since astronomy, geology and climatology have finally accepted the catastrophic history of planet Earth and its inhabitants, it is time for archaeoastronomers to recognise the findings of natural science and, where appropriate, to adopt them in their research.

Modern anthropologists who have studied ancient catastrophe traditions of the New World underline the missionary influence on particular tribal legends and mythologies. These traditions of natural catastrophes are no longer regarded as recollections of past events. Most scholars try to demonstrate that the catastrophe traditions originated from European travellers and missionaries who had read their own ideas into the religions of savage Indians. After one century of research, however, there is eventually a growing consensus among anthropologists that some of the most ancient Mesoamerican catastrophe accounts are autochthonous rather than borrowed (Horcasitas 1988).

The Aztec traditions, like most of the Mesoamerican accounts, recall the collapse of four previous World Ages (or 'Suns'). After the catastrophic end of the fourth era, the earth and the sky were "newly created" during the fifth and present age. Among the Mesoamerican peoples it was generally believed that the present era would eventually be destroyed by a natural disaster just as the previous cultures had collapsed (Peiser 1995).

A similar legend can be found in Chinese mythology. It revolves around Yi, the heroic archer, and how he saved the world from being destroyed by a cosmic conflagration. The myth recalls how in the times of Yao, ten suns rose in the sky, threatening to destroy the world. Yi, with the help of a sacred bow and arrows, saved mankind when he shot down these ten "suns" (Birrell 1993).

The historicity of these and many other legendary events are difficult to verify. Generally speaking, myths should never be interpreted literally. It is well known from many eye-witness accounts of earthquakes, the atom bomb on Hiroshima or the Tunguska impact catastrophe that survivors recall their fears of 'the end of the world.' Catastrophe myths, therefore, while they might provide rather vague data for ancient disasters, can never provide evidence about the nature and geographical extent of such calamities. Yet in view of new astronomical information, some of the widespread catastrophe myths might echo real natural events in prehistoric times (for example the breaking up of a comet into ten pieces, misperceived as ten "suns"). By comparison, most archaeoastronomers object to a natural or historical interpretation of these catastrophe legends. A legitimate reason for their general rejection of their historicity is based on the fact that these are simply legendary reports which include some sacred accounts of wondrous events. Moreover, catastrophe traditions in many of these legends are not taken seriously since they were often revised according to religious requirements or written down at a relatively late period of time. The main reason, however, for the traditional scepticism with regard to myths was the apparent absence of a scientific theory which could convincingly explain cosmic catastrophes.

Recent research, however, has confirmed that our planet and human civilisations have been repeatedly punctuated by cosmic debris during the last 10,000 years (Bailey, Palmer & Peiser 1998). These relatively new findings make it necessary for archaeoastronomers to reasses the possible historicity of some catastrophe myths and legends. According to Victor Clube and Bill Napier many of the world's catastrophe and combat myths, in which the battles of cosmic serpents and dragons lead to natural disasters, "could be describing successive stages in the break-up of a huge comet (...). The dragon of the combat myths can be seen as a description of a comet which, in a near-Earth orbit, produced impacts of fireballs" (Clube & Napier 1982). The history and development of this 'giant comet' is elaborated by Clube and Napier in terms which match the physical evolution of the observed Taurid meteorid stream (to the extent that the latter can be identified). This stream contains comets, asteroids, and meteoroids whose disintegration products are believed to be largely responsible for the observed zodiacal dust. According to both astronomers, the ascending and descending nodes of the dense core of this stream are encountered by the Earth in close succession (i.e. separated in time by about 100–200 years) every 2500 years or so (Clube & Napier 1990; Clube 1995). Most of this stream still cannot be readily observed and so its detailed structure is currently unknown.

If the 'giant comet' hypothesis could be substantiated, it would be only too reasonable to conclude that actual episodes of impact catastrophes would be most likely to have triggered the myth-making mind into the creation of 'fabulous' stories about flaming skygods and their catastrophic intervention in human affairs.

Collective Reactions to Natural Disasters

Natural catastrophes do not necessarily lead to collective responses in the form of catastrophe games or rituals. In many cases, the scope of the disaster is limited and the disaster agent is clearly identified. However, if a natural catastrophe results in extreme devastation and destruction, or if the agent of destruction is unknown, the disaster can prove to be a traumatic experience which triggers post-disaster collective reactions.

These activities stem from the desire of communities struck by catastrophe to influence and pacify the agents of destruction. Furthermore, expressive (and in most cases violent) responses are triggered by collective anger and tension following traumatic catastrophes. To avoid internal violence and civil strife, the growing aggression within a community of survivors has to be acted out in a structured, organised and generally accepted (i.e. ritualised) form.

It was particularly conspicuous in antiquity that many catastrophe rituals and festivals took place either to commemorate past catastrophes or to avoid future disasters. Although there is considerable overlap, these festivals can be classified into three different categories: i) festivals which were held in commemoration of a foregone disaster; ii) playful rituals with an apotropaic function to hinder future natural catastrophes; iii) games which dramatised and re-enacted a natural catastrophe. Most of these ceremonies were held in the form of combats, blood sacrifices and religious rituals or, alternatively, through mimetic dramatisation of the catastrophe. Right from the very start, they had the supposed function of pacifying natural forces and of hindering an apocalyptic end of the world.

The Aztec flying game volador apparently relates back to a natural catastrophe. The Aztecs believed that the Earth had been destroyed by a succession of four catastrophes (caused by devastating combats of the sky gods) leading to the present fifth age of creation. They appeared to assume that the best way to prevent another cataclysmic battle from occurring was to have the sun circle in the sky. According to Wasserman (1982/83), the Aztec flying game was a mimetic dramatisation of the Sun's movement, intended to hinder the repetition of yet another threatening disaster. The fear of a prospective cataclysm and its psychological working-off in the form of play was the proper function of this Aztec catastrophe game.

The most celebrated Aztec festival was called Toxiuhmolpilia. This ceremony took place every 52 years. On the last night of each 52-year cycle, the Aztecs extinguished every fire and proceeded to a sacred place, the Hill of the Star. Having reached the hill, they waited in terror for the final catastrophe which would mark the world's end.

Aztec astronomer-priests observed the heavenly bodies and when they saw that the stars had not changed their usual motion: "They knew that the movements of the heavens had not ceased and that the end of the world was not then, but that they would have another fifty-two years, assured that the world would not come to an end" (Coe 1975). Such apocalyptic behaviour in relation to the effects of trains of cometary debris which are circulating around the sun and intersecting the Earth's orbit is not at all unreasonable, as British archaeoastronomers conclude (Clube & Napier 1982, 1990). In fact, the common fear of the world's end and the endemic panic with regard to comets, which is common among many people, can only be explained satisfactorily if catastrophic encounters of the Earth with similar cosmic objects triggered the (only seemingly bizarre) prevalent association of comets with natural catastrophes in the first place.

Psychological Reasons for Catastrophe Rituals

Natural catastrophes always have a profound impact on survivors. The response to natural disasters, however, depends on its characteristics, its magnitude and its level of destruction. Research has shown that in severe cases, the reaction of survivors to natural catastrophes can result in a total breakdown of order and an extreme traumatisation of the survivors' psyche on a mass scale (Alexander 1992). One of the psychological consequences of natural disasters is the so-called survivor syndrome, which is characterised by extreme depression and feelings of total hopelessness.

Of utmost interest is the survivor's inner need to relive the traumatic experience of the catastrophe. The unambiguous proof for prehistoric floods, earthquakes and the total collapse of civilisations, therefore, prompts us to question the reactions of the survivors.

According to Heinsohn (1992), many sacrificial and catastrophe rituals are to be understood as a playful reaction following overwhelming natural disasters. What these catastrophes did not destroy is likely to be destroyed by the survivors themselves since they kill one another out of apocalyptic panic. As a response to traumatic crises, Heinsohn claims, that catastrophe games and blood sacrifices were introduced. Just as children try to digest a shocking experience through play, during which they relive the trauma but in so doing have everything under their own control, catastrophe games and rituals would appear to be a means of re-enacting foregone traumatic events. In this way, people were able to rid themselves of the emotional suffering inside them through the participation in the catastrophe games whose course, however, lay entirely in their own hands. In these mimetic games people changed their former role as passive victims of catastrophe into the role of active participants. In this way, they gained the control of a self-made plot which on most occasions resulted in killing, destruction or blood sacrifices.

The violence and destruction which human beings had once experienced, was acted out in an imitative form of play, so that the apocalyptic tension could be psychologically worked off and potentially self-inflicted violence could be diverted (Peiser 1994b, 1995, 1996a, 1997b). In short, if the 'giant comet' hypothesis by Clube and Napier could be substantiated, it might be able to elucidate many of the seemingly bizarre beliefs and rituals of ancient (and modern) religions.

Acknowledgments

I wish to thank Victor Clube (University of Oxford & Armagh Observatory) who read an earlier version of this paper and whose helpful information and advice certainly improved this text.

References

Alexander, D. 1992, Natural Disasters (New York).

Asher, D. J. and Clube, S. V. M. 1993, "An extraterrestrial influence during the current glacial-interglacial" Quarterly Journal of the Royal Astronomical Society 34, p. 481.

Asher, D. J., Clube, S. V. M., Napier, W. M. and Steel, D. I. 1994 "Coherent Catastrophism" Vistas in Astronomy 38, p. 1.

Bailey, M. E. 1995 "Recent Results in Cometary Astronomy: Implications for the Ancient Sky" Vistas in Astronomy 39, p. 647.

Bailey, M. E., Clube, S. V. M., Hahn, G., Napier, W. M. and Valsecci, G. B. 1994 "Hazards due to giant comets: climate and short-term catastrophism", in: Hazards due to Comets and Asteroids, Ed. T. Gehrels (Tucson: University of Arizona Press), p. 479.

Bailey, M. E., Palmer, T. and Peiser, B. J. 1998, Natural Catastrophes during Bronze Age Civilisations (Oxford: British Archaeological Reports, in press)

Baillie, M. G. L. 1995, A Slice through Time. Dendrochronology and Precision Dating. (London: Batsford).

Baillie, M. G. L. 1994 "Dendrochronology raises questions about the nature of the AD 536 dust-veil event" The Holocene 4:2, p. 212.

Birrel, A. 1993, Chinese Mythology: An Introduction. (Baltimore & London: The John Hopkins University Press).

Boulanger, N. A. 1767, L'Antiquité déveilé par ses usages (Amsterdam).

Boulding, K. E. 1992, "Punctuation in Societal Evolution", in: The Dynamics of Evolution. The Punctuated Equilibrium Debate in the Natural and Social Sciences, Eds. A. Somit and S. A. Peterson (Ithaca/London: Cornell University Press), p.171.

Cachel, S. 1992, "The Theory of Punctuated Equilibria and Evolutionary Anthropology" in: The Dynamics of Evolution. The Punctuated Equilibrium Debate in the Natural and Social Sciences, Eds. A. Somit and S. A. Peterson (Ithaca/London: Cornell University Press), p. 187.

Clube, S. V. M. 1994a, "Hazards from Space: comets in history and science", in: The Mass-Extinction Debates: How Science Works in a Crisis, Ed. W. Glen (Standford: University of Stanford Press), p. 159.

Clube, S. V. M. 1994b "Revelation and catastrophe during the Christian Era: a basis for historical interpolation and future extrapolation" Chronology and Catastrophism Review Special Issue: Proc. 1993 Cambridge Conference (Manchester: SIS), p. 66.

Clube, S. V. M. 1995 "The Nature of Punctuational Crises and the Spenglerian Model of Civilization" Vistas in Astronomy 39, p. 673.

Clube, S. V. M. and Napier, W. M. 1982, The Cosmic Serpent (London: Faber & Faber).

Clube, S. V. M. and Napier, W. M. 1990, The Cosmic Winter (Oxford:Blackwells).

Coe, M. D. 1975 "Native Astronomy in Mesoamerica", in: Native American Astronomy Ed. A.F. Aven (Austin: University of Texas Press), p. 3.

Dalfes, H. N, Kukla, G. and Weiss, H. (Eds.) Third Millennium Climate Change and Old World Collapse (Berlin/Heidelberg: Springer Verlag).

Farnell, L. R. 1919 "The value and methods of mythological study" Proceedings of the British Academy, XX, p. 30.

Gammon, G. 1980 "Bronze Age Destructions in the Near East" SIS Review IV:4, p. 104.

Gehrels, T, (Ed.) 1994, Hazards due to Comets and Asteroids (Tucson: University of Arizona Press).
Gribbin, J. and Gribbin, M. 1996 Fire on Earth: Doomsday, Dinosaurs and Humankind (New York: St Martin's Press).
Heinsohn, G. 1990, "Destruction layers in archaeological sites: the stratigraphy of Armageddon", in: Catastrophism 2000, Eds. M. Zysman and C. Whelton (Toronto: Heretic Press), p. 213.
Heinsohn, G. 1992 "The Rise of Blood Sacrifice and Priest-Kingship in Mesopotamia: A 'Cosmic Decree'?" Religion 22, p. 309.
Horcasitas, F. 1988, "An Analysis of the Deluge Myth in Mesoamerica", in: The Flood Myth, Ed. A. Dundes (Berkeley/ Los Angeles/London: University of California Press), p. 183.
Hoyle, Sir F. 1993, The Origin of the Universe and the Origin of Religion (Rhode Island & London: Wakefield).
Kilian, K. 1980 "Zum Ende der mykenischen Epoche in der Argolis", in: Jahrbuch des Römisch-Germanischen Zentralmuseums Mainz 27, p. 166
Lewis, J. S. 1996, Rain of Iron and Ice: The very real threat of comet and asteroid bombardment (Reading, Mass.: Helix Books).
Mazur, A. 1992, "Periods and Question Marks in the Punctuated Evolution of Human Social Behavior", in: The Dynamics of Evolution. The Punctuated Equilibrium Debate in the Natural and Social Sciences, Eds. A. Somit and S. A. Peterson (Ithaca/London: Cornell University Press), p. 221.
Pang, K. D. and Srivastava, S. K. 1988 "Climatic Impacts and Volcanic Eruptions: Interferences from Ice Core, Tree Ring and Historical Data" Eos 69:44, p. 1062.
Pang, K. D., Keston, R. and Srivastava, S. K. 1989 "Climatic and Hydrologic Extremes in Early Chinese History: Possible Causes and Dates" Eos 24, p. 1095.
Peiser, B. J. 1994b "Catastrophism and Anthropology" Chronology and Catastrophism Review Special Issue: Proc. 1993 Cambridge Conference, p. 130
Peiser, B. J. 1995 "Cosmic Catastrophes and the Ballgame of the Skygods in Mesoamerican Mythology" Chronology and Catastrophism Review XVII, p. 36.
Peiser, B. J. 1996a, "Catastrophe Games. Playful reenactment of traumatic events", in: Games of the World between Tradition and Modernity, vol 1, Eds. E. Pfister, T. Niewerth and G. Steins (Sankt Augustin: Akademia), p. 112.
Peiser, B. J. 1996b, "The cosmic symbolism of the Mesoamerican ballgame", in: Sport and Symbols—Symbols and Sport, Ed. F. Van der Merve (Sankt Augustin: Akademia), p. 107.
Peiser, B. J. 1997b, O-Kee-Pa: Catastrophe Myths and Legends among the Mandan Indians" Chronology & Catastrophism Review XIX, p. 34.
Raikes, R. L. 1964 "The End of the Ancient Cities of the Indus" American Anthropologist 66, p. 284.
Raikes, R. L. 1979, "The Mohenjo-doro floods: the debate continues", in: South Asian Archaeology 1977, Ed. M. Taddei (Naples: Instituto Universario Orientale, Seminario di Studi Asiatici) p. 561.
Raikes, R. L. and Dales, G. F. 1977 "The Mohenjo-doro floods reconsidered" Journal of the Palaeontological Society of India 20, p. 251.
Schaeffer, C. F. A. 1948, Stratigraphie Compare et Chronologie de l'Asie Occidentale: IIIe et IIe millénaires (London: Oxford University Press).
Schechner Genuth, S. 1997 Comets, Popular Culture and the Birth of Modern Cosmology (Princeton: Princeton University Press).

Steel, D. I. 1995, Rogue Asteroids and Doomsday Comets: The Search for the Million Megaton Menace that threatens Life on Earth (New York/ Chichester/ Brisbane/ Toronto/ Singapore: John Wiley & Sons).

Tainter, J. A. 1988, The Collapse of Complex Societies (Cambridge: Cambridge University Press).

Toon, O. B., Zahnle, K., Turco, R. P. and Covey, C. 1994 "Environmental perturbations caused by impacts", in: Hazards due to Comets and Asteroids, Ed. T. Gehrels (Tucson: University of Arizona Press), p. 791.

Velikovsky, I. 1950, Worlds in Collision (London: Victor Gollanz).

Verschuur, G. L. 1996, Impact! The Threat of Comets and Asteroids (Oxford: Oxford University Press).

Wassermann, M. 1982/83 "Myth and Cultus in Sport: The Case of the Aztec Flying Game", Stadion VIII/IX, p. 99.

Weiss, H., Courty, M. A., Wetterstrom, W, Guichard, F, Senior, L, Meadow, R and Curnow, A. 1993 "The Genesis and Collapse of the Third Millenium North Mesopotamian Civilization" Science 261, p. 995.

Weiss, H. 1996 "Desert-Storm" Sciences 36:3, p. 30.

Wolfe, I. 1994a "A Catastrophic Interpretation of Western Cosmologies" Chronology and Catastrophism Review Special Issue: Proc. 1993 Cambridge Conference (Manchester: SIS), p. 51.

Wolfe, I. 1994b "A Catastrophic Reading of Religious Systems" Chronology and Catastrophism Review XVI, p. 2.

4

Astronomical Omens

Bradley E. Schaefer
Astronomy Department
University of Texas
Austin, TX

Astronomy does affect our world in many ways, where the skies bring changes in both our physical environment and our symbolic environment. This paper examines questions concerning the power and nature of celestial omens. The source of power for signs in the sky arises since the heavens are the abode of gods, so the authorship of any message there must be divine and true believers will act upon divine authority. The opportunity for sighting impressive omens in the sky averages several times a year, so that any soothsayer can always find a celestial excuse to support any revelation. Historically, omens tend to be quite accurate in their predictions, as dictated by several reasons; interpretations are frequently made after the event, some omens result in self-fulfilling prophecies, the stated interpretations are often ambiguous, and failed predictions tend to be forgotten. Omens from transient astronomical events have a strong universal tendency to be negative in character, and this is likely the result of such events breaking the harmony of nature. When events become predictable, they lose power as omens and instead become yet a further manifestation of the harmony of nature. Signs in the sky also lose their power when they become understood, since the correct explanations are not fearsome like the whim of some powerful god.

* * *

Introduction

Astronomy is usually viewed as remote from ordinary human life. This is in contrast to the 'practical sciences' which have produced great improvements in the standard of human living. But astronomy actually impacts many areas of ordinary life for all citizens both ancient and modem. The connections between the sky and humans are through astronomical influences on our physical environment and through the sky as part of our symbolic environment. As such, the topic of astronomy and culture has many manifestations beyond alignments and folk tales.

One prominent means by which the sky changes individual's behavior and history is through celestial omens. That is, signs in the sky are interpreted as messages from divinities, which are then acted upon. I would like to address five questions relating to astronomical omens.

The Source of Power

The first question is "Why do signs in the sky have the power to impel man to action?" Here, I think that the answer is straight forward. It is because the heavens are the abode of the gods—and *only* the gods live in the sky—and usually the gods who live in the sky are among the most powerful in the pantheon. Any sign in the sky could only have been put there by the gods. Any divine message must be acted upon by a true believer. Thus, we find a strong mechanism linking events in the sky with events on Earth.

Meteor falls provide a emphatic case of this mechanism. Observed meteors for which a rock can be recovered are very rare, especially in old times of low population density. But such events are spectacular with their burning column of fire and their noisy impact. The obvious connection with the sky assures the divine origin of the exotic stone found at ground zero. If you are from a culture with an animistic religion, then the meteorite itself might be worshiped. I know of seven such cases spread out widely around the world. The better known examples are the Black Stone in the Kaaba (likely impactite from the Wabar crater), the meteorite of Einessa forced on the worship of the Roman empire by Elagabalus, and the Cybele meteorite transported to Rome for five centuries of worship. A less publicized example is the Red River meteorite which was venerated as a god of healing by the Pawnee tribe and now residing in the main entrance to the Yale astronomy department, so that I have the weird feeling of having an ex-god just down the hall from my office. Such animistic worship tends to occur early in cultures when records am poor or nonexistent, so it is likely that meteorite worship was much more prevalent in old times than these seven cases would alone suggest.

Frequency of Omens

My second question is "How often do suitable celestial phenomena happen that can be interpreted as omens?" That is, are omens rare and wonderful or common and prosaic? This question arises since on occasion a confluence of spectacular celestial events is interpreted as necessarily resulting in extraordinary reaction by societies. For example Lilley (1986) claims that a series of solar eclipses and other events provoked the Easter Islanders to build their moais, while Malville and Putnam (1989) claim that a similar series converted the inhabitants of Chaco Canyon into devoted sky watchers, while Robinson (1983) claimed that a series of winter solstice eclipses near the horizon influenced the design and orientation of Stonehenge. At the Oxford V conference, W. B. Masse claims that two separate series of solar eclipses visible across the American Southwest both individually caused changes that were previously inexplicable in the archaeological record. As a reaction to the claim by Masse, in the open discussion period, I challenged Masse to randomly pick a time and place, such that I would search for some confluence of astronomical spectacles. He chose China in the 1300's and also China in the 1400's. After review, I refined this to trying to find a spectacular series of celestial omens between the years 1361 and 1391 from the Ming capital as an explanation for the flowering of the arts at the start of the Ming dynasty. Indeed, I found one of the best apparitions of Halley's Comet. A dozen other comets were recorded from Nanking. Three total or annular eclipses crossed over Nanking. The Chinese astronomers in Nanking recorded sunspots at a rate several times higher than any other time in history. A nova was reported in 1388, and 16 total lunar eclipses were visible. But the clincher was the fantastic Leonid meteor storm in 1366! With these great visions all crammed into one place and generation, we can have no wonder at the root cause of the Ming revival. The second control time was narrowed down to the Nanking area from 1400 to 1472. There were three solar eclipses where totality passed over Nanking, while four totalities passed nearby. Of total lunar eclipses, 33 were visible from Nanking. The Chinese record 26 naked-eye comets, which include two comets that were bright enough to be seen in the daytime as well as another of the all-time best apparitions of Halley's Comet. The Chinese recorded six novae. And there was another great Leonid storm in 1466. The point from my two control studies of times and places selected randomly by Masse is that *any* time and place can be shown to have a confluence of celestial spectacles.

This high rate of celestial omens is not surprising. Masse himself found two separate half-century series of impressive omens visible from the American southwest within a several century window, and this by itself shows such series to be insignificant. Indeed, there are various times in history when many impressive events occur close together. For example, the Star of Bethlehem is restricted to a several year period, yet astronomers identify it with any of seven known events (a nova, a triple conjunction of Jupiter and Saturn, a massing of Mars, Jupiter, and Saturn in Pisces, an occultation of Jupiter by Venus, a comet, a supernova, and the stationary point of Jupiter). Similar clusters of events have been found around various times of millenialist expectations. Two other examples are given in Schaefer (1989–93).

As a more systematic approach, Schaefer (1989–93) has compiled lists of event rates for many types of transient sky phenomena. Depending on what types of events are to be considered, the rate ranges from once a decade to several times a year. In particular, the concentrations reported by Liller (1986), Malville and Putnam (1989), and Tobinson (1983) are shown to be insignificant.

All this goes to show that if modern astronomers look closely enough at any restricted time and place, they can always come up with an apparent over density of events where the rate of signs in the sky is much higher than most researchers or laymen would expect. There are two important implications for historians: First, it is dangerous to go from a perceived cluster of omens to a conclusion that this cluster motivated historical happenings. Second, omens in the sky happen so often that any seer can always find some celestial excuse to support any desired message. This relegates astronomy to the background, since the interesting question is then why the seer chose to present the given message at the given time instead of what the sky was doing.

Accuracy of Omens

My third question is "Why do omens always seem to provide accurate predictions?" That is, you never hear about portents that get it wrong. But how can this be since the physical connection is often zero?

To support this question, I have constructed a list of 60 celestial omens from the eastern and central Mediterranean regions through Roman (including Byzantine) times. Such a restricted area and time allows for a reasonable cultural continuity, yet allows enough scope to collect many cases with detailed knowledge. These 60 omens have had various properties and characteristics identified and tabulated (see Table 1). Out of 60 cases, only four did not come true. So statistically, the second question is posed with solid grounds.

In studying the cases, I have concluded that there is no one reason why omens are generally accurate. However, I have identified four separate reasons which combine to answer my second question:

The first reason is that interpretations are often made *after* the outcome is known. In such a case, the prediction will inevitably be correct. Halley's Comet and the Norman conquest of England provides a classic example of this. In the half-year between the appearance of the cornet and the Battle of Hastings, the doom foretold by the cornet worried *both* sides equally. It was only after William had triumphed that the chroniclers identified the sign in the sky as bad luck for *England*. In my census, 42 cases have a definite order to the omen and the event. Of these, seven are interpretations *after* the fact. Of the omens which did not come true, all had interpretations before the fact. So we see that this explanation is valid, but it can explain only a sixth of the cases.

The second reason is that the predictions based on the omen can work as a self-fulfilling prophecy. The classic example is during Nero's reign when a bright comet in 64 AD appeared amid popular interpretation as signaling the death of the emperor. Nero tried to divert the wrath of the comet by killing several dozen prominent Roman leaders. For my census of omens, only 2–4 cases are possible/probable examples of self-fulfilling prophecies. So my second reason is valid, but only for a minority of cases.

The third reason is that omens are ambiguous. The classic example is the oracles of Delphi, for which the statements are interpreted after the fact to correspond to what actually happened. In modem days, the classic examples are horoscopes and fortune cookies which can always be twisted to account for some facet of a day's activity. This is a generic problem for celestial omens which are intrinsically neutral, since the identity of the doomed party can always be adjusted after the outcome is known. For my census of omens, one-third of the cases are readily explained by this third reason. In particular, these include omens for which no event is identified, unclear events with no definite outcome, and omens for which the doomed side is not stated or implied.

Out of the 60 omens in my census, only 19 have the interpretation preceding the event, are not self-fulfilling prophecies, and have a definite right/wrong evaluation. That is to say, two-thirds appear correct only because of some extrinsic reason. Nevertheless, this still leaves 19 events of which 16 have correct predictions. So even after removing the forced successes, omens still have a high rate of success. I believe that this can be entirely accounted for by a fourth reason.

Table 1. Census of Sixty Celestial Omens

Year	Reporting Culture	Omen Event	Historical Association	Time Order	Positive/Negative	Come True?	Self Fulfill?
-216	Roman	comet	battle	O/E	?	yes	no
-134	Roman	comet	royal greatness	E&O/I	positive	yes	no
-87	Roman	comet	war Calamities	O/E	negative	?	no
-56	Roman	comet	political crises	O/E	negative	yes	no
-52	Roman	comet	political crisis	E/O	negative	yes	no
-44	Roman	comet	death of Caesar	O/I/E	pos/neg	yes	yes/no
-42	Roman	comet	battle	?	negative	?	no
-31	Roman	comet	battle	?	negative	?	no
-30	Roman	comet	death of queen	E/O	negative	yes	no
-12	Roman	comet	death of noble	?	negative	yes	no
9	Roman	comet	battle	?	negative	yes	no
14	Roman	comet	death of emperor	O/E	negative	yes	no
14	Roman	lunar eclipse	end of revolt	O/I/E	negative?	yes	yes
54	Roman	comet	death of emperor	?	negative	yes	no
60	Roman	comet	death of emperor	O/I/E	pos&neg	-	no
64	Roman	comet	death of nobles	O/I/E	negative	yes	yes
69	Roman	comet	rebellion	O/E	negative	yes	no
59	Roman	comet	fall of Jerusalem	O/E	negative	yes	no
76	Roman	comet	-	O/I	negative	-	-
79	Roman	comet	death of emperor	O/E	pos & neg	yes	no
200	Roman	comet	death of emperor	?	negative	yes	no
218	Roman	comet	panic	?	negative	-	no
336	Roman	comet	death of emperor	O/E	negative	yes	no
363	Roman	comet	death of emperor	?	negative	yes	no
364	Roman	comet	death of emperor	?	negative	yes	no
400	Roman	comet	?	O/E	negative	?	no
402	Roman	comet	battle	?	positive(?)	?	no
539	Byzantine	comet	invasion	?	negative	?	?

791	Byzantine	unknown	expedition	O/I/E	positive	no	no
822	Byzantine	comet	heresy	?	negative	?	no
840	Italian	solar eclipse	-	O/I	negative	-	-
904	Byzantine	comet	birth of emperor	E/I	positive?	yes	no
904	Byzantine	comet	death of emperor	O/I/E	negative	yes	no
959	Byzantine	comet	death of emperor	?	negative	yes	no
975	Byzantine	comet	-	O/I	negative	-	-
985	Byzantine	comet	Bulgar menace	O/I/E	negative	yes	no
986	Byzantine	meteor	battle	O/I/E	negative	yes	possibly
989	Byzantine	aurora	war	O/I/E	negative	yes	no
1010	Italian	solar eclipse	famine	?	negative	yes	no
1095	Crusaders	meteor storm	Crusades start	O/I?/E	neutral	yes	no
1097	Crusaders	comet	battle	?	positive	yes	no
1097	Crusaders	aurora	battle	E/O	negative	yes	no
1097	Crusaders	lunar eclipse	fall of Jerusalem	O/I/E	positive	yes	no
1097	Crusaders	meteor	battle	O/I/E	positive	yes	no?
1117	Crusaders	aurora	Deaths of princes	O/I?/E	negative	yes	no
1117	Crusaders	lunar eclipse	Deaths of princes	O/I?/E	negative	yes	no
1117	Crusaders	lunar eclipse	Deaths of princes	O/I?/E	negative	yes	no
1176	Byzantine	solar eclipse	-	O/I	negative	-	-
1176	Islamic	solar eclipse	battle	O/I/E	negative	no	no
1177	Sicily	solar eclipse	church foundation	O/I	negative	-	no
1187	Crusaders	solar eclipse	fall of city	E/O	negative?	yes	no
1187	Byzantine	solar eclipse	end of world	I	negative	no	no
1331	Byzantine	solar eclipse	death of emperor	O/E	negative	yes	no
1337	Byzantine	solar eclipse	battle	?	negative	?	no
1361	Byzantine	solar eclipse	-	O/I	negative	-	-
1431	Italian	solar eclipse	death of pope	?	negative	yes	no
1453	Byzantine	volcanic glow	savior army	O/I/E	positive	no	no
1453	Byzantine	sky lights	fall of city	O/I/E	negative	yes	no?
1453	Byzantine	lunar eclipse	fall of city	O/I/E	negative	yes	possibly
1456	Italian	comet	invasion	E/O/I	negative	yes	no

The fourth reason is that failed omens are rapidly forgotten. They don't make good stories. The classic example is that members of the Seventh Day Adventist Church forget the failed omen of the Great Comet of 1843. End-of-the-world omens are a particularly fruitful grounds for the rapid loss of interest in failed predictions. And we've all heard of the wonderful psychic readings of Jeanne Dixon, but few have heard of her abysmal track record and her very high rate of failure. Tales of fulfilled prophecy make great drama, while misguided omens are merely perplexing. It is natural that accurate omens are chosen for inclusion in history. This bias is based on a strong selection effect fundamental to human nature.

In summary, the apparent accuracy of omen predictions arises from four effects in a complex manner.

Positive or Negative?

My fourth question is "Are astronomical omens generally positive or negative in character?" This question arose when Halley's Comet last came around, with many writers claiming comets were generally negative while other writers pointed out times when comets were positive. Indeed, this question can be made more general for other classes of signs in the sky.

The general answer is clear and universal. Comets, eclipses, and meteors are bad tidings everywhere in the world. For comets, I have found ethnographic or historic evidence from a wide variety of cultures around the world, including the Incas, Romans, Tibetans, Chinese, Aztecs, Greeks, as well as all western European based societies. For eclipses, Jamaicans, Aztecs, Chinese, east Africans, Mexicans, Indians, American Indians, Greek, Mayan, and Turkish cultures all view them with dread. For meteors, negative views are prevalent in central Asia, Greece, Japan, Australia, North America, and broadly over Europe. Other transient astronomical events (supernovae, sunspots, auroras) also invoke fear throughout the world. This universal character of transient celestial omens requires a universal reason deep within the human psyche.

Why is this so? I think that it has to do with the conservative nature of old societies. That is, the cycles and rhythms of nature are comforting regularities in an otherwise mysterious universe. And these same cycles bring assurance that the Sun will dawn and spring will come. Any breaks in these harmonies of nature were bad, be they floods or droughts or deaths. The ordinary sky is epitomized by reliable cycles. But comets, eclipses, meteors, and other transient events are radical breaks in natural harmony. By analogy, these signs must also imply impending trouble.

This difference between transient and regular events is further highlighted by the common worship of regular events. That is, the objects with reliable cycles are canonized as benevolent gods, with Sun worship near the founding of all cultures and Moon worship not far behind. The regularity of the Sun's motions are enshrined in Stonehenge and religious temples of all denominations. The nature of astronomical omens is positive-or-negative based entirely on whether the event is regular-or-transient.

Despite the above arguments, there are indeed individual cases that violate the rule. Out of my census of 60 omens, 6 are definitely positive. These are a mixed bag. Three of the positive omens came from the First Crusade where the infectious optimism colored their outlook. One interpretation was made long after the event, where the positive nature of the event (the birth of Mithradites) was dictated by expedience. Political necessity has forced numerous attempts to overcome the intrinsically evil nature of certain signs. Thus, Seneca wrote a book on the good nature of comets so as to assuage Nero's fear of the comet of 60 AD, Augustus employed the first widespread political propaganda campaign to twist the evil omen of Caesar's death into a token of his deification, and a Chinese court astrologer won advancement by breaking with the traditional interpretation of calamities following the supernova in 1006.

I conclude that regular and reliable astronomical events are universally regarded as positive in character, while transient astronomical events are universally regarded as negative in character. Like all rules for human beliefs, this conclusion has exceptions, caused by forced attempts based on political expedience.

Losing Power

My fifth question is "When do omens lose their power?" In old times, portents were potent. Yet in modem times, omens are few and of little impact. (Despite pessimistic news stories, newspaper horoscopes are relegated to the comics page and scandals are created when a president's wife merely juggles her appointments schedule on astrological advice.) What has changed? I think that there are two answers:

The first answer, is that omens lose their power when the true explanation is known. The classic example is the two Athenian expeditions to Syracuse. The first was during a lull in the Peloponnesian War. When the siege of Syracuse was getting nowhere, the Athenians decided upon an evacuation before matters turned worse. But a lunar eclipse on the night of embarkation was taken as a sign to delay the retreat for a month. During that month, the Syracuse army gained the upper hand and wiped out the expedition. Ironically, the true cause of lunar eclipses had just been discovered by Anaxagoras, but he had been jailed in Athens to gag his impious explanation. However, a later expedition against the tyrant of Syracuse was made by Greeks who knew the true cause and were not frightened. When a lunar eclipse happened during the embarkation in Athens, no one worried. After all, who is afraid of a shadow that will soon pass? With the true explanation, we are no longer dealing with a whimsical god, but are instead dealing with a rational phenomenon.

The second answer is that omens lose their power when they become predictable. The omens have their power since only divine messages can appear in the sky. The classic example is Columbus' lunar eclipse, by which he frightened the native Jamaicans into supplying his shipwrecked sailors with food for many months while awaiting rescue. Columbus and his men knew of the prediction of the event in the almanac of Regiomontanus, while the event was unexpected to the Jamaicans. This difference led to fear on one side and glee on the other. A turnabout of this episode came when the chief Tecumseh used a total solar eclipse in 1806 to confirm the prophetic powers of his brother, Tenskwatawa. Here, Tecumseh's knowledge of the prediction of the eclipse was translated into religious power. A predictable omen does not break the harmony of nature, and hence is not fearful. Instead, a predictable omen *confirms* the natural order and serves as a comfort.

In modern times, correct and non-threatening explanations are widely known, and many of the signs in the sky are fully predictable. Indeed, eclipses now serve as occasions for joyous parties with special bands of devotees, while comets are a sight commended for their beauty, and supernovae create scientific excitement and research opportunities. Thus, celestial signs happen frequently, so convenient portents can always be found. Celestial omens are few and impotent in this day-and-age.

Conclusions

The omens have their power since only divine messages can appear in the sky. Celestial signs happen frequently, so convenient portents can always be found by any attentive prophet. Omens have a good track record of coming true, but only for various extrinsic reasons. Generally, the transient events are viewed negatively since they represent breaks in the natural cycles, while reliable events are viewed positively for their comforting regularity. Omens lose their power when they become understandable or predictable.

At the end of my oral presentation in Santa Fe, I ventured to make a celestial prophecy and omen of my own. I predicted that on the last night of the Oxford V field trip to Chaco Canyon (August 11/12), the heavens themselves would blaze forth with fire in the sky as a sign from above that the conference had been good. While the audience agreed that the omen would yield the correct conclusion, none were frightened by the upcoming predictable meteor shower. And indeed, I can report that on the appointed night, the skies were filled with glorious illumination.

References

Liller, W. 1986, *Archaeoastronomy*, 9, 52.
Malville, J. M. & Putnam, C. 1989, *Prehistoric Astronomy in the Southwest.* (Boulder: Johnson Books).
Robinson, J. H. 1983, *Archaeoastronomy*, 6, 124.
Schaefer, B. E. 1989–93, *Archaeoastronomy*, 11, 91.

5

Black Suns and Dark Times: Cultural Responses to Solar Eclipses in the Ancient Puebloan Southwest

W. Bruce Masse
Los Alamos National Laboratory and
Affiliate Graduate Faculty
University of Hawaii

Robert Soklow
Southwest Research Associates
San Diego, CA

The apparent subjective nature of past iconographies and traditional cosmologies has made archaeologists and astronomers reluctant to investigate the possible correlations between cultural behavior and known or reconstructible naked-eye visible temporary celestial events (e.g., novae, comets, meteor storms, eclipses). The findings presented in this paper are intended to encourage such investigations for the solar eclipses of the Puebloan portion of the American Southwest between A.D. 1 and 1540. We identify these eclipses and provide a contextual basis for suggesting that previously inexplicable events, processes, and iconography manifest at certain periods in the archaeological record are responses to specific eclipses and eclipse clusters.

* * *

Introduction

Sky as Environment

A total or high magnitude annular solar eclipse is a brief event, lasting at most a few minutes. Nevertheless, the suite of physical effects that can accompany such an event is quite impressive. In addition to the darkness created by the eclipse, the sudden appearance of stars in the daytime, and the impressive nature of the Sun's corona, effects can include noticeable changes in animal behavior, strong gusts of wind as ambient temperatures drop, clouds becoming ominously black, and the human physiological responses to anything that causes awe or fear (Allen and Allen 1987, pp. 93–109; Littmann and Wilcox 1991, pp. 51–52; Zirker 1995, pp. 183–189).

The paths of total eclipses are long and narrow, with a typical path averaging about 6,000 miles (9654 km) in length and 100 miles (161 km) wide (Zirker 1995: 30). From any given spot in the northern hemisphere a complete total or annular solar eclipse will be visible, weather permitting, on the average of only once every 320 years (Allen and Allen 1987: 40). Occasionally, however, multiple eclipses will occur in the same general area over a very short time frame, such as the total eclipses witnessed near Port Moresby in Papua New Guinea in 1983 and 1984. Our data identifies when and where multiple solar eclipses occurred within small circumscribed regions of the Puebloan Southwest between A.D. 1 and 1540. At this writing, we have focused our investigation in these areas of confluencing eclipse paths and how they can be identified in the archaeological/cultural record.

Mesoamerican and Puebloan Responses to Solar Eclipses

Recent studies of the Mexica [Aztecs] have underscored the importance of solar and lunar eclipses as major cultural events (Milbrath 1995). In Mexica chronicles only one or two events are provided each year, including major battles, floods, the death of a ruler, and solar eclipses. Solar eclipses at or near the time of death of a ruler were typically linked to that death. During lunar eclipses pregnant women put obsidian in their mouths or bosoms to protect themselves and to prevent disfigurement and disease for their unborn children (Sahagun 1953).

Solar eclipses were viewed as traumatic events because of the belief that the four preceding cosmological ages had each ended with cataclysm and the death of the Sun (Leon-Portilla 1963, 1995; Bierhorst 1992; Masse 1995, Masse and Espenak, 2003). Each new eclipse thus portended the end of the present age. Sahagun (1953), documented that prior to the Spanish conquest of Mexico, a solar eclipse would trigger a loud cry from all people, captives and people of light complexion were slain, and all people participated in blood-letting sacrifices. The noise raised by people during a solar eclipse was to scare away the *tzitzimime* demons believed to be devouring the Sun [Miller and Taube 1993; Milbrath 1995], and to prevent these demons from diving headlong from the heavens to attack the Mexica. The *tzitzimime* were visualized as constellations, individual stars, and planets transformed during the eclipse into starlike, clawed demons. This is significant since the Mexica, like most traditional cultures, viewed planets, stars, and constellations as being aspects of powerful gods or supernatural beings. In some Mexica cosmogonic myths, the transformation of these otherwise beneficent celestial gods into *tzitzimime* potentially signaled the end of the world (see also Keber 1995, p. 174, 255).

Evidence for the manner in which traditional American Southwestern cultures reacted to eclipses is limited but intriguing. The Mexica are linguistically related to the Hopi as part of the Uto-Aztecan language family, and these groups may have shared aspects of their cosmology. It is known the Hopi viewed eclipses as ominous

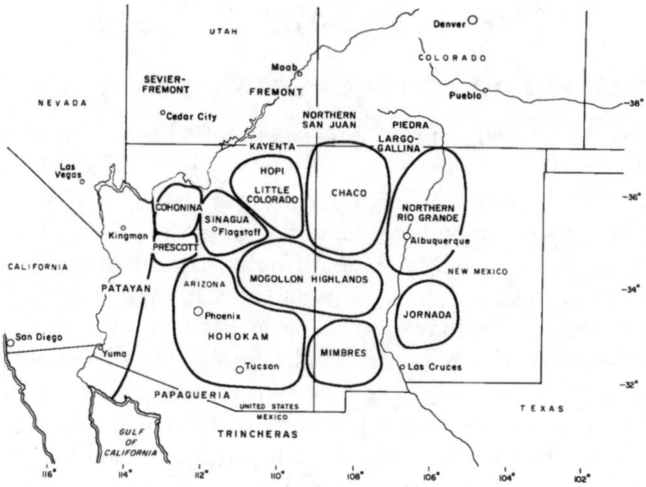

Figure 1. General map of Southwestern culture areas aound AD 1000 to 1400.

signs and believed that the Sun and Moon died (Parsons 1996: 86, 180). Disease and death were associated with eclipses, and purifying ceremonies were required by at least some groups after an eclipse. The Hopi believed that if a solar eclipse occurred immediately after sunrise, Zuni and Rio Grande pueblo peoples would die; if in the middle of morning Hopi people would die, and if at noon, people to the southwest would die. An analysis of the apparently substantive reaction to solar eclipses by the prehistoric Hohokam, likely ancestors (in part) of the modern Hopi, Zuni, and O'odham, is presented elsewhere (Masse and Espenak, 2003).

The Tewa of the Rio Grande also dreaded solar eclipses. According to their explanation of the event, when the burning shield of the Sun Father gives only a little light, he has moved away from the earth in displeasure with the people of this world who are fighting. When this happens, the Tewa fear that he will retire to his house in the underworld (just as he does at night) and never reappear (Williamson 1987:189).

Our investigation focuses on the period of A.D. 1 to 1540, when the ancestors of the Hopi and other modern Pueblo Indians occupied a vast area stretching from north and west of the Grand Canyon, south to the Mogollon Rim, and east to the Four Corners region and the northern Rio Grande valley, as shown in Figure 1. A number of subregional cultural populations were within this area, such as Chaco Canyon, Mesa Verde, Hopi-Kayenta, Piedra, Fremont, Sevier-Fremont, Largo-Gallina, the Little Colorado valley, and the northern Rio Grande. It is cautioned that the boundaries and character of these subregional populations, as depicted in Figure 1, fluctuated through time, since there is evidence, both archaeologically and in migration legends, for population movement and cultural mixing throughout the Southwest during the last 2,000 years.

Puebloan Solar Eclipses

Events, Clusters and Zones of Confluence

Table 1 presents a listing of the eclipses identified for the study period using a variety of commercially available software programs.[1] Of the 44 historical solar eclipses identified in the Puebloan region during the study period, 30 are total and 14 are annular eclipses having magnitudes of at least 95%. Table 1 demonstrates the sporadic nature of these events. We recognize the absence of noticeable eclipses for generations or even hundreds of years in many areas. At some key locations, however, two or even three events may have been observed within a single average human life span of 35–45 years. While solitary total eclipses certainly would have impressed those individuals and cultural groups who witnessed them, we are particularly interested in the locations where multiple eclipse paths crossed during a single human lifetime to create a "cluster" of events. For the purpose of our investigation, we call the area were all the cluster's events could have been observed at their maximum magnitude the "zone of confluence."

This study focuses on five of eight Puebloan eclipse clusters where sufficient archaeological information is likely currently available to investigate possible cultural responses. Cluster #1 (A.D. 149–171) and Cluster #2 (A.D. 411–436), both in the Fremont, northern San Juan and Mesa Verde zone of confluence, are omitted from present consideration. Likewise omitted is Cluster #7 (A.D. 1351–1372) whose zone of confluence is situated just northeast of Mesa Verde, and which occurred at a time when the region was largely devoid of permanent residents. Cluster #8 (A.D. 1318, 1379, 1397) warrants special and separate consideration below in "Solar Eclipses and Katsia Societies."

Eclipse Cluster #3—A.D. 711, 724, 728 Fremont/N. San Juan-Mesa Verde

The A.D. 711–728 cluster involved two annular eclipses and one total eclipse in the 17-year period of A.D. 711 to A.D. 728. Figure 2 illustrates the confluence of Cluster #3 event paths immediately north of Mesa Verde, in Montezuma Basin. The sequence began with Total Solar Eclipse (TSE)-711, an early morning, late October eclipse. This event was the first visible solar eclipse in the Puebloan region for 152 years. And, for this particular subregion, it was the first total eclipse in 275 years, when the A.D. 411–436 cluster darkened this same area. TSE-711 began at daybreak, when the sun rose as a rapidly receding crescent. By 0815 it was dark, and that portion of the morning previous occupied by the sun had been profoundly transformed. The event occurred in our modern constellation Libra, near the bright stars Arcturus and Spica, with Mercury and Venus immediately above the eclipse,

1. Four separate software programs were utilized to reconstruct likely Southwestern total or annular solar eclipses during the period of A.D. 1 and 1540 (Table 1; Figures 2–6). *Dance of the Planets 2.71©*, *RedShift Multimedia Astronomy*, *E.Z. Cosmos 4.0©*, and *Eclipse Complete 2.0©* (EC 2.0). The centerline and width of selected eclipse paths were plotted using *EC 2.0* and were checked by superimposing two to four specific centerline coordinates computed with *Dance*. In most cases, the paths produced by the two different software programs were parallel, and the majority of center lines computed by these programs coincided. The few observed discrepancies were minor and should not significantly impact our overall points and comparisons.

Table 1. Puebloan Total Solar Eclipses and Annular (>95% Magnitude) Eclipses During the Period of A.D. 1 to 1540

Date and Approximate Time of Totality			Location in Sky—Visible Planets (*=Planets Within 15°)	Altitude Above Horizon	Type and Approximate Magnitude	Eclipse Path [West to East] by General Cultural Region
047	05 Dec	10:30	**Sagittarius**—Mercury*, Saturn, Venus, Jupiter	27°	Total 102	Piedra
077	10 May	12:45	**Taurus**—Venus*, Mercury, Jupiter	61°	Total 102	Sevier-Fremont
085	10 Jun	16:00	**Gemini**—Venus/Mercury (in conjunction)	38°	Annular 96	Patayan/Prescott/Sinagua/ N. Hohokam/Mogollon Highlands/ Mimbres/Jornada
098	21 Mar	06:45 sunrise	**Pisces**—Mercury, Mars	<13°	Annular 99	N. Rio Grande/Chaco/Hopi/ Little Colorado/Sinagua/ Cohonina
128	13 Aug	10:00	**Leo**—Venus/Mercury (in conjunction); Jupiter	57°	Annular 98	Mimbres/Hohokam/Patayan
149	22 Jun	18:45 sunset	**Cancer**—Saturn*, Mercury, Venus	<10°	Total 104	Sevier-Fremont/Fremont/ N. San Juan/Largo-Gallina/ N. Rio Grande
163	15 Sep	18:15 sunset	**Virgo**—Mercury*, Venus, Mars, Saturn	14°	Total 101	Sevier-Fremont/Fremont/ N. San Juan/Piedra
171	22 Apr	17:45 sunset	Taurus—none	13°	Annular 95	Sevier-Fremont/Fremont/ Hopi-Kayenta/N. San Juan/Piedra/ Largo-Gallina/ N. Rio Grande
230	25 Aug	06:00 sunrise	**Virgo**—Venus*/Mercury* (in conjunction)	<10°	Total 103	Kayenta/N. San Juan/Piedra

Date and Approximate Time of Totality			Location in Sky—Visible Planets (*=Planets Within 15°)	Altitude Above Horizon	Type and Approximate Magnitude	Eclipse Path [West to East] by General Cultural Region
284	26 Sep	06:15 sunrise	**Virgo**—Jupiter, Venus (in conjunction)	8°	Total 103	Jornada
310	15 May	06:30 sunrise	**Taurus**—Venus, Mercury	11°	Total 105	Patayan/Hohokam/Prescott/Sinagua/ Mogollon Highlands/Little Colorado/ Hopi-Kayenta/Chaco/N. San Juan/ Largo-Gallina/Piedra
323	21 Feb	16:45	**Pisces**—Saturn	68°	Total 100	Trincheras/SE Hohokam/Mimbres/ Jornada
411	01 Dec	12:15	**Sagittarius**—Mercury, Mars, Venus	30°	Total 104	Sevier-Fremont/Kayenta/ San Juan-Chaco/Piedra/Largo-Gallina
428	28 Jun	12:45	**Cancer**—Mercury*, Mars, Venus, Jupiter	73°	Annular 98	Sevier-Fremont/Kayenta/ San Juan-Chaco/Piedra/Largo-Gallina
436	29 Jul	10:45	**Leo**—Mercury*, Venus/Jupiter (in conjunction)	67°	Total 104	Sevier-Fremont/Kayenta/ San Juan-Chaco/Piedra/ Largo-Gallina
461	27 Mar	18:15 late afternoon	**Aries**—Venus	17°	Annular 99	Sevier-Fremont/Fremont
502	24 Jan	07:30 dawn	**Capricornus**—Mercury*, Jupiter/Venus (in conjunction)	<13°	Total 103	Jornada/Mogollon Highlands
529	25 Jan	17:30 twilight	**Aquarius**—Mars*, Jupiter	<10°	Total 103	Sevier-Fremont

Date and Approximate Time of Totality			Location in Sky—Visible Planets (*=Planets Within 15°)	Altitude Above Horizon	Type and Approximate Magnitude	Eclipse Path [West to East] by General Cultural Region
559	14 Dec	15:00 late afternoon	**Sagittarius**—Venus*/Mercury* (in conjunction), Jupiter	17°	Annular 96	Papagueria/Tucson/San Pedro/Safford/Mimbres/ Mogollon Highlands/Jornada/N. Rio Grande
711	16 Oct	08:00 early morning	**Libra**—Venus*/Mercury* (in conjunction), Jupiter, Saturn	17°	Total 104	Sevier-Fremont/Fremont/N.San Juan/Piedra-Gallina
724	24 Jul	18:15 sunset	**Leo**—Venus*, Mercury, Jupiter	<14°	Annular 99	Fremont/Piedra-Gallina
728	13 May	16:15	**Taurus**—Mercury, Venus	33°	Annular 97	Sevier-Fremont/Fremont/N. San Juan/Piedra-Gallina
797	26 Aug	11:15	**Leo**—Saturn/Venus (in conjunction), Jupiter	64°	Total 105	Prescott/Sinagua/N. Hohokam/Mogollon Highlands/Mimbres/Casas Grandes
800	25 Jun	18:45 sunset	**Gemini**—Mercury, Saturn/Venus (in conjunction)	<10°	Annular 95	Sevier-Fremont/Hopi/Chaco/N. Rio Grande
804	13 Apr	14:30	**Aries**—Mercury*, Mars, Venus, Jupiter	44°	Total 107	Cohonina/Sinagua/Hopi/Chaco/N. Rio Grande
805	26 Sep	07:30 sunrise	**Virgo**—Saturn*, Mercury, Venus/Jupiter (in conjunction)	<15°	Annular 97	Sevier-Fremont/Hopi-Kayenta/Chaco/N. Rio Grande
957	29 Jul	08:00	**Leo**—Mercury, Venus, Jupiter	29°	Total 106	Sevier-Fremont/Fremont/N. San Juan

Date and Approximate Time of Totality			Location in Sky—Visible Planets (*=Planets Within 15°)	Altitude Above Horizon	Type and Approximate Magnitude	Eclipse Path [West to East] by General Cultural Region
985	20 Jul	05:30 sunrise	**Cancer**—none	<13°	Total 101	Mogollon Highlands
987	02 Jan	11:00	**Capricornus**—Mercury*, Venus, Saturn, Jupiter	25°	Total 103	Sevier-Fremont
1076	07 Mar	16:45 late afternoon	**Pisces**—Mercury*, Venus, Jupiter	<16°	Total 102	Hohokam/Mogollon Highlands/ Mimbres
1097	11 Jul	14:15	**Cancer**—Mercury, Venus, Mars, Saturn	58°	Total 107	Sevier-Fremont/Fremont/N. San Juan/ Chaco/N. Rio Grande
1156	14 Nov	16:45 sunset	**Ophiuchus**—Mercury*,Venus	<10°	Annular 96	Patayan/Phoenix Basin Hohokam/ Prescott-Cohonina-Sinagua/Chaco/ N. Rio Grande/Casas Grandes
1194	22 Apr	05:15 sunrise	**Aries**—Venus	<10°	Total 106	N. Rio Grande/Piedra-Gallina
1225	10 Mar	11:45	**Pisces**—Mars*/Venus* (in conjunction), Mercury, Saturn, Jupiter	47°	Annular 96	Sevier-Fremont
1234	01 Mar	07:00 sunrise	**Pisces**—Mercury*, Mars/Venus (in conjuntion)	<11°	Annular 96	Patayan/Prescott-Cohonina/Sinagua/ Mogollon Highlands-Chaco/ N. Rio Grande
1257	13 Jun	12:15	**Gemini**—Mars*, Mercury*, Venus, Jupiter	75°	Total 108	Sevier-Fremont/Fremont-Kayenta/ N. San Juan/ Piedra-Gallina

Date and Approximate Time of Totality	Location in Sky—Visible Planets (*=Planets Within 15°)	Altitude Above Horizon	Type and Approximate Magnitude	Eclipse Path [West to East] by General Cultural Region
1259 17 Oct 13:45	**Libra**—Mercury*, Jupiter, Venus/Mars (in conjunction)	33°	Total 105	Fremont/N. San Juan/ Piedra-Gallina-N. Rio Grande
1285 04 Jun 08:15	**Gemini**—Mercury/Mars/Venus (in conjunction)	35°	Total 101	Sevier-Fremont
1313 08 Nov 16:45 sunset	**Ophiuchus**—Mercury, Saturn	<10°	Total 104	Sevier-Fremont
1318 26 Aug 10:00	**Leo**—Mercury, Venus	50°	Total 101	Cohonina/Sinagua/mid-Little Colorado/ Mogollon Highlands/Jornada
1351 25 May 14:30	**Taurus**—Venus*, Mars*, Mercury*, Saturn, Jupiter	52°	Total 100	Sevier-Fremont/Fremont/N. San Juan/ Piedra
1372 27 Sep 09:15	**Virgo**—Mercury*, Mars*	38°	Total 101	Sevier-Fremont/Fremont/N. San Juan/ Gallina
1379 16 May 08:00	**Taurus**—Mercury/Venus/Mars (in conjunction), Saturn	37°	Total 106	Trincheras/Hohokam/ Mogollon Highlands/S. Chaco/ N. Rio Grande
1397 26 May 18:15 late afternoon	**Taurus**—Jupiter	<17°	Total 106	Sevier-Fremont/Cohonina/Sinagua-Hopi/ Mogollon Highlands-Chaco/ Jornada-N. Rio Grande

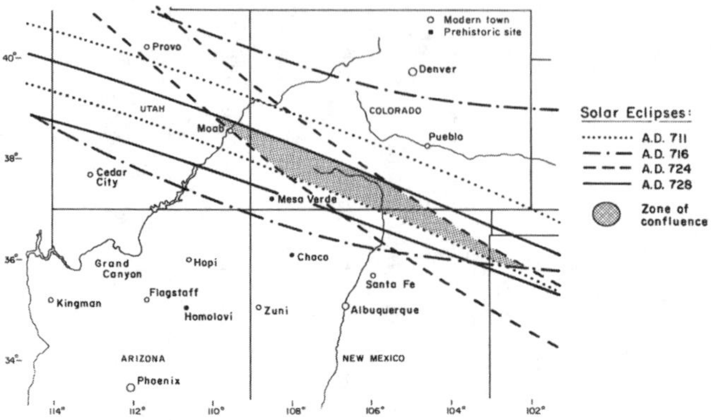

Figure 2. Reconstructed paths of the total solar eclipse of AD 711 and the high magnitude (>95%) annular eclipses of AD 724 and 728. Also depicted is the path of a lower magnitude annular eclipse in AD 716.

and Jupiter and Saturn higher in the abruptly darkened sky. Table 1 lists these celestial associations for this event, as well as for the other 43 eclipse events. Thirteen years later, Annular Solar Eclipse (ASE)-724 reached 99% totality, and would have made a startling spectacle an hour before a late July sunset. This event was followed four years later by ASE-728, yet another late afternoon annular (97%) just prior to sunset. In addition to the eclipses, several impressive and possibly portentous planetary conjunctions occurred between A.D. 709–714.

Eclipse Cluster #4—A.D. 800, 804, 805 Kayenta/San Juan

The A.D. 800–805 eclipse cluster is impressive in that it takes place over a very short period. These events included two annular eclipses and one total eclipse, and Cluster #4 formed a broad zone of confluence encompassing the Kayenta and San Juan regions (Figure 3). ASE-800 was a 95% magnitude event occurring near sunset a few days past the summer solstice. TSE-804, which occurred on an afternoon in April, was a large magnitude total eclipse, with darkness lasting more than five

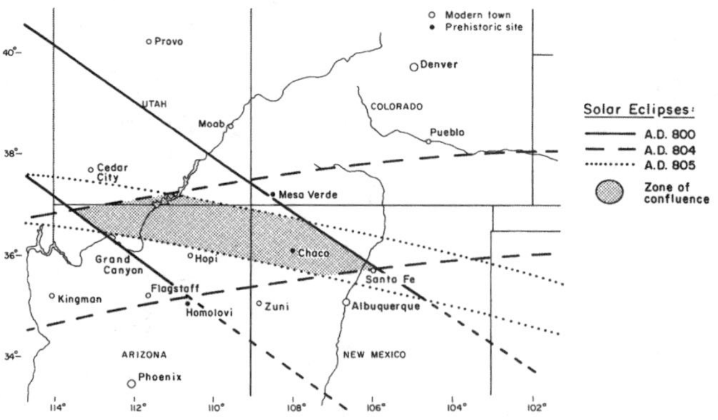

Figure 3. Reconstructed paths of the total solar eclipse of AD 804 and the high magnitude (>95%) annular eclipses of AD 800 and 805.

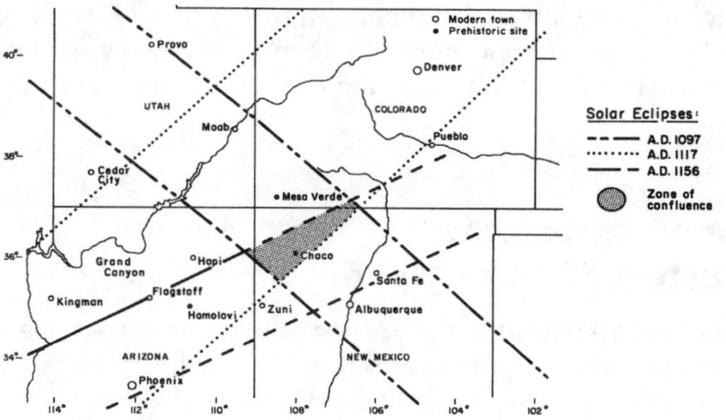

Figure 4. Reconstructed paths of the total solar eclipse of AD 1097 and the high magnitude (>95%) annular eclipse of AD 1156. Also depicted is the path of a lower magnitude annular eclipse in AD 1117.

minutes. ASE-805 was a 97% magnitude sunrise annular that occurred a few days after autumnal equinox.

Eclipse Cluster #5—A.D. 1097, 1156 Chaco Regional System

Figure 4 shows the zone of confluence created within the Chaco Regional System by the events of A.D. 1097 and 1156. TSE-1097 was an afternoon total eclipse in mid-July. ASE-1156 (96%) occurred in the late afternoon of mid-November. In addition, a lessor magnitude (93%) annular eclipse, which does not appear in Table 1, occurred at midday just two weeks after the winter solstice in A.D. 1117.

Cluster #6—A.D. 1257, 1259 San Juan and Kayenta

The zone of confluence for the A.D. 1257 and 1259 events is in the northern San Juan, Piedra, and Largo-Gallina subregions, illustrated in Figure 5. TSE-1257 occurred at high noon on June 13, in the constellation Gemini, and close enough to

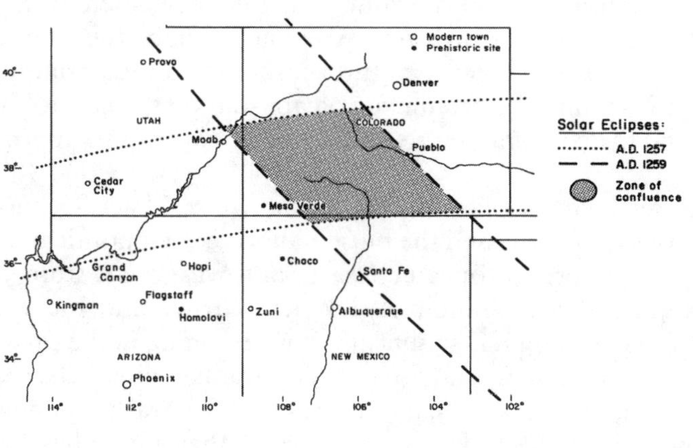

Figure 5. Reconstructed paths of the total solar eclipses of AD 1257 and 1259.

summer solstice to shake up any shaman. On this day, the Kayenta and San Juan subregions saw totality last approximately six minutes. Two years later, TSE-1259, a mid-October afternoon eclipse, occurred over the eastern San Juan, in the constellation Libra.

Cultural Implications of Eclipse Clusters #3 through #6

The A.D. 700–750 period in the area of eclipse Cluster #3 is marked by considerable transition and change in architecture. This includes the first appearance of "protokivas" containing *sipapu*, or spirit entrances to the underworld, which appear to mark a shift toward religious or ritual usage of pit-structures (Judge 1991: 23). Also potentially significant is a marked decrease in regional population in the eclipse zone of confluence after a steady climb for nearly 100 years (Dean, Doelle, and Orcutt 1994:Figure 4.1).

There are major settlement shifts in the A.D. 800–805 eclipse Cluster #4 zone of confluence between 800–850, including both rapid increases in population (Dean, Doelle, and Orcutt 1994:77; Figures 4.2–4.4, 4.6), and areas of widespread abandonment, such as that of most of southeastern Utah (Wilcox and Haas 1994:231). The cultural effects of this eclipse cluster may have been accentuated by two other celestial phenomena. The first is the uniquely near total lack of temporary celestial events (other than eclipses) between A.D. 776 and 813. This temporary celestial event lacunae, longest of the past 2,000 years, may have been disquieting to priests and shamans used to seeing such events and who typically incorporated the observance of such events into ritual and social practices (Masse 1995; Masse, Johnson, and Tuggle n.d.). Also of note is a probable northern hemisphere volcanic eruption in August A.D. 797, the effects of which greatly attenuated normal sunlight for more than two weeks in the Mediterranean (Schove and Fletcher 1984, p. 167).

The apparent coincidence of this volcanic darkness (if observable in the Southwest) with a total solar eclipse (Table 1) that traversed the southern edges of the Puebloan region and the northern portion of the Hohokam region, has been suggested as a possible portent leading to the spread of the regional Hohokam ballcourt system around A.D. 800 (Masse and Espenak, 2003). Of particular interest are purported cases of human cannibalism in the zone of confluence, such as at Teec Nos Pos (Turner 1989) in extreme northeastern Arizona. Perhaps the cannibalism and/or warfare evident at this time was a cultural response to the cosmological implications of solar eclipses and the presumed volcanic darkness, akin to the Mexica fear of *tzitzimime* and the concomitant need to sacrifice captives to appease the demons of darkness.

Eclipse cluster #5 took place during a time when Chaco Canyon and its surrounding environs had witnessed the most elaborate sociopolitical development of any region within the pre-European contact Southwest apart from Casas Grandes ("Pacquime") in northern Chihuahua, Mexico (Crown and Judge 1991; Windes and Ford 1996). Therefore, it is significant that beginning in the early 1100s the archaeological record of Chaco Canyon changes dramatically (Sebastian 1991:131). New construction greatly declines, plazas are enclosed, trash mounds are no longer used, and room spaces emphasize domestic use. Other anomalies are evident after A.D. 1100, such as a notable increase in the use of obsidian (Toll 1991), which takes on poignancy if obsidian had a similar association with eclipse events for Chacoan groups as was the case for the Mexica. We suggest a correlation between these changes and TSE-1097, which was of particularly large magnitude. Also of note is the fact that major Chacoan building phases (Windes and Ford 1996) are bracketed by eclipse events (total eclipses in A.D. 830 and 1102; and ASE-1156) whose paths crossed the Casas Grandes region. Given the possibility that Casas Grandes was a major conduit for Mexican ideas, goods, and cosmology into Chaco Canyon, responses to eclipses at Casas Grandes itself may have directly or indirectly impacted the Chacoan system.

During the period of A.D. 1250–1300 (the 'Tsegi phase' in the Kayenta region), construction of cliff dwellings and tower sites at the beginning of this period is followed by widespread conflict, the sacking and burning of villages, violent deaths including possible cannibalism, and ultimately the abandonment of the region. Scholars have long attempted to demonstrate that these behaviors were the consequence of a major drought between A.D. 1277–1299. However, such conflict was especially intensive in the 1257–1259 eclipse Cluster #6 zone of confluence area of the Largo-Gallina subregion (Wilcox and Haas 1994). These eclipses were likely accentuated by a notable lacunae in other temporary celestial events between A.D. 1245 and 1264, an absence shattered by a dramatic daylight-visible comet in A.D. 1264 whose multiple-braided tail stretched more than 100°. Perhaps the construction of cliff dwellings and towers, and the conflict as well, were stimulated by the observance of the eclipses and comet, in addition to the deteriorating environmental conditions. Moreover, features such as the striking large white circular paintings which are part of the Tsegi Painted Style rock art tradition (Schaafsma 1980:145–153) could have been produced in response to these eclipses.

Where eclipses did occur, some observers may have been impressed enough to commemorate or otherwise mark these events on the ground, particularly at sites with known celestial observation platforms. For example, Yellow Jacket Pueblo in Montezuma Basin had potentially numerous solar shrines and alignments placed in observance of solstice, equinox and other yet undetermined associations (Malville and Putnam 1991). In addition to being in the zone of confluence of eclipse Clusters #1, #2, #3 and #6, the people of this subregion were witness to three additional total eclipses in A.D. 957, 1097 and 1194 (Table 1). These three solo events occurred during the major phases of construction at Yellow Jacket Pueblo.

Solar Eclipses and Katsina Societies

Katsina Societies and Archaeology

If solar eclipses were powerful forces in the development and trajectory of Southwestern cultures, it would be expected that images of these events would appear in some fashion in rock art, kiva murals, the decoration of ceramic vessels, and religious and ritual paraphernalia. To explore this issue we turn to the 14th century, a time of great change in the Puebloan area, where many Pueblo groups aggregated from villages into large towns and there were striking innovations in terms of art and religious expression.

The 14th century witnessed the cult-like flowering of an art form believed by many scholars to be the archaeological expression of the beginnings of the katsina (also spelled kachina, katchina, katcina) societies still operant today in modern Pueblos (Adams 1991; Schaafsma 1994b). Katsinas are powerful spirits or supernatural beings of the animate world who are interceded with through the medium of dolls and masked dancers who impersonate the spirit (Parsons 1996; Titiev 1992). The symbolic aspects most closely tied to katsina ceremonialism and imagery are those of fertility and the desire for rainfall for crops, healing or curing, ancestor worship or ties with the dead, and perhaps warfare. The most important katsina ceremonies, including those marking the beginning and end of the annual katsina cycle, are tied to the solstices and solar ritual (Parsons 1996:554–590; Titiev 1992:142–154).

After detailed consideration of all available data, Adams (1991:145–148; 1994) has built an interesting case for the possible origin of at least one katsina society in the upper Little Colorado river valley (near Fourmile ruin) between A.D. 1275 and 1325, crystallizing iconographically by no later than A.D. 1325, and becoming—as he has defined archaeologically—the "Fourmile katsina cult." Fewkes (1904) noted that not only did Hopi and Zuni individuals still make pilgrimages to sacred lakes in the upper Little Colorado, but also that both groups had oral traditions of katsinas originating in the region near Fourmile ruin.

Adams (1991) suggests that katsina imagery began expanding around A.D. 1325 into areas such as the Homolovi group in the middle Little Colorado (see also Hays 1989, 1994; Cole 1992), and expanding into the Zuni and Rio Grande areas around A.D. 1375. Elaborations in katsina art and iconography on the Hopi Mesas at sites such as Awatovi between about A.D. 1375 and 1400 become codified as a separate archaeological style ("Hopi katsina cult" in Adam's terminology) at around A.D. 1400, associated with a considerable amount of rain symbolism. The Hopi katsina imagery is widespread, being recognizable at places such as Zuni, Acoma, and Rio Grande pueblos such as Pottery Mound. Adams also recognizes the "Jornada katsina cult" as a third archaeological style of katsina imagery, based on its similarities to rock art in the Jornada culture area, and which he has dated to the period of approximately A.D. 1450–1500.

There are alternative models of katsina society development, and Adams himself (1996, personal communication) acknowledges that aspects of the dating of Jornada katsina imagery are equivocal. Schaafsma (1980, 1992, 1994a) has suggested that katsina imagery was introduced into the Jornada region from Mexico between A.D. 1325 and 1350, and became manifested as the Rio Grande style of rock art which spread into other portions of the Puebloan area. Lekson and Cameron (1995) have argued that there are katsina antecedents during the 10th and 11th centuries in Mimbres and Chaco Canyon, a position also supported by Crown (1994). And Christian Downum has proposed that the Wupatki area during the 12th and 13th centuries may be an ancestral home of the katsina society (see Bates, this volume).

We suggest that all these views may be essentially correct. The occurrence and distribution of the katsina imagery noted by Adams is seemingly intimately related to the individual paths and the zone of confluence of three 14th century total solar eclipses (Figure 6), which together we refer to as Cluster #8. The earlier katsina-like images noted by Lekson and Cameron, Crown, and Downum, can be easily explained by the occurrence of solar eclipses in these regions (Table 1), assuming that at least some katsina-like imagery reflects eclipse events, as is discussed below.

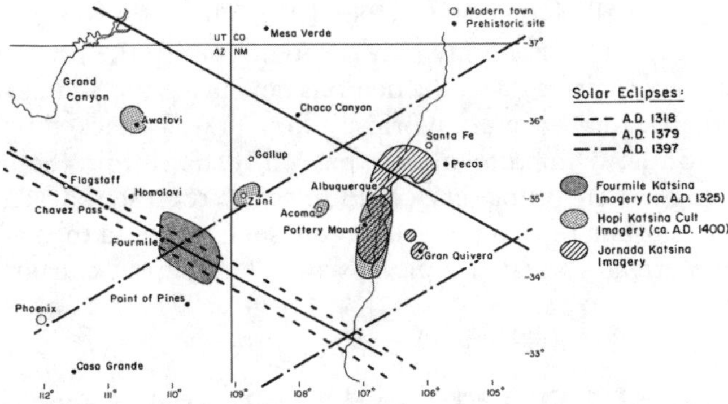

Figure 6. Reconstructed paths of the total solar eclipses of AD 1318, 1379, and 1397.

Solar Eclipse of A.D. 1318 and the Flute Clans

TSE-1318, a rare combination of annular and total eclipse, occurred near the beginning of a period (A.D. 1317–1330) when, with the exception of a single comet in A.D. 1319 and a couple of lunar eclipses, there were no known temporary celestial events. As previously mentioned for the period of A.D. 776–813, the lengthy absence of temporary celestial events can be a potentially disquieting time for priests and shamans. TSE-1318 occurred during the height of summer on 26 August 1318. The midmorning eclipse had a narrow 20 mile (32 km) wide path of totality, which transited directly through the area of the upper Little Colorado River. As shown in figure 6, the exact geographic coincidence of the TSE-1318 eclipse path with the center of Adams' (1991) reconstructed heartland of Fourmile katsina imagery at around A.D. 1325 is striking.

We suggest that the occurrence of TSE-1318 may have influenced the early rituals and calendrics surrounding the Flute ceremony of the Hopi Gray Flute and Blue Flute clans. Historically, this ceremony was held biennially (alternating with the Snake-Antelope ceremony) for the multiple purposes of securing rain, curing disease, and especially for conducting solar observances, possibly associated with the solstices (Titiev 1992:147–149; Parsons 703–708). The ceremony begins around the 16th day after the August moon is first sighted, and climaxes in the afternoon nine days later, with the altars being dismantled the next day. If conducted in 1318, the timing of this ceremony would have climaxed within two days of TSE-1318.

If the origin of some katsinas is tied in part to the Fourmile ruin area and to the path of TSE-1318, two aspects of the Flute ceremony are pertinent. The first is somewhat self-evident, as best stated by Parsons (1996: 705–706): "... this ritual, I surmise, dramatizes either the placing of the new sun in the sky or the emergence from the sunless lower world into the sunlight." This statement would certainly apply to the appearance of a "new sun" after the death of the Sun during an eclipse.

The second ceremonial aspect is the role of the locust (Kokopelli), both in the ceremony and in the Flute society itself. During the climax of the Flute Ceremony all members of the Flute society sing, shake rattles, or play flutes, with the flutes specifically being considered to represent the sound of locusts. The Puebloan "locust" is most likely the 17-year cicada (order Homoptera), commonly termed a locust, which produces a chirping flute-like sound, and which is known to burst into song during solar eclipses (Allen and Allen 1987:107). Thus, the sound produced during the climax of the Flute ceremony may be analogous to the previously noted attempt by the Mixtecs to drive away the demons devouring the Sun during an eclipse. During his emergence to the present world, Kokopelli was attacked by bolts of lightning from the clouds, which harmlessly passed through him (Schaafsma 1980: 140–141). We suggest such imagery could possibly represent an encoding of the appearance of the corona of that particular eclipse. One could then ask if the use of "locust medicine" relates to the Hopi view of solar eclipses causing human sickness and death.

Solar Eclipses of A.D. 1379 and 1397, and Jornada and Hopi Katsina Imagery

Near the end of the 14th century, TSE-1379 and TSE-1397 gave a most spectacular show. Each event had a 150 mile (241 km) wide path of totality, with darkness lasting up to 5 minutes along the center line of each path. The 1379 event reached its maximum just after sunrise on the morning of May 16, while the 1397 event reached its maximum an hour before sunset on May 26. The view west from the Hopi Mesas during TSE-1397 may have been distinctive, with the wide path of the umbra being possibly defined on the edges by the San Francisco Peaks near Flagstaff to the south and Navajo Mountain near Page to the north. The zone of confluence of the 1379 and 1397 events is relatively large, covering a diamond shaped area measuring approximately 45,000 square miles (116,550 sq. km) of west central New Mexico.

It is intriguing that the path of TSE-1379 coincides with the distribution of Jornada katsina imagery (Figure 6). We do not completely discount the possibility of a Jornada katsina society as defined by Adams at around A.D. 1450–1500, although there are no eclipses anywhere near this area between A.D. 1397 and A.D. 1531. However, given the vagaries of dating rock art and even kiva murals (such as those at Kuaua—see Dutton 1963), perhaps most of what Adams sees as Jornada katsina imagery dates instead to the period of A.D. 1380–1450. This earlier date is more in line with what Schaafsma (1980:252; 1994a) sees as the beginning of the 14th century Rio Grande style of rock art in New Mexico.

Equally remarkable with the correlation between TSE-1318 and the distribution of Fourmile katsina imagery is the fact that the distribution of Hopi katsina imagery around A.D. 1400–1450, as mapped by Adams (1991:Figure 6.5), coincides exactly with the centerline of TSE-1397 (Figure 6). Indeed, the fact that the centerlines of TSE-1318, TSE-1379, and TSE-1397 all coincide spatially and temporally with the initial localized expressions of katsina-like imagery is difficult to attribute solely to chance.

Katsina Solar Eclipse Iconography

The potential eclipse imagery that accompanies these archaeological katsina societies is striking. For example, a series of katsina images on 14th century ceramic bowls, and rock art from the Homolovi group (Hays 1989, 1994; Adams 1991; Cole 1992; Ferg 1982) depict masks which combine obvious solar characteristics (round, rayed masks) with probable lunar images such as crescents (Figure 7: j, k)

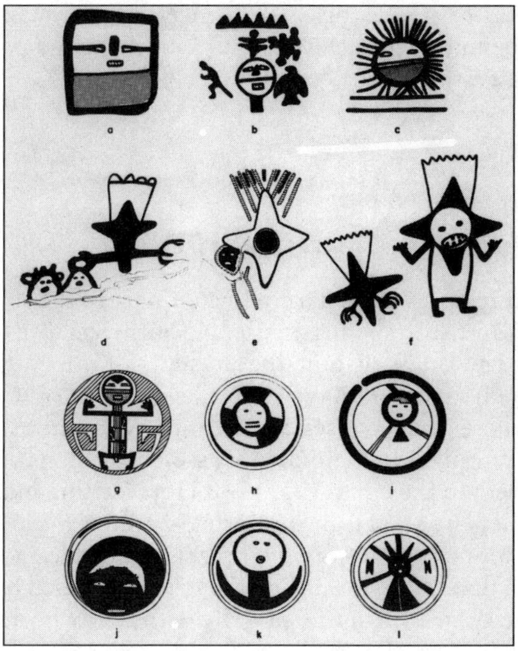

Figure 7. Potential solar eclipse or related Katsina sky imagery dating to the 14th century: (a–c) painted slabs; (d–f) rock art petroglyphs; (g–l) painted ceramic bowls.

or with darkened bands to create opposition suggesting solar-lunar symbolism (Figure 7: h, I). In two instances the masks themselves appear as actual darkened eclipse-like faces (Figure 7: j, l). Most of these bowls were likely made in the Hopi mesas, suggesting (if they are indeed eclipse images) possible association with TSE-1397, close to the estimated time that the Homolovi group is abandoned (Adams 1991; Masse and Espenak, 2003). This association suggests that the ancient Puebloan potters commemorated interactions of the Sun and Moon after witnessing an eclipse. A Fourmile polychrome bowl (Figure 7: g) and painted stone slabs from late 14th century kivas in the Point of Pines region (Figure 7: a–c) depict a different style of opposition in terms of bands of light (Sun or normal light?) and dark (Moon or eclipse shadow?) on probable solar images, perhaps a representation of TSE-1379.

Among the most convincing solar eclipse images are a series of Rio Grande style petroglyphs and rock paintings from the Galisteo basin (Figure 7: d–f), which includes the logo used for the Fifth Oxford Conference on Archaeoastronomy (see Schaafsma, this volume). These taloned, starlike-feathered images are nothing less than the central Mexican *tzitzimime* star demons of darkness who appear during solar eclipses. The distribution of these images seemingly supports an association with TSE-1379, although some may be instead associated with TSE-1397.

There undoubtedly are other solar eclipse images in the iconography of the Southwest. For example, we have previously alluded to the possibility that the widespread motif of the hump-back flute player, Kokopelli, may be associated with eclipses, as may also be the white circle motifs of the Tsegi Painted rock art style, and perhaps even the spectacular "Sun shield" images that occur in the Rio Grande area (Schaafsma 1980, 1992). As archaeologists and art historians begin to systematically factor in the reconstructed history of eclipses and other temporary celestial events in their studies of the historical sequence of Southwestern iconography, it may be possible to definitize the celestial nature of the many pantheons of symbols in the ancient Southwest.

Conclusions

We believe solar eclipses played a significant part in the drama of cultural and cognitive evolution in the ancient American Southwest. At the very least, solar eclipses were likely viewed as disquieting portents of impending (or immediately past) misfortune, such as would have been the case for total solar eclipses in the Hohokam region where such eclipses coincidently were almost always followed in one or two years by tremendous floods (Masse and Espenak, 2003). Indeed, the rain symbolism evident in the circa A.D. 1400 Hopi katsina imagery (Adams 1991) suggests similar beliefs by the Hopi, or perhaps even reflects the transmission of ideas from actual Hohokam emigrants who arrived in the Hopi Mesas following probable devastating floods in the Salt River watershed caused by unusually high levels of precipitation in A.D. 1358 and especially in the early portion of the A.D. 1380 decade (Nials, Gregory, and Graybill 1989, Gregory 1991, Teague 1993, Masse 1991, Masse and Espenak, 2003).

On the other hand, the seeming coincidence of total solar eclipses with cultural phase boundaries (see also Masse and Espenak, 2003), and the coincidence of solar eclipses with the rise and evolution of aspects of Puebloan katsina societies suggest that eclipses where not viewed simply as passive portents; they may have instead promoted the beginnings of periods of significant social change involving abandonments, migrations, and/or changes in cosmology and/or world view. We know from ethnographic evidence that kiva katsina ceremonies were part of a suite of ritualized behaviors aimed at propitiating the gods and at preventing disharmony in the world. Because of the importance of daily and annual cycles of the Sun, the appearance of solar eclipses was almost certainly interpreted as demonstrating that things were ***not*** right and that something should be done to make them right. The degree to which each cultural group did something to change things to restore harmony undoubtedly depended on a host of variables. The key fact that seems to emerge from our analysis is that things did, in fact, usually change.

The people of the ancient Southwest were keen observers of the heavens and we believe they commemorated both the known celestial features and temporary celestial events in a variety of ways. In the context of oral tradition, religion and mythology, solar eclipses and other temporary celestial events played a role in creating and preserving knowledge about both our human past and the universe itself. Continued study of the relationship between solar eclipses (and other temporary celestial events) and the archaeological record can perhaps teach us something about the historical sequence of Southwestern iconography, the beliefs of the artist at that time, and the celestial origin of at least some deities in the pantheons of the ancient Southwest.

Acknowledgments

The authors thank Rolf Sinclair, Raymond White, and George Gumerman for encouraging and supporting the presentation of a version of this paper at the Fifth Oxford Conference on Archaeoastronomy. Ron Beckwith drafted the figures. Valuable editorial comments were provided by Dorothy Washburn, John Fountain, and Polly Schaafsma. The authors take responsibility for the content of the final product.

References

Adams, E. Charles. 1991, *The Origin and Development of the Katsina Cult.* University of Arizona Press, Tucson.

———. 1994, The Katsina Cult: A Western Pueblo Perspective. In *Kachinas in the Pueblo World*, edited by Polly Schaafsma, pp. 35–46. University of New Mexico Press, Albuquerque.

Allen, David, and Carol Allen. 1987, *Eclipse.* Allen & Unwin, Sydney.

Bierhorst, John. 1992, *History and Mythology of the Aztecs: The Codex Chimalpopoca.* University of Arizona Press, Tucson.

Cole, Sally J. 1992, *Katsina Iconography in Homol'ovi Rock Art, Central Little Colorado River Valley, Arizona.* The Arizona Archaeologist 25, Arizona Archaeological Society, Phoenix.

Crown, Patricia L. 1994, *Ceramics and Ideology: Salado Polychrome Pottery.* University of New Mexico Press, Albuquerque.

Crown, Patricia L., and W. James Judge. 1991, *Chaco and Hohokam: Prehistoric Regional Systems in the American Southwest.* School of American Research, Santa Fe.

Dean, Jeffrey S., William H. Doelle, and Janet D. Orcutt. 1994, Adaptive Stress, Environment, and Demography. In *Themes in Southwest Prehistory*, George J. Gumerman (ed.), pp. 53–86. School of American Research Advanced Seminar Series, Santa Fe.

Dutton, Bertha P. 1963, *Sun Father's Way: The Kiva Murals of Kuaua.* University of New Mexico Press, Albuquerque; School of American Research and Museum of New Mexico Press, Santa Fe.

Ferg, Alan. 1982, Fourteenth Century Katchina Depictions on Ceramics. In *Collected Papers in Honor of John H. Runyon*, edited by G. X. Fitzgerald, pp. 13–29. Papers of the Archaeological Society of New Mexico 9, Albuquerque.

Fewkes, Jesse Walter. 1904, Two Summer's Work in Pueblo Ruins. In *22nd Annual Report of the Bureau of American Ethnology*, pp. 1–196. Smithsonian Institution, Washington, D.C.

Gregory, David A. 1991, Form and Variation in Hohokam Settlement Patterns. In *Chaco and Hohokam: Prehistoric Regional Systems in the American Southwest.* P. L. Crown, and W. J. Judge (eds.), pp. 159–193. School of American Research Press, Santa Fe.

Hays, Kelley Ann. 1989, Katsina Depictions on Homol'ovi Ceramics: Toward a Fourteenth-Century Pueblo Iconography. *Kiva* 54:285–311.

———. 1994, Kachina Depictions of Prehistoric Pueblo Pottery. In *Kachinas in the Pueblo World*, edited by Polly Schaafsma, pp. 47–62. University of New Mexico Press, Albuquerque.

Judge, W. James. 1991, Chaco: Current Views of Prehistory and the Regional System. In *Chaco and Hohokam: Prehistoric Regional Systems in the American Southwest*, W. James Judge and Patricia L. Crown (eds.), pp. 11–30. School of American Research, Santa Fe.

Keber, Eloise Quiñones. 1995, *Codex Telleriano-Remensis*. University of Texas Press, Austin.

Lekson, Stephen H., and Catherine M. Cameron. 1995, The Abandonment of Chaco Canyon, the Mesa Verde Migrations, and the Reorganization of the Pueblo World. *Journal of Anthropological Archaeology* 14:184–202.

Leon-Portilla, Miguel. 1963, *Aztec Though and Culture*. University of Oklahoma Press, Norman.

———. 1995, On the Meanings of the Celestial Bodies in Pre-Hispanic Mexico. In *The Inspiration of Astronomical Phenomena*, G.V. Coyne, S. J. and R. M. Sinclair (eds.). *Vistas in Astronomy* 39:451–461.

Littmann, Mark, and Ken Willcox. 1991, *Totality: Eclipses of the Sun*. University of Hawaii Press, Honolulu.

Malville, J. McKim, and Claudia Putnam. 1991, *Prehistoric Astronomy in the Southwest*. Johnson Books

Masse, W. Bruce. 1991, The Quest for Subsistence Sufficiency and Civilization in the Sonoran Desert. In *Chaco and Hohokam: Prehistoric Regional Systems in the American Southwest*. P. L. Crown, and W. J. Judge (eds.), pp. 195–223. School of American Research Press, Santa Fe.

———. 1995, The Celestial Basis of Civilization. In *The Inspiration of Astronomical Phenomena*, G.V. Coyne, S. J. and R. M. Sinclair (eds.). *Vistas in Astronomy* 39:463–477.

Masse, W. Bruce, and Fred Espenak. 2003, Sky as Environment: Solar Eclipses and Hohokam Culture Change. In *Environmental Change and Human Adaptation in the Ancient Southwest*, D. E. Doyel and J. S. Dean (eds.). University of Utah Press, Salt Lake City. (In press).

Masse, W. Bruce, Rubellite Kawena Johnson, and H. David Tuggle. n.d., *Islands in the Sky: Astronomy and the Role of Celestial Phenomena in Hawaiian Myth, Language, Religion, and Chiefly Power*. University of Hawaii Press (in preparation).

Milbrath, Susan. 1995, Eclipse Imagery in Mexica Sculpture of Central Mexico. In *The Inspiration of Astronomical Phenomena*, G.V. Coyne, S. J. and R. M. Sinclair (eds.). *Vistas in Astronomy* 39:479–502.

Miller, Mary, and Karl Taube. 1993, *The Gods and Symbols of Ancient Mexico and the Maya*. Thames, and Hudson, London.

Nials, Fred L., David A. Gregory, and Donald A. Graybill. 1989, Irrigation systems in the lower Salt River valley. In *The 1982–1984 Excavations at Las Colinas: Environment and Subsistence*. By D. A. Graybill et al. Arizona State Museum Archaeological Series, no. 162, vol. 5, pp. 59–78. University of Arizona, Tucson.

Parsons, Elsie Clews. 1996, *Pueblo Indian Religion*. University of Nebraska Press, Lincoln. [Orig. 1939].

Sahagun, Fray Bernardino de. 1953, *Florentine Codex: General History of the Things of New Spain. Book 7—The Sun. Moon, and Stars, and the Binding of the Years*. Monographs of the School of American Research 14(7), Santa Fe.

Schaafsma, Polly. 1980, *Indian Rock Art of the Southwest*. School of American Research, Santa Fe, and University of New Mexico Press, Albuquerque.

———. 1992, *Rock Art in New Mexico*. Museum of New Mexico Press, Santa Fe.

———. 1994a, The Prehistoric Kachina Cult and Its Origins as Suggested by Southwestern Rock Art. In *Kachinas in the Pueblo World*, edited by Polly Schaafsma, pp. 35–46. University of New Mexico Press, Albuquerque.

Schaafsma, Polly (ed.). 1994b, *Kachinas in the Pueblo World*. University of New Mexico Press, Albuquerque.

Schove, D. Justin, and Alan Fletcher. 1984, *Chronology of Eclipses and Comets, A.D. 1–1000*. Boydell Press, Bury St Edmunds, Suffolk.

Sebastian, Lynne. 1991, Sociopolitcal Complexity and the Chaco System. In *Chaco and Hohokam: Prehistoric Regional Systems in the American Southwest,* P. L. Crown and W. J. Judge (eds.), pp. 109–134. School of American Research Press, Santa Fe.

Teague, Lynn S. 1993, Prehistory and the Traditions of the O'odham and the Hopi. *Kiva* 58: 435–454.

Titiev, Mischa. 1992, *Old Oraibi: A Study of the Hopi Indians of Third Mesa.* University of New Mexico Press edition, Albuquerque. [Orig. 1944].

Toll, H. Wolcott. 1991, Material Distributions and Exchange in the Chaco System. In *Chaco and Hohokam: Prehistoric Regional Systems in the American Southwest,* P. L. Crown and W. J. Judge (eds.), pp. 77–107. School of American Research Press, Santa Fe.

Turner, Christy G., II. 1989, Teec Nos Pos: More Possible Cannibalism in Northeastern Arizona. *Kiva* 54:147–152.

Wilcox, David R., and Jonathan Haas. 1994, The Scream of the Butterfly: Competition and Conflict in the Prehistoric Southwest. In *Themes in Southwest Prehistory*, George J. Gumerman (ed.), pp. 211–238. School of American Research Advanced Seminar Series, Santa Fe.

Williamson, Ray A. 1987, *Living the Sky: The Cosmos of the American Indian.* University of Oklahoma Press.

Windes, Thomas J., and Dabney Ford. 1996, The Chaco Wood Project: The Chronometric Reappraisal of Pueblo Bonito. *American Antiquity* 61:295–310.

Zirker, J. B. 1995, *Total Eclipses of the Sun*. Princeton University Press, Princeton.

6

Different Astronomies, Different Cultures and the Question of Cultural Relativism

Stephen C. McCluskey
Department of History
West Virginia University
Morgantown, WV

It has become a commonplace among students of the cultural aspects of astronomies that different kinds of astronomies are practiced in different cultures. Drawing on astronomies found in a wide range of cultures, I examine a number of these astronomies, employing a number of taxonomic elements to define their significant differences both as to purpose and to scientific content. The differences in scientific content relate both to the different kinds of astronomical phenomena these astronomies consider, the different observational and calculating techniques, and the different theoretical frameworks they use to make these phenomena meaningful.

I suggest that the differences among these astronomies to address a problem which has vexed anthropologists, historians, and philosophers of science for some time, the problem of cultural relativism and the social construction of knowledge. The existence of different astronomies which have developed in different cultures, that nonetheless deal with what has, in many cultures, become the most precise kind of objective knowledge offers a way out of the false dilemma between that extreme form of cultural relativism, in which all knowledge is relative, and an extreme objectivist philosophy of science, in which there is only one kind of science that can lead to true statements about the observed world.

* * *

Introduction

Looking back over the papers presented at the Oxford series of conferences, most of them have tried to identify indigenous astronomical practices and then ask what these astronomies can tell us about the cultures in which those astronomies developed. The approach I'm taking goes in the opposite direction and asks what these different astronomies can tell us about science and how (or whether) its content is influenced by culture. This problem of cultural relativism is implicit in the notion of studying the sciences of different cultures. Science has been defined, at least according to some philosophers, by a particular method that humans have developed which has the significant characteristic of enabling us to find the Truth about nature—in the case of astronomy—to find the Truth about the heavens.

Those of us who study astronomies in cultures find ourselves faced with a particular problem. The different cultures that we study have developed a series of different astronomies. Which of these different astronomies can be said to present the scientific truth about the heavens? Are they all true? Is only one of them true and all the rest of them false? Or are they all just stories for which it is inappropriate to make judgements about their truth or falsehood?

Questions of this kind have been bubbling along for some time among philosophers, historians, anthropologists, and other social scientists trying to sort out the role that culture plays in the development of scientific institutions—and more significantly—of scientific ideas (Brown 1984; Hollis & Lukes 1982). Of late these issues have begun to take on a greater intensity as some of the more extreme advocates of the social (or cultural) construction of scientific knowledge have come to assert that since science is a product of the society in which it develops, we cannot even judge the truth or falsehood of modern science. Such extreme positions have engendered reactions from those who wish to defend the objectivity of scientific knowledge against the perceived threat of cultural relativism by claiming that modern science presents the sole way to truth about nature. Perhaps the most widely known (if perhaps not so widely read) defense of the objectivity of science is the parody of literary critical discussions of science which the physicist Allan Sokal (1996), managed to slip past the editors of the special issue of the journal *Social Text* dedicated to the "Science Wars."

This is not the place for a theoretical discussion of the merits of social constructionism or its applications to modern science. (I may note briefly that like most historians of science I take the view that scientific ideas are influenced by our observations of a real world out there, but that the questions that scientists choose to investigate, and even the form in which they express their answers to those questions, are significantly influenced by cultural factors.)

Instead, I would like to employ the findings of archaeoastronomy, defined broadly as the study of different astronomies in different cultures, to approach the issues of objectivity and the cultural construction of scientific ideas from a firmly grounded empirical perspective. We have studied a wide range of astronomical traditions; we have studied them with concern for the extent to which these astronomical traditions accurately represent celestial phenomena; and we have noted in some detail how these astronomies influence, and are influenced by, their social context.

The celestial phenomena which the skywatchers of different cultures have observed are relatively constant. Save for the changes in the rising and setting of the stars as we move from the equator to the poles, celestial phenomena are much the same throughout the world. If these relatively constant phenomena can give rise to fundamentally different kinds of astronomies, and if these astronomies can all adequately account in different ways for important aspects of those astronomical phenomena, by examining them we may arrive at a way out of the false dilemma between an extreme form of cultural relativism, in which all knowledge is relative, and an extreme objectivist philosophy of science, in which there is only one kind of science that can lead to true statements about the observed world.

The Nature of Science

Before going any further, I must deal with a central issue. Are the astronomies that we examine really scientific? To put it another way, are the skywatchers of the various societies we have studied doing science? Here there is room for disagreement. Those who focus on the progressive achievements of modern science or on the critical achievements of that science which traces its origin from Greek natural philosophy, would like to deny the scientific nature of the astronomies we study. Authors have chosen various criteria that would deny the accolade science to any natural knowledge before the rise of a distinct scientific profession and the term "scientist" in the nineteenth century, before the scientific revolution of the seventeenth century, before the rise of critical philosophy in Ionian Greece, or before the establishment of precise astronomical predictions in Mesopotamia. This lack of agreement indicates that the line of demarcation between science and non-science is indistinct and suggests that the quest to identify it is problematic, if not futile.

That many widely accepted definitions of science make no attempt at such a demarcation reinforces the notion that one may be unnecessary. In the eleventh edition of the *Encyclopædia Brittanica* W. C. Dampier Whetham (1911) succinctly defined science as "ordered knowledge of natural phenomena and of the relations between them." More recently, Marshall Clagett (1963) expanded that concept to encompass "first, the orderly and systematic comprehension, description and/or explanation of natural phenomena and, secondly, the [mathematical and logical] tools necessary for the undertaking." My research and teaching in the history of science suggests that these definitions need to be further expanded to encompass four different aspects of science: science as a process of inquiry, science as a body of results, science as methods and techniques, and science as a social institution. For our further discussion I will use the following as a working definition:

Science: (1) Any attempt by the members of a community to establish a framework that makes their observations of nature intelligible. (2) A body of accepted knowledge about nature, including both accepted observations of nature and the frameworks used to make those observations intelligible. (3) A body of exemplary practices and techniques for observing nature and interpreting such observations. (4) Groups of specialists, and the institutions which they lead, that play major roles in 1) defining and transmitting the framework and 2) directing the practices and techniques used to find intelligibility in nature.

Science and Religion

This definition is deliberately broad and, lacking a clear line of demarcation, calls upon us to distinguish science, defined in this way, from other activities such as religion. Without going into the lengthy discussion this issue requires, the chief distinction between religious and scientific inquiry is related to the different kinds of questions they ask. When faced with a natural phenomenon religion is concerned with answering the question "why" in order to find meaning in the event: for example, why does it rain on his field and not on mine, why does good fortune occur to some and misfortune to others? These questions are frequently answered by postulating the action of spiritual beings, a god or gods, who may be propitiated in various forms of rituals. Science is not concerned with "why" but with "how"; it is concerned with understanding, in some way, how nature works.

Hopi Sun-Watching

Years ago when I was working in the Southwest a Hopi sunwatcher, Ned Zeena, told me that he understood how I could look at what he was doing as science, but he looked at it as his religion. His concern reflects the fact that in most cultures and through most of human history, scientific activity has been intimately enmeshed with other activities such as religion. When the Hopi observe the rising and setting of the Sun to determine the time of religious rituals and agricultural activities, those observations are performed by the leaders of religious societies as part of their ritual preparations. The observations are coupled with the deposit of ritual offerings at precisely located shrines marking the summer and winter solstices.

Yet a careful examination of Hopi astronomy reveals that despite its religious context there is an obvious scientific component to these activities. The location of these shrines reflects the precision of Hopi astronomical practice; they are located precisely at two of the four named directions of solstice sunrise and sunset. These directions, in turn, provide a theoretical framework for organizing Hopi knowledge of a wide range of astronomical, meteorological, botanical, zoological, and social phenomena (McCluskey in press).

When the Hopi say that the Sun journeys each year between points on the horizon that mark solstice sunrise they are doing science; when they say that the Sun is an animate being who brings warmth to the Hopi and fertility to their crops in response to their prayers, they are doing religion. These activities are intimately, although not inextricably, mixed. The shrines that serve as benchmarks for observing the annual motions of the Sun are also places where offerings are made encouraging him to hurry back northward from the winter solstice and to take his time on his southward journey from the summer solstice. These religious activities are, as Zeena said, clearly part of Hopi Sun watching and influence their observations and understanding of the heavens.

The central element of this activity as science is the interaction between observations and the theoretical framework that makes them intelligible. The observations may be conducted as part of religious rituals and the resulting framework may be expressed in terms of the sacred, but the framework is logically tied to the observations which it explains. The explanatory frameworks may be of many different kinds. They may be mathematical or mythological; they may merely describe or correlate phenomena or they may provide causal accounts for their occurrence. Science is not here distinguished by any particular ideal method, but by an ongoing attempt to come to an understanding of nature. Although the definition is fairly open ended, the close tie to observations of nature leaves little shelter for extreme social constructionists (although there is plenty of room for a more moderate social study of science).

Archaeoastronomy and the Nature of Science

The many different astronomies discussed over the past years at the Oxford conferences clearly fit within this proposed broad definition of science. Sky watchers in various cultures do establish frameworks that make their observations of the heavens intelligible; they do use these frameworks to pass on the knowledge of the heavens that they have constructed; and in most cases they do observe the heavens according to well-established procedures.[1]

If we are to treat these cultures' investigations of nature with the respect that they deserve, we will need to critically examine the scientific components which exist within this matrix of religious ideas. Once we consider these astronomies as sciences, we have the opportunity to address how these sciences, which address many of the same celestial phenomena, differ, and more significantly, how these differences relate to the social contexts within which these astronomies have flourished. In a word, we have the opportunity to provide some empirical data for the question of the cultural relativism of science.

A Taxonomy of Astronomies

As a minor step forward in this task, let me propose a preliminary taxonomy of astronomies that provides a guide to the classification of diverse astronomical traditions in terms of their main characteristics.

Table 1. Elements for a Taxonomy of Astronomies

Function:	
	Theoretical / Practical
General Approach:	
	Secular / Sacred
Method:	
	Observational / Predictive
Phenomena:	
	Continuous / Discrete
Compared in:	
	Spatial Terms / Temporal Terms
Kind of Model:	
	Arithmetic / Geometric

1. In the few cases where there is little evidence for contemporary observations (e.g., medieval computus) the astronomy employs principles (here the 19 year Luni-solar cycle) that are derived from past astronomical observations.

Principal Bodies Considered:
    Sun / Moon / Planets / Stars / . . .
Reference for Observation & Model:
    Horizon / Stars / Equator / Ecliptic
Benchmark for Reference:
    Tropical / Sidereal / Lunar Phase / Irregular
Mode of Explanation:
    Descriptive / Causal
Applicability:
    Global / Regional / Local

The first two elements in this taxonomy reflect the social contexts of these astronomies by presenting dichotomies of theoretical versus practical functions and secular versus sacred approaches; the remaining eight elements present various aspects of the scientific content of these astronomies.

Western Geometrical Astronomy

To see how these elements apply in practice, let's consider the tradition of geometrical astronomy that occupied the mainstream of the history of Western European astronomy from Eudoxus to Kepler (Pedersen 1993, 57–89, 267–277). If we were to quickly work through the elements of our taxonomy, we would find this astronomy to be theoretical rather than practical in focus and to show little concern with the sacred. It does not merely observe the heavenly bodies but predicts their future positions; these predictions use geometric models to express the motions of the Sun, Moon, stars and planets as continuous functions of time; the predictions use the ecliptic as the basic reference frame using either a tropical (Ptolemy) or sidereal (Copernicus) benchmark. These models are descriptive rather than causal in nature and are not limited to local conditions but are globally applicable.

There are two main characteristics that distinguish this astronomy from the others we will be considering. The first clearly pertains to its social context; this astronomy has comparatively little direct practical utility. Despite occasional appeals by astrologers, calendar makers, and navigators, it is primarily theoretical. The second relates to its content, it tends to describe celestial phenomena in continuous, rather than discrete terms; it is concerned with continuous processes rather than with discrete events. To clarify these contrasts, we can consider a few examples drawn from recent studies of astronomies in cultures.

Puebloan and "Megalithic" Horizon Astronomies

Turning away from the mainstream of the Western astronomical tradition, we can examine two traditions that are clearly related to observations of the Sun (and Moon) against the local horizon, the puebloan astronomies of the US Southwest (McCluskey 1990; Zeilik 1985, 1986) and the so-called "megalithic" astronomy of the British Isles (Heggie 1981; cf. Ruggles in press). Summarizing their taxonomic elements, we find these astronomies to be practical in focus and apparently sacred in nature;[2] observational; concerned with discrete events; spatial; having an arithmetical (British) or a weak geometrical model (puebloan); concerned with the Sun and Moon; using the horizon as the basic framework against tropical benchmarks determined by the turnings of the Sun or Moon; descriptive (or at most weakly predictive); and local in applicability.

Perhaps the most significant scientific aspect that distinguishes these two astronomies from the geometric astronomies we have just seen is that the arrival of the Sun at discrete points on the horizon, where to the Hopi it dances for four days (Malotki 1983, pp. 435–440), is more important than its continuous day-to-day motion. Looked at another way, these astronomies, like many of the others we will consider, do not use the motion of the heavens to measure the passage of time but are concerned with marking the arrival of specific times by the arrival of the Sun at specific points. This calendric goal drives many other traditional astronomies

To carry out this goal these two astronomies use local observations to identify those days in the year that are held, for some cultural reasons, to be specially significant. This use of local observations is coupled with widely disseminated shared reference frames (the puebloan use of the solstices marking the four directions, the British use of the solstices, equinoxes, and Mid-quarter days) is not without its cultural significance. This kind of astronomy reflects a culture of local institutions sharing astronomical and ritual concepts.

Greek and Pawnee Stellar Astronomies

Besides these astronomies which focus on the Sun and Moon, there are other traditions that center their attention on the stars. Again we can consider two examples of this kind of astronomy, the astronomy described by Hesiod (1959) in his *Works and Days* and the astronomy of the Native Americans of the Great Plains that Von Del Chamberlain (1982) has documented among the Pawnee. Although both of these astronomies are primarily observational and stellar, they appear to take a different approach to their observations of the stars. Hesiod used the seasonal appearances of the stars to identify discrete times in a practical calendar used for agriculture and trade; the Pawnee note the motion of the stars at night to determine the passing of time.[3] Both treat the stars as sacred; Hesiod (1959, line 383) speaks of "the Pleiades, the daughters of Atlas" and Pawnee lore describes a pantheon of stars.

Here again the kind of astronomy, with its wide-ranging regional applicability, reflects the cultural context. A stellar astronomy, unlike the solar astronomies of prehistoric Britain or of the puebloan peoples, requires no fixed local landmarks. The Pawnee were a nomadic people, wandering widely over the Great Plains; the Archaic Greeks were a different kind of wanderers, establishing colonies throughout the Mediterranean basin. Hesiod himself notes (1959 lines 388–9) that his astronomy can be used anywhere, "whether you live in the plains or whether close by the sea."

---

2. This sacred nature is clear for the pueblos, but only inferred from later historical festivals for the British "megalithic" astronomy.

3. The lack of mention of timekeeping by Hesiod, or the lack of a documented stellar calendar among the Pawnee, cannot be taken as demonstrating a lack of concern with the stars as temporal or seasonal indicators.

Maya and Babylonian Arithmetical Astronomies

Turning from the kinds of astronomies developed within relatively small communities, we find a different kind of elaboration in the astronomies of the Maya and the Babylonians (Aveni 1980; Neugebauer 1975). These societies both supported strong states and learned priestly classes who studied astronomy extensively.

Besides the similarities of these two cultures, there are many technical similarities among these astronomies. These astronomies have strong theoretical components, perhaps reflecting the existence of a specialized priestly class, although they also served practical astrological functions. They were predictive, not observational astronomies. Although they considered the planets to be divinities, their arithmetical methods seem purely secular.

Like the "megalithic" and puebloan astronomies, they were concerned with determining the time or place of discrete astronomical events rather than with developing a kinematic or dynamic model of the continuous motion of heavenly bodies. Despite this similar concern there were striking differences in approach. When the Maya studied the planet Venus, they predicted the days when it would appear and disappear on the horizon as morning and evening star; the Babylonians, on the other hand, predicted the place among the stars as well as the day of these events, and corrected these calculations using functions of the position of the last occurrence of the same event.[4] The concern with predicting discrete events, shared by the Maya and Babylonians, was well served by their common use of arithmetical techniques.

Although both of these astronomies dealt with the Sun, Moon and planets, the Maya apparently used a horizon reference frame against tropical benchmarks while the Babylonians used a stellar reference frame with sidereal benchmarks. Despite these differences, their calculations of the appearances of the planets were both widely applicable.

Another significant scientific aspect of these astronomies is that both were merely descriptive; we have no evidence that they proposed causal models of the appearances and disappearances of the planets. It is equally significant that the Maya and Babylonians predicted the same events using arithmetical models that differed significantly from each other, and differed even more from the geometrical models that dominated Western astronomies from Eudoxus to Kepler.

Medieval Computus

The application of similar arithmetical techniques to an explicitly calendric problem appears within the medieval astronomical tradition of computus (Stevens 1995), which concerned calculating the date of Easter, which is the first Sunday after the full moon after the vernal equinox. As with the Babylonians and Maya, we see a social concern on the part of powerful agents (ecumenical and regional church councils, emperors and kings) to establish uniformity in ritual as an essential element of uniformity in society (McCluskey 1998). Since this practical calcu-

4. Babylonian predictions of an event in terms of its position in space allowed them to develop arithmetical techniques to account for nonuniformities in the spatial and temporal intervals between successive phenomena. The Maya predictions of the date of an event did not lend itself to the development of such techniques.

lation is concerned with a ritually significant event, although the method of these calculations is clearly secular, their purpose and symbolism are clearly sacred. As the Irish monk Columbanus (1957, pages 2–5) wrote near the end of the sixth century,

> It surely is impossible that at Easter any part of darkness should rule over the light, since the feast of the Lord's Resurrection is light, and there is no communion of light with darkness.

The calculations of computus were aimed at the stable regulation of a ritual calendar. Here the social concern driving astronomical techniques was not so much astronomical accuracy (although issues of accuracy were occasionally mentioned in periods of disagreement) but theological symbolism and ritual uniformity.

Computus is very similar in scientific content to the astronomy of the Maya, employing simple arithmetic to predict the date of an event, in this case the Paschal Full Moon. The positions of the Sun and Moon or the causes of their motion were not relevant to these questions; the dates of the equinox and the full moon were crucial to all medieval inquiries into the technicalities of computus.

Astronomies and Cultural Relativism

This all too brief survey of a few astronomical traditions illustrates not only the ways in which astronomies differ, but also how those differences reflect the societies in which those astronomies developed. The kinds of differences we see among these astronomies suggest not so much that they are in conflict with each other; rather they seem to be approaching the heavens in ways that are so fundamentally different that they can be said to be talking at cross purposes to one another. An astronomy that observes the arrival of the Sun at a specific point on the horizon has little to say to an astronomy that calculates the time the Moon will be opposite the Sun or the place of that occurrence among the stars. Even modern astronomy, which computes the positions of celestial bodies as complex functions of time, only indirectly calculates the time when specific events occur.

When we get to the claimed tension between the social construction of scientific knowledge and scientific objectivity, we find that this tension also largely vanishes. Although these astronomies' approaches to the heavens are formed by the societies in which they developed, each of these different astronomies does make true statements about the regularities we see in the heavens—true in the sense that those statements were tested against their observations and can be tested against our calculations. If we were to try an ahistorical comparison of the adequacy of these different astronomies in dealing with those observations, we would find that there is a real sense in which each astronomical tradition is better suited to deal with the issues raised within its social context than other traditions which could be seen as its potential rivals.

What, then, is the point of this discussion? I see two. First, and most relevant to the focus of the Oxford conferences, is that there is much to be gained in our studies by considering the astronomies we investigate as examples of sciences. They aren't just ritual or mythology or literature, although they can be looked at from those perspectives. They present a range of different scientific models for understanding the heavens that can be meaningfully compared with the models developed at other times or in other cultures. The differences among these astronomies, and the extent to which those differences reflect the differences among societies, is an important element of archaeoastronomical inquiry.

But this approach also gives our own specific concern with the comparative study of astronomies in cultures a broader significance. These different astronomies provide those who are investigating (and debating) the nature of science real examples of different sciences that use accurate, but different, models to understand the same heavens. These examples demonstrate that although different ways to do science and different scientific results do arise in different cultures, this provides little support for those who would use such differences to question the sciences' ability to provide reliable statements about the world in which we live.

References

Aveni, A. F. 1980, *Skywatchers of Ancient Mexico* (Austin: University of Texas Press).

Brown, J. R., ed. 1984, *Scientific Rationality: The Sociological Turn*, The University of Western Ontario Series in Philosophy of Science, 25 (Dordrecht: D. Reidel).

Chamberlain, V. D. 1982, *When Stars Came Down to Earth: Cosmology of the Skidi Pawnee Indians of North America* (Los Altos, Calif.: Ballena Press / College Park, Md.: Center for Archaeoastronomy).

Dampier Wetham, W. C. 1911 "Science", in *Encyclopædia Brittanica*, 11th ed. (New York: Encyclopedia Brittanica, Inc).

Clagett, M. 1963, *Greek Science in Antiquity* (New York: Collier Books).

Columbanus. 1957, *Sancti Columbani opera*. Ed. G. M. Walker (Dublin: The Dublin Institute for Advanced Studies).

Heggie, D. C. 1981, *Megalithic Science: Ancient Mathematics and Astronomy in North-West Europe* (London: Thames and Hudson).

Hesiod. 1959, *Hesiod*. Transl. Richmond Lattimore (Ann Arbor: University of Michigan Press).

Hollis, M. & S. Lukes, eds. 1982, *Rationality and Relativism* (Cambridge: MIT Press).

Malotki, E. 1983, *Hopi Time: A Linguistic Analysis of the Temporal Concepts in the Hopi Language*. Trends in Linguistics, Studies & Monographs, 20 (Mouton: Berlin).

McCluskey, S. C. 1990, "Calendars and Symbolism: Functions of Observations in Hopi astronomy" *Archaeoastronomy*, no. 15, supplement to *Journal for the History of Astronomy*, 21, p. S1.

McCluskey, S. C. 1998, *Astronomies and Cultures in Early Medieval Europe* (Cambridge: Cambridge University Press).

McCluskey, S. C. in press, "Puebloan Ethnoscience", to appear in the multi-volume *History of Science* (Rome: Istituto della Enciclopedia Italiana).

Neugebauer, O. 1975, *A History of Ancient Mathematical Astronomy*, 3 vols (Berlin / Heidelberg / New York: Springer-Verlag).

Pedersen, O. 1993, *Early Physics and Astronomy: A Historical Introduction* (Cambridge: Cambridge University Press).

Ruggles, C. L. N. in press. *Astronomy in Prehistoric Britain and Ireland* (New Haven: Yale University Press).
Sokal, A. 1996, "Transgressing the Boundaries: Toward a Transformative Hermeneutics of Quantum Gravity" Social Text, 14, p. 217.
Stevens, W. 1995, "Cycles of Time: Calendrical and Astronomical reckonings in Early Science", in his *Cycles of Time and Scientific Learning in Medieval Europe* (Aldershot: Variorum).
Zeilik, M. 1985, "The Ethnoastronomy of the Historic Pueblos, I: Calendrical Sun Watching" Archaeoastronomy, no. 8, supplement to Journal for the History of Astronomy, 16, p. S1.
Zeilik, M. 1986, "The Ethnoastronomy of the Historic Pueblos, II: Moon Watching" Archaeoastronomy, no. 10, supplement to Journal for the History of Astronomy, 17, p. S1.

7

Interpreting Rock Art: European and Anasazi Representations of Spirituality

Richard L. Poss
Humanities Program
University of Arizona
Tucson, AZ

While indigenous rock art is a rich source for anthropological and archaeoastronomical investigation, it also constitutes an important aesthetic experience in its own right. The emphasis of recent investigation (that is, the effort to demonstrate astronomical relationships present in petroglyph sites) overlooks the complex aesthetic and spiritual nourishment gained by investigators from the site itself. Discussion of these aspects of petroglyph sites should be open to the interpretive knowledge of European hermeneutic traditions of art history and criticism.

This paper examines some ways in which European art has dealt with representations of spirituality, and then compares them with two well known Anasazi artifacts at Chaco Canyon, the "Supernova" pictograph and the Fajada Butte petroglyphs. It is hoped that the juxtaposition of the two kinds of art reveals interesting insights into the nature of both traditions.

* * *

Meaning and Interpretation

Suppose chemical analysis were to demonstrate that a picture from the Italian Renaissance was painted fifty years earlier than previously thought. This discovery would cause shock waves of stylistic and interpretive reevaluation in art history, but no one would suggest that the dating of the picture closed the book on discussion about the meaning of the work itself. Rather, the dating would prompt further analysis ranging from biographical research to connoisseurship, all with a view toward synthesizing the new information with what is already known.

So it is with rock art. Identifying the community that created the work, establishing a firm date, demonstrating astronomical alignments—all these are questions of detective work, like the precise dating of a dinosaur bone, or the authorship of a Renaissance sonnet. But even the placement of a single glyph within a tradition of stylistic development involves judgements of an aesthetic and interpretive character. Like the European tradition of connoisseurship, it capitalizes on the breadth of the investigator's personal experience with the material and on individual talent and intuition. These creative and to some extent irrational qualities enable us to probe further the depths of great art, European or Anasazi.

As the new date for the Renaissance painting becomes widely accepted, the human and animal figures in the picture and their interrelationships, as well as the brushwork and values of light and shade, will change in meaning, some slightly, others dramatically. The meaning of a painting (or of any artifact) is drawn out by asking questions and engaging in a kind of dialogue or interrogation with the work. These meanings change with new circumstances, when new facts are discovered, and when new questions are asked.

How does a petroglyph mean? How do we interpret a symbolic experience made of stone, sunlight, shadows, and setting, informed by gods, mythologies, traditions and customs, of which we are largely ignorant? We can agree on what we mean by "intention" when dealing with a Renaissance court poet, and, although the situation is by no means uncomplicated, we can at least know that we share a background of assumptions about the nature of communication, art, and the processes of symbolism and literary convention. When we examine a petroglyph, we begin far removed from this relatively comfortable space, guessing about the degree of convention present and about the participation of transcendent beings in the creation of the symbol. As with many artifacts, it may or may not have anything to do with communication in its original context.

What is the significance of the location of a glyph? When is the location as important as what is drawn there? "Location" may signify not only its location culturally considered, i.e., in a canyon, near a wash, where a hero died, where a goddess gave birth, near fields bearing crops, or beside a battlefield. But also it may mean where the sun strikes, how the glyph is illuminated, what astronomical events can be sited from the location. Will the glyph greet the adult members of the community, or is it concealed from the adult members of the community? Does it greet visitors, and enemies? Is it meant to be read from the heavens, looking down at earth, or by sky watchers looking heavenward?

What happens to a petroglyph when it rains? In terms of color, hue, allusion, and metaphorical function, what are the changes, and what do they mean? Is there a change in the meaning of the glyph in different conditions of light? What is the role of acoustics at rock art sites? Does the interplay of echoes and acoustical focusing charge some rock art sites, but not others?

Are petroglyphs ever responses to other glyphs placed on the rock earlier? Are they intended to convey or communicate at all? Or are they merely what is left after the performance of a magical or shamanic rite in which more than one person is involved? Finally, how do the astronomically charged artifacts mean what they mean? What is gained when we succeed in establishing beyond a reasonable doubt that a given site is arranged consciously and intentionally as a set of markers, of solar and lunar occasions, of solstices and cross-quarter days? What is the next step? Do we put the tools back in the kit and go home, having accomplished our goal? What is the goal, ultimately?

Modes of Explanation

When we explain what something is, our explanations fall into three major categories. These three modes of explanation offer us a useful bridge between theory in science and interpretation in art.

1. Genesis: how it came into being. Suppose an intelligent visitor arrives from another planet and wants to know what various everyday objects are. It points to a car and asks, "What is this?" We explain how iron is mined and processed into steel, how rubber is collected and shipped, how the chassis is moved along the assembly line as each component is attached, until the final product is completed. When the car rolls off the line, this type of explanation is complete. We have told the story of how it came to be.

In the case of a sonnet or a play, we explain circumstances in the mind and life of the writer, and historical and literary influences which were brought to bear on the composition of the piece, including the history and meaning of the genre, and why plays or sonnets came to be and what needs they satisfied for society.

2. Internal structure, or constituent parts. This is analysis, Aristotelian and Cartesian. We break up the artifact into its components, clearly delineate each of them, then reassemble them into the whole and this gives us an understanding of what the thing is. In the case of the car, we describe the engine, the tires, the electrical system, the hydraulic system, the headlights, and the air conditioning, etc. Then we reassemble the complete car as the sum of its subsystems.

In the case of the sonnet, we describe the iambic pentameter line, the 14-line structure, the octave, the sestet, the tendency of the octave to go in one direction and the sestet in another, etc. "Petrarchan" and "Shakespearean" models and their proclivities, patterns and correspondences in imagery, sound effects, rhythm and cadence, are all aspects of the reassembled whole.

3. Function or purpose. What is the shaping principle of a car? There is the need for transportation, the distances separating home from work, food, health care, recreation. Cultural dimensions illuminate important aspects of transportation—why one person drives a Mercedes when another drives a VW bus or a pick-up truck. Nor are these accidental but essential aspects of the car, without which our visitor fails to understand what the car means in this society.

What is the purpose of a sonnet, a play, or a devotional poem? Cavalier poets wrote sonnets for seduction, patronage, and fame all at once. A devotional poem by Donne or Herbert functions as meditation, prayer, and art-making for the poet and the reader. What is the purpose of a cathedral? Or the flashing lights of a performance art piece that lasts only a few moments? These questions do not have simple answers, and we should not expect the answers to be any less complex when the questions are addressed to petroglyph sites.

"Reading" the Old Masters

When interpreting visual art from Western Europe during the period from the Middle Ages to the end of the Baroque (Duccio to Rubens), we tend to "read" the pictures iconographically. We construct meaning in the painting by building up values from the simple to the complex, using the following method:

1. Identify each object in the picture. Since the art of this period is primarily representational, we can catalog the features to register the success of the artist in the basic practice of rendering. Here is a tree, there a mountain, etc. Is the four-legged animal a wolf, a fox, or a dog? This is not always as simple a procedure as one might think. In Georgione's *Tempest* the man on the left is holding something. Is it a spear, a staff, a crook, or a pole? Each object is similar but will have widely different values in the second stage.

2. Explore the symbolic associations of each object in the picture. The spear has martial overtones and implies combat, while the staff has pastoral associations. The dog may be a symbol of fidelity, or the patron's pet, or both. The man may be a saint, a king, or Everyman. The hill surrounded by a burning cloud is virtue, a path up that hill the ascent to virtue. The clothed woman with the sword may be Justice, while the nude woman with the bow and arrow is Artemis.

These meanings are gleaned from earlier practice, contemporary accounts, and the rich mosaic of literary and artistic background to which the artist and patron would have access. There are still allegorical paintings from the Renaissance wherein both the identification of specific objects and their meaning to contemporaries remains uncertain.

3. All the separate symbolic and metaphoric associations join together into a coherent interpretation, a unified if multivalent meaning. The parts fit together into a symphonic presentation of symbolic resonances, with a clear, if complex, message or intent. The painting is a syllogism visible to the discerning reader.

Two paintings by Botticelli, The *Birth of Venus* and the *Primavera*, were once on adjacent walls of one of the Medici palaces, according to Vasari. In Botticelli's *Birth of Venus*, the graceful nude stands on a shell in the sea, while zephyrs hover nearby. We conclude from manuscript evidence (the letter from Marsilio Ficino) that the figure is Venus, in her manifestation as virtue, that she is presented as an aid to meditation on virtue itself. The *Primavera* is more difficult to assimilate. The figures are clear enough, and the symbolic associations common knowledge in his time and ours, but combining the parts is notoriously difficult. A completely convincing interpretation of this picture has yet to emerge. With these distinctions in mind, we can examine some examples of "traditional" western art.

Representations of Spirituality in European Art

Fra Angelico's *Last Judgement* (Figure 1) is a traditional subject. Explicitly religious and didactic, this composition usually shows Christ in judgement with the

Figure 1. Fra Angelico, *The Last Judgement*, 1430, tempera on wood, Museum of San Marco, Florence.

good being shepherded into heaven on one side and the evil being herded into hell on the other. Let us reverse the device of the extra-terrestrial visitor from earlier in this paper, and imagine that we ourselves are from another planet and we seek to understand the levels of representation present in this glyph from a "foreign culture." This is, after all, what we do when we look at petroglyphs.

First we notice the overall symmetry of the composition, which implies a kind of architectural order to the cosmos. Who is the figure commanding the top center of the picture? He is a god, or rather, The God, who is also manifest in other forms (Father, Holy Spirit) but who appears here as Son and who as Son was human once but is not now, but when the "event" depicted occurs he will return in the physical garment of the son but still function as all three together.

He is surrounded by spirit-beings called angels. Their differing appearances indicate their different origins. Some were human once, others were never human but were created at the birth of the universe. Others, like the faces surrounding the seated Christ, are spirit-beings but function essentially as pictorial decoration.

The stars which appear above the heads of the seated saints indicate an identification of the realm of the stars with the realm of God and the saints (although in medieval astronomy the stars were contained in one of the spheres, and heaven was the Empyrean, beyond the spheres). Picture space is not true space. The seated saint on the far right is located right next to devils which are located in hell—the rock boundary which separates them is crucial. The castle gateway through which the angels are inducting the blessed leads them into the presence of Christ who is over in the middle of the picture. The saints seated in rows at the top of the picture lived lives on earth but in different centuries in different countries speaking different languages. They are "perfected spiritual essences" but they appear visually as approximate depictions of their earthly bodies.

At the bottom right a crowd of the unrepentant (including priests, bishops, and cardinals) is whipped by devils and pushed into the cave entrance of hell, the multi-storied domain of torture chambers which, while it resembles the interior of a mountain, is understood to be under the surface of a spherical earth. We might wonder at the psychological health of this culture, with its masochistic obsession with physical torments. The horned creature at the bottom center of hell is an angel transformed through disobedience into the source of evil on earth.

The most pleasing (and most earth-like) section is the remarkable circular dance of angels and souls on the left side of the picture. The perfection of their circle is a foretaste of the perfection of heaven. It is often remarked that Renaissance art aims at the accuracy of the photograph. The simplistic notion that this picture aims to "record an event" dissolves under the multiple layers of realities present in the composition. Hell, heaven, "dead" souls reinhabiting their bodies which are springing out of the ground—all present a coherent but multivalent meaning. The only part of the picture which is an "actual event" (but an event which hasn't happened yet and whose occurrence will herald the end of this world) is the center portion where the graves are opened and the two kinds of souls move in their separate tides.

As we look at this composition from our extra-cultural distance, how much help must we have to conclude that this busy picture presents the drama of the soul's choice and its consequences, and that it was thought to apply to all humanity? It was in a sense a picture of the meaning of life, earth the "threshing floor," in terms comprehensible to anyone from the time period. To anyone not familiar with the complicated theological background, not sharing those cultural assumptions, and not having available any other textual or visual clues, the situation would be almost hopeless. But to people within the cultural context of the work, the entire composition was a conventional image treated casually. In fact, this picture was originally painted (Pope-Hennessy 1981) on the upper portion of a chair used by the priest during High Mass.

Figure 2. Donato Creti, *The Astronomical Observations: The Sun*, 1711, oil on canvas, Vatican Gallery, Rome.

Works of art confront us interlaced with spiritual traditions, mixing representation with conventions of genre, mingling "actual" with transcendent beings, in times past, present, and future, in landscapes seen and dreamed. When can we know with certainty that a rendering of a human or animal is not intended to portray a god? Immersion in the symbolic and representational matrix of the culture seems the only answer. And if that material is not available, we are at a loss, and know very little indeed.

Donato Creti's lovely painting of *The Sun* (Figure 2) is part of a series called *Astronomical Observations* from around 1711. The sun appears three times in the painting, none of them "realistically" portrayed. At the upper right is the sun writ large, with spots, as it appears through the telescope, under magnification, and larger than the sun would appear in the sky of the landscape. The telescope is not pointed at the first image of the sun, making the larger image a kind of "inset" which implies something very modern about this culture's relationship to nature. The world appears one way, but under scientific investigation it is something very different.

While the telescope is not pointed at the image of the sun, it is oriented at the optimal angle for the structure of the picture, aesthetically considered. The smaller sun on the sheet of paper held by the observer is the same as the large image at the top. The third sun is hidden behind the distant mountains of the landscape producing a sunset in the background of the painting. In this culture, depicting a sunset while the sun is high in the sky is not a problem. The sunset adds depth and color and is present in all of the "Observations." The beauty and significance of this painting comes from its "lyricalizing" of scientific investiga-

Figure 3. Piero di Puccio, *Theological Cosmography*, ca. 1390, fresco, Camposanto, Pisa.

tion into the sumptuous Bolognese style landscape. Science is not a violation of nature, nor is it reductionist or dehumanizing, but is nestled here among the traditions of western art just as the telescope is nestled among the graceful curves of the richly colored figures and the delicate leaves of the trees silhouetted against the glowing background sunset. As in the Fra Angelico picture, we have different layers of reality represented amid apparent pictorial contradictions.

The fresco by Piero di Puccio (Figure 3) in the Camposanto at Pisa depicts the entire cosmos with God as the cosmic orderer. At the center is the spherical earth, surrounded by the orbits of the moon, sun, planets, and fixed stars. These spheres are nestled within larger spheres peopled by saints, and then by angels. There is a continuity between the terrestrial, the celestial, and the spiritual in this diagrammatic description of communal belief. Pictures like this were common at the beginning of fresco cycles which traced the history of the world from creation through history to the last judgement.

To the left and to the right can be seen the hands with which God grasps and holds together the universe. This fresco is not really an astronomical diagram, although it resembles one. It is a statement of faith in order, a profound assertion of the kind of universe we inhabit. The multiple perfection of the concentric circles emphasizes the harmony of the spheres and the unity of the whole.

Figure 4. *God as Architect*, from the *Bible Moralisée*, Reims, France, ca. 1220–50, Austrian National Library, Vienna.

Together with the illustration of *God as Architect* (Figure 4) from a Reims Bible a hundred years earlier, these pictures reveal to us the twin blessing/curses of Western culture—Science and Christianity. The perfect Hebrew God removed

Figure 5. Vasily Kandinsky, *Yellow, Red, Blue*, 1925, oil on canvas, George Pompidou Center, Paris.

from nature (no longer a sun-god or sea-god or war-god but the god of everything conceivable including conceptions themselves) creates and structures nature. Being perfect, his creation is perfect in that it follows his "laws" of physics. Thus the perfect world of science is out there, waiting for humans to discover its principles. The God of the Bible Moralisée has to bend over to build this cosmos. This God is a Master Mason—He holds the mason's dividers used by the medieval master-builders who were the architects of the Gothic cathedrals. Thus engineering is the metaphor here for divine creation. There is a legend to the illustration: "Here Creates God Sky and Earth Sun and Moon and All Elements." The picture is inspiring as it incites us to go about the world "reading" God's handiwork of principles underlying all things, to, as Mandelbrot says, "take the measure of the universe" by doing science (Mandelbrot 1983).

Western, or European art history thus offers us visual presentations of multiple layers of material and spiritual and mythological material. Physical representation or rendering provides the basis for spiritual portraiture. Vasily Kandinsky's *Red, Yellow, Blue* (Figure 5) seems such a portrait, a depiction of the age of anxiety itself. A modern work, we cannot read it within the three step process outlined above. Art by this time has emerged from the traditional language of the "Old Masters."

When I look at this picture, I see the outlines of a human face on the left side of the painting, but I cannot insist that it is either intentional or "really there." How do we interpret this painting? How do we construct meanings from the pictorial elements, the black circles, the horizontal the vertical straight lines? We move into a new realm of *initial presentation* of visual material, as opposed to the traditional *representation* of forms and objects from the visible world. The large black curving line which snakes across the right side of the painting at a diagonal is reminiscent of curving lines in petroglyphs—we do not read this as "snake" or as any physical object, but as a structural element in the composition which "extends" certain spaces, and "holds together" particular elements in the picture. The thickest part of the curved line, in the upper right portion of the picture, curves around to "close off" the eye at this edge of the painting, and provides closure to the large mass of circles, rectangles, and lines and colors of that side of the

Figure 6. The "Supernova" Pictograph, western Chaco Canyon, N.M.

picture. As such it becomes a framing structural device for the corner of the composition.

The quasi-face is not entirely devoid of emotion. He looks hapless, puzzled, fragmented, his humanity deconstructed by analysis, his own terrible invention. The bright yellow and blues in the background shading make this a curiously cheerful picture, not hopeless. It inhabits a modernity charged with questions behind questions, without the reassurances (if they ever were reassurances) of religion or Newtonian science.

And yet what are the tools of this supremely cerebral and conceptual art? Circles. Parallel straight lines, horizontal and vertical. Squiggly lines. Intersecting blobs of iridescent color. The most basic elements of pictorial communication. Forms which in themselves are not so different from the lines, circles, and snaking patterns of Anasazi petroglyphs.

Representations of Spirituality in Anasazi Art

The "Supernova" pictograph and the "Sun dagger" petroglyph are two remarkable astronomically charged artifacts in the already remarkable Chaco Canyon, N.M., a center of Anasazi civilization with so many wonders and mysteries it would have planetary significance even without these two extraordinary works. The pictograph (Figure 6) is located on a narrow overhanging ledge at the western end of the canyon, below Peñasco Blanco, one of the great-houses. A general consensus places the date of this pictograph between 1000 and 1100 AD, based on the occupation of the canyon. Certain issues, such as contamination, make the reliability of rock art dating uncertain at present. An accurate dating of this pictograph, were it available, would doubtless settle many questions. (For an example see the paper by Armitage in this volume.)

The Crab Nebula M1 in Taurus was born when a star 6300 light-years from Earth became a supernova. The event was recorded by Chinese astronomers in 1054 AD (Brecher et al. 1983). Western Europe may have been under an exceptional cloud cover at the time, or was perhaps just being uncharacteristically unobservant. The new star went unrecorded in any case. Currently, the nebula is 6 light-years across and expanding. At its core is an optical pulsar, rotating 30 times a second.

How wonderful it would be if this event had been recorded by the Chacoans in 1054 in this pictograph which still survives to be rediscovered and recognized for its almost cosmic significance by American astronomers! The Anasazi generally were careful sun-watchers. (See J. Fountain, and Preston and Preston in this volume.) The hypothesis that this pictograph "records" or "memorializes" the supernova is not generally accepted, but the pictograph is still known as the "supernova," and is noted as such on maps of the canyon. Other interpretations are possible. It could depict the moon and Venus, or the star could be a symbol for the sun, and the "crescent" could represent an annular eclipse. The cultural association (of the pictograph with the supernova, whether or not it can be demonstrated) remains for the moment.

The hopeful designation of this artifact and the attempt to connect it with the astronomical event shows us something about the spiritual aspirations of scientists. It projects a connection between something scientists have accomplished and have a good understanding of (the crab nebula in its photographic glory) and an artifact from a foreign culture whose language, religion and ritual are shrouded in mystery.

The coincidence of the two, if verified, establishes an emotional connection between 20th Century astronomy and Anasazi art. It provides one more example of the observational acumen of the Anasazi (which needs no validation). From the other side it provides a kinship which humanizes the cold facts of modern astronomy with its distances and lengths of time too large for the ordinary person to grasp intuitively. For the moment, the painting of a star, a hand, and a crescent moon stands as a question mark in the history of representation. What is being revealed? There are shamanic traditions which can shed light on the hand, and there was a crescent moon very close to the supernova in July 1054 AD.

While jamb statues of kings on the west portal of Chartres cathedral are remarkable works of medieval sculpture, no treatment of them is complete without taking into account the context. That context includes the entire cathedral with hundreds of stained glass windows and sculptures all gathered into complex and interrelated schemes. Does the supernova pictograph relate to its context?

Today one comes upon the south wall of the broad low canyon after walking west from Casa Chiquita along a footpath which borders the north canyon wall. This north wall is fairly covered in places with petroglyphs at eye level, from many periods of Chacoan culture, deeply carved and thoroughly patinated. Crossing the canyon and the central wash one arrives at a location apparently no different from any other place along the south wall. To spot the pictograph is to discover a quiet image hidden in an awesome vastness. Unlike the petroglyphs crowded along the north wall, it is placed up high, out of reach, inviting comparison with the sky. Only later do we discover the large pale comet-like image, barely discernable on the wall below.

One can be impressed by the immensity of the landscape and still realize that for the Chacoans it was normal and everyday. From where one stands underneath the pictograph, you can see for miles up and down the canyon. But was this location "remote" in its original context? The image is seen looking straight up at the cliff wall at the bottom surface of a rock overhang. Up on top of the cliff is Peñasco Blanco (inhabiting, in a sense, the space above the symbols), a town whose connection with the pictograph is still not understood. Peñasco Blanco is removed from the other great-houses, strategically located at the high point of the western end of the canyon, overlooking the mouth of a tributary to Chaco Wash. It was also one of the natural doorways to Chaco, being near an incoming road. From Peñasco Blanco there is a commanding view of the landscape.

That it is painted rather than pecked may be because this location and the underside surface were desired, and the rock surfaces satisfying those conditions were not patinated, so painting may have been preferred. The pictograph can be regarded as belonging to Peñasco Blanco specifically, rather than to Chaco generally. Images specific to Peñasco Blanco may emerge from future excavations of the town, and these could shed more light on the pictograph.

That the painting has survived to our epoch is stunning. There is so little of it, it is so pure, quiet and unassuming. The Fajada Butte spirals in comparison are symphonic, announcing their cosmic relations from the grandeur of the giant butte. But the pictograph hangs patiently for a thousand years to reveal not a demonstrable relationship, but a tantalizing possibility.

Then there is the relationship between the three symbols on the horizontal surface facing downward and the much larger comet-like image on the vertical wall face. While the three symbols are almost hidden and not visible from anywhere other than from directly below, the comet in its brightly painted original state would have been visible from a considerable distance. Did this group have other members which have not survived? Were they placed here for public or state reasons, to reinforce the prestige of the community here at one of the entries into the canyon? For the Anasazi this complex might well suggest a community in close resonance with otherwise disturbing forces of nature. Any symbols placed in this location might be understood to refer to either Peñasco Blanco or to the entire canyon community, but how to distinguish the two from where we stand? If the symbols' purpose is to placate threatening forces, then would the protection that arrangement generated be seen to extend only to Peñasco Blanco, to the entire canyon community, or to all humanity?

A star, a crescent, and a hand: the three images of the pictograph are highly suggestive in their level of symbolic abstraction. The horns of the crescent point away from the hand. The star has ten points radiating outward; the hand's five fingers extend away from the star. Let us assume the ten-pointed star represents the supernova, and that the crescent stands for the moon and all the cultural and religious associations that went along with it. We have two representations which can be considered one picture, because the two were in the sky at the same time, near each other. What of the hand? We cannot extend the same sense of representation to the hand, because presumably we do not consider this to be a report that a giant hand was hanging in the sky as well. There is a tradition of hands in native American art, generally associated with shamanic practices. They are widely held to indicate contact with the spirit world.

If the hand appears "as a symbol" it still appears in some sense as being "in the sky" and its proximity to star and moon indicates an extraordinary tension. Whether the shamanic practices were addressed to fertility, warfare, the hunt, or rainfall, there should be a minimal claim which can be made for the hand image, keeping in mind that shamanic images in particular can mean more than one thing at a time. It might be that the hand indicates action upon natural or spiritual processes in order to influence phenomena in a particular direction.

The hand not a palm-print, but is painted on the surface. Is it a human hand facing heaven or a spirit hand facing downwards? It appears to this observer to be a right-hand palm facing downwards, rather than a left hand facing upwards. As human shamans initiated contact with the spirit world, the larger spirits worked new stars and other marvels in the macrocosm. It is tempting to read the hand as the gesture of a great spirit, accompanying the marvelous supernova.

Two general readings of the pictograph follow from the attitude of the hand, both of which assume the work to be a response to the supernova: one is "palm-up," the other is "palm-down." The first, palm-up, is the simplest reading. The work is the pictorial embodiment of a ritual to accommodate the new star, and to reinforce the earlier balance of cosmic forces to which the community had committed itself, with heavenly bodies following periodic patterns and a regular succession of seasons.

The second, palm-down, indicates a spirit gesturing towards humanity, accompanying the new star as a demonstration of power, an affront to the precarious order of Chaco Canyon, a provocative act requiring a palliative ritual rejoinder from the human community. To both these interpretations the fiery image of the comet on the wall beneath adds brilliance and power. The comet flies horizontally to the left, westward, toward the canyon communities.

But both these views invoke Western assumptions about gods and mortals which may be inappropriate to the Anasazi. The real cultural difficulty may arise when we try to imagine a culture where a new star does not constitute a threat to the cosmic order. It may well be that the Anasazi greeted astronomical singularities and accepted them as singular, without being threatened by them. There may have been no great need to reconcile their cosmology with such events as lunar and solar eclipses, supernovae, or comets.

The general view, however, has been that to a culture so astronomically sensitive as the Anasazi, the supernova must have been a major historical event, possibly threatening social and political upheavals. Did the paintings accompany rituals of appeasement and sacrifice, to placate angry gods and thus avert disaster? The discovery of numerous similar images in the southwest from that period (Brandt 1975) suggests they are not merely memorials, but magically involved with the event, interpreting its meaning and incorporating it into the iconography of rock art, softening the violence of its newness, and mediating its impact on community life.

While the claims for the supernova pictograph are tantalizing, the claims for the "Sun dagger" petroglyphs (Figure 7) are more substantial. Discovered by

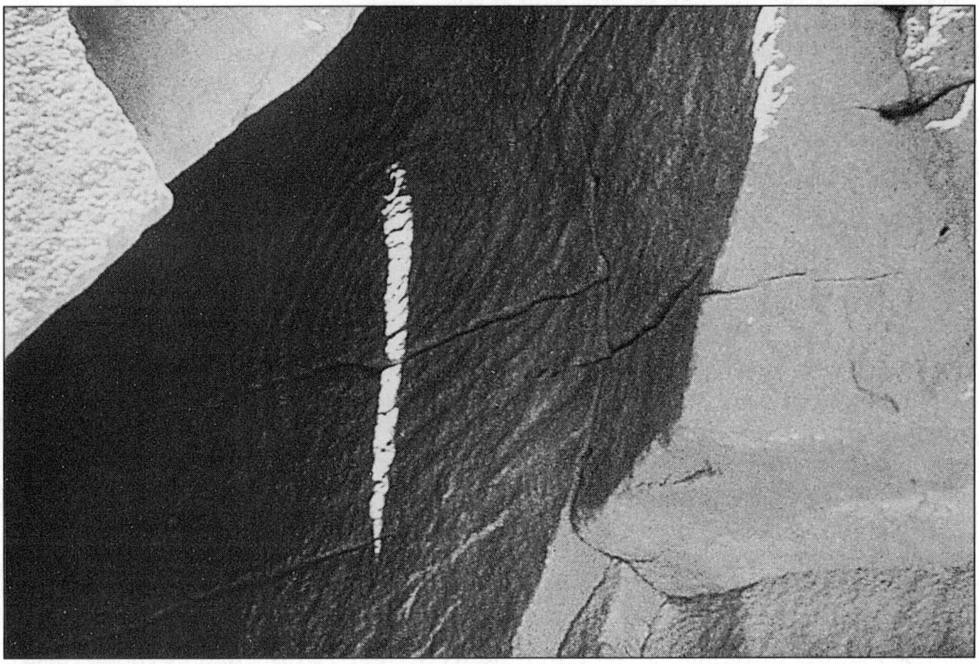

Figure 7. The "Sun dagger" petroglyph, Fajada Butte, Chaco Canyon, N.M.

Anna Sofaer and investigated by Sofaer and Volker Zinser and Rolf Sinclair (Sofaer, Zinser, & Sinclair 1979), this extraordinary complex of rock art is everything a modern scientist or humanist could wish for.

Fajada Butte towers 450 feet above the broad canyon floor, an ideal sun-watching station. Its summit seems the top of the world. There are still remains of some rooms where one can imagine Anasazi sun-priests gathered, celebrating the sacred patterns in the movements of the sun, the planets, the stars, and the moving tides of the cycling seasons.

Amongst the rock art near the top of the butte is a large spiral petroglyph carved on the wall behind three large lozenge-shaped boulders leaning against the wall. First noticed in 1977 was the intricate play of light through the stone slabs onto the petroglyphs at particular times of the year. Just before noon on the summer solstice a spot of light forms above the center of the larger spiral. It lengthens into the "dagger" shape, a sliver of light, and it moves downward through the center of the spiral. Because on the days before and after the solstice the sliver moves to the east of the spiral, it can be used to determine the solstice within a very few days.

The second sliver of light which cuts across a second smaller spiral before and after the solstice but is not present on the day of the solstice itself gives strong indication that the cluster of glyphs could indicate the summer solstice with a sensitivity of a few days. The slabs may have fallen where they are naturally in geological time, or they may have been moved in some way. The sides or edges of the slabs may have been modified, or not. The weathering is such that this cannot be determined with certainty.

Most discussion of the petroglyphs centers on the issue of how much of the site is natural and how much is artificial (Williamson 1987 and Zeilik 1984). But if we assume the case for the site being a deliberate sensitive solstice indicator (and the additional lunar indicators as well) we still have the question of purpose or function. Given their astronomical sophistication, is it plausible that this site was built by some Chacoans to inform other Chacoans of which day the solstice would be? Or could it memorialize a sacred relationship, knitting together the material of earth, sky, sunlight, and linear design into a kind of moving sculpture of light and shadow?

As Renaissance art stands against a background of Graeco-Roman mythology, without which the pictures are incomprehensible, so Fajada Butte stands intimately involved with the moving bodies of the solar system. As a sculptural "performance piece," it brings the sun and moon and their relative motions throughout the year together and incorporates them into the "installation." It is not credible that a monument like Fajada Butte stood merely to announce to history that they were capable of predicting solstice points. If that were true, we would have to say the same for Newgrange and the Egyptian pyramids. Rather, the monument of Fajada Butte seems to be a memorialization of the human community's connectedness to cosmic processes. It is often observed that a work of sculpture can alter and condition the space around it, making us perceive that space differently because of the linear thrusts in the statue. In a similar fashion, these stone slabs and spiral petroglyphs, taking advantage of the dramatic location and the leveraging function of light, bring the entire solar system into the work of art. This could be said of every solar marker, but seldom is it on so grand, so operatic a scale as at Fajada Butte.

The number of petroglyph sites where the Anasazi created or arranged shadow-casting surfaces to indicate sensitively solar interactions increases as investigators search for them throughout the southwest. (In particular see the papers by J. Fountain and by Preston and Preston in this volume.) Fajada Butte is the most impressive both in terms of its accuracy and authenticity and in terms of the grandeur of the location. It seems that, since its discovery, it has justifiably become the "Stonehenge" of solar markers. Increased scrutiny of the claims first made for this marker have strengthened the case over time. In 1989 some settling of the slabs into the ground was noticed, which altered the shape and timing of the slivers of light somewhat.

Conclusions

Archaeoastronomy has yet to come into its own in the levels of interpretation to which the artifacts are subjected. Often investigators approach a petroglyph site with a prescribed set of questions which have fairly discernable answers, which in spite of difficulties of measurement, methodology and investigator bias, are demonstrable. The questions asked now are for the most part quantitative, and can be more easily demonstrated than interpretations of an aesthetic character. It is the more difficult questions that need to be included.

"Rock art" is an admittedly inadequate term for these artifacts. What more loaded term than "art" could we choose to impose on a non-Western culture? Still, from the Western perspective, "art" is a term of praise which posits endless depths of meaning, multiple levels of interpretation, complexity, and coherence. It is a term which affirms religious, mythical, and ritual dimensions. "Art" is one term from the western tradition which is *not* reductive, but, rather, opens the artifact up to infinite dimensions of power and meaning. When we approach a rock art site we listen for what the experience is trying to tell us. We listen for indications that address our quantitative questions, but we also listen for subtle forms of music and poetry in the shape of the ground, the quality of the light, the coolness of the air, the proximity of spirit.

If rock art is art then we must admit into the conversation the hermeneutic traditions of western art history and art criticism. This is not to say that we should approach a petroglyph in the same way we approach an abstract expressionist painting. But modern non-representational art requires an openness from the viewer that would be useful in viewing rock art. The difficulties in reading a typical medieval altarpiece show the pitfalls we can encounter when we assume easy distinctions between gods and mortals, between the actual and the transcendent. We come to rock art with western classifications, and it is incumbent on us to grow beyond those cultural biases. We are in danger of oversimplifying petroglyphs if we reduce them to a strict set of linear significations. Paradoxically, it is by being aware of the complexity of our own artistic heritage that we can be more open to the full range of possibilities in Anasazi art.

Furthermore, what actually happens when Western observers visit native rock art sites is itself an experience with spiritual dimensions. The importance of the encounter of a scientifically trained Western observer with the spiritually charged artifacts of Anasazi culture has been underestimated.

Western observers point out that the Anasazi did not separate religion, science, politics, aesthetics, economics, environment, and other aspects of life into separate and discrete categories and institutions as has been the case with Western culture since the Renaissance. The encounter with Anasazi culture is one of the ways Western mankind has of encountering that underlying unity, of recognizing an environment imbued with meaning, life, and symbolic context.

If all things are related and spiritually alive as the Anasazi believed, then so is the Western observer of petroglyphs related and spiritually alive, even when bent on a mission of strictest quantification. The "meaning" of a rock art site then becomes a complex network of interlaced relationships, between the historical record, the background of the images and their traditional uses, their symbolic and religious values, and their expressive qualities evoking in us emotional and psychological responses admittedly "out of context" with the culture that produced them. The astronomical relationships inform our response to the site, but they are not the complete response. An informed appreciation of rock art is one that incorporates all the dimensions of the artifact. It may be that we are only beginning to realize how much has been left for us in the seemingly minimal figures pecked on rocks, which are gathered into landscape and leveraged into relationship with the seasons, moving in concert with the sun and moon and stars.

Acknowledgments

I want to express my appreciation to Barbara Blaha at the studio "SchwarzWeiss Labor," to Heinz Gruber and Ulrike Polnitzky at the Austrian National Library, to Tim McCarthy at Art Resource, to Russ Bodnar at Chaco Culture National Historical Park, to Joyce M. Raab of the Chaco Archives at the University of New

Mexico Library, to Laurence Bourgade of the Societe Kandinsky, and to Micheline Charton of the Centre Georges Pompidou. In addition I would like to thank John Fountain, Rolf Sinclair, and one of the anonymous reviewers—their critical insights and thoughtful suggestions helped greatly in revising the paper.

References

Brandt, J. and Williamson R. 1979, in "Archaeoastronomy" No. 1 (JHA, vol X (1979)) pp. S1–S137.

Brecher P., Fesen, R., Maran S., & Brandt J. 1983, "Ancient Records and the Crab Nebula Supernova" Ancient Records 103, p. 106.

Mandelbrot, B. B. 1983, The Fractal Geometry of Nature (New York: W. H. Freeman and Company).

Miller 1955, in "Astronomical Society of the Pacific Leaflets," Vol. VII, No. 314 (July 1955).

Pope-Hennessy, J. 1981, Fra Angelico (Florence: Scala/Riverside).

Williams R. 1987, Living the Sky: The Cosmos of the American Indian (Norman and London: University of Oklahoma Press).

Sofaer A., Zinser V. & Sinclair R. 1979, "A Unique Solar Marking Construct" Science 206, p. 283.

Zelig, M. 1984, "Archaeoastronomy at Chaco Canyon" in Noble, D. G. ed., New Light on Chaco Canyon, (Santa Fe: School of American Research Press) pp. 65–72.

Section 2

Archaeoastronomy in the United States

(Particularly in the US Southwest)

8

A Database of Rock Art Solar Markers

John Fountain
Tucson, AZ

The report by Sofaer of a rock art solar marker at Fajada Butte in Chaco Canyon, New Mexico was followed by many accounts of solar interactions with rock art on important seasonal days. A database of such observations is being maintained to assess such matters as the geographical distribution, preferred glyph type, and seasonal frequency. In spite of the 219 observations reported at 45 sites by 28 different observers, the data remain inadequate for substantial conclusions, but preliminary results are presented. Several sites have many markers at each season. Many glyphs undergo sensitive, distinctive interactions on more than one seasonal day. Few of the shadow-casting surfaces show evidence of human modification.

* * *

Introduction

In 1977 Hedges reported a distinctive sunlight interaction with a pictograph in northern Baja California on winter solstice. The well-publicized discovery by Anna Sofaer of a spectacular set of interaction at equinox and both solstices at Fajada Butte (Sofaer, Zinser, & Sinclair 1979) was followed by many reports of similar solar interactions which mark astronomically significant days of the year. These reports are often found in obscure journals and publications. The time seems right to draw this material together to examine the geographical distribution, preferred glyph type, interaction mode, seasonal preference, and temporal sensitivity of such solar markers. Such data may yield insights into the purpose for solar marker creation and the astronomical sophistication of prehistoric people of the southwestern United States. To this end, I have assembled a database of such observation, and I intend to continue to add observations as they become available.

Two classes of rock art solar markers are recognized. *Direct* solar markers are rock art images placed at a point from which the sunrise or sunset is observed to occur at a distinctive point along the horizon. *Indirect* solar markers involve a rock art image placed in such a way that light or shadow moves across a glyph in a distinctive way on a particular seasonal day. While indirect markers may result from shadows of distant rocks (Gnomon interactions) and are easily predictable, many markers result from grazing sunlight across the rock face (self-shadowing interactions), and because of the subtle undulations in the rock, are very difficult to predict.

There is much reliable ethnography to support the practice of making direct solar markers (see, for instance, Stevenson, 1985). Unaware of any literature discussing the practice of making indirect solar markers, we must currently assume the practice to be an archaic one.

Author's Observations

The plan for my observations was to observe a given site over an annual cycle of solstices and equinoxes, with occasional cross-quarter day observations. Several observers have reported distinctive, sensitive interactions of sunlight with rock art on cross-quarter days. A cross-quarter (XQ) day may be defined as a day midway in time between a solstice and an equinox.[1] Surprisingly, XQ day interactions were common at the sites I investigated, so I began spending a week at the sites on XQ days. I associate the modifier for XQ days with the following season of our calendar; thus summer XQ precedes summer solstice. It has been brought to my attention that until the 15th century, calendars in western Europe listed cross-quarter days as the beginnings of seasons, rather than equinoxes and solstice days (R. E. White, private communication, 1996). Such days are, in fact, more in tune with our sense of when the weather patterns are changing.

To date, I have investigated two petroglyph sites intensively: one in southern Nevada (Fountain, 1995) for one annual cycle and a larger site in northeastern Arizona (Fountain, 1996) for two annual cycles. The former was at the entrance to a canyon where there was a permanent water source. This is a favored environment for rock art. The second, which was studied for two years, was in a narrow canyon of importance in Zuni stories. At both sites two to twelve interactions were found at important *seasonal days*, a term used here to include solstices, equinoxes, and XQ days. Typically, a week around each seasonal day was spent observing in order to test the sensitivity of markers. While the sites were visited at other times without observing distinctive interactions, glyphs placed to mark other times of the year cannot be ruled out. Several other sites were studied less intensively.

Examples of Solar Marker Interactions

The winter solstice and spring XQ day observations in Fig. 1 show a glyph which undergoes interactions on two seasonal days. While this may seem extraordinary, with only moderate effort the author was able to identify distinctive sun-

1. A few observers have reported interactions on days when the *azimuth* of sunrise is midway between that of solstice and equinox sunrise. I refer to these days as cross-quarter days by angle bisection [XQA]. The distinction between marking a date by counting as opposed to geometrical construction may be important in considering the cultural inclinations of a people.

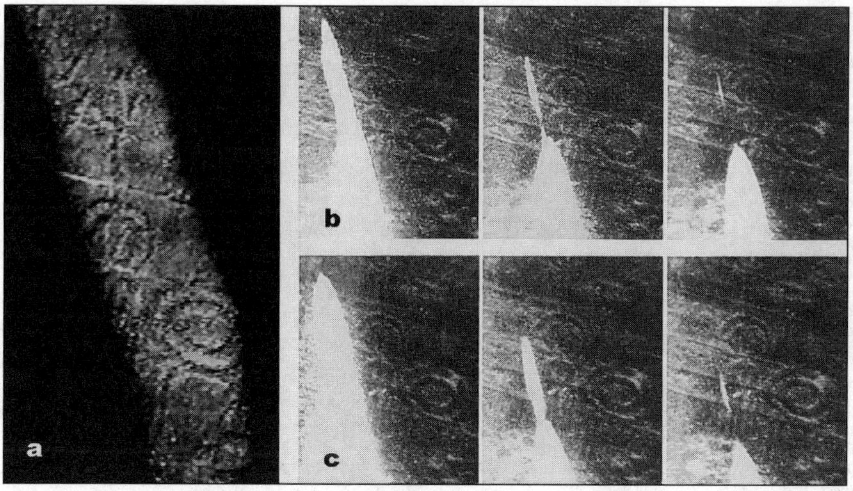

Figure 1. **(a)** At winter solstice a ribbon of light passes across two sets of concentric circles. The umbra of the shadow is tangent to the inner circle, and the penumbra is tangent to the outer circles. **(b)** At spring cross-quarter day a shaft of light narrows to an isolated sliver, disappearing tangent to the upper circle. **(c)** One day after spring cross-quarter day, the sliver of light disappears well below the upper circle.

light patterns on rocks on his property where glyphs *could* be placed to mark two different seasonal days. The spring XQ interaction shows that such dates may be marked to a precision of much better than a day.

The equinox interaction of Fig. 2 demonstrates how the shape and motion of the light on the rock face sometimes follows parts of the outline of the glyph. It is not uncommon for interactions at other times of the year, to complete important parts of the glyph form. Often glyphs are placed where the *shape* of the sunlight pattern changes significantly. This suggests that the artisan "saw" in the shape and motion of the light the glyph that he ultimately produced. This point is of interest to rock art investigators since at many sites, some rocks are covered with petroglyphs, yet others are untouched. The criteria for selection of the surface upon which to make a glyph has often been puzzling. I suggest that the motion of light and shadow across the rock face may be *one* of the factors in making such a decision. This is consistent with the substantial body of ethnography which alludes to the joining of Father Sun and Mother Earth (e.g., Cushing, 1990).

While rock art is often prominently placed at a site, some is placed in nearly hidden areas. The glyph in Fig. 3 is in a cylindrical cavity high on the canyon wall and is easily missed. We find that some of the solar markers most accurately indicating a seasonal day are less prominent. There exists a small though significant number of glyphs which undergo visually impressive interactions, while marking seasonal days imprecisely. From this one may infer that certain solar markers were produced as public commemoration of the sun. The third ring of the concentric circles (Fig. 3) is marked at both summer XQ and summer solstice. Symbolic meaning of the third ring is unclear. The summer XQ interaction is sensitive to much better than a day.

The purpose of solar markers is widely assumed to assist in agricultural endeavors. The large number (up to twelve) of solar markers for a given seasonal day

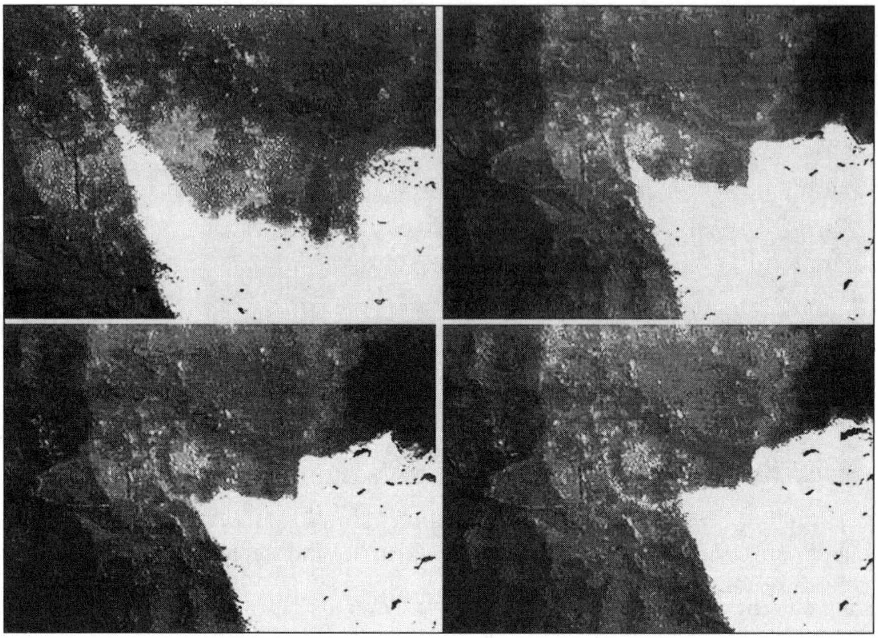

Figure 2. At equinox a shaft of light narrows and disappears tangent to the outer circle. The remaining pointer moves tangent to the inner filled circle, then moves to the outer circle and along its perimeter. Time increases left to right and top to bottom.

found at some sites makes this hypothesis unlikely. While in areas of marginal growing season, every piece of data may have been employed to decide planting times, indications from nature such as native plant budding, animal and insect activity, and bird migration may have been much more helpful in suggesting appropriate agrarian activity. With the sun being the object of great reverence by Native Americans to this day, the act of placing of rock art in such a way that sunlight interacts with it seems an appropriate memorial to that which drives virtually all activity on the face of the earth.

The Observations in the Database

Aggressive solicitation of observations and scrutiny of the literature, as well as personal observations, yielded 219 observations from 45 different sites by 28 observers. In spite of this, the number of observations in the database is inadequate. The data are replete with selection effects. First, it is impractical for an individual to monitor a site throughout the year. Expecting interactions at equinox and solstice, most observers visit sites on these days, so we have inadequate observations over an annual cycle. Even XQ day observations are rarely made, since the frequency of such interactions is not widely appreciated. A second deficit is the small number of interactions which have been examined for temporal sensitivity. Fig-

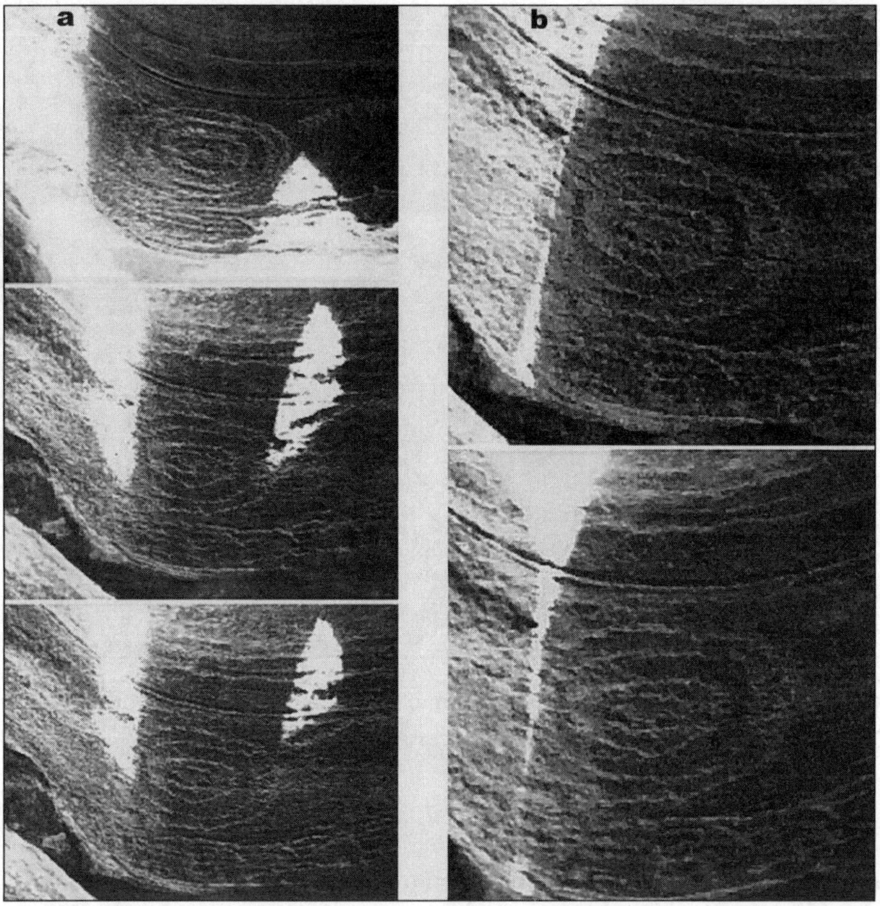

Figure 3. **(a)** Within a cylindrical alcove, two patches of light move upward, passing tangent to either side of the third of the concentric circles at summer solstice. **(b, top)** At summer cross-quarter day, a slender ribbon of light narrows and breaks tangent to the third circle. **(b, bottom)** One day after summer cross-quarter day, the ribbon of light breaks much lower.

ure 4 shows the geographical distribution of sites included in the database. This represents a minuscule sample of rock art sites in this part of the United States. It must, however, be emphasized that identifying and observing potential solar markers is a difficult and time-consuming enterprise to which few are willing to commit. Observers tend to favor sites near their residence. While some solar markers are in prominent positions, often they may be found in secretive places. It is necessary to scrutinize a site to recognize all potential solar markers. My experience indicates that where one solar marker is found, others are often present. The heterogeneous nature of the observers—their experience, interests, biases, and time commitment to the task—results in uneven scrutiny of potential solar markers at a given site.

Observations included in the database were critically examined for internal consistency of the reports, credibility of the observers, and the degree to which observations were adequately distinctive. While observations of direct solar markers are being recorded, only observations of indirect solar markers suggested to have rea-

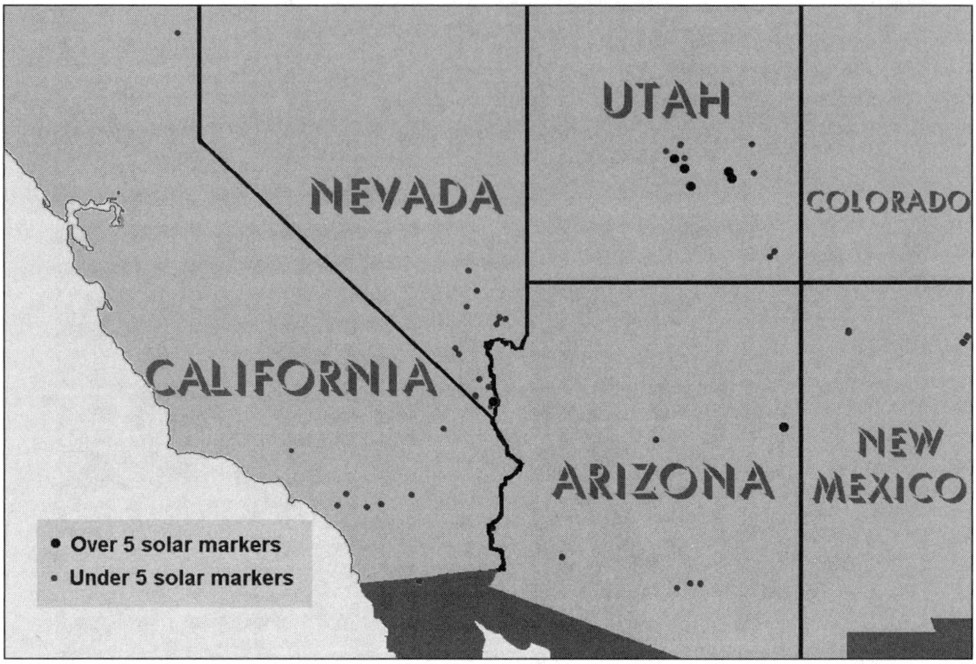

Figure 4. Map of the southwestern United States showing the locations of rock art solar marker sites in the database.

sonable temporal precision are considered in this report. The author has posted a web page to provide information on solar markers and to facilitate communication of observations. The address is http://solmar.us/.

Results from the Database

With the above qualifications, the results to date yield the following.

Table 1. Solar Markers by Season

Season	*Percentage*
Summer solstice	34%
Equinox	32
Winter solstice	14
Summer/Autumn XQ	11
Winter/Spring XQ	9

Note that particularly in narrow canyons, geometry limits the opportunities for interaction at seasons when the declination of the sun is lower. Many observers are just beginning to appreciate the potential in looking for XQ day interactions, so these events are probably more abundant than reported here.

Table 2. Solar Markers by Glyph Type

Glyph type	*Percentage*
Circles and spirals	37%
Misc.	27
Anthropomorphs	21
Zoomorphs	15

Because interactions with circular and spiral glyphs have been most publicized, some observers have given other glyph forms inadequate attention. The numbers reported here are useful only as guides for further investigation. More complete data from these and other sites will probably alter the distribution substantially.

While the database contains observations from Northern California to the Mexican border and from the Pacific Ocean to the Rio Grande River (see Fig 4), anecdotal reports indicate the likelihood that solar markers exist much farther northward and southward. The important papers by Preston and Preston (this volume) and Diane Roussel-Dupré, presented at the Oxford V Conference, form important addenda to the database as reported here.

Conclusions

Rock art placed in such a way that sunlight and shadow interacts with it in distinctive ways on important seasonal days is found throughout the southwestern United States. Such indirect solar markers may receive interactions at any time from sunrise to sunset. Midday (10:00–14:00 local apparent time) interactions appear most common, though the data on this point are insufficient.

The substantial number of glyphs over a large geographic area which sensitively mark certain seasonal days and the many glyphs which mark more than one seasonal day rules out coincidence as an explanation for most interactions. Doubtless coincidental interactions exist, but they must represent a very small percentage. The varied and complex geometries of rock art sites makes a statistical calculation of interaction probability unrealistic. This database shows that rock art was indeed intentionally place to receive sunlight interactions on important seasonal days.

The precision of the solar markers appears to follow a bimodal distribution. Most of the solar markers at equinox and XQ days distinctively mark those events to a day or better. Because the declination of the sun is changing very slowly at solstices, interactions on these dates are capable of far less precision. Some rather prominent solar marker interactions are remarkably *imprecise*. Because of inadequate data on temporal sensitivity, we can only suggest that the fraction of such imprecise solar markers to be small.

It is often suggested that solar markers were made for guidance in agrarian efforts. Clearly such a large number of solar markers at a given site is unnecessary for this purpose. Likewise, it is unclear why such precision would be necessary for agricultural concerns. Some prominently placed interactions have poor temporal sensitivity, suggesting their purpose may be more for public display.

While circles and spirals appear to be the most common glyphs for solar markers, significant numbers of anthropomorphic, zoomorphic, and other imagery are employed. The results from the current data could well be skewed by the popular impression that circular imagery is the principal form used to mark the seasons.

While some observers state that the shadow-casting surfaces of some solar marker interactions appear to have been modified by man, none examined by the author show evidence of such alteration. Most reports support the absence of human modification of the shadow-casting surfaces.

Acknowledgments

The following people contributed observations to the database. Their significant efforts were vital to this ongoing effort. An asterisk (*) indicates more than five observations.

Grace Burkholder *
Shelby Coody
Eve Cochran Ewing
John Fountain *
Cherie Freeman
Miles Gilbert
Ken Hedges
Bernard M. Jones
Reed & Norma Lance
Steve Manning
Ben Mixon
Nowell Morris
Garth Norman
John Rafter *
Nile Root
John H. Rudolph *
Clifford L. Rayl
Robert A. Schiffman
Loren Shirar
Anna Sofaer
Steve Stoney
Rory Tyler
James L. Varner
Jesse E. Warner *
Omer W. Whitman
William White
Ray Williamson

References

Cushing, F. A. 1990, Cushing at Zuni, Ed. J. Green, (Univ. New Mexico Press, Albuquerque).
Fountain, J. 1995, Utah Rock Art, 14, 75.
Fountain, J. 1996, Rock art papers, in press (San Diego Museum of Man).
Hedges, K. 1977, Fourth Annual Rock Art Symposium, Tempe, AZ.
Roussel-Dupré, D. 1996, Poster paper, Oxford V Conference on Archaeoastronomy, Santa Fe.
Preston, R. A. and A. L. Preston, this volume.
Sofaer, A., V. Zinser & R. Sinclair. 1979, Science, 206, 283.
Stevenson, M. C., 1985, The Zuni Indians (Rio Grande Press, Glorieta, NM).
Warner, J. 1993, Utah Rock Art, 12, 27.
White, R. E. (private communication 1996).

9

Consistent Forms of Solstice Sunlight Interaction with Petroglyphs throughout the Prehistoric American Southwest

Robert A. Preston
Jet Propulsion Laboratory
California Institute of Technology
Pasadena, CA

Ann L. Preston
Art Center
Pasadena, CA

At 46 prehistoric sites (≈900–1300 A.D.) throughout a 600 by 600 km. region of Arizona, New Mexico and southern Utah, 109 consistent and precise examples of interaction of sunlit images with petroglyphs (rock engravings) on the solstices have been identified. Several lines of evidence indicate that these were intentional calendrical markers, including a statistical study of the distribution of similar interactions throughout the year. Evidence also exists for the marking of the equinoxes by similar techniques. This ancient marking of the yearly cycle once existed at probably thousands of locations spread over the southwestern U.S. and can become an important tool for studying these prehistoric cultures.

* * *

Introduction

Thousands of prehistoric Indian rock art sites are scattered throughout the southwestern U.S. on cliffs, boulders, and cave walls. These images are composed of both petroglyphs (rock engravings) and pictographs (rock paintings). For the vast majority of sites, the meanings and functions of these drawings have been lost. Due in part to this difficulty in interpretation, as well as uncertainties in dating, the rock art sites have had limited value as an archaeological resource (Schaafsma 1980).

For more than fifteen years, some researchers in the southwestern U.S. have claimed that a few of these sites might have been used to mark the sun's yearly motion by the interaction of sunlit images with the rock art on the solstices (e.g., Sofaer, Zinser, & Sinclair 1979; Preston & Preston 1982, 1984, 1985, 1987; Williamson, 1984; Warner 1985; Roussel-Dupree 1997). Some of these reports also indicate that sunlit interactions may mark other dates (e.g., the equinoxes), and that moonlit interactions may mark the moon's 18.6-year period (Sofaer 1982). However, all these claims have met with some scepticism due to the lack of historical counterparts and the difficulty of determining if the events observed were chance interactions or intended calendrical markings.

In this paper we show that at least some of the types of observed interactions were intentional, and that this practice was extremely common and widespread among the prehistoric inhabitants of the southwestern U. S. during at least some portion of the time period spanning about 900–1300 A.D. Over the last 15 years we have found 109 examples of consistent forms of summer and winter solstitial sunlight displays on petroglyphs at 46 sites in Arizona, New Mexico and southern Utah. Several lines of evidence indicate these events were intentional calendrical markers including the commonality of the types of petroglyph and sunlight interaction, the frequency of occurrence of these interactions within and among petroglyph sites, and a statistical study of similar sunlight/petroglyph interactions throughout the year.

The indigenous pueblo cultures had abandoned much of the southwestern U.S. before the first European encounters in the sixteenth century, deserting large numbers of structures and rock art sites throughout the region (Cordell 1994). In the more geographically limited settlements occupied during historical times, there is no evidence that projected sunlit images on rock art were used for calendrical marking. Moreover, the making of petroglyphs was no longer as common a cultural activity (Schaafsma 1980). However, current and historical records indicate that the descendants of the prehistoric pueblo cultures monitored the motions of the sun, moon and stars, principally through direct horizon observations, to regulate ceremonial and agricultural calendars, as is common in many pre-technological cultures (Ellis 1975; McCluskey 1977; Young & Williamson 1981). It is also known from Southwestern historical records that the changing path of the sun was tracked through direct observation at rock art sites and through the motion of sunlit images on interior structural walls (Cushing 1967). Hence, it seems reasonable that the prehistoric inhabitants of the Southwest also monitored the sun's yearly motion, a hypothesis supported by probable solstitial structural alignments in prehistoric pueblos (e. g., Reyman 1971, 1976; Williamson, Fisher, and O'Flynn 1977)

Description of Sites

The 46 petroglyph sites throughout Arizona, New Mexico and southern Utah we have identified that display consistent forms of solar interaction are

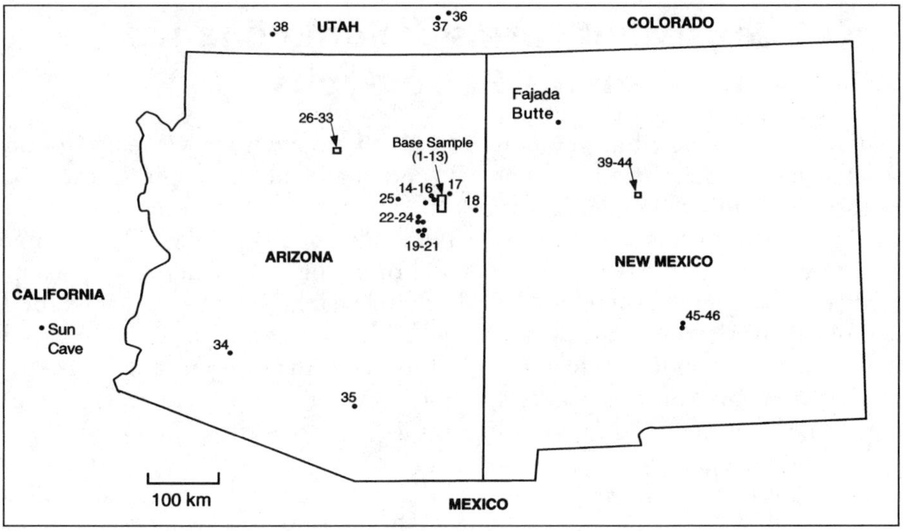

Figure 1. Map of the approximate locations of sites discussed in this paper.

composed of two groups: 1) an intensely investigated base sample of many sites within a small geographical area, which allows examination of consistent forms of sunlight/petroglyph interactions within and among petroglyph sites and their frequency of occurrence, and 2) an extended sample of sites spread across a much wider expanse of the Southwest, which allows investigation of the geographic extent and variety of the phenomenon. Figure 1 shows the approximate location of these sites, but exact locations are suppressed for site protection. The petroglyphs usually occur in concentrated groupings ("sites"), which are separated by various distances from neighboring groupings. Site dimensions generally range from about 10 to 50 meters, with each site typically containing a few to tens of separate flat rock surfaces ("panels") engraved with petroglyphs.

The base sample is composed of 13 petroglyph sites located within a 15 by 25 km region of the northeastern Arizona desert, mostly in the Petrified Forest National Park (sites 1–13 in Figure 1). The solstitial petroglyphs in the region of the base sample are probably of Anasazi origin dating roughly from 900–1300 A.D. (Pueblo II and III) (Schaafsma, 1982a). By 1450 A. D. the region had been abandoned (Stewart 1980). In the base sample region, we have searched for every petroglyph site from this epoch, and have found consistent sunlight/petroglyph solstitial interactions at 12 of the 17 identified sites. Since this is a region of sandstone mesas, some of whose rockfaces may move and erode over a millennium, it is possible that an even higher percentage (perhaps all) of these sites once displayed such interactions.

The 33 petroglyph sites we have studied in addition to our base sample range in distance up to 355 km from the center of the base sample (sites 14–46 in Figure 1), and probably date from about the same time period (although we cannot rule out slightly later dates at sites 39–44). Most of these sites are probably Anasazi in origin, but some also derive from the Hohokam and Mogollon cultures.

Consistent Forms of Sunlit Images on the Petroglyphs

The consistent solstitial sunlight/petroglyph interactions we have found at these sites involve definitive events at the centers and edges of spirals and circles. Our study indicates that identical types of interactions were used for both solstices, and that both solstices were approximately equally marked. These interactions involve either: 1) a wedge ("pointer") of sunlight or shadow whose tip either moves through the center of a spiral or circle as the sun moves across the sky, or whose side or tip traces the symbol edge without entering, or 2) a shadow line or sunlit image that suddenly forms or disappears at the symbol center or edge, or 3) a shadow line that moves up the symbol center or edge and retreats. Some examples of interactions are shown in Figures 2–6. At one site (37) a similar solstitial interaction occurs with a painted spiral rather than a petroglyph. At some historical pueblos concentric circles were used as representations of the sun (Ellis & Hammack 1968). The spiral symbols virtually disappeared from the petroglyph vocabulary by 1350 A.D. (Schaafsma 1982b), which coincides with a shift in pueblo ceremonial practices.

The precision of the observed sunlight/petroglyph interactions generally ranges from about 2 to 10 mm on the petroglyph panel, typically corresponding to an angular precision of about 0.1% to 0.2% in solar position. The observed precision of the sunlight/petroglyph interactions is generally limited by the original workmanship or subsequent movement or erosion of rocks, rather than by atmospheric refraction, long-term variations in the angle between the earth's equatorial and orbital planes, or the fuzziness of shadow edges due to the sun's angular width (although the solar width is 0.5%, the eye perceives a much sharper shadow line).

Two sites display a large number of interactions with a small number of petroglyphs, providing strong support for our hypothesis. At the Cave of Life (Site 3), 13 of the 20 interactions take place inside a rock shelter that contains three spirals and two circles. All five symbols display the described solstitial interactions, with both solstices represented and ten of the interactions occurring on one square meter of rock surface. At Alcove (Site 34), all 11 interactions occur in a smaller rock shelter on a one meter square rock surface. This panel contains two circles on which the solstitial interactions occur, with both circles marking both solstices. These two sites are separated by about 350 km, showing clearly that ideas on methods of solstitial marking were shared over a wide region.

Some petroglyphs exhibit multiple interactions at a single solstice, or mark both solstices. At least 32% of the identified petroglyphs with solstitial interactions display multiple interactions, with up to six interactions occurring on a single petroglyph. Such multiple interactions suggest they are not chance occurrences, but that significant planning was necessary before making the petroglyph, perhaps through temporary markings on the rockfaces. The rock surfaces that contain the petroglyphs and those that create the sunlit images and shadows seem to belong to natural rock formations, with no obvious evidence of alteration, although we can't rule out that smaller rocks may have been moved in a few cases.

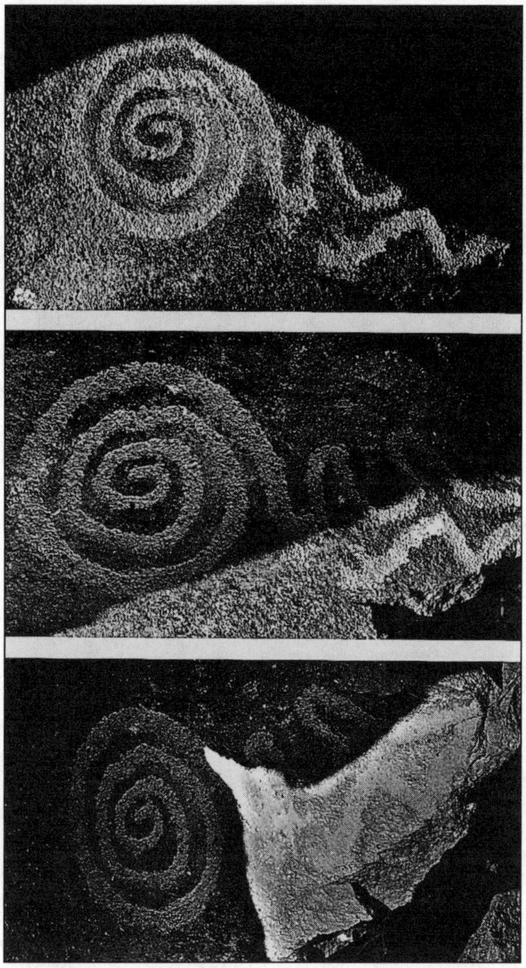

Figure 2. At a site in central New Mexico, on the winter solstice a shadow line moves upward across a spiral petroglyph until it frames the top ring of the spiral (as shown in the top photograph). The shadow line goes no higher, but reverses direction and moves downward until it is tangent to the spiral's bottom edge (as shown in the middle photograph). This is the moment of local sunset and the shadow disappears at this position. On the summer solstice the tip of a finger of sunlight, moving from upper left to lower right, traces the upper right-hand outer ring of the same spiral (as shown in the bottom photograph).

Statistical Evidence

Strong evidence that the observed sunlight/petroglyph solstitial interactions are not chance occurrences derives from a statistical study at a complete sample of 14 of our sites within a region of eastern Arizona. All spiral and circular petroglyphs at these sites were included, not just those with solstitial interactions. For all 60 spirals and 46 circles at the 14 sites, we have used Brunton compass measurements to determine all possible interactions of the types we have described that might occur over the solar year. For each possible event, the required azimuth and elevation of the sun were measured with one or two degree accuracy. Such

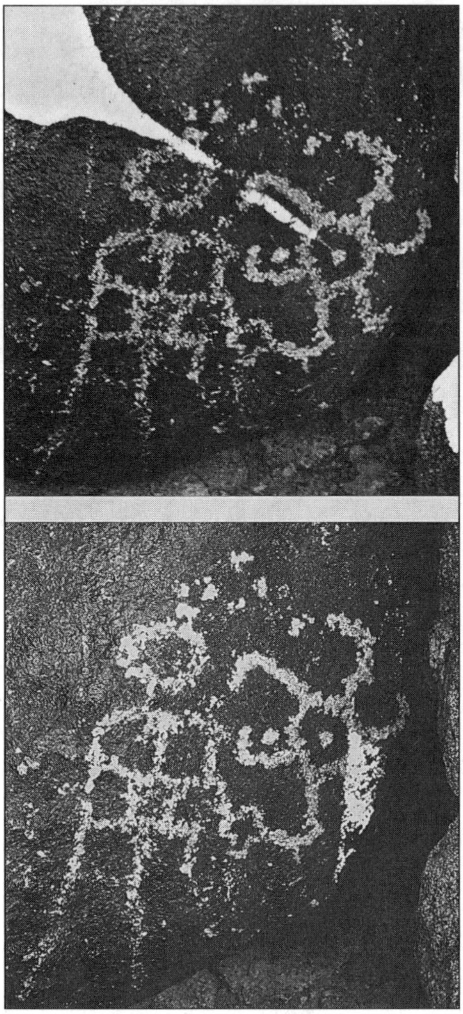

Figure 3. At a site (34) in southwestern Arizona, on the winter solstice a sunlit pointer, moving from upper left to lower right, strikes the central dot of a circle with sun-like rays (as shown in the top photo). Note that a second sunlit pointer in the upper left of the photo, also moving from upper left to lower right, is tracing the grove of another circular image. On the summer solstice, the trailing tip of a sunlit pointer, moving from top to bottom, traces the grove of the sun-like circle (as shown in the bottom photo). In total, 11 such interactions occur with the two circular images on this panel on the two solstices.

measurements of possible sunlight/petroglyph interactions are very reliable as shown by our ability to predetermine a large fraction of the solstitial interactions.

Figure 7 shows a histogram of the number of predicted interactions versus solar declination for all 14 sites combined. The plot is clearly not a random distribution, but rather shows pronounced spikes at both solstices. Interactions with spiral or circle centers or edges each individually occur most frequently on the solstices.

This study yielded an average of only 3.1 solar declinations per spiral or circular petroglyph at which interactions of these types occur throughout the year.

Figure 4. At a site (45) in south-central New Mexico, on the summer solstice a point of sunlight, moving from bottom to top, traces the right-hand edge of a circular petroglyph (as shown in the top photo). Later on the summer solstice on the same petroglyph, a new horizontal shadow line, also moving from bottom to top, rapidly converges on the original more slowly moving shadow line. The two shadow lines meet at the top edge of the circle, causing the narrowing horizontal band of sunlight between the two shadow lines to disappear at this point (as shown in the bottom photo).

Given this small average number of interactions per petroglyph, and since the precision of the interactions is high, typically only 0.1–0.2 tenths of a degree, compared to the full 47% yearly declination movement of the sun, there is only a few percent probability of a precise solstitial interaction occurring by chance on a particular spiral or circle, with a dramatically lower probability of multiple solstitial interactions on a single petroglyph. In our statistical study 22% of the petroglyphs display precision solstitial interactions.

The statistical study described above shows that a majority of the solstitial interactions are not chance interactions and that no other dates in the year are as commonly marked. Not all circles and spirals were meant to act as calendrical markers, since 61% of them among these 14 sites are located in positions where no distinct sunlight interactions can occur on any date.

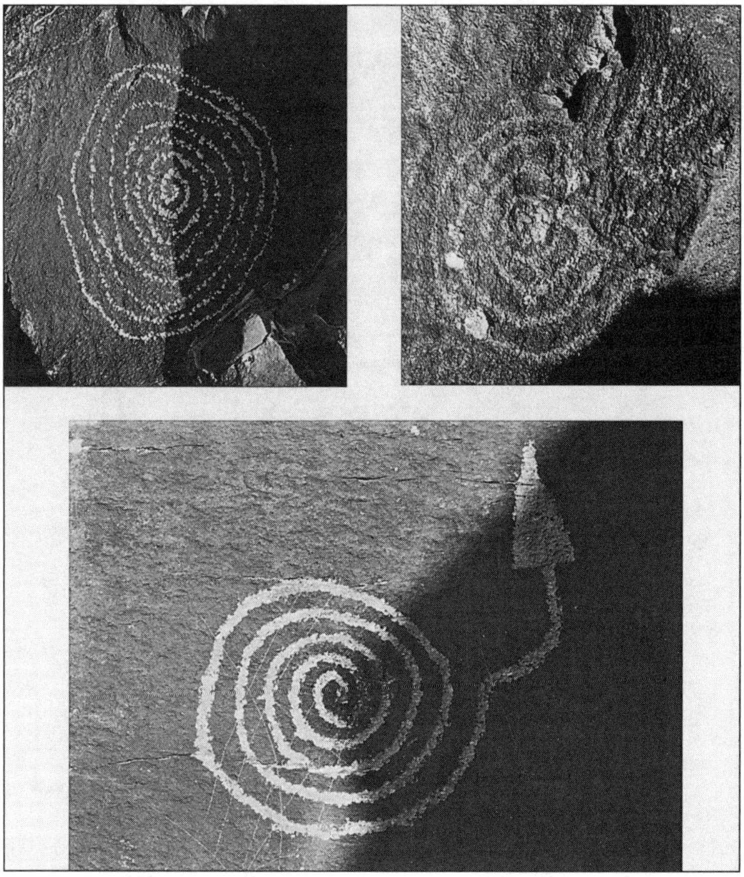

Figure 5. At a site (9) in eastern Arizona, on the summer solstice, shortly after sunrise, a fuzzy shadow from a distant rock interacts rapidly, revealing a sharp shadow line (shown in the top-left photo) through the center of a spiral. At a site (39) in central New Mexico, at the moment of summer solstice dawn, a shadow line suddenly appears tangent to (and conforming to) the outer edge of a spiral (as shown in the top-right photo). On the winter solstice a sunlit image disappears on the outer edge of the same petroglyph (not shown). At a site (1) in eastern Arizona, a shadow point, moving from lower left to upper right, passes through the center of a spiral (as shown in the bottom photo).

Conclusion

We have argued that both summer and winter solstices were intentionally marked by interaction of sunlight with petroglyphs at 46 sites ranging across Arizona, New Mexico, and southern Utah and dating from roughly 900 to 1300 A.D. Our arguments are based on the following observations: 1) the consistency of the petroglyphs and associated sunlight interactions in local regions and across a 600 by 600 km. area of the southwestern U.S., 2) the much higher frequency of these interactions at the solstices than at other times of year, 3) the fine angular precision of the interactions (typically 0.1–0.2°) compared to the large yearly declination movement of the sun and the small number of possible interactions per petroglyph throughout the year, 4) the high percentage (32%) of the identified

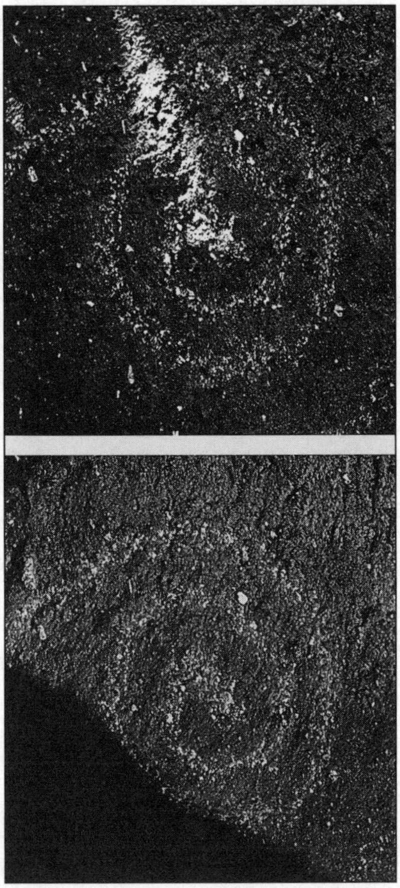

Figure 6. At a the (27) in north-central Arizona, on the summer solstice a wash of sunlight suddenly transforms into a narrow sunlit streak with the bottom end at the center of a spiral. Over the next few minutes the streak narrows and shortens (as shown in the top photo) before disappearing, all the while keeping one end at the spiral center. On the winter solstice, a shadow line moves downward on the same petroglyph until it is tangent to the bottom left-hand edge (at shown in the bottom photo), remaining at this position for 30 minutes before rising to bisect the spiral (not shown), where it reverses direction again and moves downward off of the spiral.

petroglyphs that mark both solstices or multiply mark the same solstice, 5) the high geographical density of these sites in well-studied regions, 6) the striking concentration of these interactions at some sites, 7) some consistent solstitial interactions reported by independent researchers, and 8) the precise measurement of the yearly solar cycle by historical cultures of the region.

Many of these sites still survive and function as originally intended. We have now watched several of them perform for more than a decade, precisely as they have for centuries in isolation. Our recent experience has shown that working solstice sites are easily found, almost without fail, by choosing sites with well-crafted spiral and circle petroglyphs where the rocks involved have probably not undergone significant geological movement or erosion over the last millennium. For example, Sites 26–29 are the only petroglyph sites over a

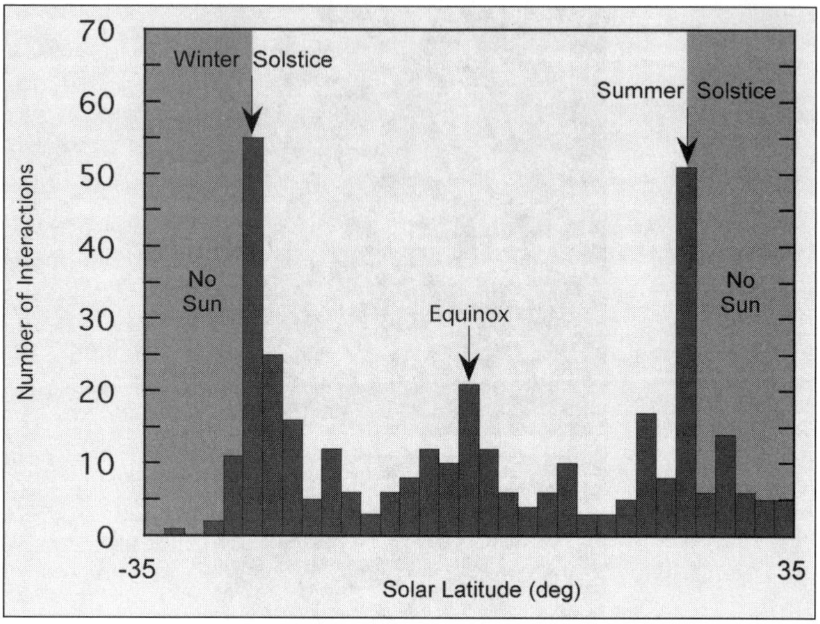

Figure 7. Histogram of the number of predicted interactions versus solar declination for all 14 sites combined.

2.5 km stretch of lava outcroppings, formations which are extremely stable geologically; 8 of 10 spiral petroglyphs among these sites display the precision sunlight solstice interactions we have described, some with multiple interactions.

Evidence also suggests that the solstitial interactions on spirals and circles act as tracers of a more complex phenomenon, with other petroglyph symbols and calendar dates being involved. In particular, strong data exist for the equinoxes being marked by similar techniques at some sites. The density of the solstice sites in the regions we studied most intensely and the broad region over which we have identified these sites indicate that thousands of such sites may once have existed throughout the southwestern U.S. Careful study of associated archeological artifacts and structures along with recent advances in radioactive carbon dating of petroglyphs raises the possibility of tracing the temporal, geographical, and cultural diffusion of these calendrical sites. The lack of memory of the purpose of these sites among the historical descendants of the petroglyph makers vividly demonstrates how easily a common cultural trait can be lost among a people lacking a written record. This interdisciplinary research, linking the disparate fields of astronomy, archaeology, and art, opens up the possibility of a previously untapped source of archaeological data in the U.S. Southwest.

Acknowledgments

We are grateful for assistance in site identification from Ferral Knight, Randy Wells, Brian Davis, and Ekkehart Malotki, for support form the staffs of the Petrified Forest National Park, the Wupatki National Monument, the Petroglyph National Monument and the JPL Photography Lab, and for funding from the Petrified Forest Museum Association and the Southwest Parks and Monuments Museum Association.

References

Cordell, L. S. 1994, Prehistory of the Southwest (San Diego: Academic Press).
Cushing, Frank G. 1967, My Adventures at Zuni (orig. pub. 1883) (Palmer Lake, Colorado: Filter Press).
Ellis, F. H. 1975, in Archaeoastronomy in Pre-Columbian America ed. A. Aveni (Austin: Univ. of Texas Press), 59.
Ellis, F. H. & Hammack, L. 1968, Amer. Antiquity, 33, 25.
McCluskey, S. C. 1977, Journal of the History of Astronomy, 8, 174.
Preston, R. A. & Preston, A. L. 1982, Bull. Amer. Astro. Soc., 14, 872.
Preston, R. A. & Preston, A. L. 1984, Bull. Amer. Astro. Soc., 16, 449.
Preston, A. L. & Preston, R. A. 1985, in Earth and Sky, ed. A. Benson & T. Hoskinson. (Thousand Oaks: Slo'w Press), 123.
Preston, R. A. & Preston A. L.1987, in Astronomy and Ceremony in the Prehistoric Southwest, ed. J. Carlson & W. J. Judge (Albuquerque: Maxwell Museum of Anthropology), 191.
Reyman, J. E. 1971, Ph.D. thesis, So. Ill. Univ.
Reyman, J. E. 1976, Science, 193, 957.
Roussel-Dupree, D. 1997, priv. comm.
Schaafsma, P. 1980, Indian Rock Art of the Southwest (Albuquerque: Univ. of New Mexico Press).
Schaafsma, P. 1982a, priv. comm.
Schaafsma, P. 1982b, priv. comm.
Sofaer, A., Zinser, V., & Sinclair, R. M. 1979, Science 206, 283.
Sofaer, A., Sinclair, R. M., and Doggett, L. E. 1982, in Archaeoastronomy in the New World, ed. A. Aveni (Cambridge, Univ. Press), 169.
Stewart, Y. G. 1980, An Archeological Overview of the Petrified Forest National Park (Tucson: Nat. Park Service Western Archeological Center).
Warner, J. E. 1985, Utah Rock Art, 4, 109.
Williamson, R. A. 1984, Living the Sky: The Native American Cosmos (Boston: Houghton Mifflin).
Williamson, R. A., Fisher, H. J., and O'Flynn 1977, in Native American Astronomy, ed. A. Aveni (Austin: Univ. of Texas Press), 203.
Young, C. J. & Williamson, R. A. 1981, in Archaeoastronomy in the Americas, ed. Williamson (Los Altos, California: Ballena Press), 183.

10

Fern Cave Rock Paintings at Lava Beds National Monument, California: Not the AD 1054 Supernova

R. A. Armitage, M. Hyman, and M. W. Rowe
Department of Chemistry
Texas A&M University
College Station, TX

J. R. Southon
Lawrence Livermore National Laboratory
Livermore, CA

C. Barat
Lava Beds National Monument
Tule Lake, CA

More than twenty western American painted and carved rock panels that include a circle or star-like symbol and a crescent in proximity have been suggested as depictions of the AD 1054 supernova that resulted in the Crab Nebula. A direct date of an image is the only way to judge the likelihood that it portrays the 1054 supernova. We analyzed radiocarbon from charcoal pigment of one such painted panel (three figures: two circles and a crescent) at Fern Cave in Lava Beds National Monument, California. The radiocarbon concentrations rule out assignment of this panel to the supernova. We suggest skepticism is warranted for other depictions until they can be dated to support or eliminate their association with the supernova. Although we have ruled out the 1054 supernova as the impetus for the Fern Cave rock paintings studied, the dates we obtained from the crescent and the upper circle are consistent with the 1604 supernova.

* * *

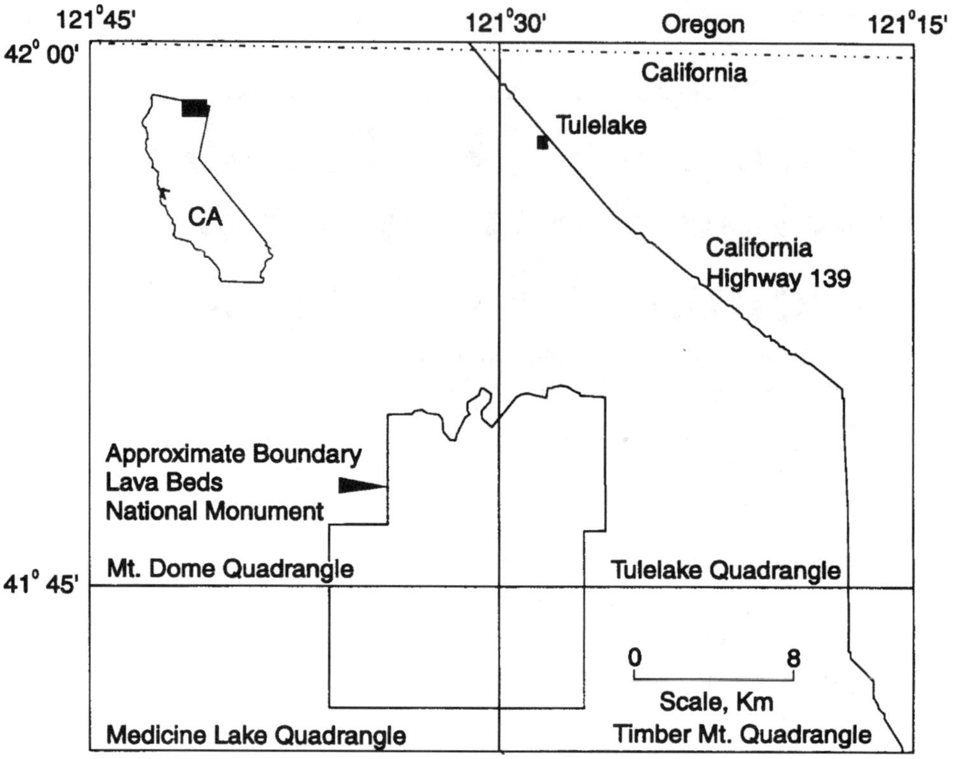

Figure 1. Map of the Lava Beds National Monument where the Fern Cave site that contains the rock paintings studied here are located.

Introduction

A new star brighter than Venus appeared in the sky on AD 1054 July 4 and remained visible for 23 days and 653 nights. This event was chronicled in at least five independent historic accounts, four from China and one from Japan (e. g., see Duyvendak 1942). Hubble (1928) suggested that the Crab nebula is the result of the 1054 supernova, a view generally accepted (Clark and Stephenson 1977). Miller (1955) was the first to postulate that specific ancient rock art in the American Southwest might represent the 1054 supernova. Both sets of rock carvings he mentions consist of a crescent and a circle: the crescent moon and nearby supernova. By the 1970s many more possible supernova depictions were suggested in North America (Brandt et al. 1973, 1975; Brandt & Williamson 1977, 1979; Eddy 1977; Mayer 1977; Mountjoy 1982; Williamson 1984). According to Mitton (1979:28), only one panel actually shows the crescent and star in correct relation. Thus as Brandt and Williamson (1979:S12) state, "...dating of the glyphs is the crucial hurdle to be surmounted before even a modicum of certainty can be ascribed to the hypothesis that the American Indians may have recorded the appearance of the Crab Nebula." Among the twenty-one possible depictions of the 1054 supernova tabulated by Brandt and Williamson (1979) are two panels of rock paintings in Lava Beds National Monument (Figure 1). We recently obtained

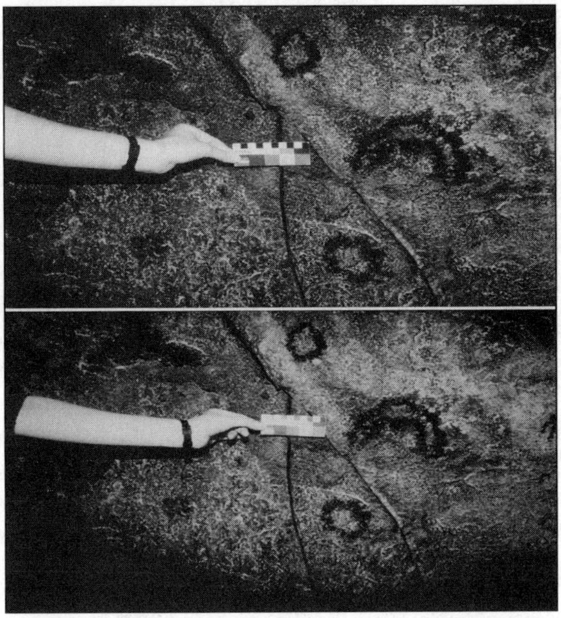

Figure 2. Figure 2A, top image, shows the "supernova" panel before the sample was taken. This is compared to Figure 2B, the lower image, that shows the panel after sample removal from the two circles to the left of the crescent and the crescent as well. It is difficult to locate where the sample was removed. The scale is 10 cm long.

three charcoal pigment samples from one of these panels located in Fern Cave that has three promising figures in close proximity: a crescent pointing downwards and two circles, one above and one below the crescent (Figure 2). Adequate samples were obtained from these three to obtain an accelerator mass spectrometric ^{14}C date on each using our plasma-chemical extraction technique. The dates do not permit interpretation of these images as the 1054 supernova.

Until a decade ago direct dating of rock paintings had never been accomplished. The advent of tandem accelerator mass spectrometry (AMS) for measuring ^{14}C greatly reduced the amount of carbon necessary for an analysis (Bennet et al. 1977; Muller 1977; Nelson, Korterling, & Stott 1977) so that the age of a small rock painting sample could be determined. Van der Merwe, Sealy, & Yates (1987) were first to directly ^{14}C date rock paintings that contained charcoal as a pigment. Since charcoal is the most commonly dated archaeological material using ^{14}C analysis, the technique for reliably dating it has been worked out in detail (Aitken 1974, 1990; Mook & Waterbolk 1985; Taylor 1987; Bowman 1990). Improvements were made in the AMS process and others followed the van der Merwe, Sealy, & Yates lead, ^{14}C dating charcoal pigments from rock paintings the world over (e.g., Ilger et al. 1995a in Angola; McDonald et al. 1990 and David 1992 in Australia; Valladas, Cachier, & Arnold 1990, Valladas et al. 1992, Clottes, Courtin, & Valladas 1992 and Ilger et al. 1995a in France; Farrell & Burton 1992 in Arizona; Geib & Fairley 1992 and Chaffee, Hyman, & Rowe 1994a in Utah; and this work in California). Other approaches to dating rock paintings have been used that also depend upon the presence of a known organic material associated with the painting, e.g., blood (Loy et al. 1990), plant fibers (Watchman

& Cole 1993) or beeswax (Nelson et al. 1995). One of the coauthor of the Loy et al. paper has raised questions about the validity of that work (Nelson 1993).

Inorganic pigments are far more frequently used than charcoal in rock paintings: reds, oranges, browns, and yellows are usually iron oxides/hydroxides and black is often manganese oxides/hydroxides (Hyman, Turpin, & Zolensyky 1996); none of these materials can be dated directly. Rock paintings using these inorganic pigments, however, can be ^{14}C dated if the following requirements are met: (1) organic matter was added to the paints initially and enough organic material has survived to yield adequate carbon to date; (2) this organic carbon can be extracted without also removing carbon from the atmosphere, limestone ($CaCO_3$ and $MgCO_3$), or mineral accretions (carbonates and oxalates) found both above and below rock paintings; (3) the extraction does not introduce significant mass fractionation; (4) carbon in the organic matter originally added to the paint does not exchange with other sources of carbon after paint application; and (5) basal rock and mineral accretions associated with the rock painting do not contain enough organic carbon (contamination) to invalidate a date. In our laboratory at Texas A&M University, we developed and tested a plasma-chemical extraction technique applicable to all archaeological materials for which a trace of organic matter remains (Russ et al. 1990, Russ, Hyman, & Rowe 1992; Chaffee, Hyman, & Rowe 1993, Chaffee et al. 1994 b; Ilger et al. 1995 b, 1996). This allows prehistoric rock paintings that do not contain charcoal to be dated by accelerator mass spectrometry ^{14}C analysis provided that the above conditions are met. The motifs studied here from Fern Cave are charcoal.

There was no difference in the macroscopic visual appearance among the three figures sampled that we could perceive. In our experience, however, it is not possible to ascertain even a relative chronology of paintings based solely on their appearances. That is true even if the observations are confined to a restricted geographical location. Differences in paint constituents, small scale micro environments, weathering and other factors may cause different rates of deterioration and fading in paintings. Although the three figures chosen for study here appeared to be visually similar, that is no guarantee that they were painted contemporaneously. Many images executed in charcoal appear in the same cave.

Experimental Procedure

Collection Protocol

Over the past few years, we have developed a standard protocol for collecting samples of rock paintings for dating. We begin by examining the selected painting to determine if any pieces seem about to spall naturally from the shelter wall. Small areas least likely to visually affect the image are selected. The selected regions are viewed carefully to assure they are free of obvious contamination—algae, rootlets, insect parts, etc. When a chosen sample passes these criteria, we photograph the motif with a camera that provides instant color photographs and one with color slide film for later development. At Fern Cave, we were interested in the panel (Figure 2) that had been suggested by Brandt et al. (1973) and others to be a probable depiction of the visible 1054 supernova.

With the permission of the Superintendent of Lava Beds National Beds National Monument, we scraped charcoal from the crescent and two circles individually, collecting three samples from that panel. Rubber gloves were worn to avoid contamination. A stainless steel dental pick was used to collect one sample and an Exacto knife with replaceable blades was used for the others. A new blade was used for each of these two samples. The removed charcoal was deposited onto a piece of aluminum foil, ~10 X 10 cm; we visually estimated when the amount of charcoal sample on the aluminum foil was adequate for a datable sample. We need only 100 mg of carbon to yield a datable sample at the AMS. Damage incurred to the three images was insignificant enough that it was difficult to determine where the sample was removed by visual inspection of the rock paintings before and after sample collection. After samples are collected, they are wrapped in aluminum foil and placed into a plastic bag. When samples arrive at our laboratory at Texas A&M University, they are placed in a desiccator filled with non-reactive argon gas to lessen the extent of adsorbed atmospheric CO_2 on the sample surfaces. Before chemically pretreating samples, we inspect them through a microscope to confirm the absence of visible contamination; any found is removed.

Chemical Pretreatment

The procedure for chemically pretreating archaeological charcoal is essentially standardized (Aitken 1974, 1990; Mook & Waterbolk 1985; Taylor 1987; Bowman 1990). First, limestone is dissolved with acid to remove carbonates that would otherwise contaminate a ^{14}C analysis. NaOH is used then to dissolve the humic and fulvic acid fractions that may be present. The sample is then re-acidified to prevent adsorption of atmospheric CO_2 caused by the NaOH. Both acid treatments have been shown to be unnecessary in our plasma-chemical technique (Chaffee, Hyman, & Rowe 1993). Limestone and other carbon-containing minerals are unaffected by the argon or oxygen plasmas and so do not interfere with our analysis. Further, argon plasmas are effective in removing adsorbed CO_2. Therefore, before the Fern Cave samples were ready for plasma-chemical extraction for ^{14}C analysis, they were treated only with 1M NaOH at 50°C to dissolve out possible humic and fulvic acids. These samples showed qualitatively the highest concentrations of humic and fulvic acids we have observed in any rock painting samples studied so far. Humic and fulvic acids make the NaOH solution appear brownish-yellow. It took eight to ten treatments with NaOH before the solutions appeared clear; we then did three additional extractions to be certain that we had removed the humic and fulvic acid components from these samples. Neutralization of the NaOH solutions with 1 M HCl produced no precipitates as expected for humic acids; thus the brownish-yellow color was more likely due to fulvic acids. Finally the samples were rinsed with doubly de-ionized distilled water, filtered and dried. The NaOH-treated charcoal samples were then ready for plasma-chemical extraction of the organic carbon.

Plasma-Chemical Extraction of Organic Carbon

We established early on the necessity for cleaning the plasma-chemical extraction system with oxygen plasmas before sample insertion to rid surfaces of organic contamination and/or adsorbed CO_2 (Russ, Hyman, & Rowe 1992). To eliminate organic contamination on chamber surfaces, oxygen plasmas were run in the empty chamber until ≤ 1 μg carbon, as CO_2, was generated during a given plasma oxidation. After the chamber is clean, a sample is inserted into the chamber via a copper-gasketed, stainless steel flange-sealed port on the chamber under a flow of argon to prevent atmospheric CO_2, or organic aerosols and particles from entering the system. Cleaning each sample with argon plasmas to remove adsorbed CO_2 before subjecting it to an oxygen plasma for organic carbon extraction was also found to be efficacious (Chaffee, Hyman, & Rowe 1993). With these precautions, negligible contamination is expected from the system or sample surfaces during plasma extraction of a rock painting sample. The following effects have also been studied and none affect our ability to date rock paintings: argon and oxygen sources; mass fractionation; calcium carbonate, magnesium carbonate, limestone and calcium oxalate decompositions; and potential vacuum leaks. Plasma treatment conditions were similar for all rock painting samples. After a sample was loaded into the plasma chamber, the chamber was evacuated to the 10^{-4} torr range and then filled with 0.2 torr ultra high purity argon (99.999%). Even though we routinely store our samples under argon, they still absorb some CO_2 onto their surfaces. Consequently, 20 to 40 watt, low-pressure (0.2 torr), low-temperature (≤ 150°C) radio frequency (RF) generated argon plasmas are performed on each sample to remove this adsorbed CO_2 by inelastic collision of the non-reactive, but energetic high-velocity argon atoms, ions and radicals with the adsorbed CO_2 molecules. We repeat this process until the amount of carbon, as CO_2, desorbed by a given argon plasma is <1 g. This amount of remaining absorbed atmospheric CO_2 should have negligible effect on the radiocarbon analysis because the background on an AMS is between 2 and 4 g carbon.

We then used RF generated 100 or 150 watt, low-temperature (≤ 150°C), low-pressure (~1 torr), oxygen plasmas coupled with high vacuum equipment, to remove organic carbon from the charcoal paint of the three Fern Cave samples, the two circles and a crescent shown in Figure 2. The low temperature of the bulk plasma is critical to avoid decomposition of carbonates and oxalates present in the sample that could seriously alter the estimation of the age of a painting. The gaseous products for oxidation of the Fern Cave samples, 250, 185 and 230 mg CO_2, were collected by freezing in a liquid-nitrogen cooled cold-finger. The amounts of CO_2 extracted were thus well above the minimum amount (100 mg) that has been routinely reliably analyzed by the AMS laboratory. See Hyman and Rowe (1997) for a review of the rock art dating from our laboratory. Samples were sent to the Center for Accelerator Mass Spectrometry (CAMS) of the Lawrence Livermore National Laboratory to be analyzed for radiocarbon.

Results and Discussion

We used the University of Washington, Quaternary Isotope Laboratory, Radiocarbon Calibration Program Revision 3.0.3a, PC Test Version #10 (Stuiver & Reimer 1993) to calibrate the radiocarbon analyses from CAMS, converting them to calendar dates. The data set used was that of Stuiver & Pearson (1993). A sum-

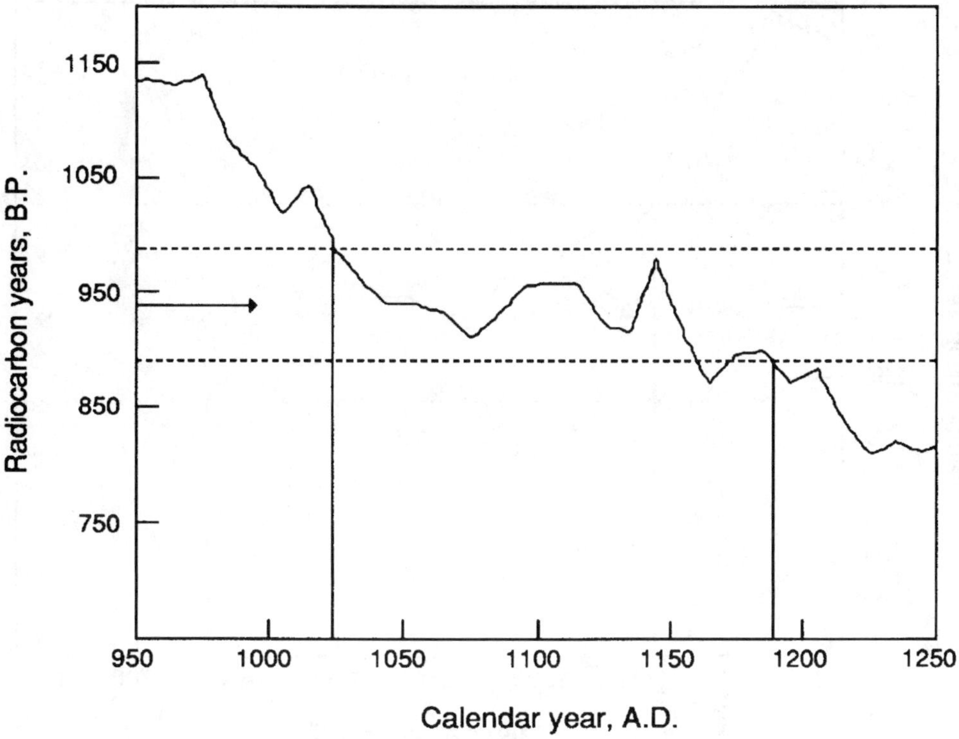

Figure 3. The calibration curve in the vicinity of the calendar date 1054 A.D. The radiocarbon date that most closely corresponds with 1054 A.D., 938 years B.P., intersects the calibration curve at five calendar dates: cal A.D. dates of 1055, 1089, 1120, 1139 and 1151, and lies within the $\pm 1\sigma$ uncertainty band of the calibration curve from 1030 A.D. to 1160 A.D. (Stuiver and Reimer 1993; Stuiver and Pearson 1993).

mary of the calibration program calculations on the three Fern Cave rock painting dates follows in Table 1. Any attempt to confirm that a particular rock painting represents the 1054 supernova is limited by the nature of the radiocarbon calibration curve in the critical area. The radiocarbon date that corresponds to the calendar date, 1054, is 938 years BP. As illustrated in Figure 3, the radiocarbon date of 938 years BP intersects the calibration curve five times. Even if one could measure the radiocarbon date with perfect accuracy, the calibrated age would be a range: cal 1050 to 1150 because of the nature of the calibration curve in the critical age range. It is not possible, therefore, to limit the age of a rock painting that was painted in 1054 to better than the range 1050 to 1150, with no way to determine more accurately the "true" date. With the additional minimum uncertainty of the ^{14}C measurement at the AMS laboratory, the calibrated uncertainty band lies between 1030 and 1160. Nonetheless, dating the rock paintings is the only direct way to confirm or deny that they are in accord with their designation as a painted representation of the 1054 supernova. A date of 1054 does not prove that a painting is of the supernova. Demonstrating that the estimated age of a painting agrees with 1054 makes it more probable that the illustration is of the supernova. However, dates that are not consistent with 1054 effectively rule out a depiction as the 1054 supernova.

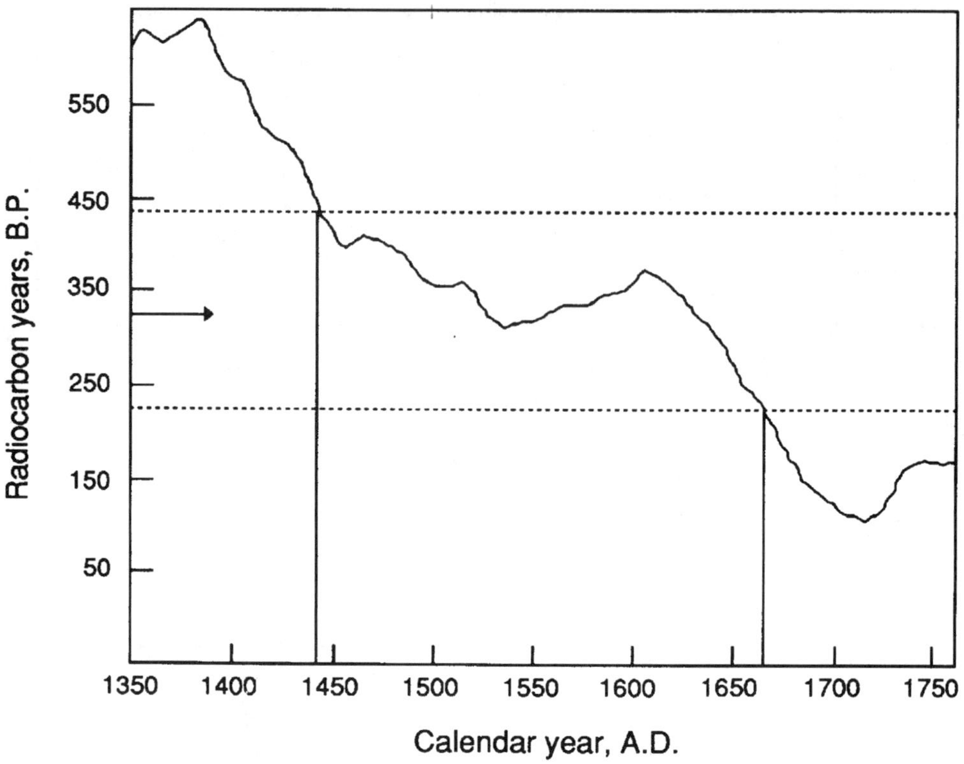

Figure 4. The calibration curve for the radiocarbon analysis of the Fern Cave crescent. The radiocarbon date, 330 ± 50 years B.P. intersects the calibration curve three times yielding cal A.D. dates of 1530, 1560 and 1630. The ±1σ uncertainty band of the calibration curve lies between 1440 A.D. to 1670 A.D. (Stuiver and Reimer 1993; Stuiver and Pearson 1993).

Table 1. Calibrated Calendar Ages Derived from the Radiocarbon Analyses

Sample	Carbon, μg	CAMS No.	^{14}C Date, years BP ±1σ uncertainty	Calibrated A.D. Date ±1σ uncertainty
Lower circle	250	27229	840 ± 70	1020 to 1290
Upper circle	185	27860	230 ± 70	1490 to 1955*
Crescent	230	27861	330 ± 50	1440 to 1670__

*1955 denotes the influence of nuclear bomb ^{14}C. The calibrated AD date overlaps with the onset of the nuclear bomb test radiocarbon levels.

The radiocarbon dates we obtained on the three components of the Fern Cave rock painting demonstrate that they are not a representation of the 1054 supernova. The lower circle, dated at cal AD 1220 (1020 to 1290) is distinctly older than both the upper circle, cal AD 1660 (1490 to modern), and the crescent, cal AD 1530, 1560 and 1630 (1440 to 1670). As can be seen in Figure 4, the radiocarbon age of the crescent crosses the calibration curve at three calendar dates; there is no way to distinguish which of the three dates is the best estimate. Radiocarbon ages on the upper circle and crescent are not significantly different from one another and may have been painted contemporaneously. Because these direct dates have shown the panel is inconsistent with it being a representation of the 1054 supernova, additional skepticism is in order for other rock graphic depictions in the American west that are considered candidates for recordings of that event.

Although we showed that the images at Fern Cave did not represent the 1054 supernova, the possibility exists that another supernova may have been represented. We did not consider other supernovae, since they were not mentioned in the previous works referenced here on supernova depictions. However, as pointed out by an anonymous reviewer, there have been other supernovae, e.g., at "A.D 185, 1006, 1181, 1572, 1604 and possibly 837, most or all of which would have been visually spectacular. In fact, the supernova of AD 1604, Kepler's supernova, which occurred near the constellation Scorpius, would have presented a scene much like that depicted in Fern Cave at around 6:45-7:15 p.m. (late twilight-early nightfall) in the evening of 25 October 1604. At that time the event was a few degrees above and to the side of a thin crescent moon (day 2-3 in the lunar cycle). The date of the Kepler supernova fits comfortably within the calibrated date ranges of the two images [the crescent and the upper circle], including if they were to be averaged as a single date." Perhaps the question of the supernova motivation of the Fern Cave images studied here is still open: only the date of the specific supernova under question must be changed from AD 1054 to AD 1604.

Acknowledgments

We appreciate discussion with Dr. Paula Reimer concerning calibration of radiocarbon dates. We appreciate the cooperation of the following at Lava Beds National Monument: Craig W. Dorman; Gary Hathaway; Michelle Moore; Chris Roundtree; and Kim Strassburg. One of us (RAA) holds a Graduate Assistantship in Areas of National Need from the Department of Education. Lawrence Livermore National Laboratory is funded by the U. S. Department of Energy under Contract W-7405-Eng-48. Partial support for this work was supplied by the Charles A. and Anne Morrow Lindbergh Foundation and the Center for Preservation Technology and Training.

References

Aitken, M. J. 1974, Physics and Archaeology (Oxford: Clarendon Press), 41-42.

Aitken, M. J. 1990, Science-based Dating in Archaeology (New York: Longman), 86-88.

Bennet, C. L., Beacons, R. P., Clover, M. R., Jove, H. E., Lievert, R. B., Litherland, A. E., Purser, K. H., & Sondheim, W. E. 1977, Science, 198, 508.

Bowman, S. 1990, Radiocarbon Dating, University of California Press: Berkeley, 28-29.

Brandt, J. C., Maran, S. P., Williamson, R., Harrington, R. S., Cockron, C., Kennedy, M., Kennedy, W. J., & Chamberlain, V. D. 1973, Bull. Amer. Astron. Soc., 5, 29 (abs.).

Brandt, J. C., Maran, S. P., Williamson, R. A., Harrington, R. S., Cochran, C., Kennedy, M., Kennedy, W. J., & Chamberlain, V. D. 1975, in Archaeoastronomy in Pre-Columbian America, ed. A. F. Aveni, (Austin: Univ. Texas Press) Chap. 3, 45.
Brandt, J. C., & Williamson, R. A. 1977, in Native American Astronomy, ed. A. F. Aveni, (Austin: Univ. Texas Press) Chap. 12, 171.
Brandt, J. C., & Williamson, R. A. 1979, Archaeoastronomy, 1, S1.
Chaffee, S. D., Hyman, M., & Rowe, M. W. 1993, in Time and Space, eds. J. Steinbring & A. Watchman, Occasional AURA Publication #8 (Melbourne: Australian Rock Art Research Association), 67.
Chaffee, S. D., Hyman, M., & Rowe, M. W. 1994a, in New Light on Old Art: Recent Advances in Hunter-Gatherer Rock Art, eds. D. S. Whitley & L. L. Loendorf, Institute of Archaeology, Monogram 36 (Los Angeles: Univ. California Press), 9.
Chaffee, S. D., Hyman, M., Rowe, M. W., Coulam, N., Schroedl, A., & Hogue, K. 1994b, Amer. Antiquity, 59, 769.
Clark, D. H., & Stephenson, F. R. 1977, The Historical Supernovae (New York: Pergamon Press), 140-160.
Clottes, J., Courtin, J., & Valladas, H. 1992, Rock Art Research, 9, 122.
David, B. 1992, Rock Art Research, 9, 139.
Duyvenduk, J. J. L. 1942, Publ. Astron. Soc. Pacific, 54, 91.
Eddy, J. A. 1977, in In Search of Ancient Astronomies, ed. E. C. Krupp (New York: Doubleday and Company, Inc.), 133-163.
Farrell, M. M., & J. F. Burton 1992, North American Archaeologist, 13, 219.
Geib, P. R., & Fairley, H. C. 1992, J Field Archaeology, 19, 155.
Hubble, E. 1928, Astron. Soc. Pacific Leaflet No. 14, 55.
Hyman, M., Turpin, S. A., & Zolensky, M. E. 1996, Rock Art Research, 13, 93.
Hyman, M., & Rowe, M. W. 1997, Techne, 5, 61.
Ilger, W. A., Dauvois, M., Hyman, M., Menu, M., Rowe, M. W., Vézian, J., & Walter, P. 1995a, Préhistoire Ariègeoise, 50, 231.
Ilger, W. A., Hyman, M., Southon, J., & Rowe, M. W. 1995b, Radiocarbon, 37, 299.
Ilger, W. A., Hyman, M., Southon, J., & Rowe, M. W. 1996, in Archaeological Chemistry: Organic, Inorganic, and Biochemical Analysis, Advances in Chemistry Series, ed. M. V. Orna (Washington, D. C.: American Chemical Society), Chap. 29, 401.
Loy, T. H., Jones, R., Nelson, D. E., Meehan, B., Vogel, J., Southon, J., & Cosgrove, R. 1990, Antiquity, 64, 110.
Mayer, D. 1977, in Native American Astronomy, ed. A. F. Aveni (Austin: Univ. Texas Press), 179.
McDonald, J., Officer, K., Jull, T., Donahue, D., Head, J., & Ford, B. 1990, Rock Art Research, 7, 83.
Miller, W. C. 1955, Astron. Soc. Pacific, Leaflet No. 314, 1.
Mitton, S. 1979, The Crab Nebula (Boston: Faber and Faber), 15-29.
Mook, W. G., & Waterbolk, H. T. 1985, Handbooks for Archaeologists No. 3: Radiocarbon Dating (Groningen: European Science Foundation), 36.
Mountjoy, J. B. 1982, Amer. Antiquity, 47, 110.
Muller, R. A. 1977, Science, 196, 489.
Nelson, D. E. 1993, Antiquity, 67, 893.
Nelson, D. E., Chaloupa, G., Chippendale, C., Anderson, M. S., & Southon, J. R. 1995, Archaeometry, 37, 151.
Nelson, D. E., Korterling, R. G., & Stott, W. R. 1977, Science, 198, 507.
Russ, J., Hyman, M., Shafer, H. J., & Rowe, M. W. 1990, Nature, 348, 710.

Russ, J., Hyman, M., & Rowe, M. W. 1992, Radiocarbon, 34, 867.
Schiffer, M. B. 1986, J Archaeological Science, 13, 13.
Stuiver, M., & Reimer, P. J. 1993, Radiocarbon, 35, 215.
Stuiver, M., & Pearson, G. W. 1993, Radiocarbon, 35, 1.
Taylor, R. E. 1987, Radiocarbon Dating: An Archaeological Perspective (New York: Academic Press), 41-47.
Valladas, H., Cachier, H., & Arnold, M. 1990, Rock Art Research, 7, 18.
Valladas, H., Cachier, H., Maurice, P., Bernaldo de Quiros, F., Clottes, J., Valdes, V. C., Uzquiano, P., & Arnold, M. 1992, Nature, 357, 68.
van der Merwe, N. J., Sealy, J., & Yates, R. 1987, South African J Science, 83, 56.
Watchman, A., & Cole, N. 1993, Antiquity, 67, 355.
Williamson, R. A. 1984, Living in the Sky - The Cosmos of the American Indian (Boston: Houghton Mifflin Company), Chap. 8, 178.

11

A Cultural Interpretation of an Astronomical Calendar (SITE #WS 833) at Wupatki National Monument

Bryan C. Bates
Coconino Community College
Flagstaff, AZ

A singular Hisatsanom (Anasazi) wall contains three portals, each directing attention across a flat eastern horizon. The sun rises center in the 1) south portal on 2–5 February, anticipating Hopi Powamu ceremony; 2) middle portal 1–4 May when soil temperatures spur seed germination; and 3) north portal 17–25 June, signaling an end to the Katsina season with preparations for Niman. Ceremonial preparations are culturally confirmed by lunar phase. One half kilometer from the calendar wall, the Crack-in-Rock community includes an enclosed plaza, a potential dance plaza of the Katsinas. A kiva near the calendar wall suggests spiritual associations and a quadruped petroglyph marks winter solstice sunrise. The petroglyph may indicate the Patki Clan, though the true identity remains an enigma. Ethnographic records indicate Patki Clan 1) announced and facilitated Powamu; 2) were farmers; and 3) assisted in Niman. The wall appears to be a cultural calendar and hierophany. Given the environmental degradation which occurred during occupation, the calendar wall supports the theory that archaeoastronomical sites may have served as a technological adaptation to environmental stress. However, such markers also allow for consistency in ceremonial observation without the necessity of pre-existing environmental degradation.

* * *

Introduction

The remnants of a cultural center (circa 1100–1300 AD) are preserved at Wupatki National Monument. These desert scrub mesas have changed little since pre-

Figure 1. The calendar wall.

Figure 2. Eastern horizon from west of wall.

historic abandonment giving us clues as to the cultural and environmental history. This chapter will present a cultural interpretation of a solar calendar developed by the pre-Columbian Hisatsanom (Anasazi).

Site Description and Surroundings

Three small portals in an independent "calendar wall" (circa 1120 AD, Figure 1) direct one's vision to specific points along a flat horizon 80+ Km to the east (Figure 2). One meter west of the wall, a rock outcrop may have allowed the prehistoric sunwatcher to observe (it is assumed) the sunrise horizon. Two flat rocks construct each portal and extend the width of the wall (Figure 3). Only one wall photograph (1971) was found in Wupatki National Monument archives, there were no records of site stabilization, and site number WS833 was assigned in 1990. Given this lack of historic documentation and the original dry laid masonry (no mortar is in the wall), it is assumed that the site has not been stabilized or significantly al-

Figure 3. Portal with flat rock on side.

tered. Map 1 details the location of Wupatki National Monument, photos and associated artifacts.

The portals track the sunrise azimuth via changes in light-shadow lines. On 2–5 February (Figure 4), the sun will rise center to the south window and may (in coordination with the moon) anticipate Powamu. From 1–4 May (Figure 5), the sun rises mid-portal forecasting soil temperatures likely to stimulate seed germination. Near 25 June (Figure 6), the sun migration reverses (as observed through the north portal) prompting the announcement and preparations for Niman (Katsina Going-Home Dance). The light pathways through each portal do not intersect at a common point.

One room block lies east of the wall and a second to the north but neither obstructs portal vision. Because few artifacts remain, the date of the calendar wall is assumed to be contemporaneous with Crack-in-Rock (e.g. late PII, early PIII) (Smith 1949, 42–48).

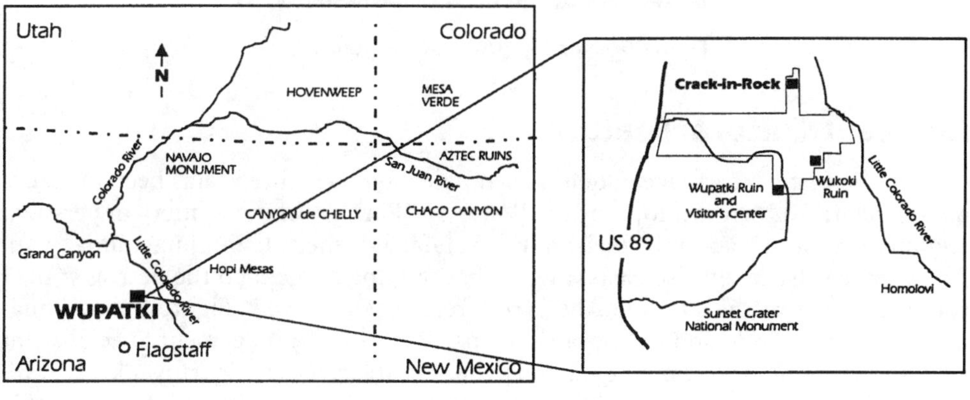

Map 1. Location of Wupatki National Monument.

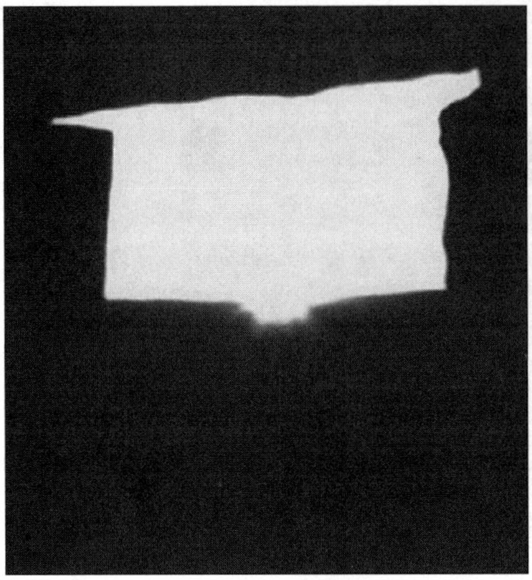

Figure 4. South portal sunrise, February 2.

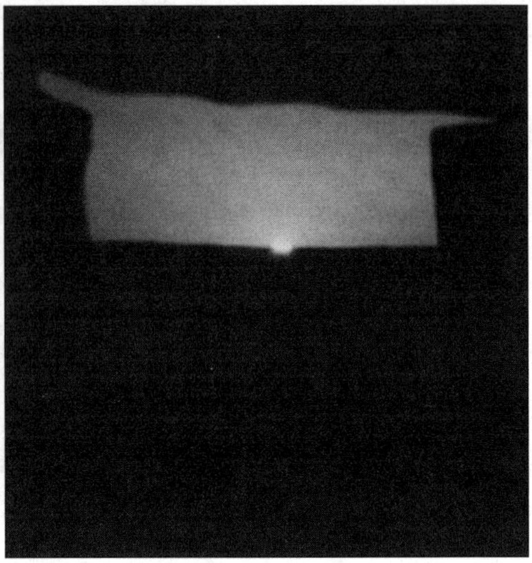

Figure 5. Middle portal sunrise, May 1.

Figure 6. North portal sunrise, June 21.

Adjacent Cultural Features

On the eastern cliff face, a quadruped petroglyph (Figure 7) has been independently identified by four Hopis as the Water or "Patki" phratry symbol in personal communications ("p.c.") (D. Nelson; E.J. Satala; S. Denett; L. Poulingyouma, p.c.). The winter solstice sunrise casts a light-shadow line tangent to the petroglyph. A room structure to the south has been referred to as a "kiva" (Figure 8) by Cinnamon (p.c.) and Anderson (p.c.), both former Resource Management Specialists at Wupatki. Wright (1992, p.c.) questions whether the structure is truly a kiva. A religious chamber would lend a religious context to the calendar wall. Smith (1952, 73) described a similar structure as a kiva (Crack-in-Rock, Room 3 in Map 1). East

Figure 7. Quadruped petroglyph.

Figure 8. Kiva.

Figure 9. Sun spiral petroglyph.

of the "kiva" is a large spiral petroglyph with rays (Figure 9) referred to as a "sun spiral" (Anderson, 1990: 9,30). From compass measurements, this petroglyph does not appear to have any anticipatory or calendrical function.

Context in Relation to Crack-in-Rock Community

The Crack-in-Rock community (NA537) was a significant dwelling with numerous smaller habitations scattered about the region (Anderson, 1990). The one-half Km distance between Crack-in-Rock and calendar wall fits research by Zeilik (1985, S8) who states "in most cases, [historical] sun-watching stations are within

Figure 10. Crack-in-Rock plaza with collapsed wall.

or close to the pueblo." Furthermore, Downum (1994 presentation and p.c.) suggests that Crack-in-Rock may be one of the original homes of the Katsina culture, a religion which "may have kept multiple cultures together." The enclosed plaza may have allowed for Katsina dancing and ceremony (Figure 10).

Similarity to Known Regional Astronomical Sites

The calendar wall is similar to known calendrical sites at Hovenweep National Monument (Williamson 1989, 116–132); however, the method of observation differs. At Hovenweep Castle and Unit-Type House, light penetrates small portals, projects to and intersects with a marked point along the back wall. The Hovenweep light patterns are more precise as the observer at the Wupatki calendar wall looks through the portal and can affect the observation with slight motions of his head (Figure 11). The erosive sandstone at Wupatki precludes determination as to whether

Figure 11. Angular projection of light shadow through portal.

Figure 12. Rock bench 1m west of wall.

observational markings were made on the rock bench (Figure 12) (Breed, 1992 p.c.), similar to the marked points on the Hovenweep walls.

Research Methodology

During first examination of the wall, a true N–S string line was laid out accurate to one-half degree via a Brunton compass, corrected for magnetic declination. Using a transit, (aligned with true N–S string) the telescope was center to each portal (Figure 13) and the back azimuth recorded. The declination of a center rise ce-

Figure 13. Measuring the back azimuth.

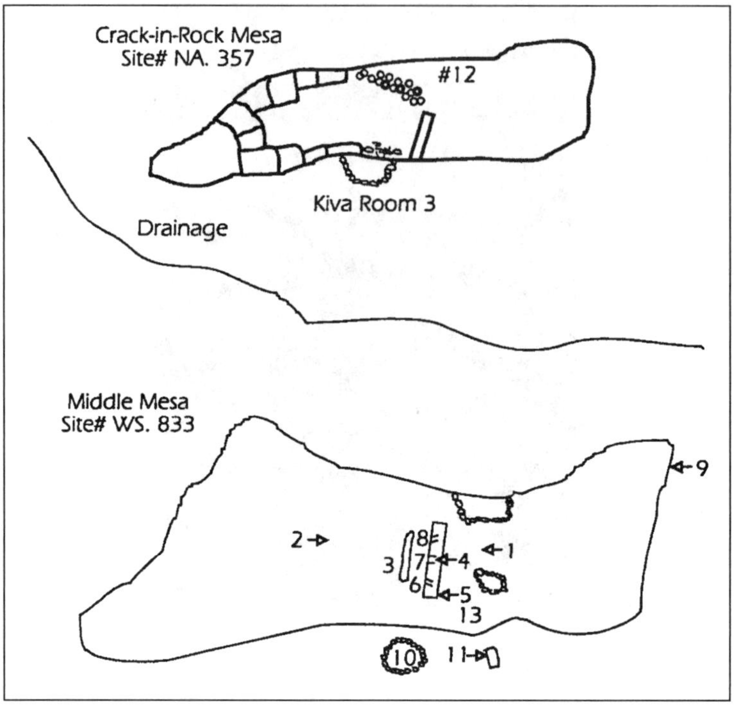

Map 2. Physical data.

lestial object was calculated for each portal. A secondary theodolite survey documented azimuth of portal perception, altitude and azimuth of the sun, and WWV time. Calculations were performed per Thom (1971) and found to be in agreement with the original survey. The physical data are detailed in Map 2. Physical measurement of the site was followed by extended library research and innumerable personal conversations with both Native and non-Native residents.

Discussion

Survival on the desert mesas of Wupatki depends on planting at a time when the seeds are most likely to germinate, mature and bear fruit, the biology of which depends on soil temperature and rain. To ensure that the rains come, that the sun returns to and from its solstice "home", and the crops grow in bounty, Hopis have created ceremonies to ask the "gods" to send their blessings. By manipulating sunlight and stone to calibrate time, the calendar wall appears to be a "hierophany" or a place "that *something sacred shows itself*" (Eliade 1959, 11; emphasis original).

Potential Hopi Clan Association

The Hisatsanom builders of Crack-in-Rock (Cinnamon, 1988; Anderson, 1990 & p.c.; Downam 1994) are likely though not exclusive ancestors of the current day Hopi (Brew 1974, 514; Downum, 1994). Tree-ring dates of the first Walpi village

(Ahlstrom 1991, 632–4) overlap in time with dates of Wupatki abandonment. The quadruped petroglyph (Figure 3) has been identified by four Hopi as a Patki phratry symbol. However, Wright (1992 p.c.) suggests the Sand phratry, a moiety partially absorbed into Snake clan with "certain prayers of the Patki clan" (Stephen 1936, 718). Fewkes (1897, 7 Figure 57) documents rain clouds as Water clan and the quadruped icon as a Sand clan. Colton (1931, 34–5) also documents quadruped as Sand clan. No symbol was documented for Patki clan. More research is needed to verify the identity of this petroglyph and the relationship between associated clans. This enigma reinforces the caveat of using historical data to interpret prehistoric ethnographic artifacts. (Binford, 1972; McCluskey, 1981).

Patki Clan Origins

"The Patki Clan came from Pala'tkwabi. No one knows just where that Redland is, but is somewhere in the far southwest." (Stephens 1936, 849; Hodge 1912, 563).

Using Hopi informants, Byrkit (1988) has traced the Pala'tkwabi Trail from the Hopi villages south to the Verde Valley. Nequatewa (1967, 134–5) agrees and notes that the clan probably split after leaving the Verde Valley. Curtis (1922, 82) states that Cloud clan (analogous to Patki clan) came from the Gila River, traveled the Pala'tkwabi trail, and settled at Homolovi (now a State Park) 50 km south of Crack-in-Rock along the Little Colorado River. However, Hodge (1912) implies a migration distance including Crack-in-Rock. The Pala'tkwabi Trail passes through Chavez Pass where the author is researching a potentially predictive October cross-quarter marking spiral petroglyph.

Clan and Ceremonial Associations

Semantics intended to distinguish often end up confusing the non-Hopi. The religious "societies" (or "fraternities") set the ceremonial calendar based on celestial observations. A phratry is a set of related clans, yet each clan may have a priest(s) who is(are) involved with a religious society and has (have) a multitude of responsibilities related to but not necessarily distinguishable as a "society" vs. a "clan" responsibility. Even this distinction maybe more "Anglo" than "Hopi". The confusion is not surprising when one realizes that the documentation of celestial observations, and the "society" vs. "phratry" vs. "clan" ceremonial responsibilities vary depending upon which Mesa (First, Second or Third), and which individual Hopi (i.e. that individual's knowledge and willingness to convey accurate information) of which society on which mesa and which clan affiliate is speaking, as opposed to which anthropologist is recording the information. As with identifying clan/phratry symbols, we are once again reminded of the caveat of using historical data quid pro quo for prehistoric interpretation (Binford 1972; McClusky 1987).

Calendar Wall and Ceremonial Associations

While there is no known connection between the words Wupatki and Patki (L. Jenkins, Hopi Cultural Preservation Office), the cultural responsibilities of the Patki Clan (Stephen 1936; Fewkes 1920) correspond with those dates predicted by the calendar wall.

Powamu

When the sun rises in/near the South portal (2–5 February) *and* the new moon is first sighted from the most elder maternal clan house, Powamu is announced in the kiva (Stephen 1936, 123–155). "That the Hopi observe both the sun and the moon, and note the relations between them is indicated by the luni-solar principles underlying Powamu" (McCluskey, 1982). Powamu is a complicated and interwoven ceremony concerning world renewal (Wright 1992, p.c.) and "a ritual designed to promote fertility and germination" (Titiev 1944, 120). The beans to be used during Powamu are planted by the chief of Powamu society at the full moon of December/ January preceding (Stephen 1936, 155). "The Sun chieftaincy [which announces Powamu] is handled within the Patki clan" (Stephen 1936, 1075) and the chief kiva in Powamu belongs to the Patki Clan (Stephen 1936, 1080). Because of the angular motion of the sunrise light shadow line across the jamb of the portal, the sun chief could anticipate the timing of the sun and associated moon for Powamu.

The first day of the 1893 Powamu ceremony (Stephen 1936, 155) corresponded with the first crescent moon (29 January), and the date of Katsina emergence from the kiva coincided with sunrise center alignment in the south portal. The 1894 announcement occurred on solar alignment (2 February) with the Katsinas going into the kiva on the first crescent moon (8 February). In 1934, Powamu began on 2 February (Titiev 1944,19), again a time of first crescent moon. The discrepancy in dates is undoubtedly the result of coordinating with the lunar aspect rather than a mistake in the anticipation of sun position. (These dates were confirmed via Voyager II astronomy computer program.)

Planting

Near the time of mid-portal center sunrise (1–4 May), the probability of a killing frost decreases as soil temperature increases. (Wupatki weather records 1950–1990). No announcement is made concerning planting (Stephen 1936, 390); however, the Patki Clan chief is the first to plant (Stephen 1936, 1082). With the largest aperture, the middle portal provides the sun watcher with greater freedom in interpretation and greater associated risk. Secondly, though no present day ceremony exist for this time period, there was formerly a ceremony geared to fertility among young Hopi called Nevenwehe. The exact dates are unknown (Wright 1992, p.c.).

Niman

The sun is first visible through the north portal near 10 June, and then rapidly slows in its angular shift. Announcement of Niman comes four days after the solstice (Stephen 1936, 61), the time when one can (with a reference system) first distinguish a change in sun azimuth direction. Following Niman, the Katsinas return to their home in the San Francisco Mountains (Nequatewa 1967), as does the smoke from their ceremonies (Titiev 1944, 172). During Niman, the Patki clan (with a large representational involvement, Stephen 1936, 539) is one of four clans to provide water for the Katsinas (Stephen 1936, 516). The chieftaincy for the "Going Home" dance is within the Patki clan (Parson, in Stephen 1936, XXXV) and the Patki clan (though not necessarily an individual chieftain) is in possession of a "tiponi" or medicine bundle (Stephen 1936, 1075). It is not clear which clan would announce Niman, though the Patki clan appears to be influential.

As the sun reverses its migration, so too does the direction of Hopi sun-watching. From the winter to summer solstice (the Katsina season) it is Hopi convention to observe sunrise, the direction in which the calendar wall is situated. From summer to winter solstice (non-Kachina season), the Hopi watch sunset (Hodge 1912, 567; Titiev 1944, 109; McCluskey 1990, 53). Sunsets are not observable through the portals of the calendar wall.

Snake and Flute Dance

If one continues observing sunrise (which is not culturally supported per above), the sun would rise mid-portal center 6–10 August, a date which may be associated with the Snake and Flute dance. Based on evidence from Eastern Pueblos, this time corresponds with the annual shift from agriculture to hunting and gathering (Wright 1992, p.c.). Harvest occurs when the crop is ready (Nelson 1992, p.c.).

Wuwutsim

Near 31 October, the sun rises center to the south portal and, with the moon, sanctifies the announcement for Wuwutsim, a preparation for darkness and initiation ceremony (Nelson 1993, p.c.). The fields are cleared and (physically and religiously) readied for next year's planting in the Above while planting commences in the Below. Wuwutsim prepares participants for added cultural and spiritual responsibilities (Stephens 1936, 1037; Nelson 1992, p.c.).

Soyal

Winter solstice is prehistorically predictable by the shift of a sunrise light-shadow line which comes tangent to the quadruped petroglyph (Figure 7). The winter solstice is not recorded by the calendar wall, posing the question "Why?". No clear answer is readily attainable yet the announcement of Soyal may be related to the responsibilities of Bear clan (Frigout, 1979, 575; Wright 1992, p.c.). Winter solstice was (and still is) carefully watched and predicted.

Other Wupatki Sites

At Wukoki (NA.203), a portal wall is found in association with a tower similar to known astronomical sites at Hovenweep (Williamson 1989, 116–132). The orientation of the four portals would indicate solar observations of summer solstice, equinox, and winter solstice. However, the wall was reconstructed in 1954 and close examination of a 1930 pre-stabilization photograph shows that the current portals are not aligned as the original portals. Consequently no definitive statement can be made as to whether the portals served an intentional calendrical function (Bates and Snow, 1996).

Only one site at Wupatki is anticipated to predict equinox (Preston, 1996 p.c.; this volume). There are no known lunar alignments, although one site has been observed and the evidence is inconclusive (Bates, 1989).

Other Regional Sites

Atop a hill near Chavez Pass, a 31 October sunrise light-shadow line bisects a spiral petroglyph. With each sunrise, the light-shadow line progressively cuts higher across the spiral and is completely off the spiral prior to 21 December. With the reversal in solar migration, the light-shadow line also reverses and leaves the spiral near the February cross-quarter date. From Patki clan history, it is anticipated that the two settlements are related. Observations will continue.

From a "shrine" (Dr. Jan Balsom, Archeologist, Grand Canyon National Park) 121 m south of Hilltop Ruin in the Colorado River corridor, one can observe the sunrise along a variegated eastern horizon with an apparent "bridge" in the Coconino sandstone aligning with winter solstice. Wright (1993, p.c.) indicates the site may be related to the Bear Clan (Bates, 1992).

Conclusion Related to the Calendar Wall

The Patki phratry are known to have brought "a high form" of sun worship to the Hopi villages (Hodges 1912, 563). Patki clan is largely responsible for Katsina season sun observations and "with the Patki clan came water, clouds and corn" (Stephen 1936, 718). The Patki clan sun chief announces Powamu (Stephen 1936, 1075) south portal, they are farmers who plant first (Stephen 1936, 390 & 1082) mid-portal, and Patki clan is integral to Niman (Stephen 1936, 519, 536 & 1075) north portal. It appears that the ceremonial and survival responsibilities of the Patki clan are documented by the calendar wall. However, Frigout (1979,575 via Eggen 1950, 130) indicates things may not be so simple.

At all primary villages, Katsina clan is responsible for Powamu in February, Patki clan is responsible for Flute dance at Oraibi and Lakon ceremony at Walpi. While Patki clan is responsible for Soyal at First Mesa, Bear clan is responsible for winter solstice on Second and Third Mesa. Stephen (1936, 1, 89) agrees, but also indicates that Patki clan has major responsibilities for Powamu and Niman. Other interpretations will not be addressed here.

Hopi Cosmology

To document light-shadow interactions across a calendrical structure and ferret out associated ceremonial relationships is far easier than to understand the cultural cosmology of a living and ancient people. Yet the two reflect each other. The construct of an observatory defines the physical limits of celestial motion, the cultural orientation to the heavens and the timing of religious activity, all of which contribute to the establishment of a *center* (Eliade 1959, 21).

Physical Limits and Cultural Interpretations

"The Hopi orientation bears no relation to North and South, but to the points on the horizon which mark places of sunrise and sunset at the summer and winter solstices" (Stephen 1936, 1190).

Beyond the extrema of solar migration is a deeper cultural concept of duality between day and night, time and space, life and death, Aboveworld and Underworld (Stephen 1936, 280 ff.; Titiev 1944, 173; Hieb 1979, 577). The sun rises in the east and sets in the west, yet culturally it passes through the Underworld to return to its rise point the following morning. Day Above means night Below, summer Above is winter Below, and the death of an individual Above marks the advent of life Below. Ceremonial elements six months hence may be embedded within an ongoing ceremony. "When Powamu shines in the Above, its counterpart Nasha'nmii is shining in the Below. Summer is yonder, winter is here. All Vegetation is mature, fruits are ripe in the Below." (Stephen, 1936: 239).

The planting and growth of beans for Powamu represents the coming of a bountiful harvest in August (Frigout 1979, 567). The names of the winter to summer solstice months are repeated with an added suffix six months hence reflecting the dual nature across time, while the timing of a month corresponds to the lunar cycle and not the solar year (Titiev 1938, 39–42). The reflective duality of a luni-solar calendar system to initiate ceremony can be witnessed throughout Hopi religion as explained above. The clearing of fields during Wuwutsim is planting in the below (Stephens 1936, 1037). Further examples of embedded ceremonial duality have been documented by Fewkes (1922), Stephen (1936), Titiev (1944) and others, and alluded to but not expressly defined by contemporary Hopi.

Katsinas

Katsinas may best be considered as reincarnated Hopis who act as spiritual messengers between the living and the "gods". Katsinas are in the Above from winter to summer solstice, and in the Underworld from summer to winter solstice thus indicating the direction of Hopi solar observation. Both Katsina spirit and prayer communication travel through the *sipapu,* the pathway of life migration from the Underworld to the Above represented as a channel in the kiva floor. Katsinas travel along the sun's path to the sun's house in the west (Titiev, 1944, 173; Stephen 1936, 280 ff.). As a reverse ceremony, those individuals who pass through Wuwutsim initiation are reserved places in the Underworld at their death and return as Katsinas to bring rain, a manifestation of cooperation between living and dead (Hodges 1912, 566 ff.; Titiev 1944, 178). Analogous to Katsinas returning "home" during Niman, a deceased individual is placed looking east whereupon at sunrise of day four, the deceased rises, follows the pathway of the sun to his new "home" and becomes a katsina (Stephen, 1936, 826; Titiev 1944, 108) returning as a cloud to bring rain (Fewkes 1923, 486; Titiev 1944, 108). The timing of ceremony and Katsina dances which may have occurred in the Crack-in-Rock plaza (Downum 1996, p.c.) may have been anticipated via the calendar wall.

Conclusion

When viewing prehistoric artifacts which have been used to demarcate time by an ancient people, we might better consider their constructs as a *hierophany*, i.e. that *something sacred shows itself to us.* (Eliade 1959, 11; emphasis original). "In the homogeneous and infinite expanse, in which no point of reference is possible and hence no point can be established, the hierophany reveals an absolute fixed point, a *center.*" (Eliade 1959, 21).

There can be little doubt that the calendar wall anticipates the times of survival activity (Planting and hunting) and sacred ceremony by the ancient Hisatsanom.

Importance of Cross-Quarter Dates

While solstices "anchor" Pueblo ritual (Zeilik 1985, S12), the cross-quarter dates indicate those times at which the rate of sun motion along the horizon is rapidly changing. Given the prominence of ceremony during the cross-quarter times, cultural awareness may have correspondingly increased.

Feb. 2	Powamu, a cleansing and reverse harvest ceremony
May 1	prepare for planting, perhaps Nevenkewe
Aug. 10	Flute or Snake dance
Nov. 1	Wuwutsim, prepare for winter and rights of passage

This hypothesis will require further testing.

Technological Adaptation or Method of Cultural Consistency

Research by Cinnamon (1988), Travis (1990), and Downum & Sullivan (1990) suggest that fluctuating environmental conditions brought on (in part) by human activity and increased population may have led to an alteration of the biophysical ability of the land to support increased human habitation. When correlated with the chronology of construction, agricultural development, and corresponding ecological degradation (Cinnamon, 1988; Downum, 1991), archaeoastronomical sites might well be interpreted as a technological adaptation to environmental stress. As crop productivity declines, precipitation patterns falter, and population stresses increase (Cinnamon, 1988; Downum, 1991), the combined ability to 1) extract more information from the natural world, 2) anticipate the latest probable spring frost and earliest probable precipitation, and 3) prepare for appropriate ceremonies, provides a culturally sensitive and scientifically reliable mechanism to anticipate the physical and religious preparation for seasonal fluctuations in an ecologically disturbed environment.

An equally viable hypothesis is that the construct of archaeoastronomical structures allow for consistency in the timing of proscribe ceremony and ritual. Whether during environmental stress or bountiful harvest, constructs which reveal the patterning of celestial motion allow for ceremonies at times which have been determined to be of value in both a cultural and biophysical realm. Thus, in a world of chaos, order is established and the annual cycle of ceremony is kept in balance with the creation and renewal of the new year. (Eliade, 1959, 20 ff.).

Prehistorically and currently, the Hopi define their orientation to and perceptions of the natural world by merging sacred time and sacred space with the physical construct of sky motion. As a mechanism to tie together the sky and the earth, life and death, material and eternal with the daily and the spiritual, calendrical use of the sun and moon serve to provide ceremonial and daily direction in Hopi life.

Acknowledgments

It is impossible to acknowledge and thank all those who have helped provide feedback and direction in the course of this research and manuscript preparation. However, in a brief and meager attempt to recognize significant parties, my heartfelt thanks to Steve Cinnamon, Barton Wright, Bill Breed, Bruce Anderson, Todd Metzger, Brent Roberts, Nancy Morgan, Bonnie Fennig, Charlie Schweighauser, Michael Zeilik, Steve McCluskey, Andy Odell, Ray Williamson, Carol Henderson, Bob

Coody, Donald Nelson, Suzanne Denet, Alan Peterson and especially (to my wife) Barbara Bates. To all others who have read this manuscript and offered suggestions and encouragement, thank you.

References

Ahlstrom, R., J. Dean and W. Robinson 1991. Evaluating Tree-ring Interpretations at Walpi Pueblo, Arizona. *American Antiquity* 56(4): 628–644.

Anderson, B.1990. Rock Art. *Wupatki Archeological Inventory Survey Project: Final Report*. B. Anderson, ed. Southwest Cultural Resources Center, Professional Paper #35, National Park Service, Santa Fe, NM.

———. 1994 Personal communication.

Bates, B. 1992. Archaeoastronomical Survey of the Colorado River Corridor, Grand Canyon National Park. Unpublished manuscript submitted to Grand Canyon National Park Service for sponsored archeological research along the Colorado River.

Bates, B. and J. Snow 1996. An Investigation of the Potential for a Solar Calendar at Wukoki Ruins (NA 203), Wupatki National Monument. Unpublished manuscript submitted to the National Park Service.

Binford, L. R. 1972. Methodological Considerations of Archeological Use of Ethnographic Data. In *An Archeological Perspective*, Seminar Press, New York, pgs. 59–67.

Breed, B. 1992 Interview and personal communication. (1960–1981 Curator of Geology and 1965–1981 Head of Geology Department, Museum of Northern Arizona, Flagstaff, AZ.)

———. Interview by author and personal communication, 1992–1996.

Brew, J. 1979. Hopi Pre-History and History to 1850. In *Handbook of the American Indian*, Vol. 9, pgs. 514–523. ed. A. Ortiz, Smithsonian Institute, Washington DC.

Byrkit, J. 1988. The Pala'tkwapi Trail *Plateau Magazine*, 59, No. 4. Museum of Northern Arizona, Flagstaff,

Cinnamon, S. 1988. The Vegetation Community of Cedar Canyon, Wupatki National Monument as Influenced by Prehistoric and Historic Environmental Change. Master's thesis, Department of Biology, Northern Arizona University, Flagstaff, AZ.

———. 1993 Personal communication. (Resource Management Specialist, Midwest Regional Office, National Park Service, Omaha, NE)

Colton, H.R. and M. R. 1931. Petroglyphs, A Record of an Adventure. *American Anthropologist* 33, pgs. 34–55.

Curtis, E. S.1922. The Hopi. In *The North American Indian*, Vol. 22. Smithsonian Institute, Washington, DC.

Denett, S. 1993 Personal communication.

Downum, C. 1994. The Origins of Socio-cultural Complexity in the Prehistoric Southwest. Grant proposal submitted to Northern Arizona University Organized Research Committee, Flagstaff, AZ.

Eliade, M. 1959. *The Sacred and the Profane*. W. Trask, transl. Harcourt, Brace & World, New York.

Fewkes, J. W. 1897. Tusuyan Katcinas. In *Fifteenth Annual Report of the Bureau of Ethnology* (1893–94), J. W. Powell, Director. Government Printing Office, Washington, DC.

———. 1897. Tusayan Totemic Signatures. In *American Anthropologist* Vol. 10, No. 1, pgs. 1–11.
———. 1920. Sun Worship of the Hopi Indians. In *Smithsonian Report from 1918*, pgs. 493–526. Government Printing Office, Washington, DC.
Frigout, A. 1979. Hopi Ceremonial Organization. In *Handbook of the American Indian*, Vol. 9, pgs. 564–576. A. Ortiz, ed. Smithsonian Institute, Washington, DC.
Heib, L. 1979. Hopi World View. In *Handbook of the American Indian*, Vol. 9, pgs. 577–580. A. Ortiz, ed. Smithsonian Institute, Washington, DC.
Hodge, F. W. 1912. Hopi. In *Handbook of American Indians North of Mexico*. Smithsonian Bulletin #30, pgs. 560–567. Washington, DC.
Jenkins, L. 1996 (June), personal communication. (Director, Hopi Cultural Preservation Offices, Hopi Tribe, Kykotsmovi, AZ.)
Kennard, E. 1979. Hopi Economy and Subsistence. In *Handbook of the American Indian*, Vol. 9, pgs. 557–563. A. Ortiz, ed. Smithsonian Institute, Washington, D.C.
McCluskey, S. 1977. The Astronomy of the Hopi Indians. In *Journal of Historical Astronomy*, Vol. 8, pgs. 174–195.
———. 1981. Transformation of the Hopi Calendar. In *Archaeoastronomy in the Americas*, R. A. Williamson, ed. Ballena Press, Center for Archaeoastronomy, pgs. 173–182.
———. 1982. Historical Archaeoastronomy: The Hopi Example. In *Archaeoastronomy in the New World*, A. Aveni, ed. Cambridge University Press.
———. 1990. Calendars and Symbolism: Functions of Observation in Hopi Astronomy. In *Archeo-astronomy*, No. 15., Supplement to the Journal for the History of Astronomy.
Nelson, D. 1992–93 Personal communications. (Coordinator of Hopi Program, Yavapai College Elderhostel and anthropology instructor, Yavapai College, Prescott, AZ.)
Nequatewa, A. 1967. Truth of a Hopi. Museum of Northern Arizona, Bulletin #8, Northland Press, Flagstaff, AZ.
Preston, R. 1996. Research presented on solstice and equinox petroglyph markers at Wupatki National Monument. Oxford V Conference on Archaeoastronomy, Santa Fe, NM. See table of contents this volume.
Poulingyouma, L. 1995 Personal communication.
Satala, E. J. 1995 Personal communication.
Smith, W. 1949. Excavations in Big Hawk Valley. *Museum of Northern Arizona Bulletin*, 4, No. 3, pgs. 42–48. Flagstaff, AZ.
———. 1952. Excavations in Big Hawk Valley II: Wupatki National Monument, AZ. *Museum of Northern Arizona Bulletin #24*, Flagstaff, AZ.
Stephen, A. 1936. *Hopi Journals*. E. Parson, ed. Columbia University, New York.
Sullivan, A. and C. Downum. 1991. Aridity, activity and volcanic ash agriculture: a study of short-term prehistoric cultural-ecological dynamics. In *World Archeology* 22 No. 3, pgs. 272–287.
Thom, A. 1971. Megalithic Lunar Observatories. Clarendon Press, Oxford University.
Thompson, L. and A. Joseph 1965. *The Hopi Way*. Russell & Russell Publishers, New York, pgs. 36–44.
Titiev, M. 1938. Dates of Planting at the Hopi Indian Pueblo of Oraibi. *Plateau Magazine*, Museum of Northern Arizona, 11, No. 5, Flagstaff.
———. 1944. Old Oraibi: A Study of the Hopi Indians of Third Mesa. In *Papers of the Peabody Museum*, Vol. 22. Harvard University, Boston.
Voyager II, Version 1.0 1995. Computer astronomy timeline program, San Leandro, CA.

Williamson, R. 1984. *Living the Sky: The Cosmos of the American Indian.* Houghton-Mifflin. Boston, MA.
Wright, B. 1973. *Kachinas: A Hopi Artist Documentary*, Heard Museum, Northland Press, Flagstaff, AZ.
———. 1992–93 Personal communications. (Curator of Anthropology, Museum of Northern Arizona, Flagstaff, AZ.)
Zeilik, M. 1985. The Ethnoastronomy of the Historic Pueblos I: Calendrical Sun Watching. In *Archaeoastronomy,* Supplement to the *Journal for the History of Astronomy*, No. 8, pgs. S1–S25.
———. 1987. Keeping the Sacred and Planting Calendar: Archaeoastronomy in the Pueblo Southwest. In *World Archaeoastronomy: Proceedings of the Second Oxford International Conference on Archaeoastronomy*, A. Aveni, ed., pgs. 143–166. Cambridge University Press.
———. 1987. Anticipation in Ceremony: The Readiness is All. In *Astronomy and Ceremony in the Prehistoric Southwest.* J. B. Carlson and W. J. Judge, eds. Maxwell Museum Press, Albuquerque, NM. Papers of Maxwell Museum of Anthropology, No. 2. pgs. 25–41.

12

The Shaw Butte Hilltop Site: A Prehistoric Hohokam Observatory

Todd W. Bostwick
Pueblo Grande Museum
Phoenix, AZ

Stan Plum
Phoenix, AZ

A three-year study of a prehistoric Hohokam fortified hill site, located in the Phoenix Mountains, Arizona, suggests the site served as a 12th or 13th century observatory, where the Hohokam marked the appearance of solstices and equinoxes with light patterns in an artificially constructed rockshelter, and by the location on the horizon of sunrise and sunset in relation to architectural features and petroglyphs. Other Hohokam sites are known which have astronomical features, indicating that the Hohokam may have had a solar calendar system.

* * *

Introduction

When Old World travelers explored the Sonoran Desert of Southern Arizona in the 17th century, they encountered numerous ancient ruins as well as native O'odham speaking (Pima and Papago) farmers living in villages scattered along the desert rivers. The O'odham observed the motions of the sun, moon, and certain constellations to enhance the success of agriculture and to establish a calendar for ceremonial cycles (Bahr et al. 1994:63). This paper addresses the question of whether or not the inhabitants of the ancient ruins of the Sonoran Desert, the Hohokam, also made systematic astronomical observations for calendrical purposes. A brief discussion of O'odham calendar systems is given, followed by a summary of published Hohokam archaeoastronomy data. Then the results of an ongoing study of a prehistoric hilltop site in Phoenix, Arizona, are presented. The paper concludes with a discussion of the possible implications of our study of the Shaw Butte hilltop site, which we argue served as a Hohokam solar observatory.

O'odham Calendar Systems

Many Southwestern agricultural groups defined their universe by the path of the sun, which provided a quadrilateral model of the cosmos (Bohrer (1994:471–473, 497). The corners of the earth in this model are marked by the summer and winter sunrise and sunset solstice locations on the horizon, with the two pairs of the solstice houses sitting where the path of the sun intersects the earth at opposite horizons. The O'odham (Piman) speaking Indians of southern Arizona, possible descendants of the Hohokam, have a similar model for their universe.

According to O'odham mythology, the sky is seen as a vault, with Earth Doctor spitting water into the sky to create stars and throwing a block of ice to make the sun, and then commanding Gray Spider to spin a web that connected the edges of the sky with the earth (Llyod 1911:29; Russell 1908:247). The stars, as well as the sun, moon, and certain planets and constellations, served as signs for scheduling festivals and establishing a calendar (Bahr, Smith, Allison & Hayden 1994:63). The O'odham believe in the omnipotence of the sun, and tracked its movement in relation to its location on the horizon at the Sierra Estrellas, south of the Salt River, and at Baboquivari Mountains and Ajo Mountains in southern Arizona (Castetter & Bell 1942:143–144). In addition, the location at dawn of the constellation of Pleiades, called The Travelers, was used as a seasonal marker (Castetter & Bell 1942:143). The observation of celestial movements by the O'odham was the function of a delegated individual, who provided information about when to plant to the headman, who then called the men of a village together six days prior to planting time to inform them of the planting schedule (Castetter & Bell 1942:143).

The O'odham's calendar is a 12 to 13 month lunar calendar that started in July, right after the ripening of the saguaro cactus fruit, which varied from location to location (Underhill et al. 1979:18). While the annual calendar was determined by the moon, year cycles were defined by the sun. The solstices were important seasonal events, with the creation myth formally celebrated during the sacred four-day winter solstice period (Castetter & Bell 1942:144). Winter solstice was called the backbone month because it divided the year in half (Underhill et al. 1979:17). Two O'odham deities are believed to live at the solstitial points on the horizon: Sinking Magician lives in his "house" at the northeast and northwest solstitial points, and South Magician lives at the southeast and southwest solstitial houses (Russell 1908:251).

Archaeoastronomy Studies

Previous studies of Hohokam archaeoastronomy have identified a few sites in which it is argued that the prehistoric inhabitants made solar observations and marked solstice and equinox events. Those sites include platform mounds, an adobe tower-like structure, and a natural rockshelter.

During the Hohokam Classic period (ca. A.D. 1150–1450) the inhabitants of the Sonoran Desert built massive, rectangular platform mounds. The long axis of the majority of those platform mounds are oriented slightly east of north or west of north; a small number of platform mounds are oriented west-east (Gregory 1987). Two of the largest platform mounds, Pueblo Grande and Mesa Grande, both have adobe rooms that apparently served as solstice markers. At Pueblo Grande, a room on top of the mound has a corner doorway that aligns with a doorway in the south wall during the summer solstice sunrise (Mixon 1989). A room at the Classic period Mesa Grande platform mound has northern and eastern doorways that align with the winter solstice (Howard 1995:120).

Another Hohokam adobe tower-like structure, the multi-story Big House at Casa Grande, has several astronomical features. The Casa Grande Big House stands more than 10 m in height and has lower walls more than 1 m in thickness (Wilcox & Shenk 1977). Nine circular holes in the walls of the two upper stories of the Big House may have been used to observe or mark solstice and equinox events, as well as possible maximum and minimum lunar excursions (Molloy 1969). For example, a hole in the west wall of the Big House is positioned to allow observation of the summer solstice sunset on the western horizon, and a hole in the east wall of the highest story allows an elliptical spot of light to fill a circular hole in a west wall of the same room, marking the fall equinox sunrise (Evans & Hillman 1981). The holes in this room also record the spring equinox sunrise, but the ray of light fills the hole several days before the equinox, anticipating instead of marking this solar event (Norton 1987:64). An O'odham legend was recorded by Fewkes (1912) about an individual called The Bitter Man at Casa Grande who was responsible for looking out of the Big House in order to salute the sun when it rose and set.

A natural rockshelter in the Papago Buttes on the north side of the Salt River, called Hole-in-the-Rock, also appears to have been used as a solar marker by the Hohokam. A natural hole in the ceiling of this rockshelter creates a ray of light that fills a man-made metate slick with light during the summer solstice, and fills a natural rock seam during the winter solstice (Mixon & White 1991:247). In addition, Hole-in-the-Rock has significant line of sight alignments with many of the platform mounds in the Salt River Valley. For example, equinoctial sunset occurs at 270 degrees, which is aligned with the platform mound sites of La Ciudad and Las Colinas, and Hohokam villages east of Hole-in-the-Rock align with the solstice sunrise (Mixon & White 1991:250).

Shaw Butte Hilltop Site

At the northern edge of the Salt River Valley is a narrow but steep range called the Phoenix Mountains. One of the highest peaks in these mountains is called Shaw Butte. On the northeastern side of Shaw Butte is a ridgetop situated about 76 m (250 feet) above the desert floor to the north. This ridgetop has two knolls which provide a clear view of the horizons to the north, west, and east; the view to the south is partially obstructed by Shaw Butte and other peaks in the Phoenix Mountains. Located on this ridgetop is a prehistoric Hohokam ruin which contains a masonry compound with rooms, petroglyphs, trails, and a rockshelter (Figure 1).

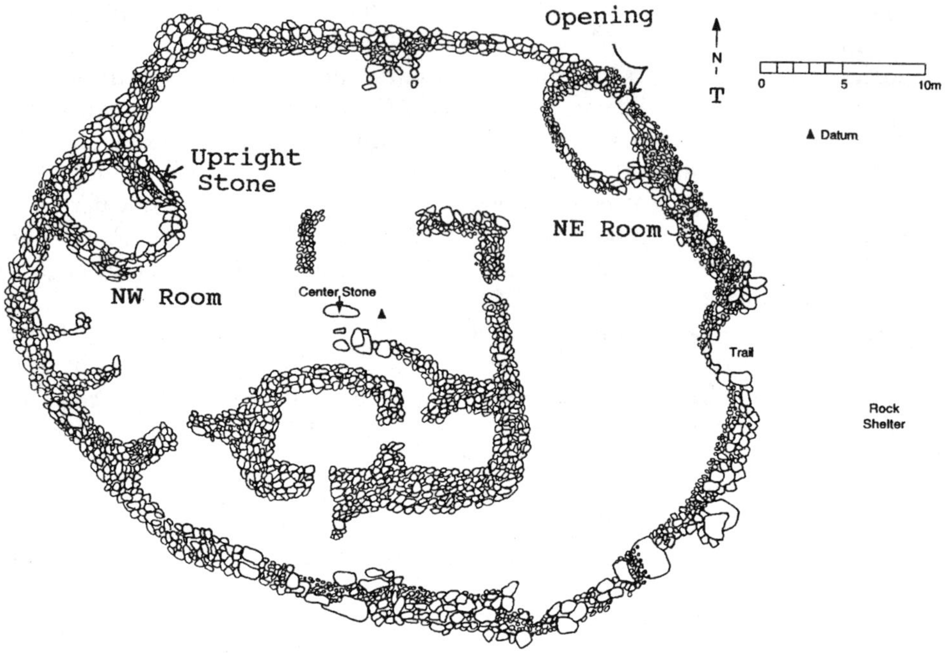

Figure 1. Schematic Plan View Map of Shaw Butte Compound.

The oval-shaped masonry compound at Shaw Butte was built with walls up to 1.5 m thick and 1 m in height, and is 42.5 m east-west and 33.75 m north-south. Some wall collapse has occurred, and the walls may have once been slightly higher. In the center of the compound the topography is elevated, creating a high point with a dramatic view, especially to the north and to the east. Several boulders containing petroglyphs are present at this elevated location, including a large, flat-faced boulder with a series of circle/dot petroglyphs; we have called this rock the *centerstone*. To the north of the centerstone is a plaza area partially surrounded by walls. Two of the largest rooms inside the compound, both with massive stone walls made of basalt boulders, are located just south of the centerstone and are adjacent to each other. The function of these two rooms is unknown since they contained few artifacts, and there is no evidence that the rooms were roofed.

The northeastern compound room is oval-shaped, with its outer wall formed by the compound wall. A wide, horizontal stone flanked by two small upright stones creates a formal opening—a door or window—in the center of the outer wall in this room. An individual sitting in the middle of this room can view the sun rising on the horizon, from behind Mt. Ord, through this door or window during the summer solstice in June. Thus, the opening in this room provides a direct view of the northernmost sunrise in the annual solar cycle.

The square-shaped northwestern room in the compound does not have an opening in its outer wall, which is the compound wall. However, there is a pointed stone in the center of the outer wall and the summer solstice sun sets over this stone. In addition, in the north wall of this room is an upright stone with a U-shaped top; a person in the center of the room can view the summer solstice sun rise over the top of this stone. This room provides a formal space for viewing, and

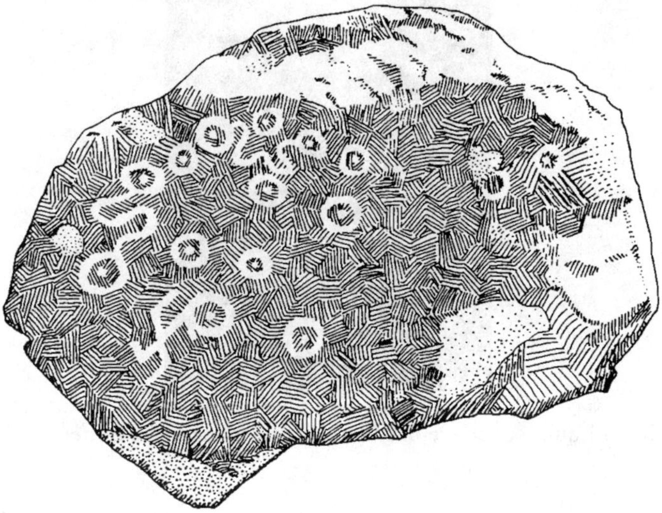

Figure 2. Centerstone Petroglyph Panel, Shaw Butte Site (circles 5 to 10 cm in diameter).

may intentionally mark, the northernmost sunrise and sunset during the annual solar cycle.

More than 75 petroglyphs are present at the Shaw Butte site, mostly geometric designs and predominately circle/dots. Although circles are common Hohokam rock art motifs, circles with dots occur less frequently. The possible symbolic meaning of the circle/dot is unclear, but its dominant presence at the Shaw Butte site may indicate it had an astronomical meaning. For some Pueblo Indian groups, the sun was represented by two concentric circles with a central dot, with the dot symbolizing the umbilicus of the sun (Ellis 1975:59). A circle with a central dot was one of the clan symbols of the Pueblo moon clan (Colton & Colton 1931). It has also been suggested that dots on Pima calendar sticks may represent the sun (Ellis 1975:61).

The importance of the circle/dot design to the builders of the Shaw Butte site is expressed in its presence on the south face of the centerstone, where 13 circle/dot designs are clustered in a swirling pattern, some connected to each other with tails (Figure 2). Another circle/dot and a circle are also present in the eastern part of this panel. The number 13 is intriguing from a calendrical perspective. The historic Hopi kept accurate solar calendars based on horizon observations that had 13 moons in an annual cycle (Parsons 1939:496–7). Some O'odham groups had 13 month calendars (Underhill et al. 1979). In Mesoamerica, the trecena calendar system that was used for more than a millennium consisted of 13 months with 28 days in each month (Edmonson 1988:111).

The Shaw Butte centerstone itself appears to be a solstice marker. This unusual boulder is 1.4 x 1.8 x. 0.75 cm in size, flat on one side and round on the other, and has a worn flat spot on its top, where items may have been placed. Differences in patination at its bottom and the outcrop it sits on indicate that the boulder has been moved from its original location, and positioned at a slight angle to create solar light and shadow interactions. At noon during the summer solstice, a ray of light shines through a flaked notch at the top of the boulder and fills the westernmost circle/dot

Figure 3. Autumnal Equinox sunrise over rock art panel near the centerstone (September 21, 1997).

on the south face of the centerstone. In addition, at sunrise during the winter solstice, a shadow splits the face of the centerstone horizontally, cutting through the same petroglyph lit up by the summer solstice.

Located 1 m southeast of the centerstone is a petroglyph panel that may be the oldest panel at the site since it pecked designs have more repatinated, or desert varnish, than other panels. This panel contains elaborate concentric circles, circle/dots, and a hollow cross; at the right edge of the panel a human with upraised arms faces the geometric designs. A viewer of this low, slightly sloping panel looks at the eastern horizon above the panel, which is aligned with the equinoctial sunrise and a panel to the east that contains a single circle/dot (Figure 3). Another petroglyph panel at the southeastern edge of the Shaw Butte compound also has a human figure wearing a headdress; a circle/dot is located to the right and above this figure's head. Do these two humans represent individuals making solar observations at the site? Interestingly, the north wall of the inner cell of one of the Las Colinas platform mounds (Mound 8), located about 14 km southwest of Shaw Butte, contained a swirling pattern of circles, some with finger indentations (Hammack & Sullivan 1981:Figure 12), giving them an appearance similar to the circle/dot petroglyph pattern on the Shaw Butte centerstone. Is this a coincidence, or were the circle/dots at Mound 8 and Shaw Butte made by members of the same social group or belief system?

The Shaw Butte rockshelter, located downslope next to a prominent trail that enters the east side of the compound, is not a natural feature, but appears to have been intentionally constructed from a naturally occurring crevice in an eroded basalt outcrop. The crevice runs along a north-south axis with an east facing opening, which was used as an entrance. Because the floor of the crevice was once irregular, basalt stones and fill dirt were brought in to level the floor. A large, patinated boulder over 1 m in size was hoisted into the crevice and wedged against the crevice walls, creating an elevated entrance approximately 57 x 86 cm in size, and about 1 m above ground level outside the rockshelter. The main area of the shelter is just large enough to fit one person comfortably: 66 cm in length (north-south), 55 cm in width (east-west), and 40 to 100 cm in height. A lower portion of the natural basalt outcropping extends out into the northeastern part of the shelter, forming a low, bench-like ledge 130 x 75 cm in size. Three short, parallel grooves have been

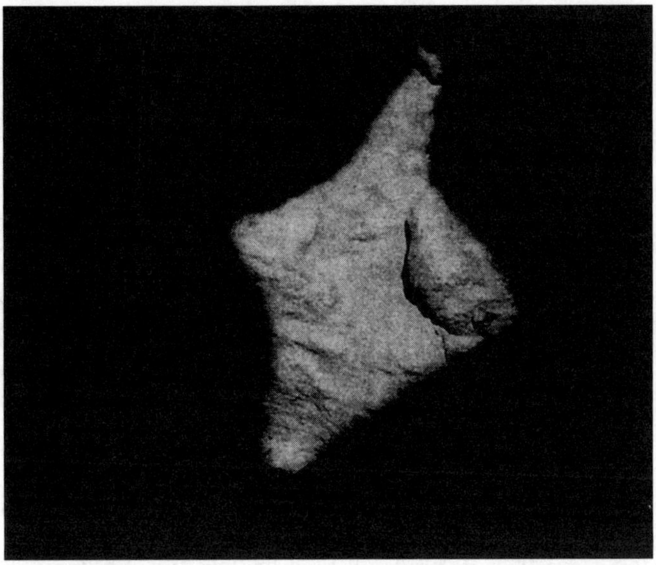

Figure 4. Four Pointed Star Light Pattern in the Shaw Butte Rockshelter (June 22, 1994, solar noon).

ground into the middle of this ledge. Outside the entrance to the rockshelter are several petroglyphs including a circle/dot, as well as a hollow cross and scratched geometric designs. The hollow cross has small circles at two of its margins, perhaps representing the sun in two of its solstitial houses.

The roof of the rockshelter was almost entirely constructed by the careful placement of several boulders, with smaller stones used to wedge the boulders in place. One of the boulders in the roof is in situ, but five of the other boulders were moved from elsewhere to construct the roof. Although the roof boulders have an apparent haphazard arrangement, three openings appear to have been intentionally created in the roof to let rays of light into the shelter. Some of the roof boulders' edges also have been flaked and/or ground in an apparent attempt to manipulate the shape and location of the light that streams into the shelter.

An upright, wedge-shaped stone covers the south end of the shelter's roof. This stone, which is flaked, creates a trapezoidal opening in the roof that allows a shaft of light into the shelter year round. In the middle of the roof, another opening is partially formed by a boulder with a flake removed that allows in a small beam of light. The third opening, located on the north side of the shelter, is formed by other boulders and the natural outcropping.

The light patterns in the rockshelter appear to have important cycles that could have provided calendrical measurements year round by marking the passage of time on the floors and walls of the shelter. During the solstices and equinoxes, dramatic light forms appear in the rockshelter. Those light forms do not appear suddenly, however, but are part of a cycle of light forms. For example, fifty-six days prior to summer solstice, a beam of light comes down from the roof and touches the interior ledge at the north end of the shelter. Over the next 56 days, the beam expands out into a four pointed star (Figure 4). This phenomena is the result of the sun's rising elevation at solar noon behind the large naturally occurring stone in the shelter's roof. There is a spot on this stone which is ground smooth, which helps define the

Figure 5. Light Beam on Notch in North Wall of Shaw Butte Rockshelter (September 19, 1993, just before solar noon)

southeast corner of the star-shaped light beam at summer solstice. As the days of summer pass, the light star withers until it vanishes 56 days after the solstice.

There is also an equinoctial light which appears in the rockshelter at spring equinox and vanishes at fall equinox. This light beam shines on a wide, flaked crack in the northern wall of the rockshelter, climbing the edge of the crack during midday and stopping as a rod of light at a small notch, where it disappears into just before solar noon (Figure 5). Twenty-eight days after appearing on this wall, the beam of light begins shining on the interior ledge. It gradually grows into a dagger/snake shape at summer solstice. Around solar noon, the dagger/snake moves across the ledge, touching the grooves that were cut into the ledge. Twenty-eight days prior to fall equinox, the light leaves the floor and shines again on the north wall. At fall equinox, the light disappears altogether.

As stated above, the main light beam in the shelter shines year round, descending the west wall and crossing the floor at solar noon. The shape of the light remains trapezoidal throughout the year until 28 days prior to winter solstice. At this time the upright, flaked stone in the south part of the rood splits the beam into two smaller light beams that form dots 3.5 cm in diameter. At solar noon, the dots move up the side of the ledge and onto the ledge, then climb the three grooves as if they were steps. The beam that creates the two dots is whole again 28 days after winter solstice.

In summary, at summer solstice, there are three distinct light patterns on the floor. These are the trapezoid, the four-pointed star in the center of the floor, and the dagger at the north end of the shelter.

Discussion

Ceramic types date the Shaw Butte site from the late Sedentary to early Classic periods (ca. A.D. 1100–1300), although repatination by desert varnish of one of the petroglyph panels inside the compound suggests that panel may be older than the compound. As discussed above, other sites dating to the Classic period have evidence that the Hohokam marked solstice events. The expanding database is increasingly supporting the hypothesis that the observation of solar cycles was important to the Hohokam. Based on our Shaw Butte research, it is apparent that other so-called forts should be studied for potential solar interactions.

The Shaw Butte site is one of the most complex Hohokam solar observatories examined thus far, equal in importance to the Casa Grande Big House. Two of the three types of calendar phenomena identified by Zeilik (1989) for the Pueblo Indians—horizon observations and light and shadow interactions—are present at the Shaw Butte site. The third calendar type, lunar phase counts, has not been a focus of study at the side but merits further investigation. According to Hardman & Hardman's (1992) theoretical model of solar observatories, the Shaw Butte site appears to fit into at least level six of their eight levels of complexity. Having several different marking devices and formal viewing locations (e.g., solstice rooms, rockshelter, and petroglyphs) may have been intended to reinforce the importance of the site, serving a similar purpose as some Hopi sun shrines—a *hierophany*, or revelation of the sacred (McCluskey 1990:S13).

In conclusion, the Shaw Butte site appears to have been a solar observatory, probably with ceremonial significance. The changing light displays inside the Shaw Butte rockshelter suggest that cycles of 28, 7 and 56 may have been important periods in a Hohokam calendar. Knowledge of a calendar could have been used by Hohokam leaders to schedule important ceremonies. Solar movements tracked at Shaw Butte may have been used to collaborate observations made at some of the platform mounds located to the south. Analysis of the Shaw Butte ceramics indicates that they probably originated from several different villages in the Salt River Valley. Thus, the site may have been tied into a larger system of ritual obligations and interactions among the various villages and other Hohokam sites in the Salt River Valley.

Acknowledgments

The authors wish to thank John Andrews and Clifford Straitor for assisting with the mapping of the Shaw Butte site, and Susan Milbrath and Henry Wallace for reviewing the first draft of this paper. Thanks also to Roger Lidman, Archaeology Section Administrator, City of Phoenix Parks, Recreation and Library Department, for administrative support.

References

Bahr, D., J. Smith, W.S. Allison and J. D. Hayden, 1994. *The Short Swift Time of Gods on Earth: The Hohokam Chronicles*. Berkeley: University of California Press.

Bohrer, V. L, 1994. "Maize in Middle American and Southwestern United States Agricultural Traditions." In *Corn and Culture in the Prehistoric New World*, edited by S. Johannessen and C. A. Hastorf, pp. 469–512. Boulder: Westview Press.

Castetter, E.F and W. H. Bell, 1942. *Pima and Papago Indian Agriculture*. Inter-Americana Studies No. 1. Albuquerque: University of New Mexico.

Edmonson, M. S., 1988. *The Book of the Year: Middle American Calendrical Systems*. Salt Lake City: University of Utah Press.

Ellis, F., 1975. "A Thousand Years of the Sun-Moon-Star Calender." In *Archaeoastronomy in Pre-Columbia America*, edited by A. F. Aveni, pp. 59–88. Austin: University of Texas Press.

Evans, J. & H. F. Hillman, 1981. "Casa Grande, An Ancient Astronomical Observatory," *Arizona Highways* 57(10):32

Fewkes, J. W., 1912. "Casa Grande, Arizona." In *Twenty-eighth Annual Report of the Bureau of American Ethnology*, pp 57–179. Washington, D.C.

Gregory, D. A., 1987. "The Morphology of Platform Mounds and the Structure of Classic Period Hohokam Sites." In *The Hohokam Village: Site Structure and Organization*, edited by D. E. Doyel, pp. 183–210. Glenwood Springs, Co.: American Association for the Advancement of Science.

Hammack, L. C. and A. P. Sullivan, 1981. *The 1968 Excavations at Mound 8 Las Colinas Group, Phoenix, Arizona*. Tucson: Arizona State Museum Archaeological Series No. 15.

Hardman, C., Jr., & M. H. Hardman, 1992. "Linear Solar Observatory Theory: The Development of Concepts of Time and Calendar," *North American Archaeologist* 13(2):149–172.

Howard, J. B., 1995. *Preserving Temples of Clay: A Stabilization Plan for the Mesa Grande*. Mesa: Mesa Southwest Museum.

McCluskey, S. C., 1990. " Calendars and Symbolism: Functions of Observation in Hopi Astronomy," *Archaeoastronomy* no. 15.

Mixon, B., 1989. "Possible Solar Alignments of Room 11 and 10 on the Pueblo Grande Platform Mound." Ms. on file. Phoenix: Pueblo Grande Museum.

Mixon, B. & R. White, 1991. Skywatchers of the Salt River Valley Hohokam. *Astronomy Quarterly* 8:245–259.

Molloy, J., 1969. The Casa Grande Archaeological Zone: Pre-Columbian Astronomical observation. Ms. on file. Tucson: Western Archaeological Center, National Park Service.

Norton, O. R., 1987. Early Indian Sun Watching Sites are Real. *American West People and Places*. August, pp. 63–70.

Parson, E. C., 1939. *Pueblo Indian Religion*. Chicago: University of Chicago Press.

Russell, F., 1908. The Pima Indians. In *Twenty-sixth Annual Report of the Bureau of American Ethnology*, pp. 3–389. Washington, D.C. (Reprinted 1975, University of Arizona, Tucson).

Schroeder, A. H., 1940. *A Stratigraphic Survey of PreSpanish Trash Mounds of the Salt River Valley, Arizona*. Masters Thesis. Tucson: Department of Anthropology, University of Arizona.

Underhill, R. D., D. Bahr, B. Lopez, J. Pancho, & D. Lopez, 1979. *Rainhouse and Ocean: Speeches for the Papago Year*. Flagstaff: American Tribal Religions No. 4.

Wilcox, D. R. & L. O. Shenk, 1977. *The Architecture of the Casa Grande and its Interpretation*. Tucson: Arizona State Museum Archaeological Series No. 115.

Wilcox, D. R. & C. Sternberg, *1981. Additional Studies of the Architecture of the Casa Grande and Its Interpretation.* Tucson: Arizona State Museum Archaeological Series No. 146.

Zeilik, M., 1990. Keeping the Sacred and Planting Calendar:Archaeoastronomy in the Pueblo Southwest. In *World Archaeoastronomy*, edited by Anthony F. Aveni, pp.143–166. Cambridge: Cambridge University Press.

13

A Sunrise Calendar at a Kayenta Anasazi Village

Von Del Chamberlain
Salt Lake City, UT

William R. Latady
Anasazi State Park
Boulder, UT

Todd Prince,
Iron Mission State Park
Cedar City, UT

A small set of ruins located on top of a butte adjacent to the Anasazi village site known as the Coombs within the city limits of the current town of Boulder, Utah suggested that the top of this small sandstone outcrop might have been used for horizon calendrics. Indeed, the view of distant mountains, both to the east and west, make this an ideal location for sun-watching. Consideration of these factors led to the interest in documentation of how this site could have been used for keeping the calendar for the people who resided at this place for about 60 years during Pueblo II into early Pueblo III times.

* * *

Introduction

The Coombs Site, located on the southern slopes of the Aquarius Plateau, within the current town of Boulder, Utah, is at an elevation of 6,700 feet. The setting is exquisite, located between the colorful canyons of the Escalante River to the south and forested Boulder Mountain to the north. Nearby streams provide some of the finest fishing in the region, which abounds with places that provide spectacular beauty: Bryce Canyon 78 miles southwest; Capital Reef 40 miles north; the Henry Mountains and numerous canyons that flow into the Colorado River (now Lake Powell) to the east and south. The landscape throughout the region is rugged, varied and colorful almost beyond description.

Located at this place, in the vicinity of a consistent stream, was a small Kayenta Anasazi village of late Pueblo II through early Pueblo III vintage, containing some 100 rooms, 10 pit-structures and one ramada in two distinct pueblo units, one L-shaped and the other U-shaped. The village was probably occupied for 50 to 60 years. Tree ring dates range from 1129 to 1169 A.D. The people survived by their abilities to grow maize, beans and squash, and to find wild game and plant foods in the surrounding region.

It is particularly interesting to note in this paper that portions of the pueblo are astronomically aligned. An L-shaped unit at the Coombs site has a string of storage rooms that are within one degree of a north-south alignment, and the walls of the habitation rooms are accurately east-west aligned. In addition, two pit structures are on the north-south alignment with the storage rooms and a series of seven pit structures are roughly aligned east-west along the south slope of the hill. The site has been described in detail in available documents (Lister 1959, Lister, Ambler and Lister 1960, Lister and Lister 1961, Morgenstein 1994, Spaulding 1994, Pollock n.d., Latady & Prince, n.d.).

The pueblo is at the stream level that runs off the Boulder Mountain. The views to east and west present somewhat elevated horizons produced primarily by buttes and mesas located near by. Just west of the village site stands a white Navajo Sandstone butte, with a mantle of red Carmel Formation, the top of which is easily and quickly reached (Figure 1). From there, one can view the valley, including the village site just below, and the distant horizons to east and west. The sounds of the valley including the flow of water in the streams and the mostly gentle sounds of life, both human and animal, from the town make this a most enjoyable place to contemplate this beautiful setting and its natural and human history. The top is relatively flat, having juniper, pinon and smaller indigenous plant life growing on it. Three prominent ledges provide platforms for observing spectacular vistas, one on the east side, one on the northeast and the other on the west. Remains of a dozen rooms are on the north end of the top of the Butte.

The butte was named "School House Ledge" by the residents of the ranch town of Boulder, because the school was located near the base. Although no extensive archeology has been done to date the ruins on top, it has architecture and artifact characteristics similar to the Coombs site. Work by two of the authors (Prince n.d., Latady and Prince n.d.) suggested that this place might have been used for astronomical observations. Indeed, there are some interesting alignments between elements of the ruins and directions to sunrise at prominent features on the distant landscape at solstice and equinox times (Prince n.d.). Thinking about this led to the desire to show how this specific location might have been used for solar observations to provide a reliable calendar for the residents of the pueblo.

Motivation for doing this was threefold. First, the ledge is a likely site for such calendric activity and the existence of the ruin there strengthens this possibility. Second, it seemed desirable to obtain observational experience for keeping such a calendar to become familiar, in detail, with the method known to be most important for calendar keeping throughout the Pueblo world (Zeilik 1985a & 1985b). Finally, photographic documentation of such a sunrise calendar would make an appropriate educational exhibit for the expanded visitor center that opened in mid-1996.

We do not claim that the site was necessarily used for calendrics. We want to illustrate that it could have been, that the ledge is a fine location for such activity, still working as nicely today as for the Anasazi people who lived there about 900 years

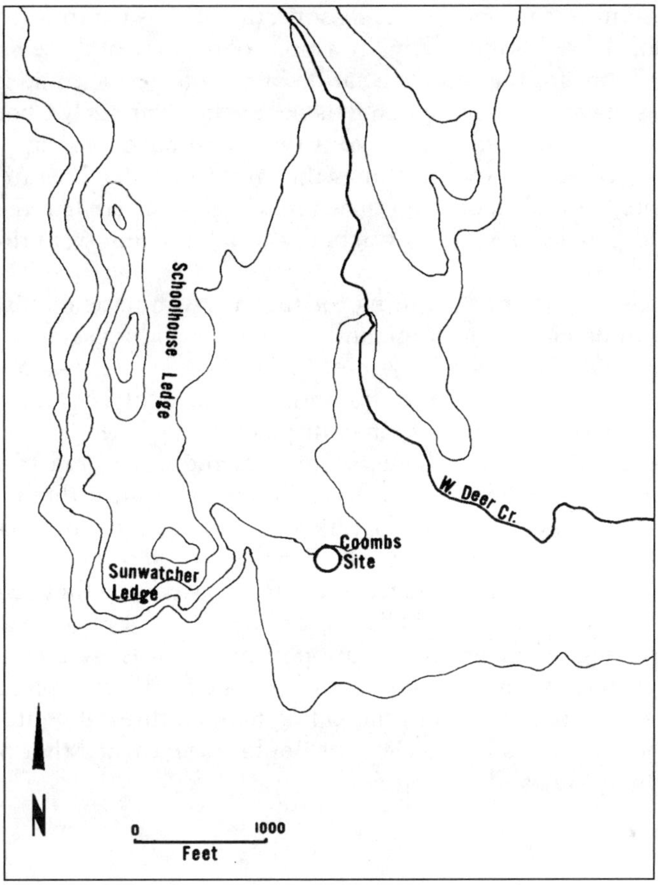

Figure 1. Topographic map showing the relationship of the Coombs Site and Sunwatcher Ledge.

ago. We will go so far as to say that because of its convenience and all the excellent factors favorable for calendrics, it seems to us highly likely that the ledge was used for such observations. Primarily, we want this to be an example illustrating for those interested in the wonderful system of solar calendaring that we know was used extensively by puebloan peoples.

Sunwatcher Ledge

Having made the short journey up to the top of "School House Ledge" dozens of times to watch the Sun both rising and setting, thinking about how others before us might have made this trip for a similar purpose, although for them the consequences of doing so would have been much more significant, we found ourselves calling the place by other names: Sun Priest House seemed appropriate for the ruin, and Sunwatcher Ledge became our favorite for the site generally. If not for the Anasazi people who might have stood there, it is a fitting name because of our own activity at the place.

Over the months, throughout the seasons, that we visited to watch the Sun rise and set, we gained insights into the many aspects of the lives of those Kayenta people: we could see reasons for the specific location of the village; we could imagine where their fields must have been on the rich flats near water courses; we could survey the mountain where they hunted, the canyons they must have frequented and the distant places they wandered to find things they needed. We deliberated such things at this grand site, and asked difficult questions: Exactly where had the people come from? Why did they leave? Did they actually watch the Sun from this location, as we watched?

Each time we stood there, waiting for the most dramatic moment of the day, we more fully appreciated the commanding view providing such an excellent, indeed accurate, calendar keeping system. From there, one looks over the nearby buttes to the distant landscape, a most important aspect of a good location for calendrics by horizon observations. A small displacement of the observer makes a significant difference in the places where the Sun rises and sets over nearby features, but little difference if the horizon is far away. It became very clear that one must climb to a place like this to have all of the factors necessary for accurate solar horizon observation. From the village site below, the Sun rises and sets during important parts of the year over features that are close and not as easily used for accurate observations of the passing days. This would have been important for planting and ceremonial timing. The growing season at that elevation is short. It was essential to know when it was safe to plant seeds and not suffer loss from late frost, yet to plant early enough so that crops would mature before the killing temperatures of winter set in. From the people who have lived in the town of Boulder we learned that planting is appropriate through May and early June.

Solar Calendar from Sunwatcher Ledge

Without a doubt the easiest and most reliable method of keeping an agricultural calendar is to monitor the directions of sunrise and sunset from a place with good contours on distance horizons to the east and west. This knowledge of solar movements could then be used along with observations of plant and animal behaviors and specific climatic factors for any particular year to decide when to plant particular crops. Calendaring by sunwatching is best done by establishing a specific observing place where horizon features can be used to mark days of interest. If such features are lacking, it is necessary to place foresights for observational reference from watching stations. In addition to, or instead of, these methods, openings in buildings can be used to observe the Sun and walls opposite these openings can be marked to monitor the play of sunlight and shadows. This latter method is more complex, but can be successful when used with thought, planning and care. It is known from records made in the latter part of the 19th century by early anthropologists living among Pueblo groups that all of these methods have been used, and it is thought likely that they were used for hundreds of years prior to that time (Zeilik 1985a, Williamson 1984).

From atop Sunwatcher Ledge both eastern and western horizons can be appropriately used for discerning the days, but the eastern contouring is best and most

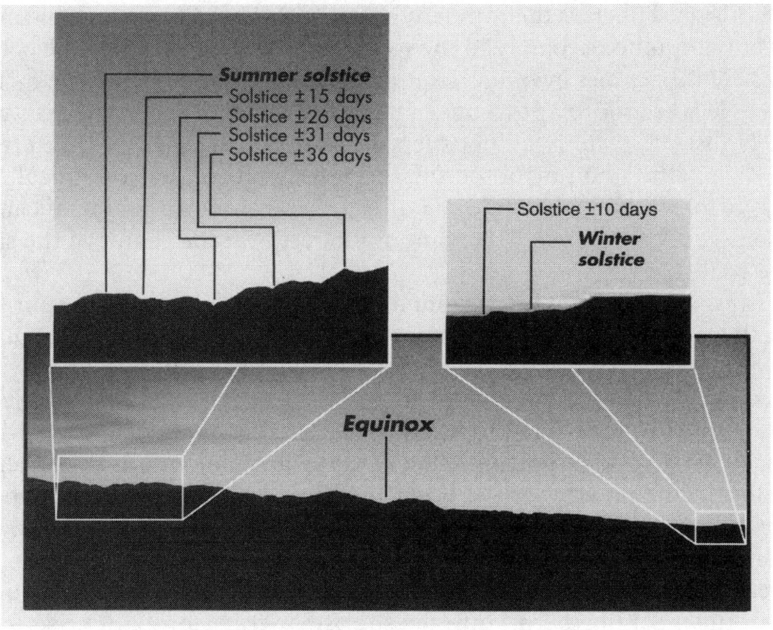

Figure 2. Eastern horizon as viewed from Sunwatcher Ledge, showing selected sunrise points between summer solstice and winter solstice. Illustration produced by Brad Greenwood through computer treatment of photographs by the authors.

of the features are farther away. The western horizon can be used for confirmation, or to substitute at critical times in case of inclement weather. Sunsets around the time of winter solstice could be especially useful because of contours in that particular direction. Because of its overall superior features, we confine this paper to discussion of use of the eastern horizon for calendrics.

Figure 2 shows a tracing of the east horizon from summer solstice to winter solstice viewed from Sunwatcher Ledge. Beginning with the winter solstice, we note that solstice sunrise occurs at a most appropriate location, the base of the second of three steps up to the top of a distant mesa. This is fortunate for solar calendar observation from this location since the horizon leading toward the winter solstice is rather flat and featureless until the steps are reached. Zeilik (1984; 1985a; 1986) has shown the importance of being able to identify anticipation points for solstices in keeping calendar by solar horizon observation. In this regard we note that the base of the first step is reached 10 days before the solstice, a fine anticipation landmark for use in preparation for winter solstice ceremonies. The actual date of the solstice can be nicely determined by observing the anticipation sunrise point both before and following the solstice. Once this is done, the actual solstice can be known as the midpoint between the dates of successive sunrises at the base of the first step, from observations made in the early years of setting up the system for this particular sunwatching station.

It is quite apparent, and well documented, that Anasazi people were interested in both solstices, but their degree of interest in equinox is less certain. Ports in walls at Hovenweep are particularly convincing that they did monitor the Sun at this time (Williamson 1984:112–150). Straight east from Sunwatcher Ledge one sees two high peaks of the Henry Mountains, the highest points of the horizon viewed from the Ledge. Mount Hillers is the southern peak and Mount Pennel is to the north. The Sun rises at the equinoxes barely to the south of the center of the saddle between the two peaks. This becomes even more interesting when we consider the fact that the Indian people were more apt to have ritualized the date when the Sun was at its "Middle House," i.e. midway between the dates of the solstices, determined by counting days rather than by observing the Sun rise at true astronomical east. This is 1.9 days before autumnal equinox or after vernal equinox, thus significantly north of the equinox sunrise point. The point of interest is at the center of the saddle between the peaks!

From vernal equinox until autumnal equinox the directions of sunrises occur along nicely contoured mountainous features. Planting would have begun no earlier than 50 or so days before the summer solstice, the beginning of May. The Sun rises at a sharp peak at the middle of May, about 36 days before the solstice, a most likely time for planting. There are several additional fine marks, but two deep dips, appropriately symbolic of putting seeds into the ground, would be especially easy to remember. The Sun rises at the bottom of a prominent dip 31 days before the solstice, and at the bottom of the deepest dip along that part of the horizon 26 days prior to the solstice.

Anticipation for the summer solstice is nicely indicated when the Sun rises at the base of the most northern butte the sunrise migrates onto. Sunrise at the base of this butte occurs 15 days prior to the summer solstice. One day later the Sun comes up along the south slope of the butte, and another day finds it rising at the south tip of the top of the butte, 13 days before the solstice. Summer solstice sunrise is slightly north of the center of the butte. Again, the Indian people could have used the north base and slope of the butte, noting when the Sun reached this point before and following the solstice, to accurately know when actual solstice dates would occur.

Conclusions

Accurate and reliable calendar keeping was important for Pueblo people generally: indeed, it was crucial for the people living at the Coombs Village. The relatively short growing season, along with ceremonial and other interests these Kayenta Anasazi people would have had leaves little doubt about the importance of calendrics at this place.

The simplest, yet most accurate method, of keeping a calendar is that of watching the Sun in its annual migrations north and south along the eastern and western horizons, a method that we can be fairly sure would have been used at this site as well as at most other places where Anasazi people have resided. Sunwatcher Ledge, as we prefer to call it, is the most convenient place with suitable factors for sunwatching. The ruins on top of the butte suggests that it was used for some significant purpose over a lengthy period of time. Although it might have served as an observation site for protection against possible invaders, sunwatching seems the most likely reason for the rooms to be located there. To be sure, this is speculative, but it is also reasonable. The view from the ledge, especially along the eastern horizon from winter to summer solstice, combined with proximity to the Coombs Village, offers an ideal opportunity for calendric sunwatching.

Putting aside the speculative aspects, this study was accomplished for the primary reason of documentation of the sunrise observation method for keeping calendar. It is an example of experimental anthropology. It was also done to obtain photographs and other information for possible use in an exhibit to inform people about the solar horizon observational method of keeping an accurate calendar.

Interesting things were learned about sunwatching during the course of this study. Actually watching the movements of the Sun through the year, day after day, with local features as guideposts, is quite different from using modern surveying equipment, computers or even a planetarium to plot out and keep a solar calendar for a particular location. Every tiny feature on the horizon can play a role, due to combinations of factors including the angle of rising or setting of the Sun at a particular latitude and the shapes and elevation of landscape features. For example, when observing sunrise at a portion of a horizon consisting of a dip followed to the south by a rise, a spot on the limb of the Sun might barely show, only for an instant, or might not quite show as the Sun angles upward and to the south in its rising. A point of the solar disk might glide along a feature rising upward to the south, then disappear behind the feature to rise again a bit later, or it might remain invisible behind such a slope for several days. A sparkle of sunlight might show at the bottom of a dip with nothing more for several minutes as the Sun is below some feature in its tilted rising movement. Many other factors and combinations are involved in actual horizon situations and these are difficult to predict with accuracy without actual observations. Adding climatic realities, variable from year to year, day to day, hour to hour and sometimes moment to moment makes the sunwatcher's role more complex, interesting and respectable than might be thought.

This study provided, for those involved, experiences which produced greater appreciation for the significance of observational calendrics as well as the knowledge, skills and persistence required for sunwatching. It also brought enhanced realization of the power of repetitive events, that are associated with tiny, constantly changing, environmental conditions that are critical to survival of whole groups of people. Playing the role of a Sun Priest, at least to a small degree, vastly increases understanding and respect for this long held responsibility which provided such fundamental information so crucial to human existence. It is hoped that this report might prove interesting and beneficial to others as well.

References

Latady, R. William and Todd Prince. n.d., The Coombs Site: Continuing Excavations at a Kayenta Anasazi Village. Paper presented at the Twenty-Fourth Great Basin Anthropological Conference, Elko, Nevada, October 1994. On file at Anasazi State Park, Boulder, Utah.

Lister, Robert H. 1959, The Coombs Site, Part One. *University of Utah Anthropological Papers No. 41* (Glen Canyon Series No. 8). Salt Lake City.

Lister, Robert H., Richard Ambler, and Florence C. Lister. 1960, The Coombs Site, Part Two. *University of Utah Anthropological Papers No. 42* (Glen Canyon Series No. 9). Salt Lake City.

Lister, Robert H. and Florence C. Lister. 1961, The Coombs Site, Part Three, Summary and Conclusions. *University of Utah Anthropological Papers No. 43* (Glen Canyon Series No. 10). Salt Lake City.

Morgenstein, Maury. 1994, Sedimentological Investigations at the Coombs Site (42GA34). Report prepared for Anasazi State Park by Geosciences Management Institute, Boulder City, Nevada. On file at Anasazi State Park, Boulder, Utah.

Pollock, Jack. n.d., Environment in Coombs Site Excavations: 1970–1991, pp. 3–7, ed. by Todd Prince. Unpublished manuscript on file at Anasazi State Park, Boulder, Utah.

Prince, Todd. n.d., Coombs Site Excavations: 1970–1991. Unpublished report submitted in partial fulfillment of work conducted under permits at the Coombs Site. On file at Anasazi State Park, Boulder, Utah.

Spaulding, W. Geoffrey. 1994, Paleoecological Investigations at the Coombs Site (42GA34). Report submitted to Anasazi State Park by Dames and Moore, Las Vegas, New Mexico.

Williamson, Ray A. 1984, *Living the Sky: The Cosmos of the American Indian.* Boston: Houghton Mifflin Company.

Zeilik, Michael. 1983, Anticipation in Ceremony: The Readiness is All. In *Astronomy and Ceremony in the Prehistoric Southwest*, ed. by John B. Carlson and W. James Judge, Papers of the Maxwell Museum of Anthropology, number 2, pp. 25–41. Albuquerque: Maxwell Museum of Anthropology.

———. 1985a, The Ethnoastronomy of the Historic Pueblos, I: Calendrical Sun Watching. *Archaeoastronomy, no. 8, Supplement to the Journal for the History of Astronomy* (JHA xvi), pp. S1–S23.

———. 1985b, Sun Shrines and Sun Symbols in the U.S. Southwest. *Archaeoastronomy, no. 9, Supplement to the Journal for the History of Astronomy* (JHA xvi), pp. S86–S96.

14

Calendric Investigations of a Complex Petroglyph Panel at a Gila River Archaeological Site in Arizona

Tom Hoskinson
1000 Oaks, CA

A rock art panel is interpreted to represent counts of various time periods, up to 24 years in length, measured in lunar months. This record of time intervals is compared with various Native American calendars.

* * *

Introduction

Many of the prehistoric Native Arizona and California desert tribal economies depended primarily on hunting, fishing and gathering rather than upon horticulture for subsistence. There is a growing body of archaeological and ethnographic evidence that systematic sky watching was (and is) an integral aspect of these cultures. Systematic sky watching, concepts of cardinal direction, navigation and calendars are fundamental to the survival of nomadic peoples and, I believe far predated the development of horticulture. The ability to navigate and to anticipate the seasons are critical basic survival tools. Without these tools, distant food resources would be difficult or impossible to locate and seasonal food availability would not be reliably predictable.

For people who spend most of their lives out of doors, the cyclic rhythms and interactions of Earth and Sky form the fundamental basis for the development of cosmologies and religions. Celestial changes above become associated with the rhythmic cycles of the Earth. Time and space are not viewed only as linear concepts, but also as concurrent, nested, continuous and cyclical. Thus Earth and Sky, in combination, are seen as being responsible for the conditions surrounding and influencing the well being and behavior of the natural world. To the Native Arizona and

Figure 1. The Sears Point Calendar Panel.

California people, skywatching was not a science, but a direct and personal relationship with the world of the sky and with the powerful forces that inhabited it.

In this paper, I concentrate on the analysis of one Sears Point Rock art panel. I propose that the panel represents a lunar calendar that counts groups of 20, 100 and 300 lunar months; 8 and 24 tropical Earth years; 1, 5 and 15 Venus Synodic cycles. In addition, the 20 month lunar cycles appear to start with the first new moon associated with the appearance of "The Hunter" constellation during the December solstice period. Each of these hypothetical cycles, would end with the new moon of the Saguaro wine ceremonial (June Solstice). This hypothetical calendar is consistent with basic Native American Indian calendric principles, and with Piman and Yuman traditions, although it is apparently a variant not previously recorded.

This panel (Figure 1) faces toward the southeast on an azimuth of approximately 120%; it faces directly into the rising December solstice Sun. The Sun casts seasonally specific patterns of light and shadow that move across and down the panel. The rock features, which cast the shadows, are 2 to 3 meters away from the panel and this produces a substantial optical lever arm. Unique light and shadow interactions with the petroglyphs have been observed and recorded at the December solstice and the equinoxes. The equinox marker has proven capable of easily discriminating the cyclic year to year movement of the shadow's position due to the tropical year effect (Hoskinson 1985: 55–64).

Systematic observations have not yet been carried out at this panel during June Solstice due to conflicting work being performed at the site during the time period of interest.

Figure 2. Arizona map showing location of Sears Point.

Sears Point (AZ:Y:3:1) (Fgure 2) is a prehistoric/historic archaeological site which occupies two small volcanic mesas and their associated drainages on the south bank of the Gila River in southwestern Arizona. The mesas are actually eroded portions of the Gila Plateau which rises about 21 to 25 meters above the Gila floodplain. The mesas and their associated terraces are covered, in part, with desert pavement and with dense basaltic ejecta from nearby volcanic cinder cones. The desert pavement is a mosaic, a single stone thick, which lies on a clay stratum. The clay stratum is saline and sterile from accumulated evaporite deposits. Pavements of this type are very stable, and the stones which make them up are covered with desert varnish, the usually glossy, dark brown to black surface deposit found on desert pavement or other desert stones. Desert varnish only covers those portions of rock exposed to air. It is composed mostly of manganese and iron salts.

In the floodplain, which starts at the eastern base of the Sears Point mesa, there is a large area which shows evidence of food gathering use over a long period of time. I believe that this area was a seasonal mesquite gathering and processing camp site. On the rocky slopes of the site are found creosote bush, brittlebush (*Encelia Farinosa*), ocotillo (*Fouquieria Splendens*), saguaro (*Cereus giganteus*) and barrel cactus (*Echinocactus Lecontei*).

The Gila River floodplain which extends to the north and east of AZ:Y:3:1 is covered by a dense velvet mesquite (*Prosopis Velutina*) bosque. Occasional ironwood (*Olneya Tesota*), palo verde (*Cercidium Microphyllum*), willow (*Salix Goodingii Ball*) and the intrusive tamarisk (*Tamarix Pentandra*) are also found in the floodplain. Gourd (*Lagenaria Vulgaris*) vines are present, as are datura (*Datura Meteloides*), squawbush (*Lycium Fremontii*) and mallow (*Sphaeralcea Emoryi* Var. *Variabilis*). In the marshy places near the river are cattails (*Typha Angustifolla L.*). The drainages of the BLM's Sears Point Archaeological District (SPAD) contain velvet mesquite, ironwood, palo verde, jojoba (*Simmondsia Californica*), catclaw (*Acacia Greggii*) and creosote bush (*Larrea Tridentata*).

Historically, Indians from two different language families, Hokan and Uto-Aztecan, have been continuously present at this site. The Hokan language family members are the Yuman speakers. The Uto-Aztecans are the Piman speakers and the Ute speakers. A large historic/prehistoric village site is located on the north bank of the Gila River immediately across from AZ:Y:3:1. This village, called Xakupï'nc ("hot water") in the Kaveltkadom (Hokan/Yuman) language, and Agua Caliente in Spanish, was occupied primarily by Yuman speakers who are believed to have belonged to the Kaveltkadom band of the Halchidhoma, (Spier 1978:1–47; Ezell 1963; Sedelmayr 1750), together with several Piman speaking families (Sedelmayr 1750). According to Garcés (Spier 1978:33), in 1775 there were 200 people living there. Agua Caliente was probably occupied for some time previous to the first recorded Hispanic contact on the Lower Colorado River in the 16th Century, through the early 19th century (Sedelmeyer 1750) (Spier 1978), (Ezell 1963). The Kaveltkadom are closely related to the Maricopa (Hokan/Yuman), and members of both tribal groups still live along the Gila River today. In addition, the Yuman speaking Colorado River Indians are closely related to the Kaveltkadom and traditionally used the southwestern Gila River area (Cocopa, Quechan, Mohave, etc.) as did the Yuman speaking Kumeyaay (Diegueño) further to the west, and the Yavapai to the north. Several groups of Uto-Aztecan Piman speakers still live along the Gila River. The closest is the Gila Bend Tohono O'odham (Papago). In addition, the Uto-Aztecan speaking Chemehuevi (Ute), who live along the Colorado River near Parker, traditionally used the southwestern Gila River area, according to tribal elders.

Currently, Sears Point is visited by Tohono O'odham people from the Gila Bend and Sells areas, and by Quechan, Cocopa, Mohave, Chemehuevi and other Colorado River people. These people regard Sears Point as a sacred site.

The Sears Point site contains some lithic archaeological evidence which indicates that the site was used by Paleoindians (10000–14000 B.P.). In addition, a variety of prehistoric cultures, including the Desert Archaic (2000–9000 B.P., lithics, campsites, petroglyphs); Patayan (proto Yuman) and Hohokam (2000–600 B.P., villages, campsites, lithics, ceramics, ornaments, petroglyphs), are believed to have utilized the site area. During the late prehistoric and post contact era, from about A.D. 1500 through the early 1800's it appears that AZ:Y:3:1 was utilized primarily by Yuman and Piman Indians.

At the Sears Point site there are about 2000 petroglyphs and perhaps 50 ground figures. Most of the ground figures in the BLM's SPAD are meanders and geometric designs created by rock alignments; several designs were formed by material removal, and others are lines formed of depressed stones that were apparently "walked" into the surface. Two of the ground figures form solstitial alignments involving a common far horizon terminal point, Granary Basket Mountain, which was recognized as a sacred mountain by the Yuman tribes of the Gila River and others (Spier 1978:23, 169–70, 252–54). These alignments and an associated analysis were documented in a previous paper (Hoskinson 1992:131–161).

Calendric Ethnographic Data

Throughout Arizona and California, the lunar month began with the new moon. The new moon was marked with its own celebration, which included shouting and prayers for good health and long life. The Wintu celebrated the new moon by having their headmen address prayers and offerings of food to Moon (Voegelin 1942:145). In northwestern California, speeches and prayers were made to the new moon, the Yuki and Sinkyone used kin terms in their prayers (Driver 1939:342–343). The Chumash extended their arms outward towards the new moon, greeting her with shouts and addressing her with prayers (Hudson and Underhay 1978:76). Many tribes believed that the Moon died monthly and that their ritual activities were necessary to help Moon become reborn (Driver 1937:88). Lunar observations were recorded, the Surprise Valley Paiute, made scratches on a stick, an extra long scratch marked the new moon (Kelly 1932:152–154). The Yokuts placed beads over a stick, feather, or bone to serve as a calendrical mnemonic device (Driver 1937:86; Aginsky 1943:426). Some of the Wintu kept their day and year counts by planting notched sticks in a row. Another Wintu method utilized a few old men who kept track of the days and years by placing pebbles in a basket (Voegelin 1942:145). The Pomo kept track of days by the use of knotted cords, as well as stick bundles (Loeb 1926:223–227).

There is evidence that some tribes recorded calendrical information far beyond what might otherwise be expected for hunter-gatherers concerned only with the solar year. The Chumash (Hudson and Underhay 1978:138, 139) were noted for recording time which extended beyond the solar year. Pomo Sun priests are known to have kept calendar records up to 512 years in length. Loeb (1926:223–227) provides the following description: "The head of the secret society watched the moons and used sticks to count them. A bundle of thirteen sticks represented the year. There were 3 bundles of 8 sticks each, which represented 8, 64 and 512 years, respectively. These sticks were inherited and used to date important events." I would speculate that it is quite possible that the Pomo were counting Venus cycles as well. Five Venus synodic cycles of 584 days make eight Earth (tropical) years (minus 2 ½ days).

Elements of the Sears Point Calendar Panel

Piman and Yuman cultures from this region associate the snake with rebirth, rain and the renewal of Earth's fertility. The snake petroglyph on this panel (Figure 3) interacts cyclically with the equinox sun (as shown in Figure 3). In fact, this vertical snake glyph symbolically divides the solar year in half. From the September equinox through the March equinox, the shadow line falls on the east side of the snake. From just after the March equinox until the September equinox the shadow line falls on the west side of the snake. On the equinoxes the shadow line touches some of the "loops" of the snake. The shadow was observed to touch different loops during the ~4 year equinox cycle.

Figure 3. "Snake" Petroglyph (4 Images).

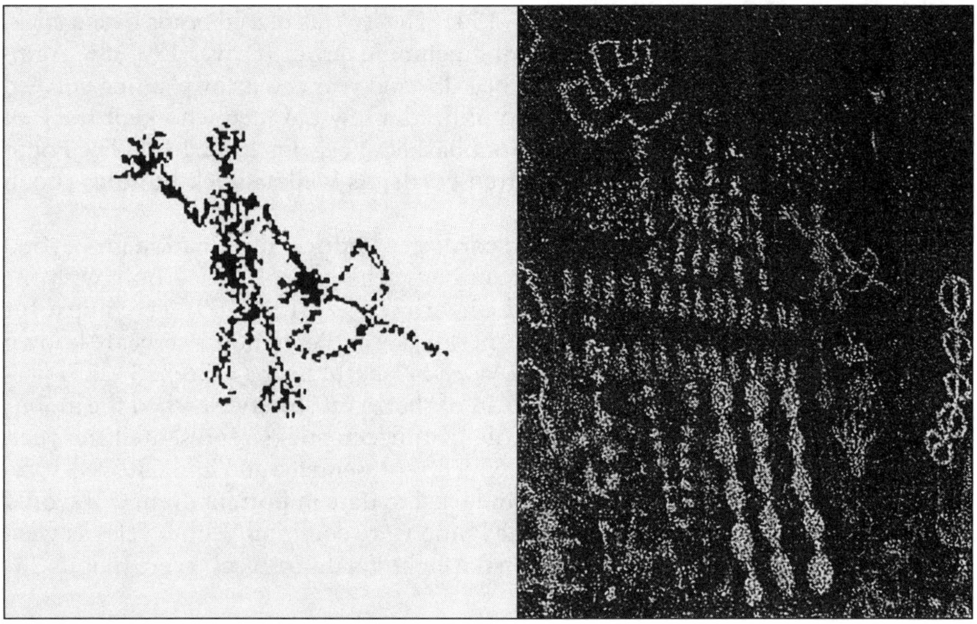

Figure 4. "Hunter" Petroglyph (2 Images).

This Sears Point petroglyph faces the southeastern horizon (Figure 4), looking along an approximate bearing of 120% true azimuth. In December, shortly after sunset, stars from Canis Major rise on this azimuth (Figure 5).

Native American Indian cultures from this region speak of a winter constellation called "The Hunter" whose main star was identified as Alpha Canis Major (Sirius) (Figure 6).

This Sears Point petroglyph (Figure 7) is an accurate representation of a One Pole Ladder. These ladders had both mundane and ceremonial uses. Shamans used them for travel to and from the world of the sky. Each circular "rung" is said to represent a transition in state of being and understanding.

At the time of the December solstice the two upper "loops" of the "ladder" are illuminated by circular spots of light just as the sun goes below the local horizon.

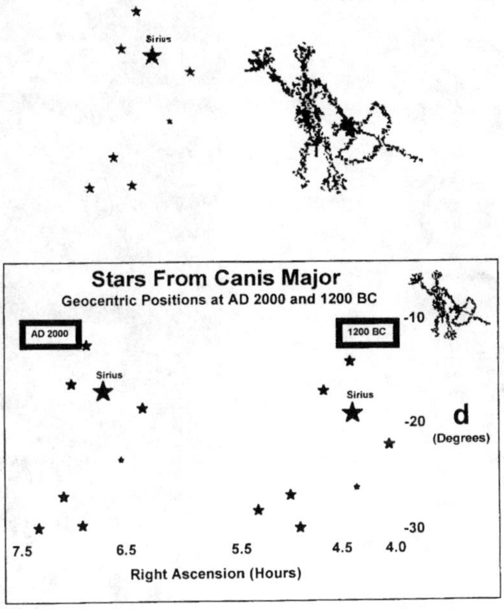

Figure 5. Stars from Canis Major and "Hunter" Petroglyph.

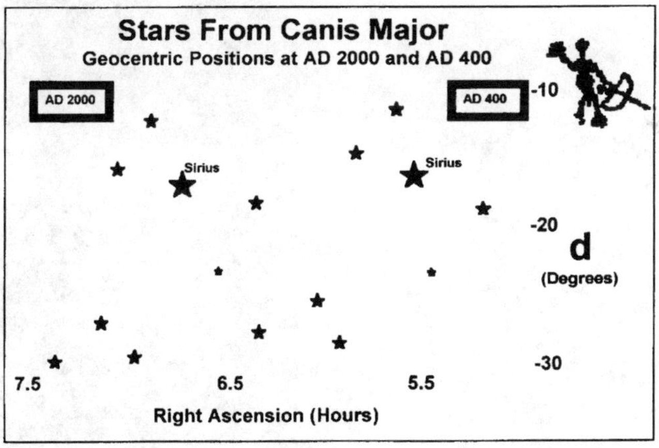

Figure 6. Stars from Canis Major at Ad 2000 and Ad 400.

This Sears Point petroglyph (Figure 8) is believed to be an abstract representation of the mythical (anthropomorphized) Saguaro cactus. The Saguaro is a Piman and Yuman metaphor for the June solstice. The Saguaro is also ethnographically identified with the northern circumpolar constellation Ursa Major.

This Sears Point Petroglyphic sequence of 29 marks (Figure 9) is believed to represent a lunar count from new moon to new moon. The two rows of 14 marks are vertically separated by a space containing a single mark which probably represents the full moon. In the native Hokan and Uto-Aztecan calendars of southwestern Arizona, each month starts at a new moon, and the month is divided in half by

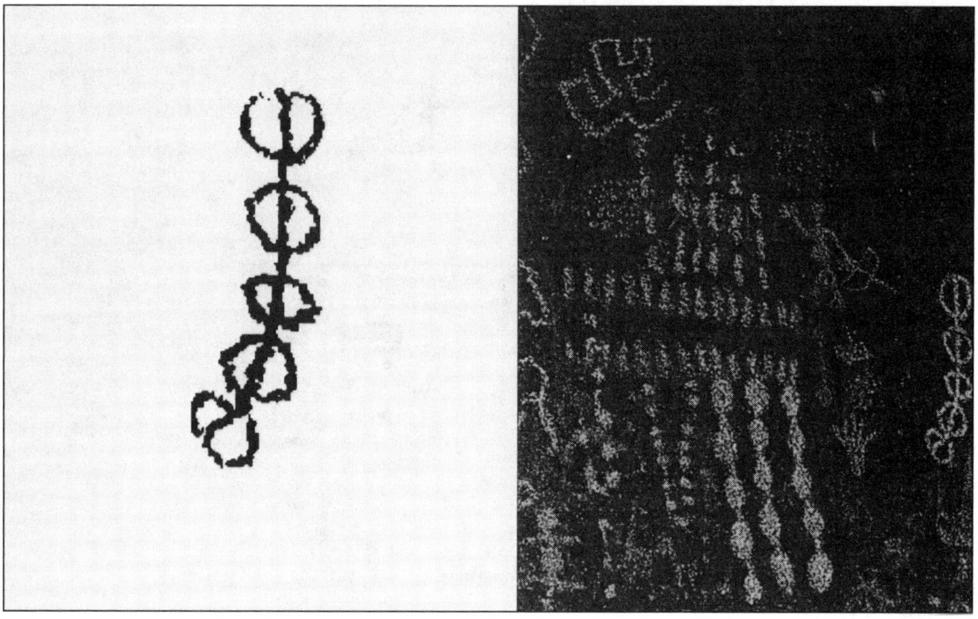

Figure 7. "One Pole Ladder" Petroglyph (2 Images).

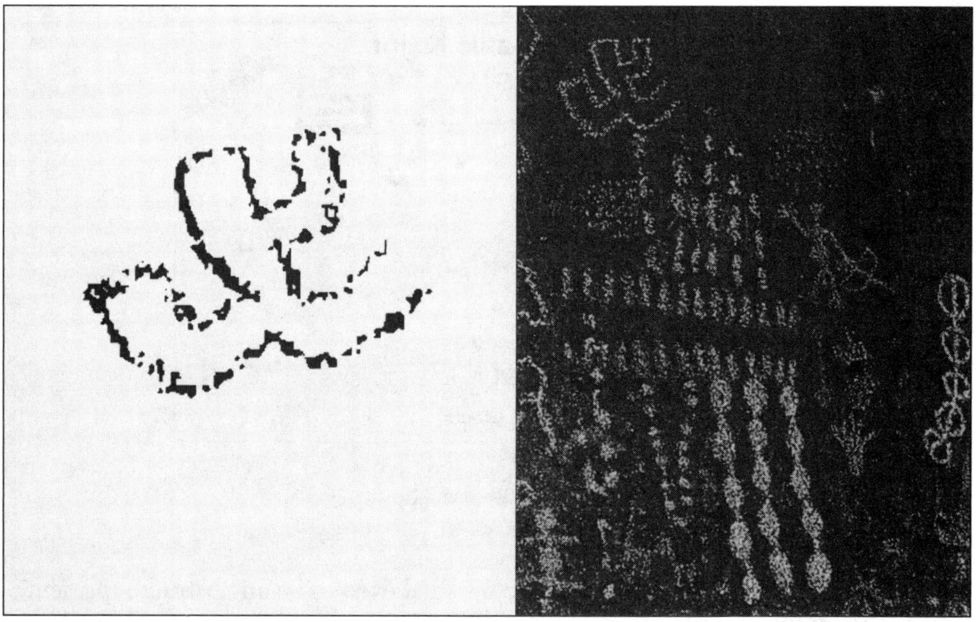

Figure 8. "Saguaro" Petroglyph (2 Images).

the full moon. At full moon, during the Autumn and Spring it is said that "Moon-looks-at-Sun."

This Sears Point petroglyphic sequence of 20 vertical marks (Figure 10) may have represented a count of 20 Lunar months of 29 days. Twenty such Lunar months

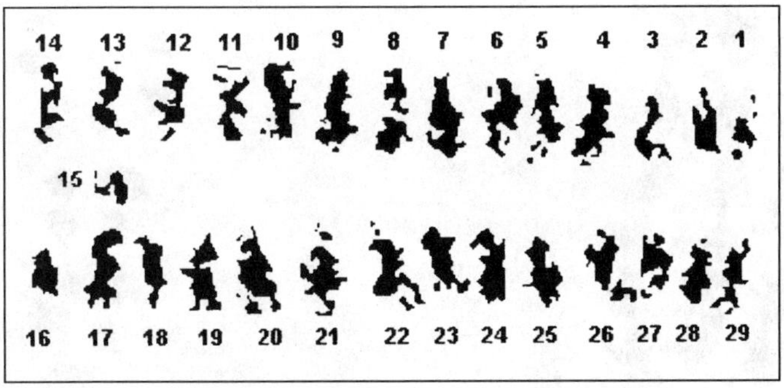

Figure 9. "Lunar Count" Petroglyph.

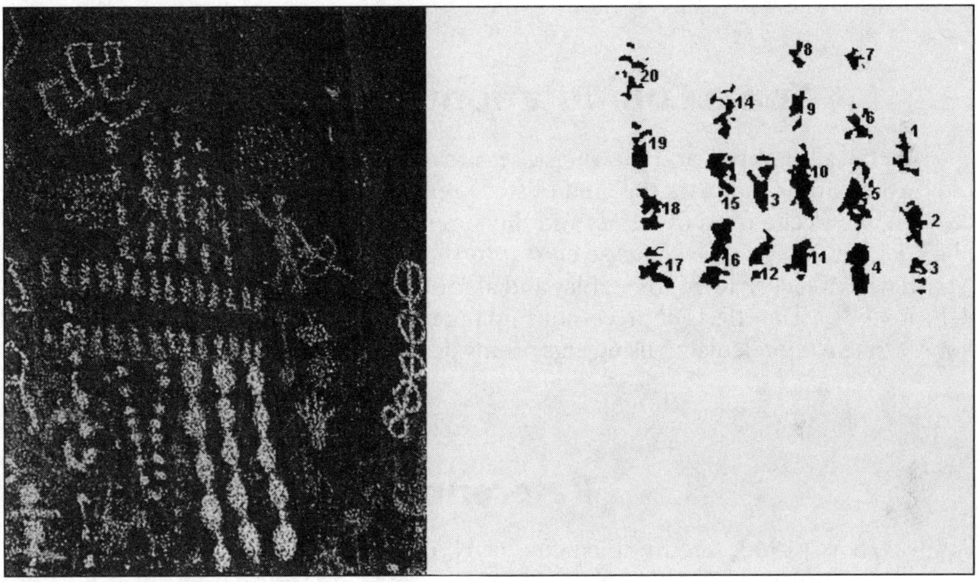

Figure 10. "Lunar Month Count" Petroglyph (2 Images).

is 580 days, which is approximately equal to 1 synodic Venus cycle. There is ethnographic evidence that both Pimans and Yumans were aware of and counted Venus cycles.

These petroglyphs strongly resemble and probably represent knotted strings (Figure 11). Knots in strings were used by many Native American Indians to represent large numbers. In this instance, I believe that each knot probably represents 20 lunar months (1 Venus cycle).

This petroglyph resembles a string of beads (Figure 12). It may provide an algorithm for tallying individual lunar month or groups of lunar month counts. For example, each bead might represent 20 months (1 "knot"), 100 months (5 "knots") or even 300 months (15 "knots").

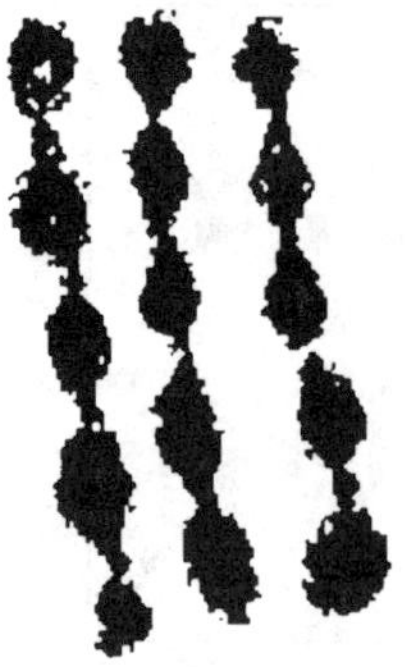

Figure 11. "Knotted String" Petroglyph.

Figure 12. "String of Beads" Petroglyph.

Some Computational Information

The data for Venus and the Sun were calculated with a computer program written by the author that uses the methods of Brentagnon and Simon (1986). Right ascension and declination of Venus and the Sun were calculated for the site longitude (113 degrees West) at the sunrise closest to the June solstice. Day of the June New Moon was calculated with the tables and algorithms developed by Jan Meeus (1983: 4-1 thru 4-36). The effects of precession and nutation on declination and right ascension for Alcyone were calculated using algorithms developed by Jan Meeus (1982: 63–75).

References

Aginsky, B.W., 1943, Culture Element Distributions, XXIV Central Sierra. University of California Anthropological Records 8(4): 393–468.

Brentagnon, Pierre and Jean-Louis Simon, 1986, Planetary Programs and Tables from -4000 to +2800, Willman-Bell, Inc., Richmond, Virginia.

Driver, Harold, 1937, Culture Elements Distributions: VI Southern Sierra Nevada. University of California Anthropological Records 1(2):53–154.

Driver, Harold, 1939, Culture Element Distributions: X: Northwest California. University of California Anthropological Records 1(6):297–433.

Ezell, Paul, 1963 The Maricopas: An Identification from Documentary Sources. Anthropological Papers of The University of Arizona, No. 6, Tucson.

Hoskinson, Tom, 1985 Cycles of The Equinox Sun. Earth and Sky, Papers From the Northridge Conference on Archaeoastronomy, pp 55–64. Edited by Arlene Benson and Tom Hoskinson. Slo'w Press, 1000 Oaks, Ca

Hoskinson, Tom, 1992 Saguaro Wine, Ground Figures, and Power Mountains: Investigations at Sears Point, Arizona. Pages 131–161 in: Earth & Sky. Visions of the Cosmos in Native American Folklore. Edited by Ray A. Williamson and Claire Farrer. UNM Press, Albuquerque, NM.

Hudson, Travis, and Ernest Underhay, 1978 Crystals in the Sky: An Intellectual Odyssey Involving Chumash Astronomy, Cosmology and Rock Art. Ballena Press Anthropological Papers 10. Ballena Press and Santa Barbara Museum of Natural History.

Kelly, Isabel, 1932, Ethnography of the Surprise Valley Paiute. University of California Publication in American Archaeology and Ethnology 31 (3): 67–210.
Loeb, Edwin, 1926 Pomo Folkways. University of California Publications in American Archaeology and Ethnology 19(2):149–404.
Kelly, William H. 1977 Cocopa Ethnography. Anthropological Papers of The University of Arizona, No. 29, Tucson.
Meeus, Jean, 1982 Astronomical Formulae for Calculators, Second Edition. Willmann-Bell, Inc., Richmond, Va.
Meeus, Jean, 1983 Astronomical Tables of the Sun, Moon and Planets. Willmann-Bell, Inc., Richmond, Va.
Sedelmayr, Jacobo, 1749 Entrada a la nación de los yumas gentiles.... Tubutama [Sonora, Mexico], January 15, 1750. MS in Arizona Pioneers' Historical Society, Tucson, Arizona
Spier, Leslie, 1978 Yuman Tribes of the Gila River. Unabridged republication, originally published in 1933, by The University of Chicago Press, Chicago, Illinois. Dover Publications, New York.

15

A Petroglyph Panel of Astronomical Significance

R. P. Olowin
Department of Physics & Astronomy
Saint Mary's College
Moraga, CA

A petroglyph panel located near White Rock, New Mexico, appears to contain a host of what might be considered astronomical imagery. The images include a crescent superposed upon a concentric set of circles; a circle with radial dot-like rays; several nearby concentric circle-sets; and three linearly disposed disks. A few other images (anthromorph and ambiguous) decorate the panel, but it remains uniquely celestial in character. An astronomical hypothesis is proposed to suggest that the panel represents a record of actual astronomical events: SN1054 and the appearance of a "Great Comet."

* * *

Location and Particulars

The petroglyph panel is located near White Rock, New Mexico, on traditional Tewa lands, now belonging to the Los Alamos National Laboratory, near the lip of Water Canyon, GPS:N3548-W10614, USGS Los Alamos Quadrangle. Positioned on the base of a small cliff, on a terrace on top of talus, the site is basically south facing and consists of several panels and outliers with images pecked into the desert varnish of the surrounding rock; there is little deterioration of the panels and no overt paint traces or associated tools can be found. Identified as site LA-49944 for the Laboratory of Anthropology of the Museum of New Mexico by K. J. Ewing on 8 May 1983, it is labeled "Nebula Site" (Ewing 1983).

Knowledge of the site was obtained from LANL Archeologist, B. Larson (Larson 1996), through C. Boliek (Boliek 1996), late of Ghost Ranch National Conference Center in Abiquiu, N.M. and P. Mahon (Mahon 1996), of Star Hill Inn in Las Vegas, NM.

The site consists of several panels, some rich with petroglyphs, but separated both in distance and, apparently, style. The "Nebula" panel (Figures 1 and 2) stands in isolation some three meters from a rich display of complicated and polymorphic

Figure 1. "Nebular" Petroglyph, Water Canyon, New Mexico.

representational petroglyphs. Although large enough to accommodate much more imagery compared to adjacent panels, the "Nebula" panel appears mostly devoid of images other than those described here. There are several concentric sets of pecked circles near the top of the cliff face, a little over a meter above the "Nebula" image, which are occasionally cut by the shadow of an adjacent overhanging rock. That these present a possible shadowgraph is currently under study (see Figure 3), as there are several such shadowgraph solstice markers in the immediate vicinity showing similar features. The panel images conform to those of the Anasazi Rock Art Styles and are thus considered pre-Columbian.

The dominant feature of the site, however, remains the so-called "Nebula." The central images of the panel are a superposed pair composed of a pecked filled-

Figure 2. "Nebular" Petroglyph Site, Water Canyon, New Mexico, after Ewing.

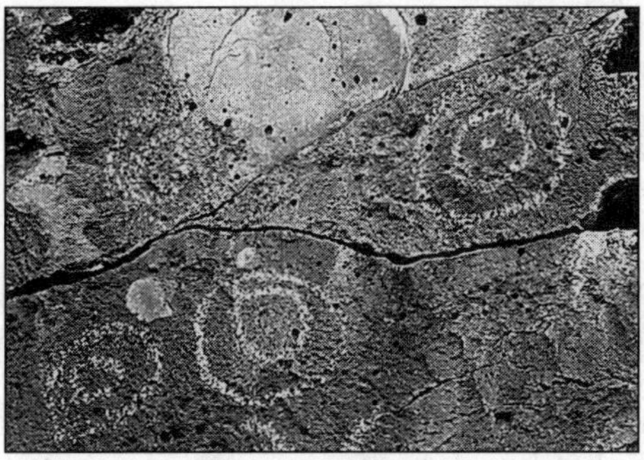

Figure 3. Possible Shadowgraph Elements at the "Nebular" Petroglyph Site.

crescent and two concentric circles, the latter being somewhat the larger; above these is a circle with five major radial series of pecked full-dots radiating outward to the ten-o'clock position. About a meter to the left of this cluster is a set of concentric circles; about 20-cm to the left of the cluster at the 7-o'clock position lies a linear row of three pecked full-dots, each over a centimeter in diameter, spanning over 10-cm. An outlier completes the imagery and consists of a concentric set of circles with an umbilicus. These elements are shown in a photographic enlargement of the petroglyph cluster and can be seen in Figure 1.

It is interesting to point out that the images in this cluster are congruent to those seen at Chaco Canyon, at Peñasco Blanco. There, the pictograph images of a crescent and an asterisk are found well preserved on an overhang above and perpendicular to an image consisting of a series of concentric circles, dressed in ochre and filled with yellow, surrounded by a faded ochre envelope and long right-facing trail. The former of these has been described as SN1054, the latter sometimes as comet C1066-Halley. These possible-identifications have provided the impetus to suggest that the imagery seen in the Water Canyon panel are likewise astronomical and may be related to actual events in the sky.

The Astronomical Hypothesis

The suggestion that the images on the " Nebula" panel are indeed astronomical must be supported by evidence that the petroglyphs "provide a number of possible relationships of form and position, which must then correspond to the relationships found in the sky (Mayer, 1977)." That this is a reasonable case for the panel in question will be decided by how well the images represent the sky.

Discussion

The main cluster elements of the panel consist, a filled crescent superposed with a somewhat larger set of concentric circles. Assuming that the latter represents a bright object and the former a lunar crescent the pair suggesting a waxing moon in appulse with a planet, bright star, nova, or supernova. It is further assumed that the images are coeval and are associated with the same event; obviously, with increasing numbers of petroglyphic elements there is great difficulty in mapping them with single events in the sky.

Above these elements lies a circle with radial dotted lines reminiscent of the classical descriptions of a comet. The articulation of this "cometary" petroglyph, because of the style, depth and character of incisions, suggests an application at a different time and most certainly, performed by a different author (Shaafsma 1996), and so may represent an *independent* record of a cometary appearance.

The main panel might also offer some clues to the understanding of the images and their relationship to themselves and to the sky. Recall the set of concentric circles containing an umbilicus, somewhat apart from the cluster, as well as the linear array of three dots. The former image is often associated with the sun while the latter, in an astronomical connotation, may represent Mintaka, Alnilam, and Alnitak: the three stars in the belt of Orion. It is generally accepted that kiva ceremonies were timed by the positions of the stars: for instance at the prehistoric pueblo of Kuaua, near Bernalillo, New Mexico, ceremonial images identified as the Pleaides and other star groups still remain (Waters 1963). Taken as a group, then, the images on the "Nebula" panel could depict the circumstances of the heavens on the morning of 5 July, 1054, with the heliacal rising of the horizontal—perpendicular array of the Belt of Orion, and the Taurus Event in the twilight morning sky. Troublesome, however, is the lunar phase if the date above is chosen as accurate. The petroglyph shows a lune in waxing phase, whereas the lunar disk of the Taurus event was waning. For a discussion on the importance of this matter, the reader is referred to the many articles debating its merits (e.g. Braudt & Williamson 1979).

Astronomical Search

Given the possibility of the astronomical content of the panel, a search has been made to suggest possible candidates for the petroglyphic images. The time-frame of the search was restricted to the pre-Columbian period, 500–1500 AD, spanning the Southwestern periods Basketmaker III and Pueblo I-IV. As candidates, lunar conjunctions with planets and bright stars or other objects, the appearance of "Great Comets," and transient phenomena such as novae and supernovae, as well as eclipses of the sun and moon, are all likely.

Comets

Within the selected time frame, 59 "Great Comets" are found in the historical record that could have been witnessed in the American Southwest. By "Great" is meant that the comet has sufficient brightness and angular extent to be easily witnessed: magnitude, $-1 < m_V < 4$, tail > 5 degrees. These have been identified in Appendix 1.

Of the comet candidates, several stand out as remarkable: C837(Halley), C989 (Halley), C1066(Halley), C1080, C1230, C1245, C1299, C1301, C1345,

C1366(Temple-Tuttle), C1491I, and C1491II. A curious note is that the latter comets were both visible near the horizon in the western sky, near Venus, on January 22, 1491. Though the moon was close to full (97%) it represents a unique situation of two comets being visible in close proximity at the same time.

Sky simulations for observations in the American Southwest suggest three interesting candidates for the comet glyph. C837(Halley) was the closest approach of Halley's comet in human history with the object spanning some 92 degrees across the sky and being recorded as brighter than Venus. However, the timing appears too early for our purposes. Another Halley apparition in 1066 is more attractive since it was observable as a morning object during the interval April 3–22 and, after conjunction, as a western evening object from April 24 through June 6. The comet was last observed 77 degrees from perihelion, suggesting an unusually bright post-perihelion apparition. Finally, the "Great Comet" of 1366, C1366 (Temple-Tuttle), was unusually bright and spanned many 10's of degrees in the sky during the evenings of October 25–26 and was seen in the northern sky near Uma/Mi and Draco.

Planetary Conjunctions

A series of planetary conjunctions were simulated for visibility in the region with the Moon and Venus, Mars, Jupiter and Saturn. Over 500 of these per object were calculated and particular attention was paid to those in appulse with a waxing or waning lunar crescent. Given that the brightest circumstances occur with the planet Venus (–4 to –3 magnitudes) and the large number of occurrences of this type of conjunction, it seems much too common an event to warrant considerable attention. These conjunctions are therefore excluded from consideration as candidates for the bright object appearing on the petroglyph panel. The frequency with which a waxing crescent of the moon in appulse with Venus was observable in the American Southwest in pre-Columbian times is obvious by consulting the appendices.

Solar and Lunar Eclipses

Of the 229 Solar and 843 Lunar eclipses seen in the centuries 500–1500, those seen in the American southwest probably do not account for what appears to be depicted on the "Nebula" petroglyph panel. Certainly the apparent lunar phase rules out this possibility. It is noteworthy, however, to point out that several total solar eclipses have ground paths through the American Southwest giving near occlusion of the solar disk during daylight hours: 711, 804, 1097*, 1318, 1372, 1379*, and 1397*. Figure 4 shows an example using the 1097 event that passed with a swath of totality through Chacoan territory on July 11 (17, Gregorian) of that year and the appendix provides the particulars.

Galactic Novae and Supernovae

As notable transient objects, the galactic novae and supernovae rank as the most remarkable. Of these, 37 appearances could have been observed in the American southwest from 800–1500. Three of these occurred in the year 837, the year of the greatest appearance of comet Halley. The longest duration began in May of that year and

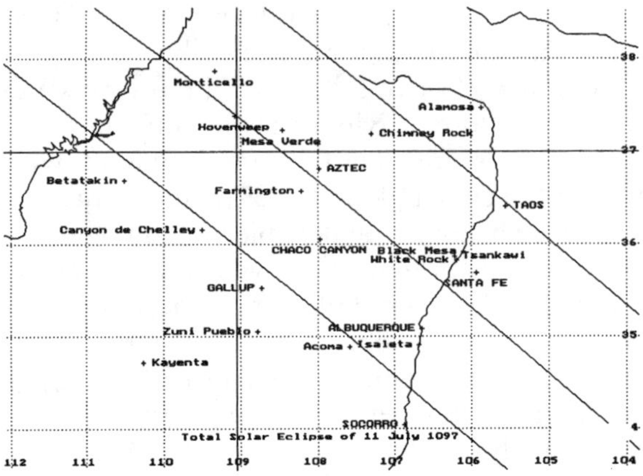

Figure 4. Total Solar Eclipse Path (1097) in the Chacoan Domain.

was seen for some 75 days in the constellation Virgo. C837 (Halley) was too far south to be seen along with this object. The prime candidates appear to be SN1054 and the guest star of 1181 seen in the constellation Cassiopia. The lunar circumstances strongly suggest SN1054 as the prime candidate, seen in heliacal risingon the morning of July 5, 1054 in appulse with the waning crescent moon in Taurus.

Conclusions

If the hypothesis that the "Nebula" panel represents a visual mnemonic of actual astronomical events, the celestial candidates include the SN1054 event and the *separate* appearance of a comet, whose date remains uncertain due to the current inability to date petroglyphic impacts on desert patination. It appears, therefore, that the petroglyph panel may represent distinct episodes: the Taurus event and C1066(Halley), a convenient pair because of temporal proximity; or, again, the Taurus event and another of the cometary candidates, the best of which is C1366(Temple-Tuttle), albeit a much later appearing candidate at the terminus of the Chaco Phenomenon.

Table 1. Pre-Columbian Galactice Novae and Supernovae*
Observable in the American Southwest

Date	Duration	R.A.	DEC	CONST
537 Jan/Feb	-	-	+65	Circum
541 Feb/Mar	-	-	+65	Circum
641 Aug 6	25 days	12 20	+20	Com
684 Dec/Jan	2 weeks	03 40	+25	Ara
722 Aug 19	5 days	01 00	+60	Cas
829 Nov	-	07 50	+15	CMI/Gem
837 Apr 29	22 days	07 00	+10	CMi
837 May 03	75 days	12 10	+05	Vir
837 June 26	22 days	18 00	-25	Sco
877 Feb 11	-	23 50	+20	Peg
891 May 12	-	16 50	-20	Sco/Lib
900 Feb/Mar	-	17 00	+10	Oph/Her
911 May/Jun	-	17 10	+15	Her
1006 Apr 03	2+ years	15 10	-40	Lup
1011 Feb 08	-	19 20	-30	Sgr
1035 Jan 15	-	01 20	+05	Psc
1054 Jul 04	22 months	05 40	+20	Tau
1065 Sep 11	-	09 20	-25	CMa/Pup
1069 Jul 12	11 days	18 10	-35	CrA
1070 Dec 25	-	02 40	+05	Cet/Ari
1073 Oct 09	-	00 10	+05	Psc
1074 Aug 19	-	00 10	+05	Psc
1138 Jun/Jul	-	01 50	+20	Psc
1139 Mar 23	-	14 10	-10	Vir
1163 Aug 10	-	17 30	-20	Sgr
1175 Aug 10	5 days	15 40	+50	Her/Dra
1181 Aug 06	185 days	01 30	+65	Cas
1203 Jul 28	9 days	17 10	-40	Sco
1224 Jul 11	-	17 10	-40	Sco
1240 Aug 17	-	17 10	-40	Sco
1356 May 03	-	05 50	+30	Aur
1388 Mar 29	-	00 10	+20	Peg
1399 Jan 05	-	18 50	-20	Sgr
1404 Nov 14	-	19 50	+30	Cyg
1430 Sep 09	26 days	07 30	+05	Mon/Ori
1431 Jan 04	15 days	04 50	-10	Lep/Ori
1437 Mar 11	14 days	16 50	-40	Sco
1460 Feb/Mar	-	11 30	-15	Cra

* Clark, D. H. & Stephenson, F. R., *The Historical Supernovae,* Pergamon, 1977.

Table 2. Pre-Columbian Comets: 500–1500
Observable in the American Southwest

Comet	peri(AU)	Σ	i^0	Ω^0	omega^0	peridate
530	0.575590	0.96871	163.3940	41.9740	97.5820	530.09271
539*	0.160000	1.0	19.0000	33.0000	246.0000	539.11060
565	0.832000	1.0	121.0000	180.0000	79.0000	565.07150
568	0.870000	1.0	4.0000	302.0000	35.0000	568.08277
574	0.730000	1.0	54.0000	155.0000	342.0000	574.03250
607*	0.580830	0.96804	163.4760	43.2610	98.7990	607.03155
684*	0.57958	0.96815	163.4180	43.8000	99.1490	684.10028
760*	0.581840	0.96785	163.4430	44.6870	99.9970	760.05207
770	0.580000	1.0	117.0000	111.0000	88.0000	770.06058
837*	0.582320	0.96781	163.4470	44.9300	100.1010	837.02283
868	0.420000	1.0	65.0000	321.0000	277.0000	868.03040
905	0.200000	1.0	140.0000	70.0000	50.0000	905.04260
912*	0.58016	0.96807	163.3110	45.6460	100.7770	912.07187
961	0.630000	1.0	119.0000	2.0000	85.0000	961.12280
989*	0.581910	0.96789	163.3990	46.5610	101.4840	989.09057
1014	0.560000	1.0	117.0000	174.0000	84.0000	1014.04060
1018	0.620000	1.0	35.0000	185.0000	197.0000	1018.08270
1066*	0.574500	0.96887	163.1120	47.6240	102.4730	1066.03209
1080	0.680900	1.0	6.9000	322.9300	73.6000	1080.09109
1092	0.770000	1.0	124.0000	114.0000	62.0000	1092.02220
1097	0.30200	1.0	41.0000	352.0000	298.0000	1097.09223
1110	0.830000	1.0	137.0000	321.0000	358.0000	1110.05180
1132	0.735500	1.0	106.3000	213.2000	114.2000	1132.08307
1145*	0.574790	0.96879	163.2240	49.0540	103.7040	1145.04186
1147	0.120000	1.0	110.000	282.0000	300.0000	1147.01280
1222*	0.574210	0.96884	163.1920	49.3040	103.8490	1222.09288
1230	0.860000	1.0	16.0000	304.0000	181.0000	1230.12288
1240	0.667700	1.0	75.3000	135.1000	331.4000	1240.01216
1245	0.50000	1.0	20.0000	180.0000	87.0000	1245.04010
1264	0.824900	1.0	16.4000	151.0400	159.7100	1264.07203
1293	0.780000	1.0	30.0000	89.0000	313.0000	1293.10280
1299	0.317930	1.0	111.0000	116.9400	103.8900	1299.03318
1301*	0.572710	0.96893	163.0760	50.1520	104.5000	1301.10256
1305	0.650000	1.0	99.0000	106.0000	56.0000	1305.01190
1337	0.749000	1.0	143.6000	97.6000	79.6000	1337.06148
1340	0.13620	1.0	176.4800	187.6300	32.4400	1340.05136
1345	0.890000	1.0	23.0000	150.0000	210.0000	1345.08230
1351	0.010000	1.0	7.0000	264.0000	165.0000	1351.11190
1362	0.300000	1.0	155.0000	291.0000	348.0000	1362.02250
*1366***	0.976230	0.90640	162.2390	225.4300	164.4810	1366.10185
1368	0.480000	1.0	131.0000	217.0000	90.0000	1368.05050
1376	0.550000	1.0	76.0000	301.0000	81.0000	1376.07317
1378*	0.576200	0.96837	163.1130	51.0200	105.2950	1378.11107
1385	0.790000	1.0	103.0000	289.0000	182.0000	1385.10240
1402	0.380000	1.0	55.0000	126.0000	91.0000	1402.03210
1433	0.492770	1.0	103.9790	104.2550	189.3920	1433.11083
1439	0.120000	1.0	81.0000	192.0000	140.0000	1439.05090
1449	0.32742	1.0	155.6790	268.8090	356.6910	1449.12099
1456*	0.579700	0.96800	162.8900	51.8660	105.8350	1456.06096
1457I	0.703360	1.0	13.2510	257.5290	194.5970	1457.01185
11457II	0.769910	1.0	7.3350	187.5510	183.2810	1457.08081
1458	0.490000	1.0	69.0000	117.0000	258.0000	1458.11070
1462	0.250000	1.0	130.0000	316.0000	137.0000	1462.08040
1468	0.850000	1.0	138.0000	107.0000	91.0000	1468.10073
1472	0.485890	1.0	170.8660	292.8730	245.7280	1472.03014
1491I	0.761000	1.0	73.4000	280.2000	164.9000	1491.01089
1491II	0.850000	1.0	9.0000	166.0000	316.0000	1491.01227
1499	0.950000	1.0	16.0000	329.0000	42.0000	1499.09090
1500	1.110000	1.0	99.0000	305.0000	352.0000	1500.04300

Legend: **BOLD** = Candidate *=Halley *ITALIC*=Great Comet

References

Baxnes, F. A. 1982, Prehistoric Rock Art, Salt Lake City: Wasatch.
Bettis, C. 1989, Legacy in Stone: Images from the Past, Portland: (RARE).
Boliek, C. 1996, private communication.
Brandt, J. C. and Williamson, R. A. 1979, Archeoastronomy, 1, 10, Sl–S35.
Clark, D. H. and Stephenson, F. R. 1977, The Historical Supernovae, New York: Pergammon.
Ewing, H. 1983, New Mexico Laboratory of Anthropology Site Record, Santa Fe: LA-49944.
Griffin-Pierce, T. 1992, Earth is My Mother, Sky is My Father: Space, Time and Astronomy in Navajo Sandpainting, Albuquerque: UNM Press Hadingham, E. 1984, Early Man and the Cosmos, Norman: OU Press.
Krupp, E. ed. 1978, In Seaxch of Ancient Astronomies, 137–141, New York: McGraw Hill.
Larsen, B. 1996, (Staff Archeologist, Los Alamos National Laboratories), private communication.
Ligon, T. R. 1994, Dance of the Planets, Lovelend, CO: ARC Softwaxe.
Mahon, P. 1996, private communication.
Malville, J. M. and Putnam, C. 1989, Prehistoric Astronomy in the Southwest, Boulder: Johnson.
Mayer, D. 1979, Archeoastronomy, 1, 10, S52.
Patterson, A. 1992, A Field Guide to Rock Art Symbols of the Greater Southwest, Boulder: Johnson.
Russeau-Dupre, D. (LANL) 1996, private communication.
Shafsma, P. 1996, private communication.
State of New Mexico 1993, New Nexico Cultural Resource Information System, Santa Fe: ARMS93-101.
von Oppolzer, T. R. 1887, Canon of Eclipses, tran. Gingerich, 0. 1962, New York: Dover.
Waters, F. 1963, Book of the Hopi, New York: Penguin.
Williamson, R. A. and Farrer, F. A. 1992, Earth & Sky: Visions of the Cosmos in Native American Folklore, Albuquerque: UNM Press
Williamson, R. A. 1984, Living the Sky: The Cosmos of the American Indian, Boston: Houghton Mifflin.
Yeomans, D., Rahe, J., and Freitag, R. 1989, Halley's Comet in History, Scientific American.
Yeomans, D. 1992, private communication.

16

Feathered Stars and Scalps in Pueblo IV

Polly Schaafsma
Research Associate, Museum of Indian Arts and Culture
Laboratory of Anthropology
Santa Fe, NM

The four-pointed star and feathered star represented in the art of Pueblo IV is a multivalent symbol embodying several interrelated meanings and implications of war. Knife-Wing, Beast god of the Zenith, and scalps and fertility, linked specifically to the growth of corn, are among the concepts conveyed by this figure, that also is called Morning Star.

* * *

Introduction

A star with four points expanding around a circular center is a significant visual metaphor in Pueblo art during the Pueblo IV period (A. D. 1325–1680) (Fig. 1). This star may be further elaborated with a face and a fan of feathers and talons. Such stars are unevenly distributed in the archaeological record from Hopi to Pecos and from the Piro country through the northern Tewa, with a primary focus among the Rio Grande Pueblos (Fig. 2). Feathered and taloned stars are notably well represented in the southern Tiwa petroglyphs of Albuquerque's West Mesa and in southern Tewa rock art in the Galisteo Basin (Schaafsma 1992a: Fig. 130, 132;1992b: Fig. 10). Tompiro rock paintings portray detailed star kachinas in full anthropomorphic form with a star as a mask (Fig. 3), and these and other variations on Tompiro stars are described in detail by Baldwin (1992). Feathered stars are featured on certain layers of the Pottery Mound kiva murals (Hibben 1975; Pottery Mound mural slides, Maxwell Museum, Albuquerque), but scarce and unfeathered at Awatovi and Kawaika-a at Hopi (Smith 1952). At Pecos on the eastern extreme of the Pueblo domain they are incised on clay pipes (Kidder 1932: Figs. 138, 139, 145, 147 and p.180).

What is the metaphysical message of this icon that appears to carry many layers of meaning? To grasp the symbolic implications of this star, that is often a conflated image combining the attributes of both star and bird, one needs to examine

Figure 1. Oxford V logo, feathered star with face, and talons, southern Tiwa, Petroglyph National Monument, Albuquerque.

its various attributes and contexts as it appears in Pueblo art. As observed by Troike (1982: 181), searching for the meaning of gesture in Mixtec codices, meaning can be inferred from the various environments, or in our case, symbolic contexts, in which a given figure occurs. Each context provides a further opportunity for refining our understanding of what any given figure is all about. As with the codices, stars in Pueblo IV art are the work of different artists, often in widely different places. Therefore each star varies accordingly, and subtle differences in meanings may be implied, according to the understanding of each artist and purpose of each representation. Nevertheless, there seems to be a general cultural consensus of the star's general, if multiple, significance. In addition to visual connotations, an un-

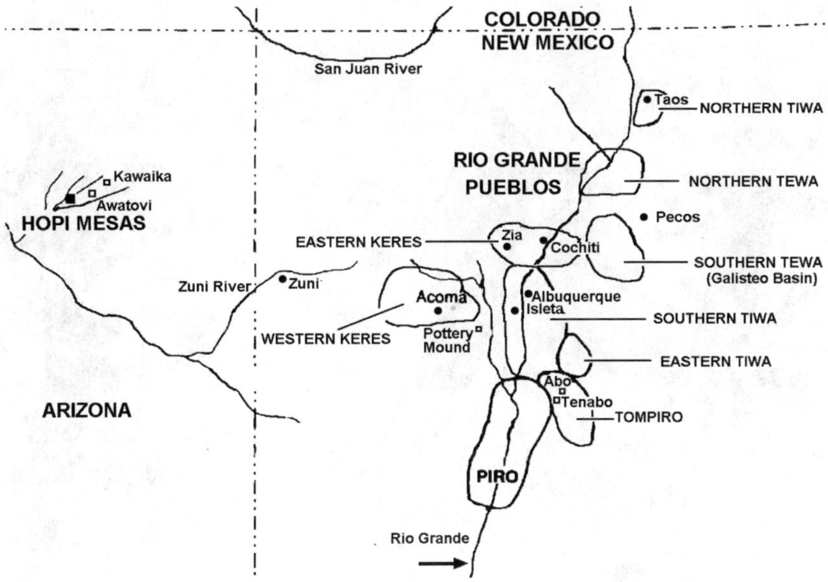

Figure 2. Map of the Pueblo Southwest indicating places noted in text.

derstanding of Pueblo cosmology and the metaphoric language of mythology is basic to comprehending this image.

To begin with, the star under discussion has a face, and it is clearly a person—in the Pueblo sense of the word—a *significant* person (like a kachina), culturally and ritually defined and given symbolic expression (Hieb 1994: 24). Some stars have an arrow as part of the headgear beside the feather fan (Schaafsma 1992a: Fig. 130). In painted examples, the face is usually dark. Stars and star kachinas are sometimes represented in pairs, and the fully defined anthropomorphic kachinas hold weapons in their hands (Fig. 3)

There is a patterned association of stars with other elements in both the rock art and kiva murals. Stars, usually without avian attributes, are very commonly found on round war shields, the latter pictured as discrete elements or carried by warriors (Fig. 4; and Schaafsma 1992a: Fig. 147; 1992b: Fig. 1). Stars on shields may also be represented by simple equilinear crosses, and five or six pointed asterisks are not unknown. Stars also are juxtaposed or found in nearby association with other ritually defined persons such as birds with extended wings (Fig. 5), dogs or coyotes, snakes (Fig. 6; and Pottery Mound murals, kiva 8, layer 14, Maxwell Museum slide), horned serpents (Hibben 1975: Fig. 34), and Shalako kachinas (Schaafsma 1992a: Fig. 132). There are occasional renditions of snakes and Shalakos with stars for faces (Figs. 7 and 8).

Ethnographically, all stars are said to have an anthropomorphic divine character (Parsons 1939: 182) and to bestow blessings or punishment (Parsons 1939: 54n, 133). All stars are associated with war, including falling stars and comets, and among the Taos and Tewas, stars were prayed to in war (Parsons 1939: 937). The Zuni scalp chief was said to pray to all the stars, "our night fathers, our night mothers" (Parsons 1939: 182). Elsewhere, however, Parsons (1929: 266; 1936: 181) specifies that it was Morning Star that was asked for help. The significant role played

Figure 3. Morning Star kachinas in black and red with scalloped kilts in Tompiro rock paintings. a) This pair was painted on a thin layer of plaster that subsequently flaked off destroying most of the head of the right-hand figure, as indicated by dashed lines. Both hold yucca whips painted in blue-green in their right hands, and bows and arrow bundles in the left. (see also Schaafsma 1994, Pl. 5). b) A tentative rendition of a third star kachina at the same site wearing a scalloped kilt or possibly scalloped apron over the kilt. (For alternative reconstruction see Baldwin 1992, Fig. 1).

by stars in war is confirmed by the many shields adorned with stars in Pueblo IV petroglyphs and kiva paintings. Further, the Pueblo shield was much more than just a protective device, but was believed to be animated by spirits (Parsons (1939: 197) and to possess magic power *derived from its design*, that was, in turn, transmitted to its owner (Wallis and Titiev (1945: 555). The Morning Star as an empowering element was especially popular on shields during Pueblo IV and has persisted into recent times (Wallis and Titiev 1945: Plates 14, 16, 20; Wright 1976).

The Role of Morning Star in Pueblo Cosmology

Do stars in the rock art represent stars in general or a specific star? Baldwin (1992: 5) argues that this four-pointed star in Tompiro rock art references the Morning Star and that simple dots represent true stars. Morning Star plays a prominent role in Pueblo cosmology, and is (was) important to both hunters and warriors. Coyote-like animals found in conjunction with stars in Pueblo IV rock art may symbolically denote this association with hunting (Young 1992: 84–85). Since it is the

Figure 4. Warrior with body shield embellished by large central star. He carries a quiver of arrows and holds a bow in his left hand and an arrow in the right. Southern Tewa petroglyph, Galisteo Basin, New Mexico.

Morning Star's connection with war, however, that is most explicit in Pueblo IV iconography, I will concentrate on that aspect.

While Evening Star was sometimes thought of as an old woman (Harrington 1916: 49; Parsons 1939: 181), the Pueblos associate the War Gods or Twins with Morning Star, the messenger and guard of the sun (Young 1992). At Zuni, Morning Star is regarded as the Sun's war chief, never far from the sun and going before him at dawn (Stevenson 1904: 27). In times of war, Morning Star was addressed as "Dark Star Man" by the Tewa (Parsons 1939: 181), an appellation that may underlie the concept of this figure as a warrior with darkened skin and account for the black-faced stars in Pueblo IV paintings. In Tompiro rock paintings the bodies of Morning Star kachinas are dark or blackened, and scalloped kilts are worn by some (Fig. 3 and Schaafsma 1994: Pl. 5).

Stars are impersonated in various kachina forms. Star-faced Morning Star kachinas in the rock art may carry a war club or a yucca whip in the right hand and a bow in the left (Schaafsma 1992a: 133), in the manner of other kachina warrior impersonators (Bunzel 1932: 872). Their occasional appearance as twins (Fig. 3)

Figure 5. Knife-Wing and taloned star petroglyphs, southern Tewa, Galisteo Basin.

Figure 6. Three stars, a snake and concentric circle that may represent a coiled snake. The star on the right has eyes and a large mouth full of teeth. Southern Tewa petroglyphs, Galisteo Basin.

suggests the dualism implicit in the aspects of the planet Venus (see below) as well as that of the Pueblo War Twins. A-turn-of-the-century representation of star kachinas by a Hopi artist (Fewkes 1903: XXVIII) depicts a pair of these figures known as Coto. Coto wears a mask with a single large star, is outfitted with lots of eagle feathers, and his paraphernalia is characteristically red in color (Fewkes 1903: 119). In one hand he carries a yucca whip, in the other a bell, the sound of the

Figure 7. Snake with star face and canine whose tail becomes an arrow, southern Tiwa, Petroglyph National Monument, Albuquerque.

Figure 8. Star-faced Shalako with crown of feathers. The significance of the large footprint is unknown. Southern Tewa, Galisteo Basin.

Morning Star rising (Parsons: 1940: 38). Sohu, a Hopi star kachina described by Wright (1977: 44) also carries a yucca whip and is listed as a guard. Wright (1973: 125) describes this kachina as wearing a fringed hunting shirt and a skirt of turkey feathers and notes that he never appears singly. He also describes red-stained warrior feathers on the tips of the stars on his headdress. A Chasing Star or Planet kachina is also pictured by Colton (1959: 55) who notes that this kachina appears in pairs.

That stars in Pueblo IV rock art may occur with the Shalako kachina was mentioned previously. The so-called Zuni Shalako is a giant kachina with a bird-like beak and conical body, that today appears before the winter solstice at Zuni, bringing seeds and blessings of abundance in all things to the pueblo. More rarely in the rock art, the Shalako itself is represented with a star face. One of the lesser known attributes of the Shalako is that it is also a warrior or guardian, and Shalakos are regarded as protectors as well as bringers of good luck (Ed Ladd, 1989; Wright 1985: 37). Two dancers are required to alternately sustain the giant effigy as it is "danced." These impersonators are called "elder brother" and "younger brother," appellations that incorporate the symbolism of the warrior twins. When unmasked, these dancers carry yucca whips, and the stone axe in the belt and warrior's cap of white buckskin are apparent. The latter is described as a "cover of thin clouds," (Bunzel 1932: 764) a reference to water and rain similar to that of scalps (see below).

A complex Southern Tewa petroglyph of a star-faced person embodying themes of war and fertility is shown with arrows emerging from his feathered headdress. In one hand he holds a plant, perhaps corn, and in the other an arrow that turns into a snake (Fig. 9). Another arrow, is suspended from his elbow, and a bird appears on his body. This figure has all the attributes of Sotuqnangu, the Hopi Heart-of-Sky God, whose symbol, or face (Brody 1991: Pl. 39), is the Morning Star (Wright 1977: 65). Sotuqnangu is a warrior deity in charge of summer thunderheads, lightning, and rains. He is described as "lightning shooting arrow-points from his fingertips" and as a scary war god "who kills and renders fertile" and who initiated the practice of scalping (Parsons 1939: 178, 181).

Figure 9. Star-faced deity with projectile points and Knife-wing with attributes of Sotuqnangu, a Hopi warrior god in charge of lightning and rain. Southern Tewa, Galisteo Basin.

Scalps, Stars, and Knife-Wing

A theme explicit in the fearsome aspect of Sotuqnangu is the linkage of Morning Star with scalping. Morning Star himself as a scalper appears in a Taos narrative in which he kills seven Redheads so that the Corn Girls can bring their heads into the pueblo on poles. Redheaded enemies are described in several Taos stories (Parsons 1939: 937). Implicit in this story is a linkage between Morning Star, scalps, rain (see following discussion), and corn. At Taos red pigment, possibly symbolic of blood, is offered to stars and scalps, and the latter are painted red inside (Parsons, 1939: 299). This relationship was dramatized in a Zuni ritual in which the act of scalping was reenacted during Bow Priesthood initiations "just as the Morning Star appeared" (Young 1992: 85).

Among the Hano, southern Tewa descendants at Hopi, and some northern Tewa on the Rio Grande, scalps themselves are called "Morning Star" (Parsons 1939: 181–182). According to Hibben (1975: 134), Acoma informants described the field of star faces in the Pottery Mound murals (Hibben 1975: Figs. 106–107) as "soul-faces," an interpretation that is reminiscent of the Aztec association of stars with dead warriors and sacrificial prisoners (Nicholson 1971: 426). If stars, in fact, are equated with the dead killed in war, then it is easy to make the extension that they are simultaneously in a general sense equal to scalps.

The association of Morning Star with death in general and scalps specifically is made clear in several other Pueblo tales. One describes the origin of Morning Star and Big Star, the latter in this case probably another planet, as occurring when a man was chased by his dead wife (Parsons 1926: 23). In the story of how Circle kachina and his wife became stars, Circle kachina tells his wife to meet him at the place where the scalps are dressed (Voth 1905: 65–71).

Although in another Taos story, stars are described as birds with bright breasts like those of hummingbirds (Parsons 1940: 170), the taloned and feathered stars in the rock art embody the attributes of Knife-Wing, Zuni and Acoma eagle-like beast god of the zenith and patron of scalping (Kelley 1964). This deity has a tail and wings of knives and tears scalps from the enemy with his claws (Bunzel 1933: 132; Parsons 1939: 187). A comparable figure at Hopi is Kwataka (Fewkes 1919: 67–68), or Kwatoko (Parsons 1939: 178). The conflation of the star/Knife-Wing images are explained in a Zuni story that when Knife-Wing was killed, his head and heart were thrown into the sky where they became the Morning and Evening Stars (Benedict 1935: 97). The Isleta assessment that "stars are mean" (Parsons 1932: 342), is in keeping with this image.

In rock art, the figure of a blunt-tailed, spread-eagle bird is frequently represented near stars and shield-bearers. Although these may resemble swallows, the ethnographic data, as well as their iconographic contexts suggest strongly that they are Knife-Wing (Marc Thompson, 1992). The previously described person with attributes of Sotuqnangu has such a bird on his chest (Schaafsma 1992b: Fig. 6). Images of this monster eagle on altars show him wearing a terraced cloud cap and with arms or wings that hang with eagle feathers, lightning lines, or flint knives. In addition to his knives, one of Knife-Wing's weapons is a lightning arrow.

War, scalping, rain, lightning, and the subsequent fertility and growth of corn are the involved and culminating themes of this star complex. An identity between lightning and corn in both Mexico and the Southwest is discussed by Taube (1986: 74). According to a Zuni myth, perhaps one of the most important results of Pueblo warfare was the obtaining of scalps "which bring good fellowship between the deceased enemy...and the A'shiwi [Zuni] and therefore much rain" (Stevenson 1904: 39). Much more than a war trophy, a scalp, once obtained and ritually brought into the pueblo is transformed into a rain-bringing fetish. A quote from a Zuni text regarding a Navajo scalp illustrates this point: "For, though in his life, the enemy was a worthless lot, now through the Corn Priest's rain prayers...he has become a rain person" (Bunzel 1933: 133). There are references in Zuni ritual poetry and other texts to the scalp as a "water-filled covering" (Bunzel 1932: 676; 1933: 131–134). Reference was made earlier to the Shalako dancer with his close-fitting warrior's cap of white buckskin, viewed symbolically as a cover of thin clouds. As a War God impersonator, the Shalako dancer by extension is affiliated with the Morning Star, scalping and ultimately rain, and as the masked Shalako he also brings seeds, including corn, to the pueblo. The two aspects of this complex figure embody a fundamental duality in Pueblo life.

The Question of Venus as Morning Star

With the exception of the clustered star faces at Pottery Mound previously discussed, where there are presumably too many of them to be a single supernatural, a relationship between the various manifestations of the feathered star in Pueblo IV and the Morning Star as a mythological and symbolic entity seems clear. The question remains as to whether the conflated bird/star image that incorporates the meanings of Morning Star, as described in Pueblo stories and other ethnographic

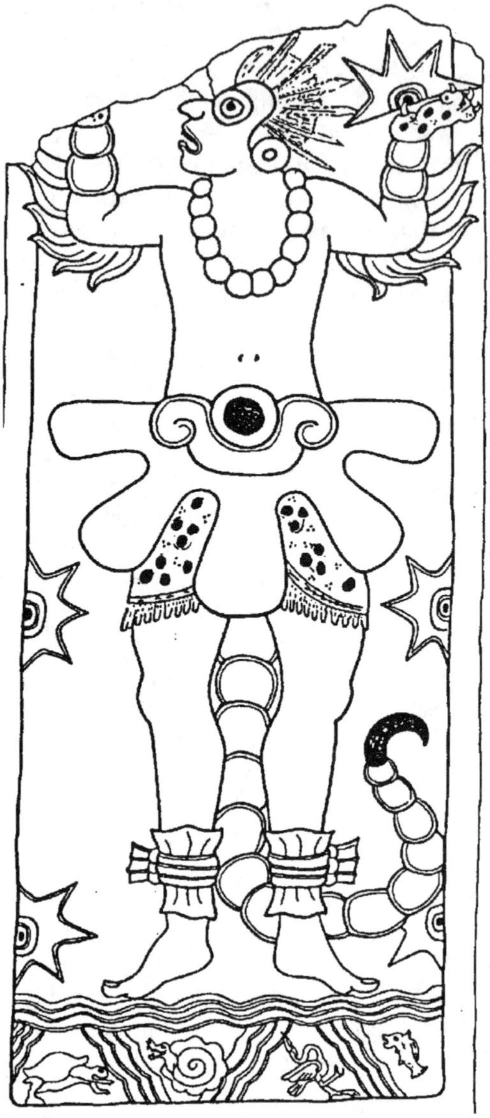

Figure 10. Venus Scorpion Man from Cacaxtla, Mexico, ca. A. D. 700–900). (After Carlson 1991: 8d, with permission).

sources, is also the planet Venus. The strength and complexity of the association of Venus specifically with hunting and war in Mesoamerica (Baird 1989; Carlson 1991; Nicholson 1971: 426; Freidel *et al.* 1993) and the symbolic parallels between the Morning Star complex in the Southwest and the Mesoamerican Venus cult raises this issue. A complicating factor may exist in that in spite of a predominance of Venus symbolism in Mesoamerica, Mayan battles were tied to the cycles of both Venus and Jupiter (Freidel *et al.* 1993: 296). Likewise in the Southwest, although the Morning Star and Evening Stars are observed by the Zuni for the purposes of timing ceremonial events (Bunzel 1932: 536; Parsons 1939: 182; Stevenson 1904: 130–131), Mars and Jupiter as well as Venus can have this role (Stevenson 1904:

Figure 11. Warrior from Chichen Itzá from a fresco, Temple of the Warriors. (After Carlson 1991: 12d, with permission).

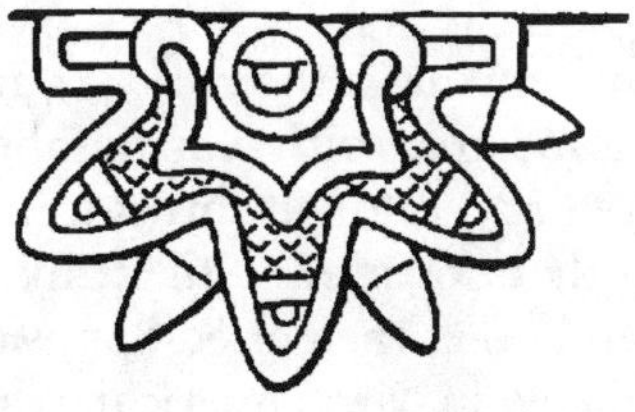

Figure 12. Venus glyph with knife-like elements from the Aztec Calendar Stone, early 16 century, Tenochtitlán. (After Carlson 1991: 1f2, with permission).

26; Young 1992: 76). Typically the Pueblo identity of Venus *in particular* as Morning and Evening Star is inconsistent or unclear. Harrington (1916: 49) claims that "The Tewa did not know planets other than the Morning Star and the Evening Star. The latter are now one planet, now another, but they did not know it." In other words, any bright planet, perhaps Jupiter, could function as the Evening or Morning Star, especially in the absence of Venus. Venus, however, would often qualify for the brightest star seen in the morning, a male divinity, also known as "big star" (Harrington 1916: 49). Only Venus, as both Morning and Evening Star, due to its proximity to the sun embodies the concept of a guard of the Sun and has the aspect of twins, hence the War Gods. It was earlier noted that paired stars occur in the rock art and that star kachinas are also paired. Reagan (1917: 46–47) reports that the Morning and Evening stars were created as two brothers, who are now the sun's guards. Thus it would seem that the basic metaphors ascribed to the Morning Star would have derived originally from observations of Venus itself.

There are additional Pueblo iconographic elements associated with stars with Mexican analogs linked with Venus. Among these is the strong association between stars and snakes, or mythological serpents, a relationship that in the Mexican case has well established Venus connotations associated with Tlahuizcalpantecuhtli, or Quetzalcoatl as Venus. Star-faced serpents in Pueblo rock art unify the star/serpent concept and the twinned aspect of some of these figures (Hibben 1975: Fig. l05) would seem to reinforce the Venus association. Another tentative parallel may exist between the scalloped kilts or aprons of some star kachinas (Fig. 3) and the Venus glyph apron at Cacaxtla and Venus warriors from Chichen Itzá (Figs. 10, 11). Mexican Venus glyphs with knives (Fig. 12) are symbolically reminiscent of the star/Knife-Wing conflation described here. Unfortunately lack of space precludes an investigation of these tantalizing parallels at this time.

In conclusion, it is possible that iconography and a symbol system regarding Venus originating in Mesoamerica would have been borrowed but simplified among the Pueblos. There are no indications to date that the Pueblos followed the Morning Star/Evening Star cycle of Venus, much less schedule events according to any planetary calendar. Although Young (1992) states that Morning Star and Evening Star served as heralds for various ceremonial and agricultural activities, this seems to have been on the basis of nightly appearance, and any bright planet could have served the same purpose.

Summary and Conclusions

The Pueblo IV star as an element in the rock art and kiva murals is a multivalent symbol predominantly associated with war. Symbolic of the Morning Star and the War Twins, in its feathered and taloned forms it incorporates aspects of Knife-Wing, Beast God of the Zenith associated with scalping. Via the symbolic aspects of the scalp, Pueblo warfare embodies the paired opposites of killing (bloodshed) and fertility, a theme closely paralleled in Mesoamerican war ideology and sacrifice. The data suggest that a desired outcome of Pueblo warfare may have been acquisition of scalps as rain fetishes especially valuable for the growth of corn. Venus as the observed Morning Star is loosely implicated.

Acknowledgments

This paper is an expansion of my presentation at the Oxford V Conference explaining the significance of the conference logo. I would like to thank Rolf Sinclair for inviting me to participate and Marc Thompson for his discussions on Knife-Wing. I am also grateful to John Carlson for his permission to use his illustrations from Cacaxtla, Chichen Itzá, and Tenochtitlán.

References

Baird, Ellen T. 1989, Star and Wars at Cacaxtla. *Mesoamerica After the Decline of Teotihuacan, A. D. 700–900.* Edited by Richard A. Diehl and Janet Catherine Berlo, pp. 105–122. Dumbarton Oaks Research Library and Collection, Washington, D. C., 1989.

Baldwin, Stuart. 1992, Evidence for a Tompiro Morning Star Kachina. *The Artifact* 30(4): 1–14. El Paso Archaeological Society, El Paso.

Benedict, Ruth. 1935, *Zuni Mythology*, 2 vols. Columbia University Contributions to Anthropology 21. New York.

Brody, J. J. 1991, *Anasazi and Pueblo Painting.* A School of American Research Book. University of New Mexico Press, Albuquerque.

Bunzel, Ruth L. 1932, Zuñi Ritual Poetry, Zuñi Katcinas: An Analytical Study. *47th Annual Report of the Bureau of American Ethnology, 1929–1930.* Pp. 612–1086. Washington.

———. 1933, Zuni Texts. *American Ethnological Society Publications* 15, G. E. Stechert and Co. New York.

Carlson, John B. 1991, Venus-regulated Warfare and Ritual Sacrifice in Mesoamerica: Teotihuacan and the Cacaxtla "Star Wars" Connection. *Center for Archaeoastronomy Technical Publications No. 7.* College Park, Maryland.

Colton, Harold S. 1959, *Hopi Kachina Dolls with a Key to their Identification.* University of New Mexico Press, Albuquerque.

Fewkes, Jesse Walter. 1903, Hopi Katcinas Drawn by Native Artists. *Twenty-First Annual Report of the Bureau of American Ethnology, 1899–1900.* Washington, D. C.

———. 1919, Designs on Prehistoric Hopi Pottery. *33rd Annual Report of the Bureau of American Ethnology, 1911–12*, pp. 207–284. Washington. (Dover edition, 1973, New York).

Freidel, David, Linda Schele and Joy Parker. 1993, *Maya Cosmos: Three Thousand Years on the Shaman's Path.* Quill, William Morrow, New York

Harrington, John Peabody. 1916, The Ethnogeography of the Tewa Indians. *Twenty-Ninth Annual Report of the Bureau of American Ethnology, 1907–1908*, Washington, D. C.

Hibben, Frank C. 1975, *Kiva Art of the Anasazi at Pottery Mound.* KC Publications, Law Vegas.

Hieb, Louis A. 1994, The Meaning of *Katsina*: Toward a Cultural Definition of "Person" in Hopi Religion. *Kachinas in the Pueblo World*, edited by Polly Schaafsma, pp. 23–34. University of New Mexico Press, Albuquerque.

Kelley, David H. 1964, Knife-Wing and other Man-Eating Birds. *35th Congreso Internacional de Americanistas: Actas y Memorias* 1, pp. 589–590. Mexico.

Kidder, A. V. 1932, *The Artifacts of Pecos* Papers of the Southwestern Expedition 6. Published for Phillips Academy by Yale University Press, New Haven.

Ladd, Ed. 1989 (Personal communication).

Nicholson, Henry B. 1971, Religion in Pre-Hispanic Central Mexico. *Archaeology of Northern Mesoamerica*, Part 1, Gordon F. Ekholm and Ignacio Bernal, volume editors. Handbook of Middle American Indians, Vol. 10. University of Texas Press, Austin.

Parsons, Elsie Clews. 1926, Tewa Tales. *Memoirs of the American Folk-Lore Society* 19. New York.

———. 1932, Isleta, New Mexico. Pp. 193–446 in *47th Annual Report of the Bureau of American Ethnology, 1929–1930.* Washington.

———. 1939, *Pueblo Indian Religion*, 2 vols. University of Chicago Press, Chicago.

———. 1940, *Taos Tales.* American Folk-Lore Society. J. J. Augustin, N. Y.

Reagan, Albert B. 1917, The Jemez Indians. *El Palacio* 4(2): 25–72.

Schaafsma, Polly. 1992a, *Rock Art in New Mexico*. 2nd edition. Museum of New Mexico Press, Santa Fe.

———. 1992b, Imagery and Magic: Petroglyphs at Comanche Gap, Galisteo Basin, New Mexico. In *Archaeology, Art, and Anthropology: Papers in Honor of J. J. Brody.* The Archaeological Society of New Mexico No. 18, edited by Meliha S. Duran and David T. Kirkpatrick, pp. 157–74. Albuquerque Archaeological Society. Albuquerque.

Schaafsma, Polly, editor. 1994, *Kachinas in the Pueblo World,* University of New Mexico Press, Albuquerque.

Smith, Watson. 1952, Kiva Mural Decorations at Awatovi and Kawaika-a. *Papers of the Peabody Museum of American Archaeology and Ethnology,* 37. Harvard University, Cambridge.

Stevenson, Matilda Cox. 1904, The Zuñi Indians: Their Mythology, Esoteric Fraternities, and Ceremonies. *Twenty-Third Annual Report of the Bureau of American Ethnology, 1901–1902*, Washington, D. C.

Taube, Karl A. 1986, The Teotihuacan Cave of Origin. *Res: Anthropology and Aesthetics* 12, pp. 51–82.

Thompson, Marc. 1992 (Personal communication).

Troike, Nancy P. 1981, The Interpretation of Postures and Gestures in the Mixtec Codices. *The Art and Iconography of Late Post-Classic Central Mexico*, Elizabeth Hill Boone, editor, pp. 175–206. Dumbarton Oaks, Washington, D. C.

Voth, Henry R. 1905, Traditions of the Hopi. *Field Museum of Natural History Publication 96, Anthropological Series 8*, Chicago.

Wallis, Wilson D. and Mischa Titiev. 1945, Hopi Notes from Chimopavy. *Papers of the Michigan Academy of Science, Arts, and Letters* 30, pp. 523–555. University of Michigan Press, Ann Arbor.

Wright, Barton. 1973, *Kachinas: A Hopi Artist's Documentary*. Northland Publishing with the Heard Museum. Flagstaff.

———. 1976, *Pueblo Shields: From the Fred Harvey Fine Arts Collection.* Heard Museum, Phoenix.

———. 1977, *Hopi Kachinas: the Complete Guide to Collecting Kachina Dolls.* Northland Press, Flagstaff.

———. 1985, *Kachinas of the Zuni.* Northland Press in cooperation with the Southwest Museum, Flagstaff.

Young, M. Jane. 1992, Morning Star, Evening Star: Zuni Traditional Stories. In *Earth and Sky: Visions of the Cosmos in Native American Folklore*, Ray A. Williamson and Claire R. Farrer, editors, pp. 75–109. University of New Mexico Press, Albuquerque.

17

Evidence for Early Navajo Horizon Astronomy in Chaco Canyon

Carol W. Ambruster
Department of Astronomy
Villanova University
Villanova, PA

Tony Hull
Rock Art Archive
Institute of Archaeology
University of California at Los Angeles
Los Angeles, CA

A site in eastern Chaco Canyon, New Mexico, provides the first comprehensive evidence that the Navajo once marked the solstices and equinoxes by means of horizon astronomy. Three large boulders, containing significant amounts of sacred rock art dating to the 18th century, indicate the place from which to observe the solstice or equinox sunrise at the base of a cliff on the distant horizon. The three cliffs involved are among the few clearly discernible features on a gently sloping horizon. The winter solstice event, however, is extraordinary: the Sun rises in a notch formed by a rock-art-covered foreground boulder and a 1 km-distant cliff on the horizon. By a remarkable coincidence, the angle of the sloped ridge of the boulder is the same as the angle of the rising Sun's path in the sky: therefore the Sun appears to rise along the ridge of the boulder over the course of 2–3 hours. Both the tight constraints on the location of the observer and the rock art itself, which contains drilled versions of the traditional Navajo constellations for November and December, argue that the Navajo were aware of the significance of this boulder. The published journal from the Simpson expedition of 1849 contains a drawing of this same winter solstice boulder and suggests that several features were added after 1849. If so, a tradition involving sacred rock art and horizon observations may have continued here for over 100 years.

* * *

Introduction

The fact that the Navajo were deeply aware of the sky is proven by their elaborate set of traditional (i.e., non-Puebloan) constellations (Haile 1947; Griffin-Pierce 1992). On the other hand, there has been little or no ethnographic evidence that the Navajo practiced horizon astronomy, i.e., the use of rising or setting positions of the Sun on the horizon to mark a yearly calendar. This is somewhat surprising given the Navajo's extensive contact with Pueblo peoples dating back to before the Pueblo revolt of 1680, and the well-known fact that horizon calendars have long been utilized among the Pueblos and their ancestors, the Anasazi (e.g., Williamson 1984; Zeilik 1985).

In this paper, we present evidence from an ongoing survey of site 29SJ1655 and vicinity in Chaco Canyon, NM, that challenges the apparent absence of horizon astronomy among the Navajo.

The Site

Background

A remarkably complex panel of Navajo rock art at 29SJ1655 is often referred to as the 'Shabikeshchee Sun Shield' (Figure 1). There are no clear ethnographic reasons for the term 'Sun Shield'; in fact, Ramona Begay-Martinez, a Navajo Park employee, whose ancestors were displaced from Chaco earlier this century, told us that the panel seemed to her more reminiscent of a sand painting (private communication, July 1996).

There are two suggestions in the literature that the 'Shabikeshchee Sun Shield' area may, in fact, have a cultural association with the Sun. The first is Ellis' (1975) report that an observer standing in front of the 'Sun Shield' at Equinox sees the Sun rise at the base of a cliff on the eastern horizon, at which time the Sun's rays fall on the 'Sun Shield'. The second, and perhaps most tantalizing, is that 'Shabikeshchee' is translated variously as 'Sun picture place' (Roberts 1929:1); 'Sun's path across a petroglyph' (Van Valkenburgh 1941, 33); or 'Sun is engraved on it' (Young & Morgan 1987:691). (If, however, the correct Navajo word is 'Tsebikeshchee', the translation is simply "Graven Rock").

The Site

Site 29SJ1655 and vicinity consists of a field of scattered, naturally placed, large and small boulders at the base of a talus slope up against the south wall of eastern Chaco Canyon.

The surface structural remains in the immediate vicinity of the 'Sun Shield' are overwhelmingly Navajo. Although there is archaeological documentation of Basket Maker III through Pueblo II habitation, these horizons, with minor exceptions, are either on top of Chacra Mesa or underground (Dabney Ford 1994, private communication). The main surface evidence of Anasazi occupation is scattered greyware and black-on-white potsherds; lithics are also occasionally found, but cultural attribution is more difficult.

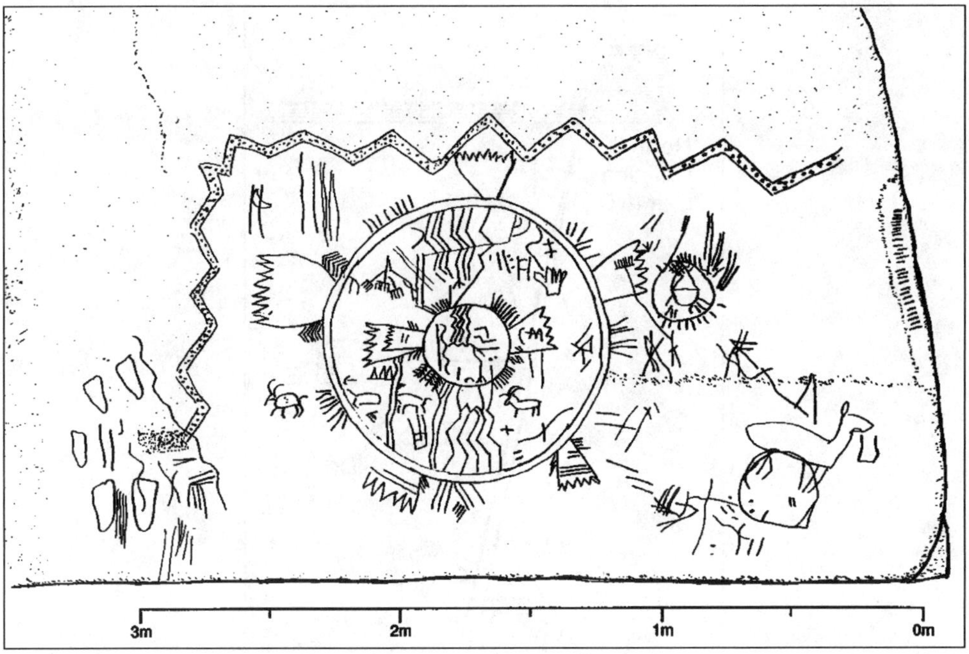

Figure 1. Drawing of the Shabikeshchee 'Sun Shield' (Panel E1), showing many of the fainter features that are only visible under optimum light conditions. Almost all possible rock art techniques are demonstrated on this panel: incision, abrasion, pecking, painting and scratching. The outer circle is made by shallow abrasion 5 cm in width; its (outside) diameter measures 103 cm both horizontally and vertically. The inner circle is slightly elliptical and is defined by an abraded line 1.3 cm wide; the 34 cm major axis is oriented roughly horizontally. Around the outer circle is an abraded, then pecked, zigzag. To the left of the shield are 5 scratched moccasin outlines.

Navajo cultural remains include several hogan foundations, two or three Dinetah gray sherds, one Gobernador polychrome sherd (D. Brugge, private communication 1998), a handful of large boulders with major rock art panels (see below), and a few other, generally smaller, boulders with less extensive rock art. While some of the rock art is pecked Anasazi, most is a mixture of deeply incised and lightly scratched Navajo; some panels contain both Anasazi and Navajo glyphs. The high percentage of early, high-quality, deeply incised, sacred Navajo rock art suggests the area was used for more than just habitation.

The archaeological record documents two windows of Navajo occupation of Chaco Canyon: 1750–1820, and 1880–1945 (Brugge 1988:98). Deeply incised ceremonial subjects (Yei figures, shield-like figures, etc., which are common in our area, are associated with the 1750–1820 period (Schaafsma 1992:37; Brugge 1988:80). However, there is evidence that sacred Navajo rock art in the early style was added after 1849 (see 4.2 below), suggesting that the site may have been culturally and ceremonially active for over 100 years.

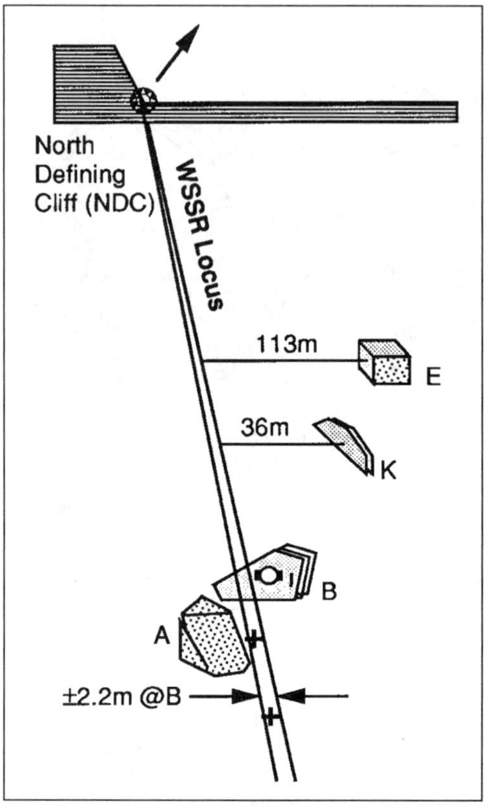

Figure 2. Schematic representation of the winter solstice sunrise geometry. Rock E is approximately 260 m from Rock B, while the North Defining Cliff is 1030 m distant from Rock B. To observe what we see, the observer must be in a narrow wedge which is 4.4 m wide at Rock B. The two authors stood at the '+' symbols. Not to scale.

Survey Methodology and Nomenclature

To avoid over-interpretation of the rock art panels during our survey, we named the boulders 'A', 'B', 'C', etc. in our field notes; for example, Rock or Panel E1 is the 'Sun Shield' (rock art panel E2 is on the south side of the same boulder). The letters were assigned initially with 'A' at the western end of the region, increasing to the east; but as additional boulders were included, either for survey purposes or because new rock art was discovered, the pure west-east order is partially lost. The schematic layout of the site is given in Figure 2.

The three rock art panels discussed in this paper ('B', 'E', and 'K') are on three different large sandstone boulders (several meters long, 1.5–4 m high) spread out over 260 m; the only human modification of the rock besides the glyphs themselves was a preparatory smoothing of some of the surface of each rock face.

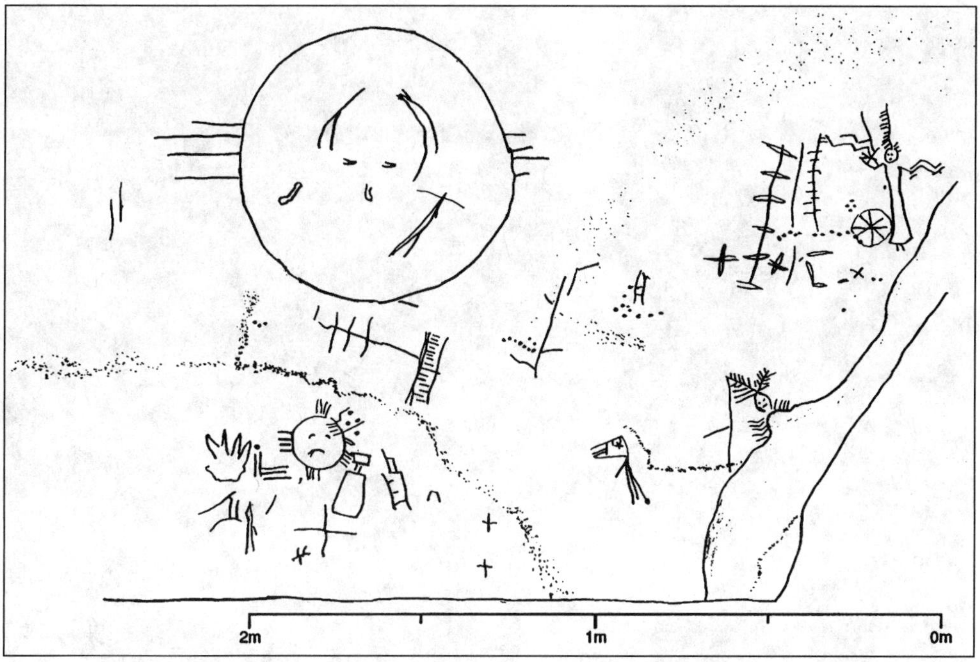

Figure 3. Scale drawing of the incised and drilled rock art on Rock B.

Sunrise Observations

Winter Solstice Sunrise

One face of Rock B contains a rich variety of incised rock art (Figure 3), including two shields, two Yei figures, a circle with spokes, a hand, and several drilled patterns, at least some of which seem to be constellations, as we discuss later.

We observed sunrise under clear skies from rocks 'A-B' on five mornings in December 1994 (December 16, 17, 19, 20, and 21 [winter solstice]). The maximum change in the solstitial sunrise point on the horizon over these five days, about 1/8 degree, is too small to be perceptible to the naked eye or to affect any of the arguments in this paper. Photographic sunrise sequences were taken from two locations near Rock A (Figure 2). The winter solstice sunrise sequence is shown in Figure 4.

As our earlier compass and clinometer measurements had predicted, the winter solstice Sun rose at the most conspicuous feature on the southeastern horizon, the approximate right angle formed where the 1 km distant north canyon wall descends to meet the flat valley floor (hereafter called 'North Defining Cliff' or 'NDC'). Furthermore, from both observers' positions, the base of the foreground Rock B, which slants up from the ground to the south at an approximately 50 degree angle, intersected the distant horizon at the base of the same NDC. As sunrise continued, the winter solstice Sun ascended the slanted ridge of Rock B (Figure 4), i.e., this unworked ridge happens to rise from the ground at the same angle as the path of the winter solstice Sun. About 5/8 of the way up Rock B, the Sun dips below this ridge, but can easily be brought back into view by small shifts of the head or camera positions. In other words, without moving his feet, an observer could watch the Sun

Figure 4. Winter solstice sunrise sequence. Stacked images show the sunrise as seen by an observer from the western-most observing position indicated in Fig. 2. Time increases downward from the top photograph. In the left foreground is the edge of Rock A. On the right, intercepting the base of the NDC, is the sloping edge of Rock B.

ride up the entire ridge of Rock B. Alternately, since it is neither particularly comfortable nor safe to gaze at the Sun, the observer could easily keep most or all of the Sun's disk barely below the ridge of Rock B, adjusting his position occasionally to check on its progress. In either case, the necessary small shifts in eye point would not detract from the drama or importance of the phenomenon to a Sun watcher.

There is a distinct practical advantage to the prolonged journey of the Sun up the ridge of Rock B around winter solstice: in the event of clouds at the horizon, or early morning clouds which later clear, the arrival of the solstice can be confirmed up to three hours after sunrise (depending on the observer's distance from Rock B), as long as the Sun can be seen along the ridge of Rock B.

Finally, the winter solstice can be anticipated by noting when the Sun rises at the top of the NDC (about 22 days before winter solstice), as well as at intermediate positions between the top and bottom of the cliff.

Figure 5. Vernal equinox sunrise. Standing in front of Boulder E, the 1997 vernal equinox sunrise (20 March) occurred at the base of one of the few cliffs at the eastern horizon.

Equinox Sunrise

Sunrise observations from Rock E at the 20 March 1997 Vernal Equinox confirmed Ellis' (1975) report (Figure 5). The eastern horizon as seen from 29SJ1655 is a fairly level mesa top with few distinct features; none are as conspicuous as the NDC. The Equinox Sun rises at the base of one of the few cliffs on the horizon. The cliff is 1030 m from Rock E (Table 1), sufficiently distant that the observer can stand anywhere in front of E and still see the Sun rise at the base of the distant cliff. Rock E, therefore, marks the place from which to observe, or verify the arrival of, astronomical Equinox.

Table 1

Object	Distance along Wedge from NDC (Figure 2)	Width of Wedge	Perpendicular Distance to Wedge from	Az NDC minus Az WSSR
Panel E	796 m	±1.7 m	112.8 m	−8.1°
Panel K	909 m	±2.0 m	36.2 m	−2.3°
Panel B	1030 m	±2.2 m	0.0 m	0.0°
NE Corner of Wijiji Pueblo	3070 m	±6.7 m	24.5 m	−0.5°
N Canyon Wall beyond Wijiji	3161 m	±6.9 m	—	—

There is little ethnographic evidence that either astronomical Equinox, or the mid point in time between solstices, had calendric significance for the Navajo, and it is possible that the serendipitous location of Rock E merely provided a convenient calendric marker for Spring and Fall. On the other hand, there is very little ethnographic information on 18th c. Navajo culture, particularly in remote areas like Chaco Canyon. And, in the present context where both solstices are marked as well, the simplest interpretation of the facts appears to be that a horizon calendar was implemented at Shabikeshchee by early Pueblo-influenced Navajo. Since the Navajo took advantage of the position of a naturally occurring boulder, the question of whether astronomical Equinox or the mid-point in time between solstices was important probably cannot be answered at the present time.

Finally, we find that the entire rock face containing the 'Sun Shield' is uniformly sunlit at dawn, not only at the Equinox but also at both solstices and throughout the year. Thus the 'Sun Shield' itself is never involved in any sunrise calendrical light-shadow displays.

Summer Solstice Sunrise

Standing in front of Rock K, an observer sees the summer solstice Sun rise at the base of another of the few cliffs on the eastern mesa top horizon (Figure 7). Because this cliff is about 20% closer than the Equinox cliff (800 vs. 1030 m), the constraints on the observer's position are a little tighter, barely extending the length of Rock K.

Of the three boulders ('B', 'E', and 'K'), carved plants only appear on Rock 'K', where they are the dominant motif. This certainly is suggestive of summer associations, including fertility; the carved plants may also symbolize the summer form of Changing Woman (Haile 1947:24–26). Interestingly, the carved plants are not similar to known renditions of corn or other plants known from Navajo ceremonials either in 18th c. Dinetah or in more contemporary sandpaintings (H. Rogers, private communication, 1997; D. Brugge, private communication, 1998).

The confirming sunrise observations were made on 19 and 20 June 1997, one and two days before summer solstice. Since the Sun rises at the same place on the horizon, to naked-eye precision, for several days on either side of the actual solstice, our observations were equivalent to viewing on the solstice itself.

Discussion

What is remarkable about the Shabikeshchee site is that the fortuitous location of three naturally placed, large boulders within 260 m of one another on the Canyon floor provide views of sunrise against distinct horizon features for all solstices and equinoxes. There are at least half a dozen other large boulders in the survey region that have smooth faces well suited for rock art, but which have not been inscribed. But from these boulders there are no natural topographic features on the horizon that would mark Equinox or solstice sunrise positions. That is, in this limited area, the only three canyon-floor boulders having significant symbolic or sacred rock art are the only ones that view horizon features clearly marking the solstice or Equinox sunrise positions.

There is further compelling evidence that the winter solstice sunrise event at Rock B was intentional: 1) there is a narrow wedge within which an observer can

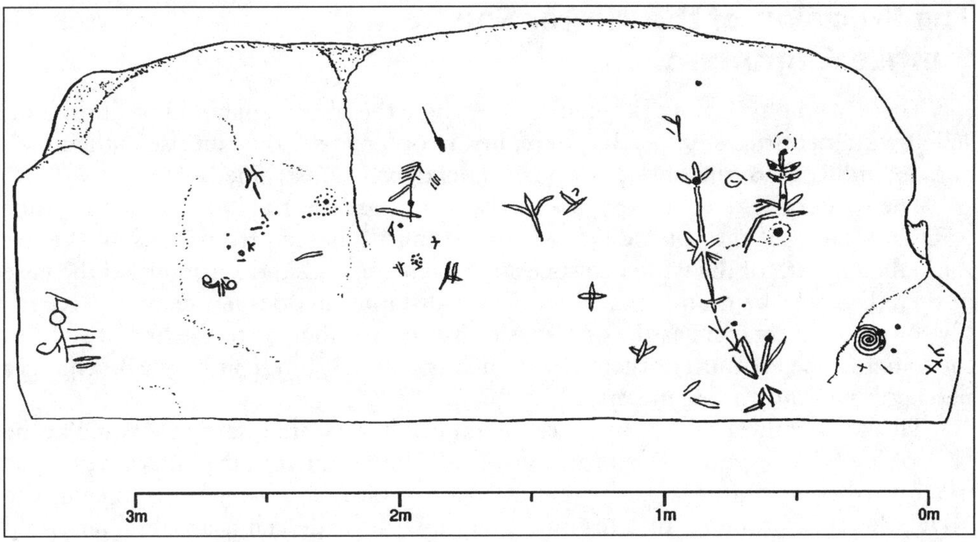

Figure 6. Scale drawing of the incised and drilled rock art on Rock K. Note incised plants, deeply drilled holes, and an eroded Anasazi spiral at the lower right.

Figure 7. Summer solstice sunrise. Standing in front of Rock K, the 1997 summer solstice sunrise occurred precisely at the base of another of the few distinct features on the eastern horizon.

watch the full event; and 2) rock art, which is often frustratingly hard to interpret, in this case provides strong references to the Sun and to winter months.

The Precision of the Winter Solstice Sunrise Alignment

There is no physical marker indicating where the observer should position himself at winter solstice sunrise. But, in reality, in order to see the Sun rise in the notch and ascend Rock B, the observer's position is tightly constrained.

The winter solstice Sun appears to rise at the base of the North Defining Cliff (NDC) because of the coincidence of two azimuths as seen from Rock B (Figure 2): 1) the azimuth of the winter solstice Sun at sunrise, and 2) the azimuth of the base of the NDC. The azimuth of the 1.5×10^8 km distant Sun does not change perceptibly for many kilometers as the observer moves perpendicular to the line of sight to the Sun, but the azimuth of the NDC, which is only 1030m from Rock B,[1] changes perceptibly in only a few meters.

There is a range of 0.25° in sunrise azimuth that would qualify as 'rising at the base of the NDC' (Figure 8). At the latitude of Chaco Canyon, the Sun moves 0.25° northward of its winter solstice azimuth in 7 days. At the ~1 km distance of the NDC, this 0.25° range translates into a band 4.4 m wide. Thus an observer could see the winter solstice Sun rise at 'the base of the NDC' only from within a wedge 4.4 m (0.25°) wide at Rock B.

However, the observer's actual position is further constrained by involving Rock B as the right side of a notch. Since Rock B is only a few meters away from the observer, he needs to move only a few centimeters perpendicular to the southeast sunrise direction, i.e., roughly south (toward Rock B) or north (away from Rock B), before Rock B either: 1) occults the rising Sun, or 2) appears against the flat horizon to the south of the NDC. In either case, the effect of the 'notch' is lost.

As shown in Figure 2 and Table 1, neither Rocks E nor K fall within this wedge, i.e., an observer standing in front of them would not see the Sun rise at the base of the NDC on winter solstice.

Although the wedge within which an observer can see the winter solstice Sun rise in the notch is very narrow, it is also long: it eventually terminates across the canyon, just east of Wijiji ruin about a kilometer away. There are few, if any, naturally occurring objects of any distinction in this wedge, and no man-made constructions. It is not particularly likely that a boulder suitable for rock art would be located within it, let alone one whose ridge rises at the same ~50° angle as the winter solstice Sun. That Rock B was recognized, and made into a rock art site of apparent sacred significance, is not surprising.

Astronomical Elements of the Rock Art on Rock B

Schaafsma (1992: 38) has noted that there is little ethnographic evidence for Navajo horizon astronomy. However, it is known that the Navajo did use the sky

---

1. We determined the distance between Rock B and the NDC to be 1030m by tying in local features, which we measured with compass, clinometer, and a metric surveyor's tape, to identifiable features on the USGS topo map (such as the NDC). We then carefully scaled the distance of the NDC off the topo map. We estimate a 3-σ error of ±50m, which is too small to affect any of our results. Distances to various features of interest obtained in this way are given in Table 1.

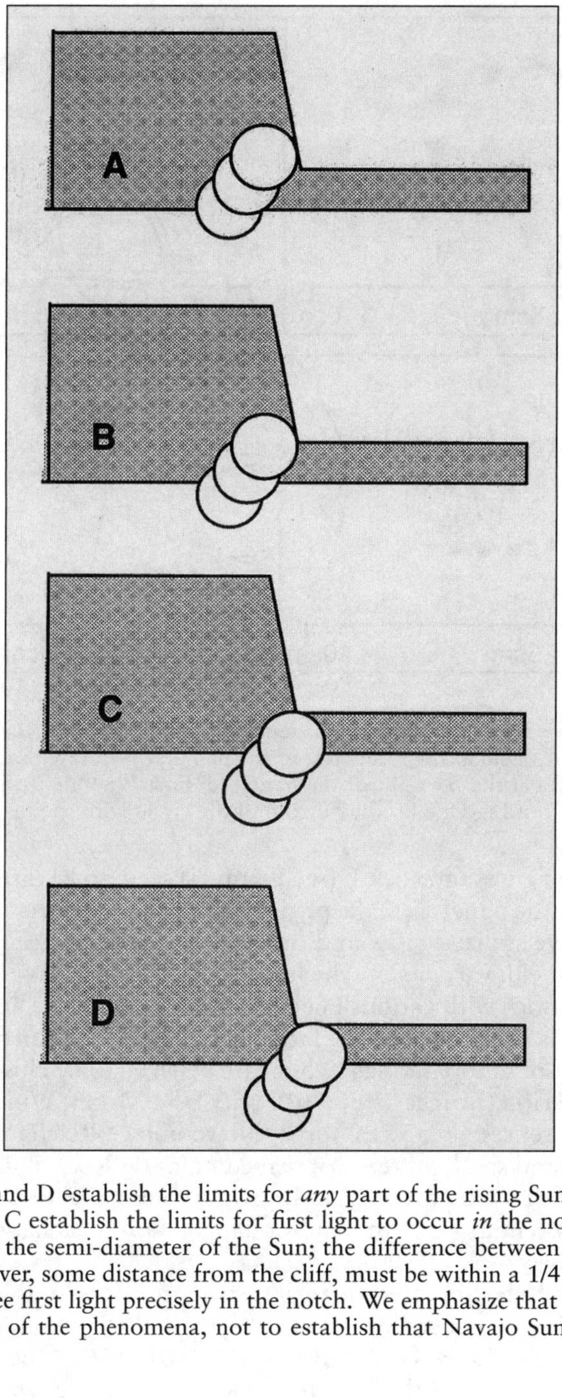

Figure 8. Cases A and D establish the limits for *any* part of the rising Sun's disk to touch the notch. Cases B and C establish the limits for first light to occur *in* the notch. The difference between B and C is the semi-diameter of the Sun; the difference between A and D is the full diameter. The observer, some distance from the cliff, must be within a 1/4% wedge (the Sun's semi-diameter) to see first light precisely in the notch. We emphasize that it is our purpose to define the precision of the phenomena, not to establish that Navajo Sun watchers required that precision.

in a variety of ways to set a yearly calendar: for example, most months have an associated constellation, one that rises heliacally in that month (Chamberlain 1983; O'Bryan 1956: 16), and there are documented cases of the use of the changing length of shadows and daily height of the Sun to determine time of year (Williamson 1983).

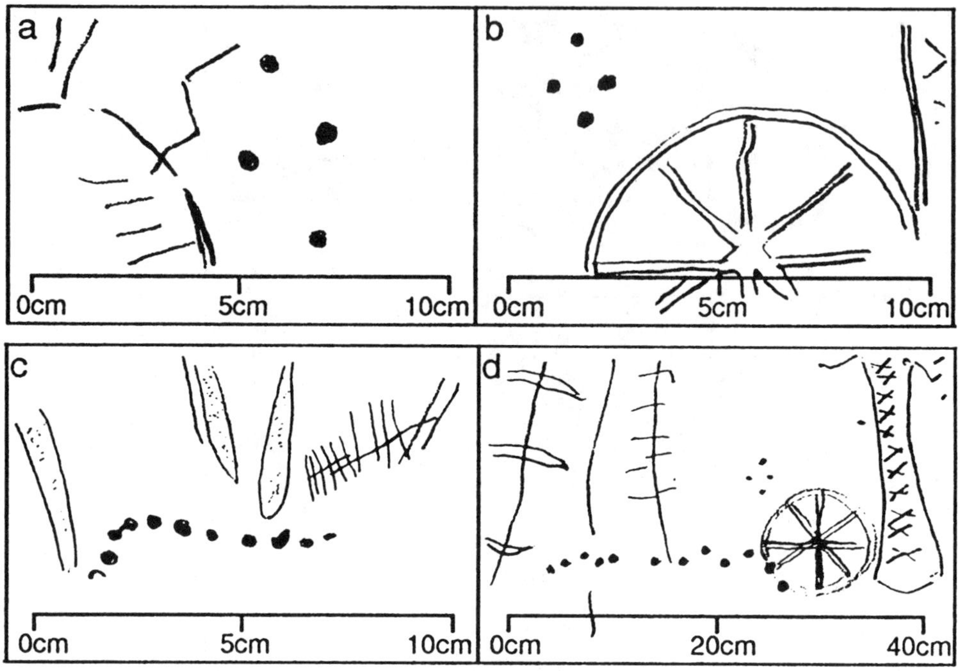

Figure 9. Drawings of probable drilled constellations on Rock B. a) Man with Legs Ajar (to the right of the small shield at the lower left of the panel); b) Man with Legs Ajar (just above the circle with spokes at the far right of the panel); c) First Big One (from the crevice on the side of Rock B); d) First Big One (at the bottom of the circle with spokes).

Several of the figures on Rock B (see Figure 3) seem to be directly related to the sky. On the left of the panel are 2 deeply incised circles with sets of lines to the 4 directions. Both have stylized faces and one has horns. They are reminiscent of the famous Zia Sun shield, variants of which the Navajo would have been familiar with from their associations with various Pueblo villages both before and after the Pueblo revolt of 1680. It is probable that the larger shield truly is a Sun shield or Sun symbol: in the context of sandpaintings (which are believed to be closely related to rock art of this early period (Brugge 1988:80)), only Sun, Moon, Winds, Earth, and Sky are allowed to be represented with horns (Newcomb & Reichard 1975).

There are several small patterns of drilled holes on Rock B, made with either a metal or a stone drill (Dabney Ford 1994, private communication). Among them are patterns matching renditions of the traditional Navajo constellations for November and December.

Figures 9a and 9b show two quadrilateral patterns that appear to match one of the 8 major Navajo constellations, Hastiin Sik'ai'i, or Man with Legs Ajar in Corvus (Chamberlain 1983; Griffin-Pierce 1992: 85, 166). One is just to the right of the small shield, and the other just to the left of the circle with spokes at the far right of the panel (Figure 3).

Man with Legs Ajar rises heliacally in November, and is the constellation the Navajo traditionally have associated with that month (Chamberlain 1983). It is also one of the more common Navajo constellations to appear in sandpaintings: Figure 10 shows an identified Man with Legs Ajar in a drawing of a Shooting Chant sand-

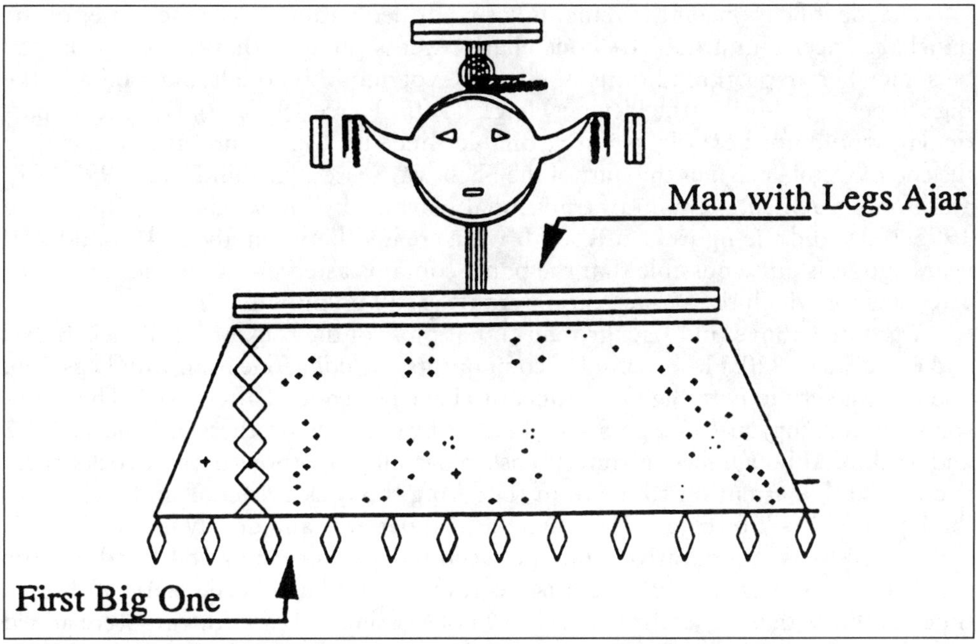

Figure 10. Man with Legs Ajar and First Big One are indicated in a drawing of a Shooting Chant sandpainting (Newcomb & Reichard 1975:58).

painting from Newcomb & Reichard (1975:58). It is possible that the presence of 'Man with Legs Ajar' indicates an anticipatory observation for the winter solstice.

Rock B appears to contain at least one, and perhaps two, renditions of another of the 8 major constellations, 'Atse'Etsoh, or First Big One, the front part of Scorpius (Chamberlain 1983; Griffin-Pierce 1992: 85,166). This is the traditional Navajo constellation for December, the month of its heliacal rise. The more probable First Big One, located in a crevice on one side of Rock B, is drawn in Figure 9c, and is also shown in a sandpainting context in Figure 10. Figure 9d shows the less certain First Big One on the front of the panel, tangent to the base of the circle with spokes (Figure 3). Most published versions of this constellation that we have seen, whether on ceremonial gourd rattles (e.g., Chamberlain 1983) or in sandpaintings, have a sharper angle at one end than does the version Figure 9d. In fact, the First Big One of Figure 10 is the straightest of any we found, and is closest to the pattern on the front of Rock B. On the other hand, there is a good deal of variation between versions of the same constellation done by different chanters, so it is quite possible that the long, scraggly line of stars in Figure 9d is indeed another First Big One. The repetition of both Hastiin Sik'ai'i and 'Atse'Etsoh may reflect the importance of multiples of two in Navajo culture (H. Rogers, private communication 1997).

It is interesting and perhaps significant that both Corvus and Scorpius are constellations of the far southern sky as seen from the latitude of Chaco Canyon, and both rise at about the same place on the horizon as the winter solstice Sun:

	Approximate declinations
'Man with Legs Ajar':	−16% to −23%
'First Big One':	−18% to −28%
'Winter Solstice Sun':	−23.5%

The identifications of the remaining two drilled patterns near the center of the panel are uncertain; if they are constellations, it is possible that one or both may be somewhat atypical renditions of either Revolving Male (Nahookos Bika'ii: the Big Dipper), or Dilyehe (the Pleiades) (e.g., Griffin-Pierce 1992:84). Dilyehe, in particular, would not be entirely unreasonable, since its appearance at twilight near the end of October signals the start of the Nightway season (Griffin-Pierce 1992:165). Since only fragments remain of traditional Navajo sky knowledge (Griffin-Pierce 1992:103), and the figures on Rock B were created between about 100 and 250 years ago, it is quite possible that the panel contains asterisms no longer in use, or asterisms for which the Western counterpart is still unknown.

We stated above (in "Background") that most of the rock art on Panels B, E1, and K predates 1820. However, some components, including one Man with Legs Ajar, and the less certain First Big One, appear to have been added after 1849. The Simpson journal (Simpson 1852) contains sketches by the Kern brothers of Panels E1, E2 and Rock B. Although there is some license in overall proportions of these rocks, there is considerable attention to detail in rendering the rock art icons and figures on these panels. Rock B (Figure 3), with its drilled patterns and deeply incised lines, is well suited for a comparative study between what was seen in 1849 and what is seen 147 years later in 1996. Because features would have been clearer in Kern's time, it is likely that any conspicuous icons not seen in the 1849 drawing were added later, even if the content and style are similar to that of an earlier period. We also note that while there are some features seen now and not then, there are very few features seen then and not now. The only conspicuous loss is the bottom half of the Yei at the lower right, which presumably fractured off in the past and is now lost.

If this panel has been added to, it suggests continuity in the production of sacred rock art from an earlier to a later habitation period, and the use of rock art as a continuing medium. Figure 11 depicts the panel on B as seen in 1996, and as seen in 1849 (Simpson 1852).

Summary

In Chaco Canyon, in the 18th and early 19th c. at least, the Navajo appear to have been familiar with horizon astronomy. In our survey area, only the three boulders marking observing positions for Equinox or solstice sunrises have significant and clearly sacred rock art panels. Although the Sun rises at the same type of horizon feature (the base of a cliff) as seen from each boulder, each boulder utilizes a different cliff, each 'works' for a different season, and each has distinctly different icons and artistic styles. Only Rock B actively participates in a sunrise; Rocks E1 and K merely mark where the observer should stand.

The comparison of Rock B in 1849 and 1996 implies that horizon astronomy was practiced near Shabikeshchee for at least 100 years, and that changing trends in secular rock art styles did not affect post-Fort Sumner additions to the panels, that is, an older tradition was maintained, at least for sacred rock art.

There is only one other report, to our knowledge, of Navajo horizon astronomy: the winter solstice sunrise event at the Wijiji petroglyph site (Williamson 1983; Zeilik & Elston 1983). Here, also, are found drilled constellations of Navajo origin, and other sacred Navajo iconography. It is interesting that these two sites are al-

1849

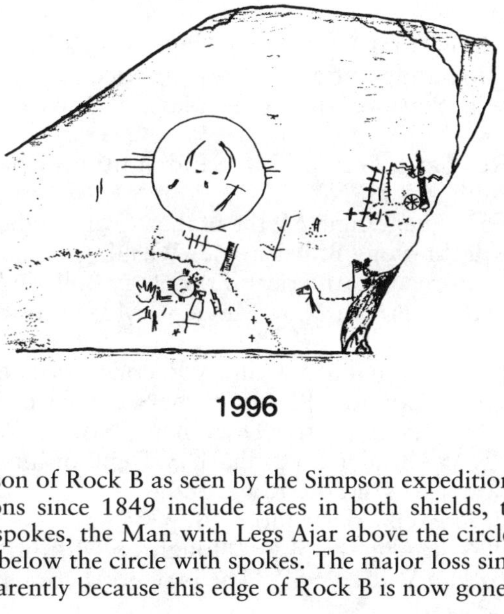

1996

Figure 11. A comparison of Rock B as seen by the Simpson expedition in 1849 (top), and in 1996 (below). Additions since 1849 include faces in both shields, the horns in the large shield, the circle with spokes, the Man with Legs Ajar above the circle with spokes, and the possible First Big One below the circle with spokes. The major loss since 1849 is most of the bottom Yei figure, apparently because this edge of Rock B is now gone.

most directly across the canyon from each other, although the Wijiji site is not actually visible from our survey area. We are currently investigating possible connections between the two sites.

Acknowledgments

We thank the National Park Service for research permits, and also Chaco archaeologist Dabney Ford for providing encouragement and access to data, and for numerous discussions from her comprehensive knowledge of the Canyon. We are indebted to Ray Williamson who suggested this research project to one of us (CWA), to Von Del Chamberlain and Ed Krupp for discussion and constructive comments, and to archaeologist E. R. Jewell who has listened to our approach, visited the site, and provided many helpful and constructive comments. We also thank Bill McCarty and Stacey Clemensen who assisted with the figures in this paper. Several references were made available courtesy of the Braun Research Library, Southwest Museum.

References

Brugge, D.M. 1988, 'The Historical Archaeology of Chaco Canyon', in 'Archaeological Surveys of Chaco Canyon, NM', A.C. Hayes, D.M. Brugge, & W.J. Judge, U. New Mexico Press: Albuquerque (reprint of 1981 work).

Chamberlain, V.D. 1983, Archaeoastronomy 6, p. 48.

Ellis, F.H. 1975, 'A Thousand Years of the Pueblo Sun-Moon-Star Calendar', in 'Archaeoastronomy in Pre-Columbian America', Ed. A.F. Aveni, Austin: University of Texas Press.

Griffin-Pierce, T. 1992, 'Earth is My Mother, Sky is My Father', Space, Time and Astronomy in Navajo Sandpainting, U. New Mexico Press, Albuquerque.

Haile, B., O.F.M. 1947, 'Starlore Among the Navajo', Museum of Ceremonial Art, Santa Fe, NM.

Newcomb, F.J., & Reichard, G.A. 1975, 'Sandpaintings of the Navajo Shooting Chant', Dover Publications, NY

O'Bryan, A. 1956, 'The Dine: Origin Myths of the Navajo Indians', Smithsonian Bureau of American Ethnology Bulletin 163, Washington, DC.

Roberts, F.H.H. 1925, Bureau of American Ethnology Bulletin 92, Washington, DC.

Schaafsma, P. 1992, 'Rock Art in New Mexico', Museum of New Mexico Press, Santa Fe, NM.

Simpson, James H. 1852, Journal of a Military Reconnaissance from Santa Fe, New Mexico to the Navajo Country. Report of the Secretary of War to the 31st Congress, 1st Session, Senate Executive Document, No. 64, Washington, DC.

Van Valkenburgh, R.F. 1941, Dine Bikeyah, (U.S. Dept. of Interior, Office of Indian Affairs, Navajo Service: Window Rock, AZ).

Williamson, R.A. 1983, Archaeoastronomy 6, p. 59.

Williamson, R.A. 1984, 'Living the Sky', (Houghton Mifflin: Boston).

Young, R.W., and Morgan W., Sr. 1987, 'The Navajo Dictionary', (UNM Press, Albuquerque), p. 691.

Zeilik, M. 1985, Archaeoastronomy Supplement to the Journal for the History of Astronomy, S1 & S86.

Zeilik, M., & Elston, R. 1983, Archaeoastronomy 6, 66.

18

Tracking Stars in Dinétah: Astronomical Symbolism in Gobernador Phase Rock Art

Von Del Chamberlain
Salt Lake City, UT

Hugh C. Rogers
Salt Lake City, UT

Cosmological interests among Navajo ceremonial practitioners have been known to anthropologists for over one hundred years and astronomical symbolism in dry paintings and on ritual objects has been noted and recorded. Although Sun, Moon and star symbols have been noticed in rock art and a few publications have speculated about identification of constellations in rock art panels, only recently have convincing examples of star patterns been found. The Largo-Gobernador area of New Mexico is referred to by Navajos as "Dinétah," their traditional homeland. Pueblo people passed through these canyons, living with the Navajos for a time following the Pueblo revolts of the late 17th century. The rock art of this region, mostly of Navajo origin, but with apparent Pueblo influence, is unusually rich in astronomical symbolism.

In this paper we document numerous examples of Dinétah star symbols, star ceilings, constellation patterns, and other astronomical motifs with some comparisons to Navajo symbolism in other media.

* * *

Introduction

Astronomical symbolism in rock art is not uncommon. Indeed, rayed disks and other motifs resembling images of the Sun are prolific in rock art and crescents, the most likely identifying emblem for the Moon are sometimes noted as well. Star symbols vary from equal-armed crosses to asterisks and dots. Planet symbols would be expected to be the same as and not distinguishable from stars except when ethnographic information exists to suggest that certain motifs represent specific planets, such as Venus as an element of war in Pueblo and Aztec iconography (Schaafsma 1980:295–298). Symbols that could represent comets, meteors and fireballs are difficult, if not impossible, to identify in rock art, although possibilities exist. Finally, symbols for atmospheric phenomena are common rock art motifs.

The authors and others have reported on star ceilings at Navajo occupation sites (Britt 1975; Chamberlain 1974; 1978; 1989a; 1989b; Chamberlain and Schaafsma 1993, In Press; Grant 1978; Jett 1984; Schaafsma 1963; 1966; 1972; 1980). One of us has also reported finding star pattern depictions in Navajo rock art that can be substantiated by comparison with star patterns on dry paintings, gourd rattles and other artifacts (Chamberlain 1983). During the past few years additional examples of both of these astronomically-inspired features have been found in the Largo-Gobernador region of New Mexico. This is the area that most traditional Navajos accept as their ancestral homeland, the region they refer to as "Dinétah," interpreted to mean "among the Diné" (Diné is the name Navajo's use for themselves, "The People").

Dinétah contains some of the most sacred places of Navajo tradition, and it has numerous early Navajo archaeological sites as well as classic Navajo ceremonial imagery, Anasazi rock art and pictographs and petroglyphs with Pueblo IV influence (Copeland and Rogers 1996). The region (Figure 1) can be approximately outlined as running from the Colorado-New Mexico border on the north, going south to Chaco Canyon, and from the Animas River on the west going east to the Continental Divide (Copeland and Rogers 1996).

Archaeological evidence suggests that the Navajos might have been in the region by 1500 A.D., and Navajo tradition would put them there much earlier. It is also known that Pueblo people lived among them during the period of the Pueblo Revolts, 1680–1696. Apparently most Navajos stopped living in the area around 1750, probably moving south and west. Those interested can find details on the archaeology of Dinétah in *The Archaeology of Navajo Origins*, edited by Ronald H. Towner (1996).

Current Navajo tradition does not permit the manufacture of ceremonial imagery as rock art, especially not the rendering of sacred figures of the type that are abundant in Dinétah. It is clear, however, that placing such figures on rock surfaces was an important part of Navajo ceremonialism during the major Navajo occupation of Dinétah. In addition, a few scattered figures in Canyon de Chelly and other places are found depicting Navajo deities, known to them as yeis, but it seems that the practice of putting such images on the rocks stopped about the time the Diné left Dinétah. It is thought that the use of dry painting succeeded generation of sacred rock art. This same line of thinking suggests that the Dinétah rock art was probably produced and used in rituals that might have been similar to, yet predated, recent and contemporary Navajo ceremonies that typically use dry painting. This concept has considerable foundation, including the fact that the Dinétah rock art figures are strikingly similar, and sometimes even identical, to sand painting figures seen today (Olin 1984).

Both Dinétah rock art and present day sand paintings include symbols representing Sun, Moon and stars. Other sky symbols found in both forms are: rain and rainbows, lightning, clouds and possibly such things as eclipses, comets, fireballs and meteors. Although it is difficult, if not impossible, to identify them in the rock art, one can find examples in dry painting representing such things as sundogs. One very interesting form of sky symbolism is its placement directly on or as part of deities: stars on the mask of Black God, or on the clothing of an anthropomorphic

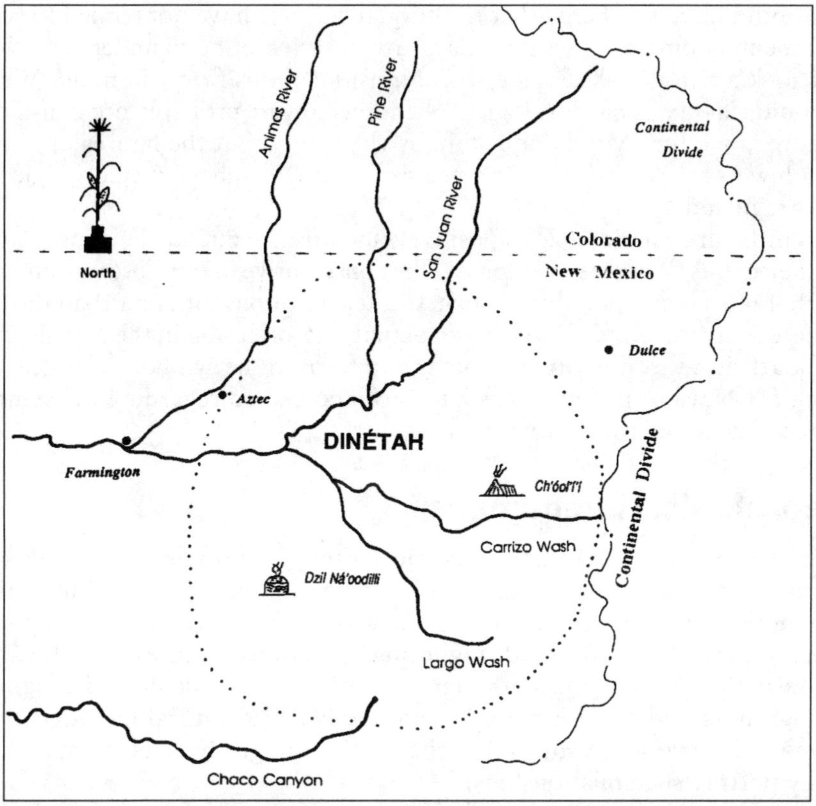

Figure 1. The Largo-Gobernador region referred to as Dinétah (within dotted area) is considered by most Navajo people to be their ancestral homeland. *Ch'óol'í'í* (Gobernador Knob) and *Dzil Ná'oodilli* (Huerfano Mesa) are two of the most sacred features of Dinétah.

figure; Sun or Moon as Holy People with elaborate head dresses artistically revealing the essence of these luminous bodies.

In this paper we present the most up-to-date information we have on astronomical symbolism of Dinétah. We are primarily interested in depiction of stars painted beneath rock overhangs and in recognizable patterns, but we include a few examples of other sky symbolism as well. Having not critically studied the directional orientation of sites nor the astronomical significance of place, we do not comment on these aspects of archaeoastronomy in this treatment.

Sky Symbols in Dinétah Rock Art

The number of sites and details of the rock art that suggest sky symbolism are far too numerous to individually discuss in this paper. Instead, selected sites are described which collectively contain a representative sample. The focus will be on star ceilings and star pattern sites with just a few examples of other sky symbolism included.

Even though the main subject of the paper is astronomical symbolism, the survey would be incomplete without including other obvious representations of the sky. It should be remembered that traditional cultures have not tended to separate environmental domains as we do today. Anyone attempting to understand the context of rock art in the observable world must keep this firmly in mind. What science would classify as meteorological phenomena were probably not considered to be remote from Sun, Moon, and stars by those who put the figures on the rock. Thus, a few examples, from the many existing in Dinétah, of meteorological symbolism are included.

It is difficult or impossible to positively identify the intended meanings of these rock art elements. We have attempted to be conservative in our observations and we know that it is always possible to suggest interpretations different than those indicated here. Fortunately we are closer to actual interpretation in the Navajo case for the rock art shows continuity in content and form to the symbolic elements of dry painting for Navajo traditionalists have been consulted regarding the symbolism they recognize in the rock art imagery.

Meteorological Phenomena

Some of the rock art symbols are more easily identifiable than others based upon their consistency with symbols that have been used within the cultural tradition in more recent times, i.e., since scholars and others have had the opportunity to observe the cultures involved. Lightning depictions (Figures 3, 7, 10), for example, usually seem obvious, although it is sometimes difficult to distinguish between lightning and snakes: indeed, there are Navajo cultural factors that relate lightning and snakes and sometimes the rock art might have been made to symbolically portray such relationships.

The typical Dinétah cloud symbol consists of sets of stepped tiers (Figures 3, 10). As with lightning, these abound in this area. This paper includes only a few examples of meteorological symbolism that occur on panels in association with other elements of astronomical significance. Many additional examples, including such things as rain and rainbows, exist throughout Dinétah.

Sun and Moon

Some depictions of the Sun seem certain, but even these could possibly represent other things. More difficult is the interpretation of rock art that might depict an eclipse, either of the Sun or Moon. There are no examples with strong enough possibilities for eclipses to speculate at this time, and there are far too many Dinétah representations of the Sun to include them in this paper.

One must be careful with crescent shapes. In Pueblo-related rock art a crescent can be the body of a bird with head and legs missing. It does seem appropriate, however, to interpret some Dinétah crescents as most likely representing the Moon. Identifying portrayals for the full moon is much more difficult because so very many things can be symbolized by a circle or a disk. Although they are not described here, there are several clear representations of the Moon in Dinétah.

Stars

Stars are consistently and repeatedly imaged in Native American symbolism as four-pointed figures or simple crosses (plus signs). This is well established in Navajo tradition. The cross can, however, represent other things such as fire or the four directions. It is significant to note that the five-pointed star symbol is recent in America and that the truest Native American star depiction is four-pointed. The other Navajo traditional way of representing stars, especially when they are in patterned groups, is the use of dots and holes. The star patterns formed by pits in rock are very consistent with the way star patterns are made on sand paintings, gourd rattles and other artifacts (Figure 2).

There is little doubt that crosses on the ceilings of rock shelters or ledges represent stars. The question of why they were put there and their ceremonial context is, however, a different story. Since this has been discussed elsewhere (Chamberlain and Schaafsma 1993 and In Press), we will not dwell on it here. This work has been limited to identifying and briefly describing each of the Dinétah star ceiling sites. All of our study, spread over many years, convinces us that the star images on the ceilings have no intended patterning. Instead, they seem to represent stars in a more general context.

On the other hand, we are confident that we have identified star patterns as sets of dots in Dinétah rock art. We know of few other examples in all the world where one can clearly identify rock art depictions of specific stars. Since this is one of the most important parts of our study, we will treat it as we do star ceilings, as an independent titled section.

In the section on identifiable star patterns we will be dealing with a few groups of stars which are particularly important to Navajos and they will be referred to as Navajo constellations. Since they are encountered over and over, some more than others, it is appropriate to briefly introduce the constellations. Greater detail about these stars can be found in a few published sources (Chamberlain 1983; Griffin-Pierce 1992).

Black God is the source of fire, including stars, in Navajo cosmology. It is said that when Black God and other Holy People had made the Sun and Moon they had bits of crystal left over. They decided to place these in the sky to represent principles the Diné should live by. In the process of carefully arranging the stars into appropriate pairs of patterns, but before they had set all of them into the sky, Coyote came along, quite unhappy because he had been denied participation in this work. When the others were not looking, he seized the buckskin upon which the remaining crystals rested and with one swift motion flung them into the heavens. Thus, there are a few important patterns that represent the laws of life, but most of the stars are random, just as Coyote put them there.

The important stars that have been clearly identified, the ones dealt with here, are as follows, with their Navajo names. Polaris is the **"Fire Star,"** *Náhookos bokhóh*. It represents the central fire of the heavens and the fire in the hogan. On one side of the Fire in the sky is found *Náhookos bika'ii*, the **"Revolving Male,"** known to most as the Dipper in Ursa Major. On the other side of the Fire Star, about equally spaced opposite Revolving Male is *Náhookos ba'áadii*, **"Revolving Female,"** stars in Cassiopeia. This pair represents family life principles: being at home with the family around the fire. The revolving pair have also been referred to as parents of the other stars.

"Dilyéhé," the Pleiades, are referred to as Black God's stars. When he first entered the creation hogan, Black God wore them on his ankle. He stomped his foot upon the ground and they jumped to his leg. Again he stomped and the stars flew to his thigh. On the third stomp they landed on his shoulder. The fourth and final stomp placed them upon his face. The mask of Black God often includes this cluster of stars on the face or temple and has a crescent moon as eyebrows (Figure 2).

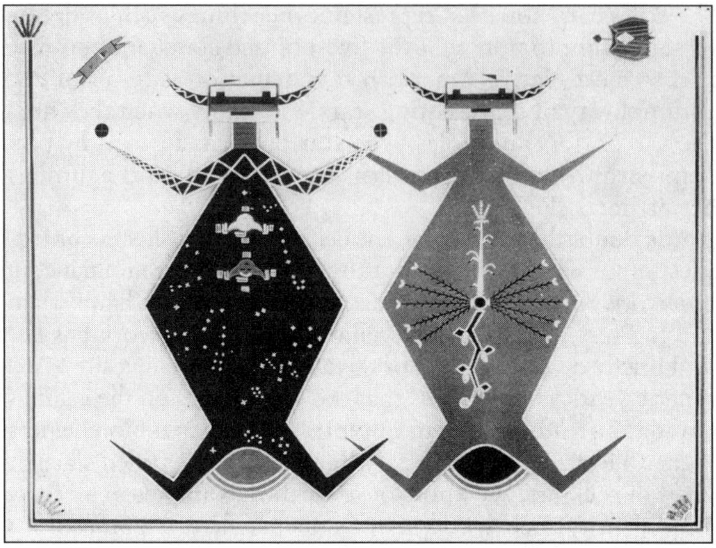

a

Figure 2. Examples of sand painting, gourd rattle and Black God mask with identifiable star patterns. (a) The following patterns can be identified on the sand painting of Father Sky: left center is Revolving Female (Cassiopeia) with Fire Star (Polaris) just beneath and Revolving Male (Big Dipper) still lower; right side below center is Man With Legs Ajar (Corvus) above First Slim One (Orion) which is above Dilyéhé (group of six dots); right just above center is First Big One (front of Scorpius) with Rabbit Tracks (tail of Scorpius) at top right; single star at the top center is likely Venus and the crosshatching along the arms and shoulders represents the Milky Way; the two upside-down horned figures are Moon (top) and Sun (bottom). (b) The seven holes at the top of the gourd rattle represent Dilyéhé and the six at the bottom are First Slim One shown as belt and sword of Orion. (c) Black God's mask includes seven dots representing Dilyéhé and crescent Moon at the top.

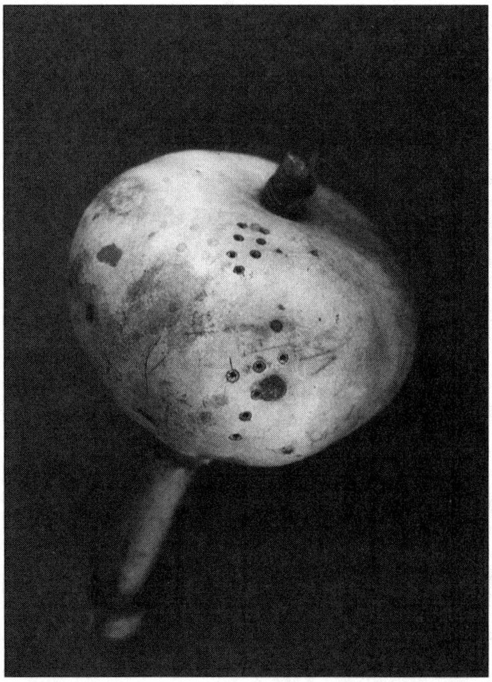

b

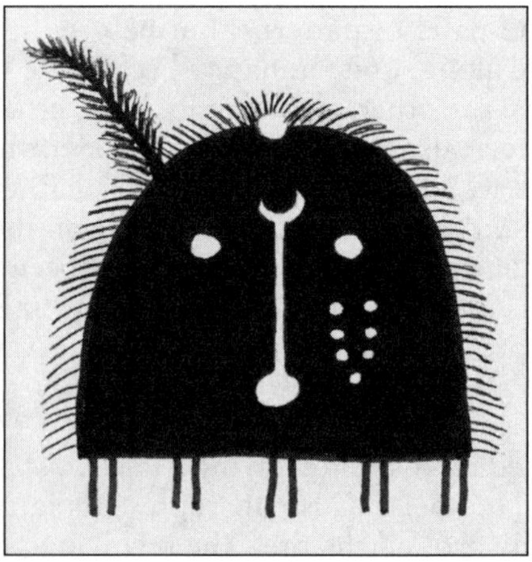

c

They are usually pictured as three side-by-side pairs, one beneath the other, with a final seventh star at the bottom, but variations occur. Usually there are seven, sometimes six, and occasionally eight or more. *Dilyéhé* can be used as a shorthand notation for all of the stars that Black God created. They were placed near the stars known as *Atsé'ets'ózi*, **"First Slim One,"** stars in Orion. This pair is especially important for agriculture. It is said, for example, "never let *Dilyéhé* see you plant." Putting seeds into the ground after the Pleiades have disappeared in the early May evening sky should assure that plants will not be lost to late frost. They must, however, be put in before late June when *Dilyéhé* is back in the early morning sky: after that time will be too late to assure successful harvest. Both of these constellations are also important for ceremonies. Most ceremonies are not held until *Dilyéhé* is in the autumn evening sky, nor are they held after *Dilyéhé* has vanished in the spring.

Hastiin Sik'ai'í, **"Man with Legs Ajar,"** stars of Corvus, is visualized as an old man and his mate is *'Atsé'etsoh*, **"First Big One,"** which includes the front part of Scorpius. This figure is visualized as an old person with a cane and relates to old age and the process of always seeking harmony and balance in order to enjoy long and happy life.

Gah heet'e'ii, **"Rabbit Tracks,"** stinger and tail of Scorpius, lets the Diné know when it is appropriate to hunt wild game by its position and orientation in the evening sky. It is thought of as being mate to the Milky Way, named ***Yikáísdáhi***, meaning "Awaits the Dawn." The principle attached to the Milky Way is that one should rise before the dawn to walk in first light, spreading pollen or corn meal in a motion that resembles the Milky Way, and contemplating the harmony and balance necessary for long and happy life. The Milky Way is often pictured as cross-hatching across the arms and shoulders of Black God and Father Sky. As of yet, we are unable to identify rock art depictions of the Milky Way in Dinétah.

Venus is also important and generally shown on sand paintings of Father Sky, along with Sun, Moon and stars. There are other stars which are mentioned in Navajo literature and dialogue with Navajo people, but they are more difficult to identify and do not seem to be as important to contemporary Navajo traditions.

Dinétah Star Ceilings

"Star ceiling" is the term for stars painted, usually as crosses, on the ceilings of rock shelters and undersides of overhangs. They have also been called "planetarium sites."

Previously Studied Dinétah Star Ceilings

In her study of rock art of the Navajo Reservoir District, Polly Schaafsma (1963) documented two star ceiling sites along the San Juan River that are now under the waters of the reservoir, one where Cottonwood Canyon enters the San Juan and the other where Todosio Canyon enters the Pine River.

The Cottonwood star ceiling had crosses said by Schaafsma to have colors that were "brilliant and clearly or precisely depicted," in orange, orange-red and red colors with two being dark gray or black, all located on an overhang about ten feet wide and thirty feet above ground level. Schaafsma described other impressive Navajo rock art in association with the star ceiling.

The star ceiling on the Pine River at Todosio Canyon was also related to other Navajo rock art, including painted deities and a petroglyph of clouds and lightning. The ceiling contained a clustering of crosses and flower-forms in red paint with possible other colors.

Just outside the town of Dulce, New Mexico, there is a rock art site that was reported in recent work by Chamberlain and Schaafsma (1993 and In Press). A small rock shelter, open to the west, has a large sun shield painted in red, black and white colors on the back wall. The wall to the south has a red queue symbol and 26 star-crosses, all red. Some of the stars are negative images enclosed in red disks, while others consist of red crosses. The ceiling over the sun shield has several hand prints and one red star.

Schaafsma (1972:44–51) described a star ceiling located west of Cuba, New Mexico, and Chamberlain and Schaafsma (1993 and In Press) have designated this as the Penistaja star ceiling. The large rock shelter overlooks an active spring that has been used for domestic stock. The overhanging rock contains many black and red star-crosses and a few bird forms. Tiny lance-bearing men on horseback and the image of a bear are lower on the slanting back wall.

Eight more star ceilings have been located directly in the Dinétah canyons. These are briefly described below.

"Star Gazer Caves," Gobernador Canyon

Gobernador Canyon runs basically east and west and is located just south of the Navajo Reservoir. Along one of the washes of this canyon there are two star ceilings within a pair of well-formed rock shelters. The smaller of these, "Star Gazer Cave," (Figure 3) has 12 star-crosses on the ceiling rock: seven are gray, two are black and three are red. A double cloud altar is painted on the back wall. The larger shelter, called "Second Star Gazer Cave," has three red stars on the ceiling.

"Star Holding the Rock at Rabbit House," San Rafael Canyon

One of the most interesting star ceilings is located at a ruin site named "Rabbit House," located in San Rafael Canyon. Near the ruin stands a great boulder supported only by a narrow stem of rock. Underneath the boulder, over some grinding slots, there is one red star-cross. This provides strong support for one of the theories that has been proposed as partial explanation for some star ceilings, the concepts that the star symbols might have been placed at some of these locations for ceremonial protection, specifically for keeping the rock from falling.

"Star Springs" and "Two Red Circles High Up," Jesus Canyon

Jesus Canyon has two star ceiling sites. One is in a rock shelter containing an active spring overlooked by a star ceiling. Twenty-one yellow star-crosses have been

Figure 3. Star ceiling site named "Star Gazer Cave" in Gobernador Canyon. The top view shows the rock shelter. At the bottom is a portion of the star ceiling with one prominent cross in the center and others more difficult to see scattered over the photograph.

counted, some faded and difficult to see. The stars easiest to make out are on more substantial rock, while the faded ones are on badly deteriorating surfaces and a few are literally pealing off the rock. It is interesting to note that mineral deposits at the edge of the spring seemed to have the same color as the pictographs.

In another part of Jesus Canyon there is a very large shelter with high ceiling that has two small red circles very high up. After studying many star ceilings throughout Navajoland, we believe that this is similar enough to belong in the inventory.

"Stars Watching Over Painted Yeis," Delgadito Canyon

A rock shelter in Delgadito Canyon is very well known for its painted yeis, including some that were cut from the cliff under controversial circumstances. Extremely faint star crosses are on the slanting cliff above the paintings: two orange ones are revealed by the four points. There are four circles and the remains of a good-sized cross.

Figure 4. Petroglyph panels at Big Star in Adolfo Canyon. The portion of the panel at left shows a pattern of seven dots (right of second to top bird track) that could represent Dilyéhé and another W-shaped pattern under the plant growing out of cloud symbol at the extreme right of the group. This could indicate Revolving Female (Cassiopeia). The main panel shown in the right portion has several dot patterns that could represent stars along with four pointed stars and other figures. The large divided circle at the left side of the group might represent a shield with Dilyéhé drawn as small circles. A protecting shelf (not shown here) above the main panel is a star ceiling with one prominent black painted cross and possible traces of others. Drawing by James M. Copeland.

"Big Star," Adolfo Canyon

The most impressive set of sky symbols known in Dinétah is the site in Adolfo Canyon called Big Star (Figure 4). A rock shelter was formed when a portion of a large rock outcrop fell away, leaving a flat surface upon which the Diné placed a wonderful petroglyph panel. Detailed later in this paper, this site has many star depictions ranging from pecked four-pointed stars, star-crosses, star pattern dots, and a set of circles. It also has a star ceiling above the primary panel. One large black star-cross adorns the ceiling formed when the rock separated to form the shelter. A smaller black star is also apparent and there are hints of others more difficult to see.

"Tse'yah so', Stars Under the Rock," Largo Canyon Rincon

Located in a small unnamed rincon running off the north side of Largo Canyon, there is a star ceiling that contains more crosses than any other known in Dinétah. The site is a good-sized shelter facing out to the north west with a low ceiling. Because of the overall characteristics of the place we refer to it as "Tse'yah so'," meaning "Stars Under the Rock." Most of the crosses, 53 by count, are black, with five red ones.

Summary

The Dinétah star ceilings have considerable variation one from another. The one at Rabbit House Ruin has a single painted star appearing to hold up a large balanced boulder, whereas Tse'yah so' has more than fifty stars. Some of the ceilings, such as Big Star and the ones destroyed by the Navajo Reservoir, overlook impressive rock art panels, while others like Star Springs watch over places that are of obvious importance for other reasons.

Chamberlain and Schaafsma (1993 and In Press) have suggested that the star ceilings of the Gobernador region probably predated those of Canyon de Chelly. Strictly from visual observation the Dinétah star ceilings do appear to be older than most others we have studied. They tend to have considerable weathering and mineral encroachment can be seen upon some of the star-crosses. Multiple factors, including the mineralogical properties of the rock and weather trends, make such observations subjective. We emphasize that no site-dating information for any known star ceilings confirms relative age hypotheses. In addition, it would be interesting to study such things as the pigments used to paint the stars and their relationships to color symbolism among the Navajo. Further study is required to apply the best available and appropriate techniques and technologies for determination of age and other parameters within the Navajo and general American Southwest cultural framework.

Identifiable Star Patterns

Another way in which stars are shown in Dinétah is as dots or pits in the rock. Under other circumstances, in other places, we would be very hesitant to think that such collections of holes might represent stars. Everyone familiar with rock art knows that dots are very common nearly everywhere rock art is found. What is of central importance for this study is the fact that it has been shown quite convincingly that Navajos have made stars in such patterns, shown often as dots, on sand paintings and as holes punched into gourd rattles. Some of the patterns are readily recognized, while others are more difficult.

Bullet scars are sometimes associated with rock art and could be confused with star dot patterns. Fortunately, not much of the Dinétah rock art suffers from this problem. Having carefully studied, by experimentation in proper places, the imprints made by bullets, we are confident that we can recognize such scars and distinguish them from dot patterns that are truly part of petroglyph panels. We are certain that no bullet scars are included in our list of potential star pattern panels.

Listed below are the dot patterns we are most certain about. We continue to study others that, like star patterns on dry paintings, are more difficult to positively identify.

Frances Canyon

A boulder at Frances Ruin has badly weathered petroglyphic features that include queues, or hourglass, symbols known to represent the hero twin Born for Water. Next to these there are sets of drill holes that probably represent stars. This whole set of motifs is strikingly similar to one found on a boulder near Wijiji in Chaco Canyon. The Chaco site has patterns that are readily identifiable as the Revolving Male and Rabbit Tracks. Even though the patterns at Francis Ruin are in-

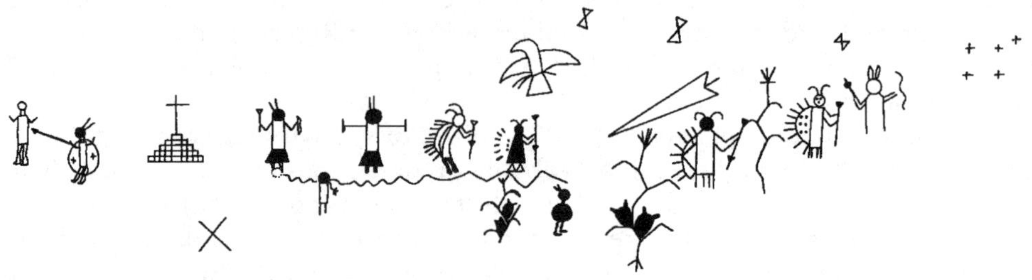

Figure 5. "Yeis-in-a-Row" located in Encierro Canyon. Notice shield figure with pair of cross stars on shield at left. The pattern of five crosses at the extreme right could represent Man With Legs Ajar (Corvus). Drawing by James M. Copeland.

distinct, the similarity of the set of motifs (drill holes along with Born for Water symbols) with those near Wijiji helps convince us that they intend stars.

"Yeis-in-a-Row," Encierro Canyon

High on a large freestanding boulder with a pueblito on top, a row of petroglyphs includes some of the more important deities, making this panel one of many that represent early Navajo ceremonialism (Figure 5). At the left (north) end of the panel there is a shield-bearing human figure with star-crosses on each side of the shield. At the other end there is a set of five crosses in a pattern that is a good representation of the Navajo constellation known as Man-with-Legs-Ajar: stars in Corvus. This is the only recognizable Navajo constellation pattern we have found where the stars are crosses, rather than dots.

Delgadito Canyon

Standing as if to guard the intersection of Cereza and Delgadito canyons, a prominent petroglyph panel stands high on the north rock face. Zigzag lines surround a cloud tier, along with other figures. Most interesting are four tiny sets of rabbit track dot configurations, two vertical and two horizontal. The smallness of the dots suggest that these probably represent the stars of Rabbit Tracks rather than tracks on the ground. Not far away, however, near another set of lightning lines and running along a break in the rock, there is a line of rabbit track sets formed by larger dots that look more like representations of ground tracks. This combination of depictions of both the ground tracks and the stars representing them is found at other sites not reported here. Indeed, the rabbit track motif is quite common in Dinétah and many of them could represent stars.

Down at ground level, not far from the above described panel, there is another star pattern representation. A petroglyph panel is marked by a pair of hand prints and just left of these there are many pits in the rock (Figure 6). One set of pits is a very fine rendering of Revolving Male, consisting of the seven stars of the Big Dipper. What makes this identification unusually convincing is the small eighth star

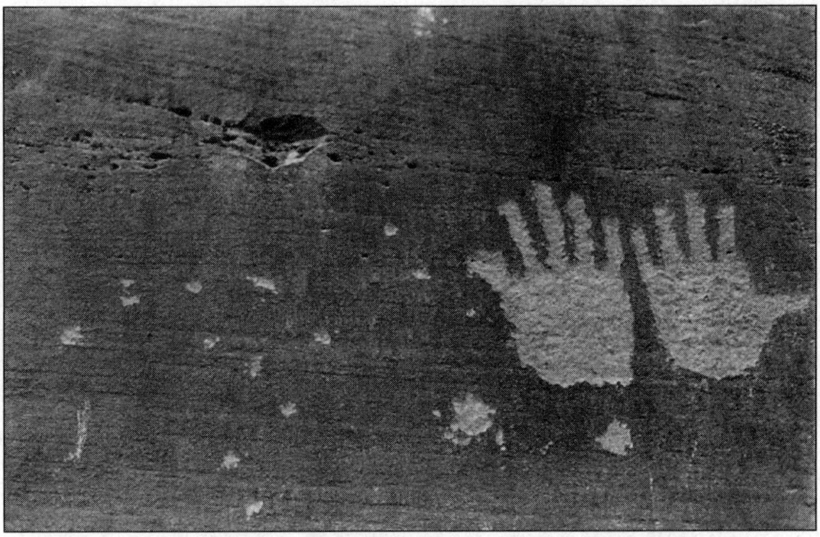

Figure 6. Petroglyph panel in Delgadito Canyon with Revolving Male dot pattern just left of the pair of hands. Note the dipper shape complete with the stars Alcor and Mizar, the pair next to the end of the handle of the Dipper.

next to the one at the center of the dipper handle, just as we see them in the sky. At the extreme left of the same panel (not shown in figure 6) there is a second dipper, this time with the bowl at the top. This pattern also has the double star in the handle of the dipper and an extra star-dot at the end of the handle.

"Hunter Tracking Stars," Cereza Canyon

In Cereza Canyon, just before entering Encierro Canyon, a site contains a petroglyph that most graphically makes the point that the stars known as Rabbit Tracks are, for Navajos, "the Hunter's Guide." A star-cross stands next to a rabbit track motif with each print of the tracks being composed of cat paw tracks (Figure 7). Considering that the cat is the ultimate hunter, how could one possibly say it better. This is, indeed, the clincher for the significance of these stars in Navajo cosmology.

Martin Canyon

A petroglyph panel in Martin Canyon contains a horizontal line of figures that include a foot print, hand print, a figure resembling an owl with lightning bolts coming out of both sides and a cluster of eight dots in one of the known Navajo variations for *Dilyéhé*, the Pleiades (Figure 8). Usually there are seven dots in this pattern, three in a horizontal line, two below the first line, and two more vertically beneath the second-level pair, but the entire pattern on this panel is so much like it that we feel certain it does represent *Dilyéhé*. The presence of the eighth dot does not really diminish the identification, for it is possible to see up to a dozen stars with naked eyes when observing the Pleiades under favorable conditions. This vari-

Figure 7. "Hunter Tracking Stars" in Cereza Canyon. Next to the star-cross is the Rabbit Tracks pattern, but with each of the tracks transformed to a cat paw print, a provocative portrayal of the Rabbit Tracks group of stars as the "Hunter's Guide."

ation is seen on Navajo artifacts as well as in other Dinétah rock art. The Pleiades are the most common stars shown on artifacts throughout Native America, on

Figure 8. Petroglyph panel in Martin Canyon with moccasin prints, Owl, foot print, hand and variation of *Dilyéhé* (left of foot print). Drawing by James M. Copeland.

shields and tipi covers, for example, and they occur in many variations, always as a tight cluster of dots, circles or other depictions having six or more stars.

Adolfo Canyon

One of the most interesting astronomical sites in Dinétah is a petroglyph panel located in Adolfo Canyon. Since it contains a large pecked four-pointed star and several smaller ones, we named it "Big Star" (Figure 4). This site has many sky symbol features, including clouds and lightning, but of special interest are the many dot patterns and one pattern of circles probably representing specific stars.

A circle with vertical division has a set of seven small circles on one side. This could symbolize day and night and the pattern of circles likely represents *Dilyéhé*. Throughout Native America there are many known shields with clusters of dots or circles representing the Pleiades. It is suggested that this is a rock art example.

This panel also contains a cluster of seven dots in a pattern that is a variation for *Dilyéhé*, the same pattern found in Martin Canyon: three side-by-side dots with another pair below and two additional dots in a vertical line below the second level pair. This place has another set of dots that could be a variation for *Dilyéhé*, consisting of six dots in a circle with a seventh dot in the center.

At least three of the dot patterns on this impressive panel are clear representations of Rabbit Tracks. Another pattern on this panel, three dots in a row with another three in a row angled to the row above is a well-known representation for the stars Navajos call First Slim One. Finally, a set of five dots in a "W" pattern, placed under one of the "sky holes" (cloud symbol), is a fine rendering for the stars known as Revolving Female. This sky hole has an incompletely rendered plant growing out of it and it stands along side another sky hole with a corn plant growing out. Considering the fertility significance of the cloud and corn figures, it seems appropriate for the Revolving Woman grouping to be there too. This is the clearest representation of Revolving Female we are aware of in Dinétah rock art, although it is common on sand paintings.

The final touch at this site, making it a treasure for this study, is the fact that overlooking the entire panel, as though to protect it in ways stars have been thought of by Navajos as protecting all on Earth, there is a star ceiling.

Crow Canyon

One of the best known and most frequently visited rock art sites of Dinétah is located in Crow Canyon. Several interesting petroglyph panels are found along the base of the cliffs just above ground level. One of two possible renderings of the mask of Black God, the fire god who was, according to Navajo tradition, responsible for the creation of the stars is located here. One of the magnificent panels has two sets of star-dots representing the star group known as First Slim One, stars in Orion. These are most frequently shown as holes in gourd rattles having three or four holes in a row with three or four more in another row intersecting the first row at an angle, very much like the belt and sword of Orion in the sky. It is in this format that they are found also in Dinétah rock art. The case in point has four dots in the top row with three in the intersecting bottom row.

A nearby panel consists of seven bows, symbols for the Hero Twin known as Monster Slayer, and seven queues, symbol for the other twin known as Born for Water. Six of the queues are solidly pecked and run horizontally underneath the bows. The seventh queue is drawn with lines and has a set of dots, in classic Rabbit Tracks motif, integrated into it. The side-by-side pair of dots is at the top of the queue with the other two dots at the right side of the queue.

A site in upper Crow Canyon has a set of dots in a distorted dipper shape. It even has the double dot at the center of the dipper handle where the double star is in the sky.

Blanco Canyon

The "Hoop Man Panel" (Figure 9) has a male figure with wand in one hand holding a large circular hoop in the other. There is a smaller circle left of the Hoop Man, and under this lies a set of dots in the perfect dipper shape of Revolving Male. Even the smaller star next to the one in the middle of the handle is shown. It seems possible that the man holding the circle might have some symbolic relationship to the movement of Revolving Male in the sky around the "Fire Star," Polaris.

Proceeding along the panel, past many interesting figures including an arching set of rabbit tracks, the rock wall makes a right angle turn then angles back again leaving a flat surface perpendicular to the panel just described. Here are six sets of dots (Figure 10), a star panel showing several important Navajo constellations. It is strikingly similar in motif as well as technique for rendering stars on gourd rattles used in the Nightway ceremony.

Beginning at the upper right there are two sets, one below the other, representing First Slim One, drawn as the belt and sword of Orion. Slightly to the lower left is a rendering of *Dilyéhé*, the Pleiades. Still lower and a bit to the right is a Rabbit Tracks group. Back at the top, just left of the upper First Slim One set, there are two joined pits. It is suggested that these represent stars known as the "Pinching Stars," sometimes pictured as wrestling boys. These are said to be one of the pairs of stars in the Hyades, or perhaps they could be the two stars at the top of the "V" shape of the Hyades. A final set of dots on the left center side of the panel is a circle of dots. The identification of these is uncertain, but they could represent Man with Legs Ajar, stars in Corvus.

Without a doubt, this is the finest star panel known in Navajo rock art studies. It was the one which first drew attention to rock art portrayals that can definitely be identified as specific stars.

Largo Canyon

Located on the north side of Largo Canyon is a petroglyph panel that has been called the "Coyote Panel" containing two dot clusters. One is a very nice depiction of *Dilyéhé* and the other is a probable variation for these same stars (Figure 11). For the Navajo, *Dilyéhé* can be thought of as representing all of the stars of the sky. The panel has an image of Coyote along with a pecked disk and crescent. There is also a single star cross on the panel as well as a cloud symbol and lightning. All

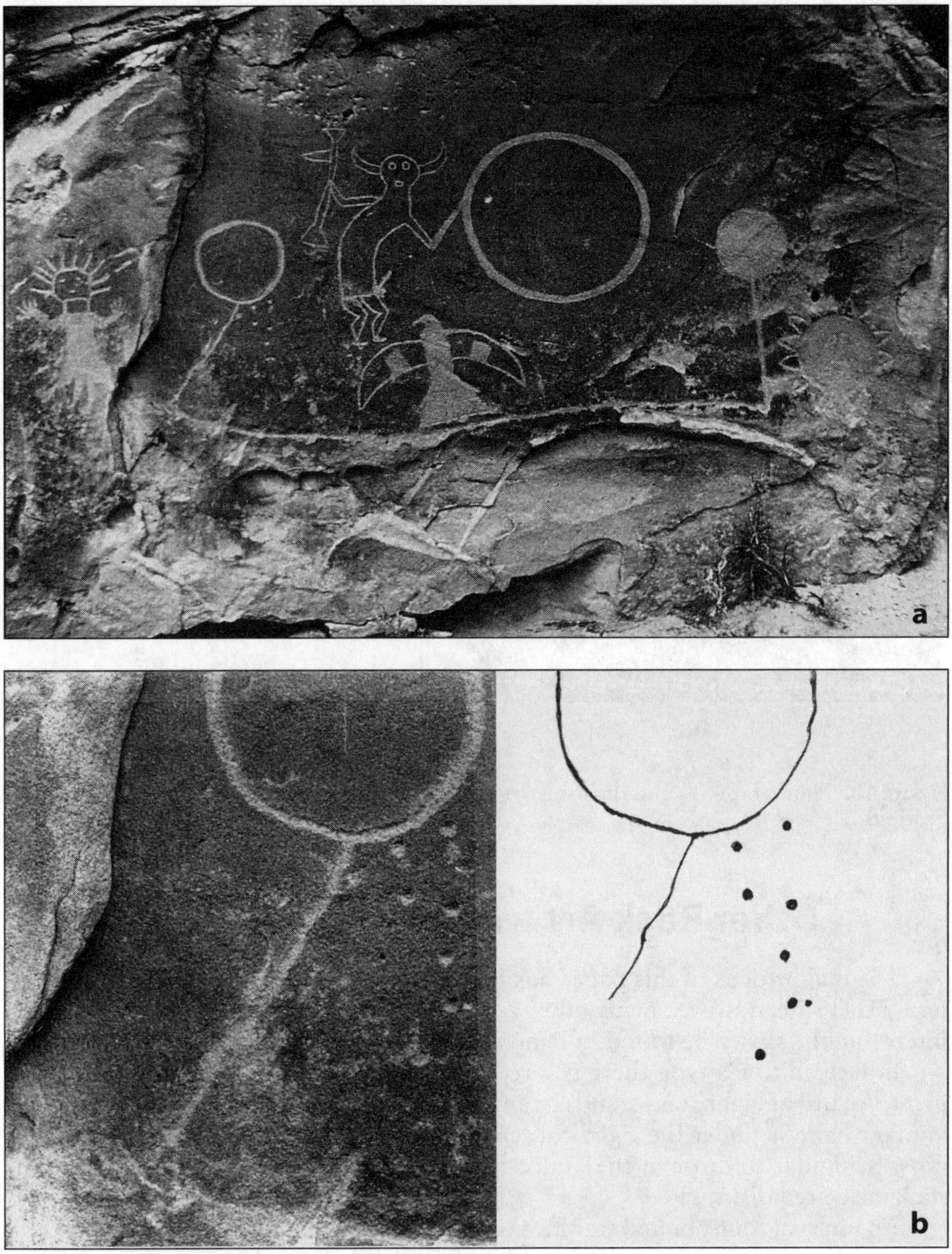

Figure 9. "Hoop Man panel" in Blanco Canyon. **(a)** Entire panel shows a male figure holding a wand in one hand and large hoop in the other, eagle and other figures. **(b)** Under the smaller circle is a fine rendering of Revolving Male formed by eight pits in the form of a dipper, complete with a small pit next to the second star in the dipper handle. The drawing shows the pattern of pits that are difficult to distinguish, in the photograph, from natural features in the rock.

of the elements of a night scene are there: nocturnal animals, full Moon, new Moon and stars.

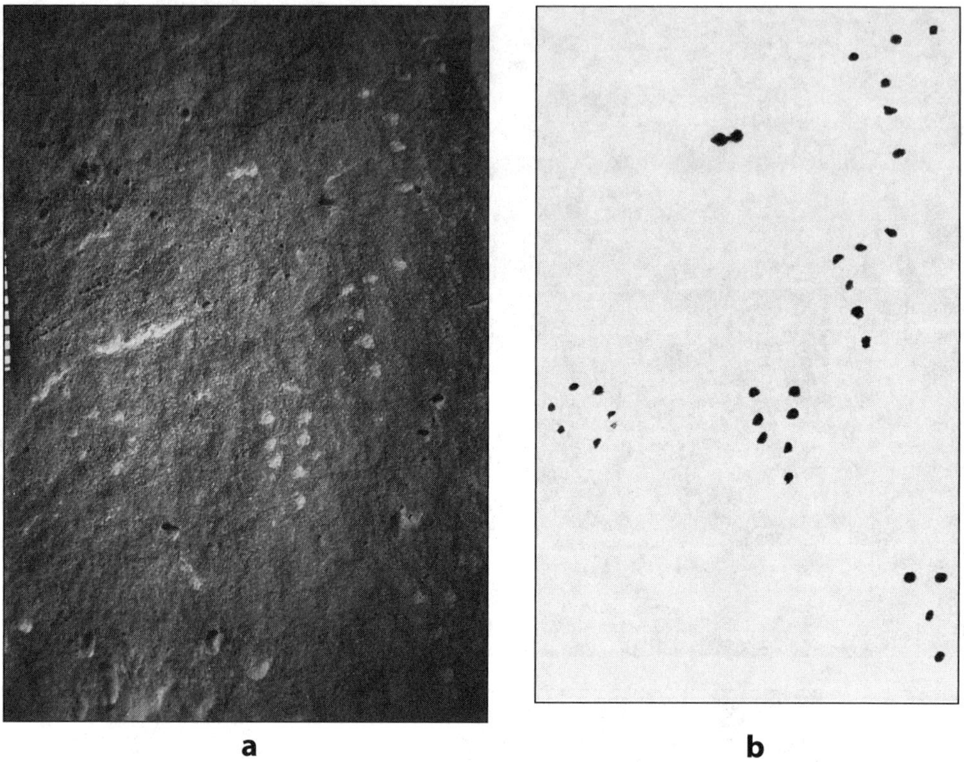

Figure 10. Photograph (a) and drawing (b) of the Blanco star panel. Refer to text for identifications.

Other Rock Art with Sky Associations

The major focus of this paper has been star ceilings and recognizable star patterns. There are, however, many other very interesting manifestations of early Navajo interest in the sky to be found in Dinétah rock art.

In Delgadito Canyon there is a rectangular horned mask, painted in red pigment. An arc of eight stars stands over it, a pair of hand prints are to the right and another hand is under the right corner of the mask. The star images are negative crosses, similar to those at the Dulce star ceiling site, crosses with the hue of the rock inside red disks.

An unusual four-pointed star face is located in an unnamed canyon of the Gobernador drainage. The face is formed by a square portion left unpecked except for the round eyes and triangular mouth. A rain symbol (rake) is attached to the top point of the star. We think this is Navajo rock art, strikingly similar to the many star Katcina glyphs found around the pueblos to the south, except that all of those we know of have round faces.

Scattered around the canyons are found individual star crosses, both pictographs and petroglyphs. One star-cross is on the body of a quadruped, another is on the body of a horse, and one shows a whole herd of horses with a pair of four-pointed stars at one end of the panel.

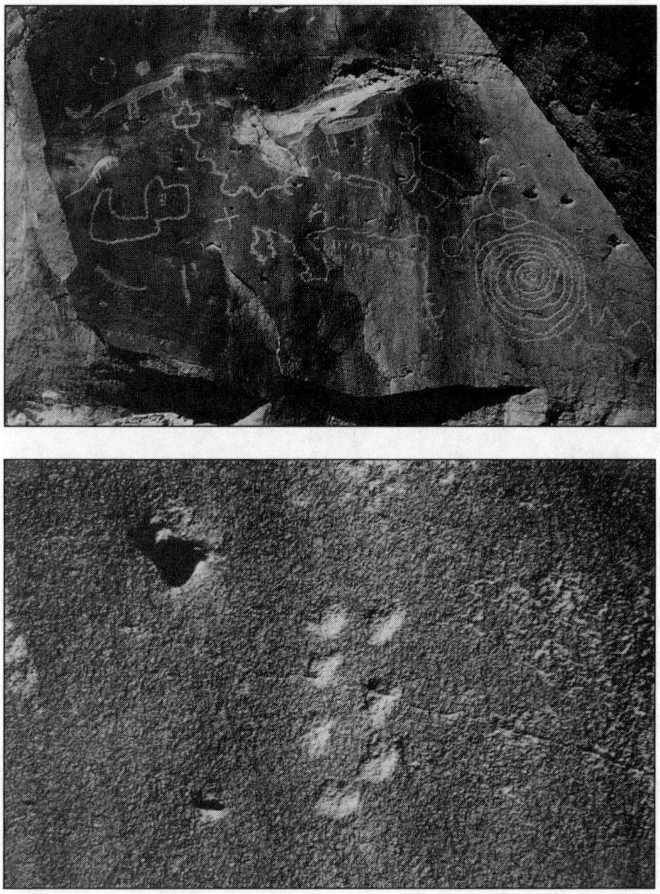

a

b

Figure 11. Coyote panel in Largo Canyon. (a) Entire panel showing coyote, crescent and full disk, cloud and lightning, star-cross, deer and other figures. Two sets of dots might both represent *Dilyéhé* (Pleiades): one is inside the irregular enclosure beneath coyote; the other is just left of the deer with antlers. (b) Closeup photograph of the dot pattern that most likely depicts *Dilyéhé*.

In Crow Canyon a horned and feathered mask has a clear division vertically down the center and dots on the right side, with a faint star-cross on the other side. Another site has a horned mask, this time without feathers, a vertical line split with a star-cross and a pair of dots on the right side, a possible Black God figure.

One of the most magnificent and interesting of all the rock art panels, in the southern part of Largo Canyon, is known as "Biih Yaazh." Many painted deities are in a long row within a large rock shelter. The vividness of the colors and crispness of detail is astonishingly beautiful. Other figures are pecked into the rock, and this panel makes it clear that many of the petroglyphs of Dinétah were once painted, for a number of the figures here have both carved and painted elements. One Ganaskidi, the Humpback God, has a star-cross on its chest. Here also is found a badly weathered but convincing petroglyphic Black God mask, complete with crescent at the top of the forehead and star dots on the temple.

One female figure at Biih Yaazh is painted mostly in white with touches of red here and there. The white skirt has more than a dozen red four-pointed stars. We nearly lost this one, for it and two companion figures are on the rock where someone had sawed completely around them. At least one figure was destroyed in this cutting and a whole slab of rock had been removed from the panel right next to these figures (O'Keefe 1996).

One more Biih Yaazh figure will be described to end this section. A large round headed figure with carefully painted red feathers resembles the appearance of the Sun. Carved on the face are the symbols of both Hero Twins, the bow and queue. Looking at this figure, one is easily reminded of a critical episode in Navajo history when two boys, children of Changing Woman and the Sun, went to their father and obtained weapons which they used to slay monsters. No longer boys, the two were hence forth known as the Slayers of Alien Monsters.

Conclusions

Dinétah abounds in sky symbolism. We have documented a dozen star ceilings, more than 30 dot patterns that probably represent stars at a dozen different sites and many other examples of sky symbolism. In this study we have not attempted to locate and list meteorological symbols, but merely to point out a few examples alongside the main topic of astronomical symbolism.

Seventy-eight star ceilings have now been studied. One star ceiling, located at San Cristobal, has definite Pueblo origin and we have reports of others. Canyon de Chelly and nearby places have 79% of the known star ceilings, 14% are in Dinétah and 7% are elsewhere. Dinétah, then, is second only to the Canyon de Chelly region for density of known star ceilings. It is now possible and important to make critical comparisons between those in the heart of the region Navajos consider to be their place of emergence and the other areas. It will be especially interesting to know how the Dinétah star ceilings relate to the one at San Cristobal and any others of Pueblo origin that might be found. Clearly, the process of understanding the star ceilings has just begun.

Star pattern recognition began with the star panel in Blanco Canyon. It is such a simple rendering, consisting only of tiny pits that resemble bullet scars on other rock art panels, that it easily could be considered unimportant by the uninformed. Instead, it should be appreciated, treasured among American rock art, and protected. It was the key to our studies of star patterning in rock art. Once it had been recognized, noting its striking similarity to the figures on Navajo sand paintings and gourd rattles, we began to recognize the other star-dot patterns in Dinétah and other places where Navajos have lived and followed their traditional ways.

Finding so many other star patterns scattered throughout Dinétah is exciting. Several can be recognized in illustrations from Schaafsma's (1963) publications on the rock art of the Navajo Reservoir (her figure 22 has a possible *Dilyéhé*, figure 31 shows definite *Dilyéhé* and Revolving Male). Only a few other recognizable star patterns are known at this time: one at Chaco Canyon; one in Alcove Canyon near Round Rock, Arizona; three in southeastern Utah; and a couple of likely cases in Canyon de Chelly.

Most claims for star identification in North American rock art studies (e.g., Mayer 1975) involve great uncertainties. There was a rock art panel, described by Stevenson (1904:444–445, plate CVIII) near Zuni that was somewhat like the ones in Dinétah, but it was removed and taken to the Smithsonian Institution. This, too, had hole patterns said to depict stars and, in spite of other identifications stated by Stevenson, we believe that two of the patterns at Zuni are the same as we have been discussing for the Navajo: one pattern is a rendering of what we would call First Slim One, the belt and sword of Orion; another is a fine depiction of what we would label Rabbit Tracks. Other hole patterns in this Zuni example are uncertain. One additional case of Pueblo origin is a Pleiades pattern on an Anasazi panel near Bluff, Utah, and it would not be surprising to find other Pueblo cases. These Pueblo examples suggest interesting relationships that should be explored. Much remains to be done for better understanding of the roles of stars in Native American ceremonial systems, along with the significance of the patterns in rock art that we have treated here.

The rock art of Dinétah is a key to understanding the evolution of the Navajo culture in all its manifestations. Certainly it is fundamental to comprehending the roles of the objects and phenomena of the sky in the cosmology of the Diné and important in the world views of indigenous American Southwest people generally.

Acknowledgments

We express appreciation to those who have participated, in one way or another, in this work. Years ago Harry and Sally Hadlock, now both deceased, spent immense amounts of time documenting Dinétah rock art and introduced one of us to the canyons. Harry and Anna Walters have been good advisers and friends in many ways, including visiting the sites. Sally Cole first informed us of the existence of one of the star ceilings and introduced us to Margaret and Glenn Stone who took us to this site. Since then we have been with the Stones on other occasions to share the joy of learning through wandering around. Our colleagues, archaeologists David Brugge and James Copeland have offered companionship in the canyons as well as insights regarding everything about Dinétah. Along with the many others we do not mention by name, we thank our families who have encouraged and supported our interests and work.

References

Britt, Claud. 1975, Early Navaho Astronomical Pictographs in Canyon de Chelly, Northeastern Arizona, U.S.A. In *Archaeoastronomy in Pre-Columbian America,* ed. A.F. Aveni, pp. 89–107. Austin TX: University of Texas Press.

Chamberlain, Von Del. 1974, American Indian Interest in the Sky as Indicated in Legend, Rock Art, Ceremonial and Modern Art. *The Planetarian* 3: 89–106.

———. 1978, Sky Symbol Rock Art. In *American Indian Rock Art. Papers Presented at the Fourth Annual American Rock Art Research Association Symposium*, vol. 4, pp. 79–89. El Toro CA, American Rock Art Research Association.

———. 1983, Navajo Constellations in Literature, Art, Artifact and a New Mexico Rock Art Site. *Archaeoastronomy* (Center for Archaeoastronomy) 6:48–58.

———. 1989a, Navajo Indian Star ceilings. In *World Archaeoastronomy*, ed. A.F. Aveni, pp. 331–340. Cambridge: Cambridge University Press.

———. 1989b, Rock Art and Astronomy: Navajo Star Ceilings. In *Rock Art of the Western Canyons*, ed. J.S. Day, P.D. Friedman and M.J. Tate, pp. 30–45. Denver Museum of Natural History and Colorado Archaeological Society (Colorado Archaeological Society Memoir, no. 3).

Chamberlain, Von Del and Polly Schaafsma. 1993, The Origin and Meaning of Navajo Star Ceilings. In *Archaeoastronomy in the 1990s*, ed. C. Ruggles. Group D Publications Ltd., Loughborough, UK.

———. In Press. The Origin and Meaning of Navajo Star Ceilings. In *Songs From the Sky*, publication of the First International Ethnoastronomy Conference held at the Smithsonian Institution in 1983. To be published by the Center for Archaeoastronomy.

Copeland, James M. and Hugh C. Rogers. 1996, In the Shadow of the Holy People. In *The Archaeology of Navajo Origins*, ed. by Ronald H. Towner. Salt Lake City: University of Utah Press.

Grant, Campbell. 1978, *Canyon de Chelly: Its People and Rock Art*. Tucson, AZ: University of Arizona Press.

Griffin-Pierce, Trudy. 1992, *Earth Is My Mother, Sky is My Father: Space, Time and Astronomy in Navajo Sandpainting*. Albuquerque, NM: University of New Mexico Press.

Jett, Stephen C. 1984, Making the 'Stars' of Navajo 'Planetaria.' The Kiva 50:25–40.

Mayer, Dorothy. 1975, Star-Patterns in Great Basin Petroglyphs. In Archaeoastronomy in Pre-Columbian America, ed. Anthony F. Aveni. Austin, TX: University of Texas Press.

O'Keefe, Joanne. 1996, Stealing Our Cultural Heritage, *New Mexico Magazine* 74 (6):50–57.

Olin, Caroline B. 1984, Early Navajo Sandpainting Symbols in Old Navajoland. *Papers of the Anthropological Society of New Mexico*, vol. 9, pp. 43–74.

Schaafsma, Polly. 1963, *Rock Art in the Navajo Reservoir District*. Santa Fe NM: Museum of New Mexico Press (Papers in Anthropology, no. 7).

———. 1966, *Early Navaho Rock Paintings and Carvings*. Santa Fe, NM: Museum of Navaho Ceremonial Art.

———. 1972, *Rock Art in New Mexico*. Albuquerque, NM: University of New Mexico Press (published for the Cultural Properties Review Committee, State Planning Office).

———. 1980, *Indian Rock Art of the Southwest*. Santa Fe and Albuquerque NM: School of American Research and University of New Mexico Press.

Stevenson, Matilda Coxe. 1904, The Zuni Indians: Their Mythology, Esoteric Fraternities, and Ceremonies. In *Twenty-third Annual Report of the Bureau of American Ethnology for the years 1901–1902*: Washington: Government Printing Office.

Towner, Ronald H. 1996, *The Archaeology of Navajo Origins*. Salt Lake City: University of Utah Press.

19

Celestial Imagery in Lakota Culture

Mark Hollabaugh
Department of Physics and Astronomy
Normandale Community College
Bloomington, MN

The traditional Lakota (Sioux) culture of the American great plains flourished from the early 18th century until 1891. These semi-nomadic people lived between the seemingly never ending expanse of the prairie and the endless dome of the sky. In the last 25 years, anthropology tended to look at the Lakota with an "ecological" mindset, and to understand the Lakota's interaction with their environment, in particular the plant and animal world. However, allusions to the sun, moon, planets, stars, comets and meteors frequently appear in Lakota stories and art. Indeed, the Sun Dance is the most important ceremony in their culture. The question to ask is "Why are celestial objects common in Lakota stories and art?" An answer may lie in their understanding of the concept of *wakan*, which is often translated "sacred" or "mystery." This word also embodies the notion of "that which is different." For example, the pole star is *wakan* because, unlike the other stars of the little dipper, it does not move. Lakota astronomy is not separable from Lakota spirituality or religion, and their astronomy must be understood in this cultural context.

* * *

Introduction

"One star never moves and it *is wakan.* Other stars move in a circle about it. They are dancing in the dance circle," Ringing Shield, May 1903 (Walker, 199 1, pp. 114–115). That comment by a Lakota elder about Polaris was the genesis for this paper. The Lakota, or Sioux, lived for over 100 years on the great plains of what is now Nebraska, South Dakota, North Dakota, Montana and Wyoming. They are perhaps the best known of the plains Native American tribes due to their persistent appearance in popular history of the American west. Much of the interpretation of their life from about 1750 through December, 1890, has focused on their interactions with their great plains environment. What has often been missing from the "standard" treatments of their culture has been an investigation and exploration of their relationship to the celestial world beneath which they roamed the plains. The purpose of this paper is to explore this celestial dimension in the context of their culture. It will not be possible to examine all of the different kinds celestial images that were important to this people, nor to fully discuss their meaning. Rather, this is an overview and a survey of those areas that might be fruitful grounds for further investigation.

Lakota History and Culture

The term Sioux is itself derived from an Ojibwe term, *nadouessioux,* which means "little snakes" or "enemies." This usage reflects the often hostile relationship between the Ojibwe *(Anishinaabe,* also know as Chippewa) and Sioux. It is thought the Sioux people originally inhabited the Minnesota area, living a farming, hunting life, and were displaced southward by the Ojibwe when the Ojibwe migrated west into Minnesota following the westward progression of the fur trade. Some anthropologists, based on linguistic patterns, have suggested the Sioux originally lived in the mid-Atlantic area of the east coast of North America. In this paper, I generally will not use the somewhat pejorative term "Sioux" to refer to these people, but their word, namely *Lakota.* Lakota is derived from the verb *lakolya,* "to be friendly with" (Beuchel, 1983, p. 322). An English-Lakota dictionary translates "friendly Indian" as *Lakota* (Karol, 1974, p. 29).

The Lakota Nation is broken into three distinct cultural and linguistic groups. The eastern Sioux speak the Dakota dialect, the central Sioux speak Nakota, and the western Sioux speak the Lakota dialect. The dialect name has become the common name for the people as well. If you asked a member of the *Mdewaka to wa* band in Minnesota the name of her people, she would respond, "Dakota." But, if you asked a man from the *Oglala* band at Pine Ridge, South Dakota, the same question, he would say, "Lakota." (The Lakota do not have a "d" sound in their dialect, and hence in Beuchel's *Lakota-English Dictionary,* "E" entries follow "C" entries.) These language groupings are also known by their common names, the Santee, Yankton, and Teton. The four Santee groups, two Yankton groups and the Tetons were said to comprise the seven "council fires." [Note: The Greek η used in transliterating Lakota to indicate nasalized vowels, *aŋ*, *iŋ*, *oŋ*, and *uŋ*. Usually I will follow Beuchel's spelling, and translations, for Lakota words.]

The Teton Sioux or Lakota are the focus of this paper. The name Teton is derived from *Títo wa* , meaning "to live in community" in Lakota, and has nothing to do with the Teton mountain range in Jackson Hole, Wyoming. The Teton are further subdivided into seven subdivisions including the *Oglala* ("To scatter one's own") who are found today mostly at the Pine Ridge Reservation in South Dakota and the *Sicá gu* or *Brulé* ("Burned" in French) Lakota who live today on the Rosebud Reservation.

Several excellent histories give further details of the history of the Lakota. George E. Hyde, born in 1882, has been called the dean of Indian historians. While late twentieth century scholars have taken issue with some of his interpretations and his lack of direct references, nonetheless Hyde's work on the Lakota serves as a primary survey of their nineteenth-century life (Hyde, 1961, 1975, 1993). Other books give a late twentieth century interpretation of these events and people (D. Brown, 1962; D. Brown, 1970; Hassrick, 1964; Utley, 1984).

Sources of "Celestial" Information

In any research into ethnoastronomy, it is important to consider the sources from which the data collected. The focus of this paper is on nineteenth century Lakota celestial thought and hence, nineteenth century sources must be utilized. Some images are available in museum collections and archival ethnography. The most readily available material was complied by a man who became a trusted friend of the Lakota.

James R. Walker was a physician at the Pine Ridge Reservation from 1896 to 1914. Walker, who tried to treat tuberculosis among the Lakota at Pine Ridge, decided to work with traditional Lakota healers instead of against them. This cooperative spirit lead to a relationship of trust between Walker and the medicine men. Health conditions improved and the Lakota shared their stories, beliefs, and rituals with Walker. His three books, *Lakota Belief and Ritual, Lakota Myth,* and *Lakota Society* are primary sources on the late nineteenth century Lakota. Because of a pledge of secrecy he made, these books have only been available in the last two decades of the twentieth century.

Walker's compilations are the primary source for this paper. This limits the scope of this exploration to the insights given by the people Walker interviewed late in the nineteenth century. Generally, contemporary Lakota have a high regard for Walker and the veracity of the accounts he recorded. Vine Deloria, Jr., for example, gives high praise for Walker and his methods, but notes the informants were all "late" in Lakota history and little is known about the origins of Lakota beliefs about the stars. He further claims that Walker asked the "wrong" questions for the information to be of value to archaeoastronomy (Deloria, 1982). 1 would disagree with Deloria's last assertion because I believe Walker has left a wealth of information, which, if *combined with other historical materials,* could yield rich insights. Another criticism of Walker is that he tended to rely on just a few translators.

There is an important epistemological point to be made. There is no such thing as one Lakota "system' ' of "orthodox" belief. Unlike Christianity, for example, which has dogmatized its beliefs into a set of doctrines, the Lakota are much less rigid. Lakota belief can even be very "fluid" and exhibits diversity (DeMallie, 1987, p. 43). One cannot say, "This is what the Lakota believe about the stars." It is possible to say what a *specific* Lakota person says about the stars. Two non-celestial examples will illustrate this lack of systemization. A common visual image in contemporary Lakota life, apparently going back to the famous Lakota spiritual leader Black Elk, is the red, white, yellow and black "medicine wheel." Generally the colors are fairly constant but the quadrant locations of red, white and yellow seems to fluctuate. In one image I found, the symbol of Sinte Gleska University at Rosebud, blue replaces black! For a modern Lakota example, consider the character Ten Bears (portrayed by Dakota actor Floyd Westerman) in *Dances With* Wolves. Whenever there is a consultation among the band's leaders, they ask for Ten Bears' opinion. He always follows his comments with "This is what I think." He is listened to because he is respected and his counsel in the past has always proved valuable and wise. Although written into a late twentieth century movie script, this is an excellent example of a very Lakota attitude. What this means for this discussion is that Walker's interviews and stories will enable us to understand what the Lakota at Pine Ridge believed about the stars. And although we may not be able to directly understand the origins of those beliefs, these accounts are a starting place,

The goal of this paper is to explore the celestial images of the Lakota and attempt to understand their meaning. This effort will proceed in three steps. First, I will give specific examples of celestial images including comets, meteor showers, and the sun and moon. Second, I will examine the Lakota concept of *wakan,* or the incomprehensibleness of the universe. Finally, I will demonstrate that it is necessary to understand *wakan* in order to understand the meaning of the celestial images.

Examples of Celestial Imagery

I turn now to specific examples of celestial images from the nineteenth century Lakota that have been preserved on winter count hides and ledger books. It is not possible to include every image they made and indeed I continually discover new images. What I will attempt here is to give a sampling of those images commonly available. Most of the examples are reproduced from Walker's books. Von Del Chamberlain (1984) has compiled an excellent list of Plains Indian winter count hides and ledger book entries that specifically refer to astronomical events. One of the most readily accessible hides, the Big Missouri Winter Count, is on display at the Sioux Indian Museum in Rapid City (Cheney, 1979).

Garrick Mallery, a late nineteenth century expert on winter count hides, explained the use and function of the winter count hide.

> ...The Keeper of the Count was responsible for the perpetuation of the history...with this counsel of the old men of his tribe, he decided upon some event or circumstance which should distinguish each year as it passed, and marked what was considered to be its appropriate symbol or device upon a buffalo robe kept for the purpose. The robe was at convenient times exhibited to other Indians of the tribe, who were thus taught the meaning and the use of the signs as designating several years (Maurer, 1992, p. 275).

For a people who relied largely on oral tradition, the hide was a visual prompt for the teller of the people's stories.

Meteors

In 1822, a brilliant meteor blazed across the sky. This was recorded on September 20, 1822, at Fort Snelling at the confluence of the Minnesota and Mississippi Rivers in Minnesota (Hyde, 1975, p.318; Chamberlain, 1984). A Lakota couple saw this and several months later named their new son Red Cloud *(Mahpiya* Luta; the implication of luta is the scarlet red of the morning sky), supposedly in recognition of the event. This is only one example of a meteor making an impression on the Lakota. In November, 1833, the Leonid meteor shower put on a spectacular display. Estimated rates are 100,000 to 240,000 per hour (Rao, 1995). Using "Moon-Times for Windows" (Zephyr Services, 1995), one can compute that the waning gibbous moon would have risen in the late evening for those observers in what is now western South Dakota. The Lakota noticed this spectacular celestial display.

Figure 1. 1833 Meteor Shower.

> In November, 1833 when Spotted Tail was ten, the stars fell. The entire sky was streaked with fire as myriads of meteorites flashed across the heavens, and the frightened Indians thought that the world was coming to an end (Hyde, 1961, p. 29).

A winter count hide by No Ears made note of this Leonid meteor shower (Walker, 1982, p. 138). His drawing of the shower is reproduced in Figure 1. Additional images of the 1833 Leonid shower and other showers were also noted by Mallery (1886, p. 116). Several of those images are reproduced in Figure 2. Chamberlain notes the universal appearance of this event on plains Indian winter count hides (1984). These events made an impression on the Lakota. What appears in any Lakota discussion concerning a comet or meteor, is that the apparition is considered to fit into the larger picture of how the universe works. The meaning of the event is intricately connect to what is *wakan* or sacred.

Walker offers an insight into the impression such meteors made and the meaning of the concept *Wakan.* He reports a conversation with Finger, an *OgIdla* Lakota holy man who spent an entire night instructing Walker in the sacred ways of the Lakota (Hirschfelder and Molin, 1992, p.88). This passage also indicates how Walker approached his task of determining the meaning to the Lakota of these concepts and events.

> A few days before leaving the Pine Ridge Agency, I had an interview with Finger, an old and conservative Oglala, which was of much interest to me, and of much value relative to the mythology of the Oglala.

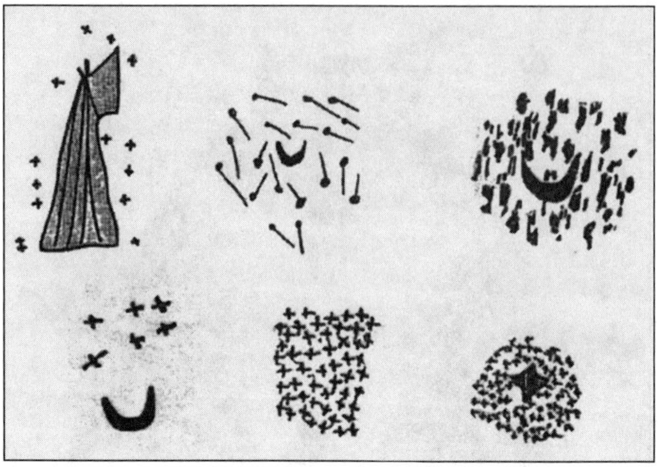

Figure 2. Lakota Winter Count Meteor Shower Images.

> It came about in this way: I was at the house of Finger in the evening, and when starting for the agency, all were out in the gloaming, and a very brilliant meteor fell. Finger exclaimed in a loud voice, "*Wohpe. Wohpe-e-e-e.*" He then harangued for a short time and the women built a fire and when it had burned to coals, Finger burned a quantity of sweet grass on it, evidently with forms and ceremonial mutterings.
>
> I asked him the meaning of this, but he would tell me nothing. I then offered him pay, and he agreed to come to my office and tell me what he knew of the mythology relative to his performance. About thirty days afterwards on the 25th of March, he came for this purpose and I secured an interpreter of unusual ability for grasping the concept of the Lakota language and translating it. The interview lasted nearly all night, and I believe the old man tried honestly to give the concepts of the shamans relative to the matters discussed. The most of the matter discussed was relative to *Taku Škanškan*, or *Škan*, to *Wohpe*, and to the immortality of the *Wakan*. I left the Agency on the first of April, so had no opportunity of reviewing the matter with Finger or of submitting it to others of the Oglala for their discussion

Wohpe, the Lakota word Finger uses for the meteor, derives from the verb *wohpa*, "to make fall by shooting," or "to shoot down" (Beuchel, 1983, p. 598). *Wohpe*, also associated with the White Buffalo Calf Woman, is a prominent personage in the Lakota creation story. The burning of sweet grass is a characteristic Lakota ritual. "Good" spirits like the smell of sweet grass, and the burning of the grass is a means Finger employed to bring the good spirits to him. Note that Walker attempted to corroborate the information he received from Finger with other Oglala Lakota. This was typical of his approach. He would ask several people to give their opinions. Walker goes on to mention how he checked his translations with George Sword, one of his principal interpreters.

> The information I got from Finger clears up much that was obscure, especially relative to *Taku Skanskan*. Perhaps you will remember that I said

> that I could not give a translation of *Skan,* which is the shamanistic term for *Taku Skanskan* and that according to the best information I had, *Skan* meant the sky I so translated it with the approval of several Indians, including George Sword, though each and all declaring that *Skan* was the sky, and was also a spirit that was everywhere and that gave life and motion to everything that lives or moves. Every interpreter interpreted *Taku Skanskan* as "What Moves-moves," or that which gives motion to everything that moves. From the information given by Finger it is evident that his concept of *Taku Skanskan* or *Skan is* a vague or nebulous idea of force or energy. Recalling attempts of other Oglala to define the word I am sure that they had the same kind of a concept of *Skan. I* am now surprised that this did not appear to me before talking with Finger.
>
> Another proposition that Finger gave utterance to is that "Anything that has a birth must have a death. The *Wakan* has no birth and it has no death. The spirit, the ghost, and the familiar of man are not born with him but are given to him at the time of his birth. They are *wakan* and therefore will never die "
>
> He cleared up other things that were obscure to me. I should have very much liked to have talked with him about the Sun Dance, but did not have the opportunity.
>
> My interpreter was an educated man, and well informed, but he had never before this interview heard the words *Skan, Wohpe, sicun,* or *ton* used to convey the concepts as given by Finger though he recalled having heard them used with an allusive sense.

Taku Skanskan could also be translated as the "energy of the universe." Beuchel translates the phrase "a power working, moving things secretly" (Beuchel, 1983, p. 476). In Finger's mind, the concept is closely related to the *Wakan.* Walker was not only aware of who his sources were, but also who his interpreters were. Although he himself learned some Lakota, he relied on native interpreters to convey the meanings of the elder's words. In the following passage, Walker relates several Lakota concepts.

> Finger's discussion of *Wakan Tanka* agreed with that given in that part of my paper on the Sun Dance as submitted to you [i.e., Walker, 1917], except relative to *Skan* and the relative existence of the four superior Gods. For instance, he gave *Inyan,* the Rock, as the first in existence and the grandfather of all things. *Maka,* the Earth as next in existence and the grandmother of all things. *Skan* next in existence after the Earth because He gave life and motion to all things: Wi, the Sun, as the last in existence but as the most powerful and August of *Wakan Tanka,* being *Wakan Tanka Kin,* The *Wakan Tanka.* He also said that the Associate *Wakan Tanka, Wi* (the Sun), Wi *Han,* The Moon; *Tate,* The Wind, and *Wakinyan,* The Winged, and *Wohpe* were as the other self of the four Superior Gods; that is, that Wi and *Wihan* are as one: *Skan* and the Wind are as one; The Rock and The Winged are as one; and that the Earth and *Wohpe* are as one; that while there are eight personalities that are *Wakan Tanka,* four Superior and four Associate, they are all as one and there is but one *Wakan Tanka.* This is The Great Mystery known only to the wisest shamans

> Discussing Wi and *Skan* Finger said that while The Sun was the superior and most powerful of the Gods, yet He derived His power from *Skan;* that many of the Lakotas believed Wi and *Skan* to be one and the same personalities; but the Wi was a *Wakan Tanka* visible in the sky only half the time while *Skan was* the *Nagi Tanka,* the Great Spirit, everywhere at all times and invisible except his color which was the blue seen in the sky at all times.

Walker has now introduced the concept of *wakan,* and beings that are *wakan.* He is appropriate in recognizing that a simple translation will not do and he attempts to understand the interrelatedness of the ideas Finger expressed. In this section he suggested the translation of "Great Mystery" for *Wakan Tanka.*

In a concluding comment, Walker acknowledges his own limitations. It is interesting that he says his paper is "constructive." Apparently he was well-aware he was putting his own thoughts into the report.

> I am fully alive to the sense that my paper is based entirely upon information given by others, and that it is in part constructive; that I may have been misinformed either intentionally or because of the difficulty in getting correct translations of the language of my informants. But the intention of the paper is to give such information as I have received (Walker, 1983, pp. 9–10).

Comets

The Lakota also noticed comets and apparently distinguished them from meteors. Two hides in the collection of the State Historical Society of North Dakota illustrate comet art (Maurer, 1992, pp. 274–275; Chamberlain, 1984). One winter count hide from the Upper Yanktonai Nakota band depicts events from 1785 to 1913. It clearly shows an early nineteenth century meteor pictograph and also a comet. A second hide, is from the *Hutlkpapa* Lakota. It chronicles their history from 1797–98 to 1911–12. One entry depicts Comet Halley's 1910 apparition. Maurer notes the *Hunkpapa* remembered the years 1911–1912 as the "time when children had measles and a bright comet appeared in the sky." This illustrates the important point that common Gregorian civil calendar dates may not always coincide with dates given for a particular pictograph.

Sun and Moon

A discussion of the role of the sun and the moon in Lakota life would encompass an entire book. For example, one could study how they reckon time. We take it for granted that the solar and lunar cycles can provide a calendar. Although the Lakota use solar and lunar cycles, they have a very different concept of time than Americans or Europeans. Times of the day, for example, are reckoned by what the sun or moon are doing in the sky. I will not in this discussion, undertake an in-depth analysis of the Lakota "calendar" and their nomenclature of the months and seasons of the year. Walker provides a convenient summary of the Lakota names for the divisions of time, which also give insights into how they viewed time.

> The day is divided into *anpao*, before sunrise; *anpa*, sunshines, and *htayetu*, fading-time (evening).... Midnight is when the first stars seen in the east are in the middle of the sky, *hancokaya*, night-middle-made.

> A moon was the time from the first appearance of the new moon in the west after the sun had set until the last appearance of the old moon in the east before the sun rose. The name of the moon was *hanwi*, night sun, which was commonly abbreviated to *wi*, sun.

There were thirteen moons and four seasons.... There were either four or five moons in the winter season.... The Discernible-not moon [*Tanin sni wi*] sometimes appeared and sometimes did not, and no one could tell for certain whether it had appeared or not until the seasons showed that it had appeared. The year began with the first moon of the first season which ordinarily was when the ducks return together in flocks [*Magaksica agli wi*], but if there was a Indiscernible moon, the year began with it.

> The winter season completed the year, and in speaking of the past, a winter meant a completed year. For this reason the Oglalas designated the past by winters instead of years. And the nearest that they could specify a particular time was of the moon of a season of a winter. (Walker, 1982, pp. 122–123)

Winter count hides were the visible representations of Lakota history and functioned as an annual calendar (Chamberlain, 1984). Although the Lakota use repeating cycles of nature to give names to the passage of time, I believe they have a very linear, continuous, view of time. Their history seemingly would never end, at least as long as the buffalo were on the earth. In attempting to understand the Lakota concept of time, I find the following assessment to be useful:

> In Lakota culture time was not conceived of as a causal force; history was not directed nor did it embody that notion of progress and change which is so fundamental to European culture. Instead, the universe was perceived as existing in harmonious balance. As Ella Deloria once put it, 'You see, we Indians lived in eternity.' (DeMallie, 1987, p. 3 1)

I have come to think of the Lakota as being suspended in time. There was no indication their situation would ever change until the white man (*wasichu*) began "putting down roots."

Living in an eternity where the sun was a persistent feature in the sky led to giving it a place of importance, and there is one ritual where the Sun's central role is best exhibited. The paramount ritual in Lakota life, then and now, is the Sun Dance. Beginning with Walker, many scholars have written volumes about the Lakota Sun Dance. Again, it is not possible to include a lengthy discussion of this important ritual in this discussion. It is, however, important to say that the Sun Dance is *not* Sun worship.

The Sun Dance is practiced by many of the Plains Indians and the Lakota call it *Wi Wa yag Wachipi*, or "Dance looking at the Sun." It is one of the seven sacred rites of the Lakota. These seven rites are a part of the "Sacred Pipe" as explained by the legendary Lakota elder Black Elk (Brown, 1989). The purpose of all seven rituals is to strengthen the *tiyospaye* ("community" or "extended family") and sense of *mitakuye oyasin* ("You are all my relatives").

The Sun Dance ritual includes fasting, purification, and dancing. At the center of the dance circle is a sacred tree around which the participants conduct themselves. Sometimes, even in modem times, the Sun Dance includes piercing of the flesh of male dancers who wish to provide a personal sacrifice. The sun is all-powerful. When a dancer participates in the Sun Dance, it is a participation in the sacred and mysterious. The Sun Dance serves primarily as an agent to strengthen the dancer's sense of self and spirituality, and strengthen connections to the community (*tiyospaye*).

In a contemporary reference, the Lakota at the Sun Dance are described as practicing a kind of sun watching reminiscent of the southwestern pueblo peoples.

> The old-man-who-counts watches the sunset, cutting notches in his stick, even though with calendars this is no longer necessary. The old man watches until the sun sets on a well-known landmark. When finally it reaches the appropriate place, the time has come to make the sacred lodge (Amiotte, 1987, p. 78).

I am uncertain how wide-spread this sun watching practice is. The mention of it, however, is a good reason to undertake further research into Lakota sun watching. This is the only reference to Lakota calendar sticks that I have found. This may be a late-nineteenth century or early twentieth century practice that became practical after confinement to the reservations and the Lakota no longer traveled great distances. However, at the Northern Plains Tribal Art Show in Sioux Falls, on September 28, 1996, Arthur Arniotte told me sun watching was practiced at the time of the Winter Solstice. He mentioned that although nomadic, the Lakota did return to the same location each year for their "winter camp" and hence there would be familiar horizon landmarks. Amiotte also said he knew of no Lakota calendar sticks in existence today. In addition to Walker's nineteenth-century records (Walker, 1917), an excellent modem, personal account of Lakota Sun Dancing is Thomas E. Mails' *Sundancing at Rosebud and Pine Ridge* (1978).

The sun and moon often appear together in Lakota images. I found an example of a *heyoka* warrior's body ornamentation (Walker, 199 1). The *heyoka* was a contrary who used his spiritual powers to satisfy the *Waki ya* or thunder beings of the west. On the great plains summer thunderstorms (and winter blizzards) approach from the west. The *heyoka* would dress warmly in hot weather, and wear few or no clothes in the dead of winter. The *heyoka* were considered *wakan*, and are an example of "that which is different."(Hirschfelder and Molin, 1992, p. 119) This particular drawing from a ledger book shows both a sun and moon symbol on the warrior's body. In a sense the sun and moon are opposites or contraries, and hence they are a sort of celestial "joke" when they appear together on the *Heyoka*.

Of course, the sun and moon are together during an eclipse. Solar eclipses were noted by the Lakota. An 1888 winter count hide pictograph by Short Man (Figure 3) reports *A pa wi wa te*, literally, "the sun died." (Walker, 1992, p. 151) However, a note dating of this eclipse on January 1, 1889, is problematic. Using *TheSky* 2.03 (Level III), I determined the dates of solar eclipses visible from western South Dakota from 1883 to 1892. Using "SunTimes for Windows" (Zephyr Services, 1995), I checked the sunrise and sunset times for the eclipse dates. Although there was an eclipse on 22 December 1889, the sun was not up in that area during the eclipse. The closest eclipse dates to 1888 to which Short Man could have alluded are 16 March 1885, 6 June 1891, and 10 October 1892. (Computational Note: I generally use the location of Wall, South Dakota, for these calculations for two reasons: It is centrally located in the area populated by the Lakota in the nineteenth century and there is no doubt on the part of any software, or this user, that Wall is

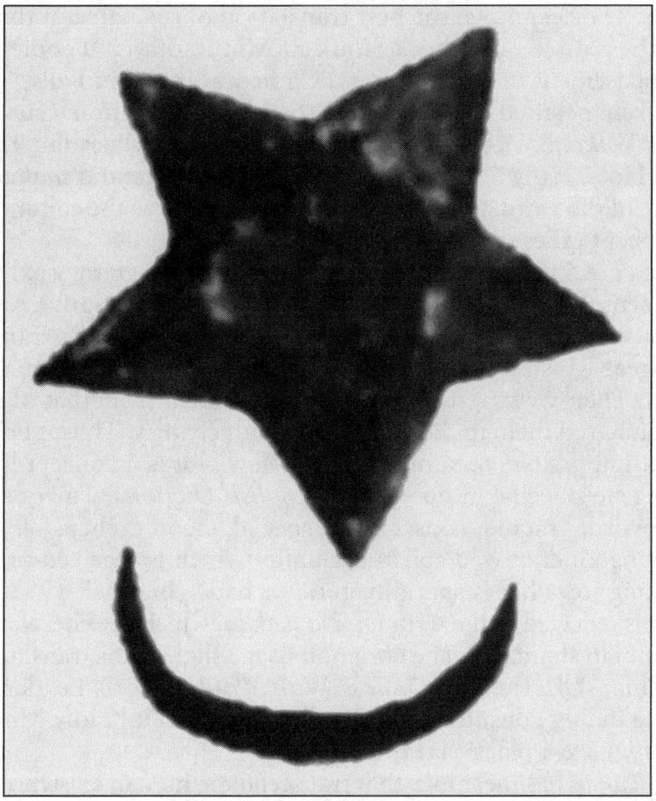

Figure 3. Solar Eclipse, 1888.

in the *Mountain* rather than *Central* time zone.) It also could be that the hide dates given for the event are slightly off. An event that occurred, for example, in the winter of 1888–1889 could be recorded in either 1888 or 1889 (Chamberlain, 1984).

Wakan — That Which Is Different

Having discussed some examples of Lakota celestial images, I will now attempt an interpretation of these images in terms of the Lakota concept *Wakan*. The Lakota did not exist apart from nature, but were very much a part of it. In this sense, the traditional ecological focus of the Lakota's relationship to the land and the animals, especially the bison (*tata ka*), that inhabited the land is correct. There has, however, been little interpretation of the meaning of the *celestial* events and phenomena that have been discussed in the previous section of this paper. It is my thesis that to understand the meaning of these celestial events and images one must attempt to understand *wakan* as the nineteenth century Lakota understood it.

I will draw on sections from Walker's conversations with Lakota elders to interpret the meaning of *wakan* itself. The most common translation of *wakan* is "sacred" or "holy." For example, the best translation of the name of the *Mdewaka to wa* band of the Santee or Dakota Sioux in Minnesota is "People of the Sacred Lake." In the basement of the Episcopal Cathedral in Sioux Falls, South Dakota, there is a small chapel used by a Lakota Christian congregation. Carved on an altar are the words *Wakan, Wakan, Wakan*, alluding to the Christian Trinitarian formula, "Holy, Holy, Holy." The best approach to understand *wakan* is to understand the usage of the word, and it is important to recognize the cultural context from which this concept takes its meaning.

The Lakota are a profoundly spiritual people. Although they exhibit one of the most genuine senses of humor and light-heartedness of any Native Americans, they are very serious about their relationship to *wakan ta ka* (often translated "Great Spirit" or "Great Mystery"), the created world, their fellow human beings, and their inner self. Their view of the world is very different from that of the white men who first encountered them in the early nineteenth century. Thus, I believe, it can be difficult, if not impossible, to summarize in a few words a concept like *wakan*.

Father Eugene Buechel in his *Lakota-English Dictionary* gives several connotations to the word: "sacred, consecrated, special, incomprehensible, possessing or capable of giving an endowed spiritual quality which is received or transmittable to beings making for what is specially good or bad" (Buechel, 1983, p. 525). This latter meaning is reflected in the term *wicasa wakan*, which Buechel translates "priest, formerly the Indian shaman." The common usage slightly mistranslates that phrase as "medicine man." In the film *Dances With Wolves*, Doris Leader Charge, who was the Lakota dialog consultant for the film, rendered it "Holy Man." The Black Hills (*Paha Sapa*) are a place that is *wakan*.

However, *wakan* has meanings that transcend the human existence of the Lakota on the great plains. James Walker offers valuable insights into the meaning of this concept to the nineteenth century Lakota. In the following lengthy excerpts from his writings, he uses Lakota elders' explanations of the meanings. As always, it is important to realize these are the opinions of the people he interviewed. First, he quotes from Good Seat:

> *Wakan* was anything that was hard to understand. A rock was sometimes *wakan*. Anything might be *wakan*. When anyone did something that no one understood, this was *wakan*. If the thing done was what no one could understand, it was *Wakan Tanka*. How the world was made is *Wakan Tanka*. How the sun was made is *Wakan Tanka*. How men used to talk to the animals and birds was *Wakan Tanka*. Where the spirits and ghosts are is *Wakan Tanka*. How the spirits act is *wakan*. A spirit is *wakan*...

The *Wakinyan* (Thunderbird) is one. The *Tata ka* (Great Beast) is one; the *Unktehi* (One Who Kills), *Taku Skanskan* (Changes Things), *Tunkan* (Venerable One), *Inyan* (Stone), *Heyoka* (Opposite'to Nature), *Waziya* (Of the North), *Iya, Tate* (Wind), *Yate* (North Wind), *Yanpa* (East Wind), *Okaga* (South Wind), *Iktomi* (Spider-like). These are all *wakan*. The sun, the moon, the morning star, the evening star, the north star, the seven stars [i.e., Pleiades], the six [sic] stars [i.e., Big Dipper], the rainbow—these are all *wakan*... (Walker, 1991, pp. 70–72)

At first glance, it seems as if *everything* was *wakan* for Good Seat. In one sense that is true, because so much of the Lakota world is extraordinary and "*Wakan* was anything that was hard to understand." But notice that the *wakan* or "mysterious" things Good Seat describes are things people cannot create or control. Good Seat's list of these objects, animals, and phenomena are things that have kept the best scientists and theologians busy for centuries. That is, they are extraordinary. *Wakan* things are those experiences, events, or objects that are worthy of wonder. Something that is *wakan* does not reach the analytical mind, but rather catches the spirit of a person and causes him or her to reflect.

Following his transcription of Good Seat's conversation, Walker adds his own interpretation concerning the meaning of *wakan*. Again, in this summary, he draws on his conversations with many Lakota, allowing them to interpret one another.

> The old Lakotas also believed that each thing except the *Wakan* and mankind had something like a spirit. This something they called a *nagila* (spiritish). These *nagipila* (spirits-ish) were *wakanpila* (*wakans*-ish). Ordinarily, the people call a *wakanla, wakan. Wakan Tanka* (Great *Wakan*) and *Wakan kin* mean the same. In former times the term *Wakan Tanka* was seldom used but now it is used more often than *Wakan kin*. The younger Oglalas mean the God of the Christians when they say *Wakan Tanka*. When a shaman says *Wakan Tanka*, he means the same as *Wakan Kin* as used in former times. This means all the *Wakan* and the *Wakanpila*, both of mankind and of other things, for the old Oglalas believed that these were all the same as one. This is *kan* (incomprehensible, an incomprehensible fact that cannot be demonstrated).
>
> Things other than the *Wakan* are called *wakan* by the Lakotas because they amaze as the *Wakan* do. A *wicasa wakan* (*wakan* man, shaman) is so called because he has marvelous power and wisdom so that he can speak and do as the *Wakan* do. *Mini wakan* (*wakan* water, intoxicating liquor) is so called because its effects are like those of the *Wakan*. *Maza wakan* (*wakan* iron, a gun) is called *wakan* because a shot from it is like the act of the *Wakan*...
>
> When an Oglala is amazed by anything he may say that it is *wakan* meaning that it is *wakanla* (like *wakan*). It appears from the above information that "divine" is the proper interpretation of *wakan* (Walker, 1991, pp. 72–74).

It is important to note that these characterizations are based on traditional Lakota beliefs from the nineteenth century as recorded by such white men as Walker. To give the meaning of *wakan* as "divine" shows a decidedly Christian influence. Encounters with whites, and in particular Christian missionaries, has had an effect on modifying the traditional mid-nineteenth century beliefs. Indeed, two important Lakota sources were Christian: George Sword was an Episcopal deacon and Black Elk was a Roman Catholic catechist. Comments from twentieth century Lakota elders may be "colored" by such encounters. Good Seat hints at this when he said, "I am an old man. I know what my father said. I know what his father said. In the old times, the Indians knew many things. Now they have forgotten many things. The white men have made them forget that which their fathers told them" (Walker, 1991, p. 70). Walker alludes to this modification when he comments "The younger Oglalas mean the God of the Christians when they say *Wakan Tanka*." Hence, the missionaries used *Wakan Tanka* to convey the Christian idea of God and usually translated the phrase "Great Spirit."

We who are white and born into a western way of thinking may have a difficult time understanding these meanings of *wakan*. In a white American (or European) way of thinking, there are distinctions between the natural and supernatural, between the secular and sacred, between the human and the divine. These distinctions are blurred or even non-existent in the Lakota mind. Such dichotomies do not exist. The important distinction for the Lakota was between the ordinary and the extraordinary. More than anything what characterizes the universe for the Lakota is its incomprehensibleness, its uniqueness, and its wonder. Thus, I believe the best interpretations of *wakan* are "incomprehensibleness," "that which is different" and "worthy of wonder." The best interpretation of *wakan tanka* might be "the great incomprehensibility." The Lakota were able to share in this wonder, mystery and incomprehensibleness through their rituals (DeMallie, 1987, p. 28) and their partnership with the natural world. Interestingly enough, although Lakota *belief* shows many variations, the *rituals* are generally more uniform. This may be because the rituals are a collective, community effort, serving to strengthen kinship ties.

Wakan and the Stars

Having discussed the meaning of *wakan*, I will now briefly explore how this important Lakota concept influenced their understanding of the stars. Walker records a conversation with Ringing Shield from May, 1903, in which the stars are said to be *wakan*.

> A wise man said this. The stars are *wakan*. They do not care for the earth or anything on it. They have nothing to do with mankind. Sometimes they come to the world and sometimes the Lakotas go to them. There are many stories told of these things. No medicine can be made to the stars. They have nothing to do with anything that moves and breathes. A holy man knows about them. This must not be told to the people. If the people knew these things, they would pull the stars from above. There is one star for the evening and one for the morning. One star never moves and it is *wakan*. Other stars move in a circle about it. They are dancing in the dance circle. There are seven stars. [Seven stars is a Lakota name for the Big Dipper.] This is why there are seven council fires among the Lakotas. Sometimes there are many stars and sometimes there are not so many. When there are not so many, the others are asleep. The spirit way is among the stars. This moves about so that bad spirits can not find it. *Wakan Tanka* keeps the bad spirits away from the spirit way. The spirit way begins at the edge of the world. No man can find it... The stars hide from the sun. They must fear him. So mankind should not try to learn about them. It is not good to talk about them. It is not good to fight by the light of the stars. They must be evil for they fear the sun (Walker, 199 1, pp. 114–115).

This passage is packed with both insights and contradictions. First, consider the North Star. The Lakota name for Polaris is *Wicahpi Owa jla*—"Star that always stands in one place." What is especially interesting about Ringing Shield's words is that the pole star is *wakan* because it does not move. This suggests that *wakan* connotes "that which is different" about Polaris compared to the other stars. Then Ringing Shield goes on to mention the morning star and evening star. He equates the seven stars of the dipper with the seven council fires of the Lakota. These statements about the stars may seem to be in conflict with the statement that "... mankind should not try to learn about them [the stars]. It is not good to talk about them." He says, in essence, "Don't even think about the stars," yet he, and the Lakota as a whole, have names for stars, legends about the stars, and record celestial events and phenomena on hides! In a similar discussion Little Wound says "The stars are *Wakan Tanka*, but they have nothing to do with the people on earth. Mankind need pay no attention to the stars" (Walker, 1991, p. 70).

There is a contradiction, or at least an ambiguity, here. If mankind should not even talk about the stars, or pay any attention to them, why should celestial events be recorded on hides, or appear on a *heyoka*, or become a part of story to be retold? I believe what Ringing Shield and Little Wound were saying was that the stars and their motions, or non-motions, are a part of the "great incomprehensibility" and hence mankind cannot fully understand them or their meanings. None the less, the sun, the moon, and the stars permeate Lakota thought and culture. Why?

It is simply because meteors, comets, the sun, the moon, and the stars *are wakan* that they were noted, recorded, and re-lived in stories and ritual. The stars are worthy of wonder, they are mysterious, they are incomprehensible. As fruitless as it may seem, reaching for the stars brings the great mystery a little closer to earth. Although the celestial images preserved on winter count hides, in ledger books and on clothing serve an ornamentation function, they too are a part of the belief and ritual of the Lakota. Each Lakota ritual serves a purpose: It reflects and transmits belief and tradition. Thus, Lakota celestial art is an integral part of the beliefs of the people who constructed it. And, the center of Lakota belief is the incomprehensibleness of the world in which they live. The stars, although extraordinary and *wakan*, are a integral part of that world.

Discussion

There are at least three areas for future research that would reveal more insights into the Lakota and their astronomy. First, I have briefly discussed meteor showers, eclipses, and comets. It would be helpful to continue the work of Von Del Chamberlain (1984) and to correlate the same events observed by various Native Americans on the great plains as recorded in ledger books and on winter count hides. Art historians might be able to shed some light on the Lakota's use of five-pointed stars. It may be that they began to use this symbol after encounters with the American flag. Second, I have observed aurora while camping in the Badlands National Park (about latitude 43° N). Are there any instances of aurora in Lakota art or stories? How did they interpret them? Finally, Amiotte's reference to calendar sticks and horizon sun watching is very intriguing. An examination of this aspect of Lakota astronomy would allow for comparison with the pueblo people of the southwest. A sensitive and respectful inquiry might reveal more about any use of these sticks today and the relative importance of sun watching for calendrical or ritual purposes.

More importantly, understanding *wakan* and how it relates to the stars must be the focus of any research into Lakota astronomy. That is, understanding their culture is the key to understanding their astronomy and their use of celestial images. There is a danger that meaning could be "read into" Lakota ideas. It is possible to avoid this by employing an approach from the "historical-critical method" of biblical hermeneutics and allow Lakota sources to interpret and explain other Lakota sources. There are ethnographic materials other than those of Walker which might reveal other insights. Also, it would be very interesting to ask contemporary Lakota elders and *wiscasa wakan* to comment on the monologues Walker recorded. However, it is important to recognize such commentary would represent the views of late twentieth century Lakota, many of whom have been strongly influenced by both Christianity and the twentieth-century writings of Black Elk.

The passion to be a part of "living the sky," to use Ray Williamson's term, was a central facet of Lakota life. Raymond DeMallie offers a final assessment that may help explain the ambiguity the Lakota have about the stars and their *wakan* nature:

> In the nineteenth-century Lakota system of belief, the unity of Wakan Tanka embraced all time and space, together with the entirety of being, in a universe where the place of human beings was minor but well-defined. Because this universe was most fundamentally characterized by incomprehensibility, it was beyond humanity's power ever to know it fully, and perhaps it was this futility that made the quest for understanding of the *wakan* the driving force in Lakota culture (DeMallie, 1987, p. 32).

Ultimately any attempt at understanding the Lakota's use of celestial images in their culture must also come to terms with their understanding of *wakan*. It may for us, as for the Lakota, be futile to come to any final conclusions that can satisfy a twentieth century analytical, scientific mind.

Acknowledgments

I am grateful to the librarians at Normandale Community College for procuring numerous reference materials for me. Katharine Salter-Godell, Department of Anthropology, Normandale Community College offered valuable suggestions and insights on this paper. Dr. Von Del Chamberlain, former director of the Hansen Planetarium, Salt Lake City, brought to my attention his important work on Plains Indians calendars. I am deeply indebted to my good friend, and Lakota mentor, Dr. Martin Brokenleg, Augustana College, Sioux Falls, for his patience with my endless questions.

References

Arthur Amiotte, "The Lakota Sun Dance: Historical and Contemporary Perspectives," in *Sioux Indian Religion*, ed. by Raymond J. DeMalhe and Douglas R. Parks, University of Oklahoma Press, Norman, 1987, pp. 75–89. ISBN 0-8061-2166-1.

Dee Brown, *Bury My Heart at Wounded Knee*, Henry Holt and Company, New York, 1970, ISBN 0-8050-1730-5.

Dee Brown, *The Fetterman Massacre*, University of Nebraska Press, Lincoln, 1962, ISBN 08032-5730-9.

Joseph Epes Brown, ed., *The Sacred Pipe: Black Elk's Account of the Seven Rites of the 0glala Sioux*, University of Oklahoma Press, Norman, 1989. ISBN 0-8061-2124-6.
Eugene Buechel, S.J., *Lakota-English Dictionary*, ed., by Paul Manhart, S.J., Red Cloud Indian School, Holy Rosary Mission, Pine Ridge, South Dakota, 1983.
Roberta Carkeek Cheney, *The Big Missouri Winter Count*, Naturegraph Publishers, Happy Camp, California, 1979, ISBN 0-87961-081-6.
Von Del Chamberlain, "North American Plains Indian Calendars," *Archaeoastronomy: Supplement to Journalfor the History of Astronomy*, Vol. 15, No. 6, 1984, pp. S1–S54.
Vine Deloria, Jr., *Red Earth, White Lies: Native Americans and the Myth of Scientific Fact*, New York: Scribner, 1995; ISBN 0-684-80700-9.
Vine Deloria, Jr., "*Sundancing and Stargazing*," Archaeoastronomy, Vol. V, No. 3, July–September, 1982, pp. 33–35.
Raymond J. DeMallie, "Lakota Belief and Ritual in the Nineteenth Century," in *Sioux Indian Religion*, ed. by Raymond J. DeMallie and Douglas R. Parks, University of Oklahoma Press, Norman, 1987, pp. 25–43. ISBN 0-8061-2166-1.
Don Doll, S.J., *Vision Quest: Men, Women, and Sacred Sites of the Sioux Nation*, Crown Publishers, New York, 1994, ISBN 0-517-59904-X.
Sven G. Froiland, *Natural History of the Black Hills and Badlands*, Center for Western Studies, Augustana. College, Sioux Falls, South Dakota, 1990, ISBN 0-931170-47-8.
Royal B. Hassrick, *The Sioux*, University of Oklahoma Press, Norman, 1964, ISBN 0-8061-2140
Arlene Hirschfelder and Paulette Molin, *The Encyclopedia of Native American Religions*, Facts on File, New York, 1992, ISBN 0-8160-2017-5.
George E. Hyde, *A Sioux Chronicle*, University of Oklahoma Press, Norman, 1993, ISBN 08061-2483-0.
George E. Hyde, *Red Cloud's Folk*, University of Oklahoma Press, Norman, 1975, ISBN 0-80611520-3.
George E. Hyde, *Spotted Tail's Folk*, University of Oklahoma Press, Norman, 1961, ISBN 08061-1380-4.
Joseph S. Karol, *Everyday Lakota: An English-Sioux Dictionary for Beginners*, Rosebud Educational Society, St. Francis Mission, St. Francis, South Dakota, 1974.
Lakota Home Page, http://maple.lemoyne.edu/~bucko/lakota.html
Lakota Star Knowledge: Studies in Lakota Stellar Theology, Sinte Gleska University, Rosebud, South Dakota, 1992.
Thomas E. Mails, *Sundancing at Rosebud and Pine Ridge*, Center for Western Studies, Augustana College, Sioux Falls, South Dakota, 1978, ISBN 0-931170-01-X.
Garrick Mallery, *Pictographs of the North American Indians*, Smithsonian Institution, Bureau of American Ethnology, Annual Report 4, 1886, pp. 1–256.
Evan M. Maurer, *Visions of the People: A Pictorial History of Plains Indian Life*, The Minneapolis Institute of Arts, Minneapolis, 1992. ISBN 0-295-97229-7.
Joe Rao, "The Leonids: King of the Meteor Showers," *Sky & Telescope*, Vol. 90, No. 5, November, 1995, pp. 24–31.
Karen Strom, *Native American History and Archaeology Resources on the Internet*, 1996, http.//hanksville.phast.umass.edu/misc/NAhistory.html.
Robert M. Utley, *The Indian Frontier of the American West 1846–1890*, University of New Mexico Press, Albuquerque, 1984, ISBN 0-8263-0716-7.
James R. Walker, *Lakota Belief and Ritual*, University of Nebraska Press, 1991, ed. by Raymond J.
DeMaille and Elaine A. Jahner, ISBN 0-8032-9731-9.
James R. Walker, *Lakota Myth*, University of Nebraska Press, 1983, ed. by Elaine A. Jahner, ISBN 0-8032-9706-8.

James R. Walker, *Lakota Society*, ed. by Raymond J. DeMaille, University of Nebraska Press, Lincoln, 1982, ISBN 0-8032-9737-8.

James R. Walker, *The Sundance and Other Ceremonies of the Oglala Division of the Teton Dakota, Anthropological Papers of the American Museum of natural History*, Vol. 16, Part 2, 1917, Lakota Books, Kendall Park, New Jersey (reprint).

20

Effigy Mound Archaeoastronomy

Frank D. Stekel
Physics Department
University of Wisconsin-Whitewater
Whitewater, WI

Effigy mounds are earthworks that include representations of birds and animals. These were constructed from about 700 to 1,300 AD by the people of the Effigy Mound Culture in the area centered around southern Wisconsin. Most of these mounds have been destroyed by agriculture and construction, but good surveys were made of many of these features in the late 19th Century by T. H. Lewis. Statistical analysis of Lewis' data for effigy mound surveys indicates that the orientations of long, linear features are not randomly distributed. Some, but not all, alignments appear to be associated with astronomical phenomena. There are indications that alignments with the cardinal directions exist. There may also be correlations with solstice sunrise locations.

* * *

Introduction

Effigy mounds are distinctively shaped earthworks. They were often constructed in stylized shapes of birds and animals. The mounds have heights up to 2 meters above grade and range in size from less than 10 to more than 100 meters in linear dimensions. Effigies are often found in groups along with round, conical mounds and linear mounds. Estimated ages indicate their construction during the period from about 700 A.D. to 1300 A.D., with a wider range being from as early as 300 A.D. and extending up into the 17th century. These earthworks are found across what is now southern Wisconsin and the adjacent border areas of Iowa, Minnesota and Illinois. It is currently accepted that they were built by the native inhabitants of the area, the people generally referred to as the Effigy Mound Culture. (Peterson, 1979; Hurley, 1975; Ritzenthaler & Goldstein, 1985; Rowe, 1956; Lapham, 1855) 20,000 is often used as an estimate of the original number of effigy mounds in Wisconsin (Petersen, 1979). Since the settlement by Europeans, most of these earthworks have been destroyed as a result of road or building construction and agricultural cultivation. (Petersen, 1979; Northwestern Archeological Survey, 1895; Lapham, 1855)

There is no single explanation for why these earthworks were constructed. Some, but not all, mounds have burials within them. Mounds are often found near bodies of water, but these are the areas that also provided food resources for the people who constructed them. It has been suggested that the construction of the mounds was an important social activity. (Hurley, 1986; Goldstein, 1987; Mallam, 1976) It is evident that the construction of the mounds represented a substantial effort and significant amounts of time and labor were needed.

Archaeoastronomy includes the study of surviving artifacts to provide evidence of astronomical practices in the ancient world. Good evidence has been obtained for this type of activity at old world sites such as Stonehenge (Hawkins, 1965; Fernie, 1990). Sites investigated in North American have included those of the Maya in Mexico (Aveni, 1980), the "medicine wheels" of the northern plains (Eddy, 1974) and others in the USA (Williamson, 1984). The earthworks of the Effigy Mound Culture appear to offer an interesting problem for an archaeoastronomy investigation: Can a relationship be established between the features of Effigy Mound earthworks and astronomical phenomena?

Cultural, ethnographic information describing activity that includes astronomical observations is often a productive guide to archaeoastronomical investigation. (Carlson, 1990; Williamson, 1984) Of the Native American tribes present in Wisconsin at the time of the arrival of Europeans, the HoChunk or Winnebago have been identified as descendants from the Effigy Mound culture (Radin, 1923). However, not all evidence identifies the Effigy Mound People as the HoChunk ancestors, and the prevailing archeological view has been that "the Effigy Mound peoples have yet to be firmly attached, via direct historic methods, to any modern tribe in the Upper Great Lakes region." (Hurley, 1986) Ethnographic data may prove useful for supporting or refuting such a linkage, but I am not aware of any astronomical applications beyond recognition of cardinal directions among recorded HoChunk activities. (Radin, 1923)

Procedure

Consequently, one must begin with the earthworks and determine if an astronomical application is reasonable. Some caution is required. "When archaeoastronomy was new, enthusiasts rushed to plumb, align, and measure anything that looked halfway plausible—and tended to find significant celestial alignments everywhere." (Carlson, 1990) If enough sight lines are considered, sooner or later, one of them will be found that points to something. Effigy mounds are often found in groups along with conical and linear mounds. If sight lines defined by pairs of points are considered, using points for all mounds in a group and several points to define a large animal or bird effigy, a large number of lines is generated from even a small group of mounds. A more restrictive procedure seems appropriate for an initial investigation.

For the first phase of this investigation, long straight features of the earthworks were used to define azimuth directions. The set of directions that was established was then examined to see if any correlation with astronomical phenomena was evident. The precision of the angles that were defined by this process depended upon the length, L, and W, the uncertainty in the location of the center of the linear feature. When L is much greater than W, the uncertainty in the angle can be estimated by:

$$\Delta\Theta = [\, W / 2 \pi L \,]\; 360°$$

An arbitrary choice of 80 feet was made as the minimum length for the consideration of a straight, linear feature as a directional indicator. This 80 foot criterion could be satisfied by a linear mound as such or by a straight section of a large effigy such as a bird wing or an animal tail. Feet, not meters, are used as units since these are the units for the historic survey data that are available. (Northwestern Archeological Survey, 1895; Stout & Skavlem, 1908) These are earthworks and their edges are not crisply defined, so the location of the center will also be somewhat uncertain. If feature width, 2W, is about 1 meter or 3 feet, this means that $\Delta\Theta$ is about 1°. Note that this precludes any excessive claims of angular precision; one need not worry much about 0.10 degree errors due to factors such as atmospheric refraction (Weiss, 1990) if the intrinsic uncertainty is more like ±1°. If some sighting aids such as stakes or poles (Williamson, 1984) had been used, more precise directions would be defined. However, I am not aware of any such features having been reported as associated with linear or effigy mounds.

The historic surveys were made using magnetic compasses to define north. (Northwestern Archeological Survey, 1885). A correction of 5° was added to each of Lewis' azimuth values so that they correspond to azimuth positions for astronomical phenomena at these sites in southern Wisconsin.

The data acquisition and analysis procedure must consider these uncertainties as well as statistical processes that will be involved. From a given location, one can see a full 360 degree horizontal circle around that point barring obstruction by terrain or vegetation. However, using two points or even a straight, linear feature establishes only a line and not a compass azimuth. Consequently, only 180 degrees of azimuth values were considered, with each value corresponding to both the azimuth and the opposite direction: 90 degrees refers to both 90 and 270 degrees, not just east but more properly east-west. No attempt was made to determine which direction an earthwork was "facing," only the line along which it was directed was recorded. These direction lines were recorded from microfilm copies of T. H. Lewis' survey notebooks. (Northwest Archeological Survey, 1895) In most instances the line of interest was present in the field data. When the most obvious linear direction was not specified by the survey points that had been recorded, a scaled sketch of the earthwork was drawn and the azimuth of this line was measured manually with a protractor.

The analysis of the set of azimuth values followed the same procedure as had been applied in the first phase of this project. (Stekel, 1990) This procedure had generated 67 azimuth values using published maps for effigy mound groups around Lake Koshkonong in southern Wisconsin. (Stout & Skavlem, 1908) Ten 18 degree increments were again used as the categories in which the data were grouped and a chi square test was used to determine if the number of values in each 18 degree increment was significantly different from the uniform number that a random distribution would predict. (See Table 1)

Table 1. 18 Degree Azimuth Increments & No. of Features Present

Lewis' Values	Azimuth	TOTAL	Animals	Birds	Linear
166.5 to 180 and/or 0 to 4.4	172 to 180 and/or 0 to 9	65	22	16	26
4.5 to 22.4	10 to 27	80	25	26	27
22.5 to 40.4	28 to 45	70	21	27	18
40.5 to 58.4	46 to 63	75	16	37	18
58.5 to 76.4	64 to 81	68	16	25	25
76.5 to 94.4	82 to 99	81	21	33	27
94.5 to 112.4	100 to 117	84	21	24	37
112.5 to 130.4	118 to 135	75	18	35	16
130.5 to 148.4	136 to 153	52	8	24	16
148.5 to 166.4	154 to 171	54	9	27	15
Sum	704	177	274	225	
Mean	70.4	17.7	27.4	22.5	
Chi Square	15.3	15.8	12.5	20.0	

All of these data are from Lewis' records. (Northwestern Archaeological Survey, 1895) The chi square test was used to determine if the population in the listed 18 degree increments is significantly different from the uniform values that could be expected from a random distribution. For the total set of 704 features, $0.10 > P > 0.05$. This was at the one chance out of ten level, so it was unlikely to be a random distribution. For the animal features, $0.10 > P > 0.05$. This was also unlikely to result from a random distribution. For the 274 bird features, $0.20 > P > 0.10$. The level of significance was not as great here, so interpretation here must be more cautious. For the 225 linear mounds, $0.02 > P > 0.01$. This was very unlikely to be a random distribution.

If the values were not randomly distributed, there should be a pattern. Note the graphs in Figures 1 through 4.

The data sets with the greatest statistical significance were most apt to be a non-random distribution for their azimuth values. Regions with the most features in them were compared with the azimuth values for phenomena listed in Table 2.

Examination of the distribution of mound features' orientations implies some recognition of the cardinal directions. North-south has more than an average number of mounds for both the animal and linear mounds' data. East-west is about equally evident for the total and all of the individual mound types.

There may be some recognition of the solar system phenomena. The 46–63 region, representing a line northeast-southwest, has more than an average number of mounds in it due to the large number of bird features found here. This could be an alignment with northernmost rises, such as the sun at summer solstice, or southernmost setting positions, such as sun at winter solstice or moon/planets at south extreme. The 118–135 region, representing a line southeast-northwest, has fewer mounds than average in it, but there are more than an average number of bird features here as well. This would fit with solar system objects' southernmost rises and northernmost setting positions. Only the bird features seem to be associated with the sun, moon or planets in this simple, linear features analysis.

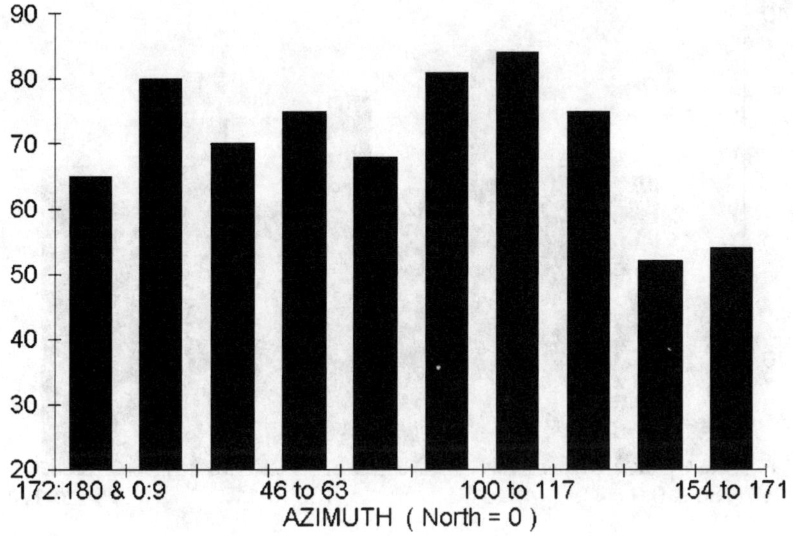

Figure 1. LINEAR features associated with effigy mounds in Wisconsin (Total = 704).

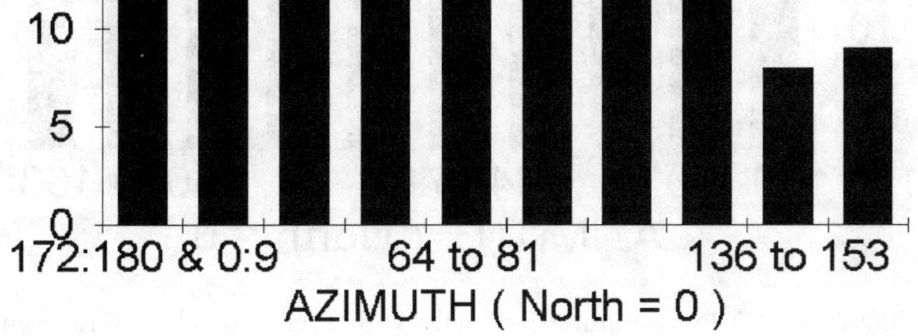

Figure 2. ANIMAL features associated with effigy mounds in Wisconsin (Total = 177).

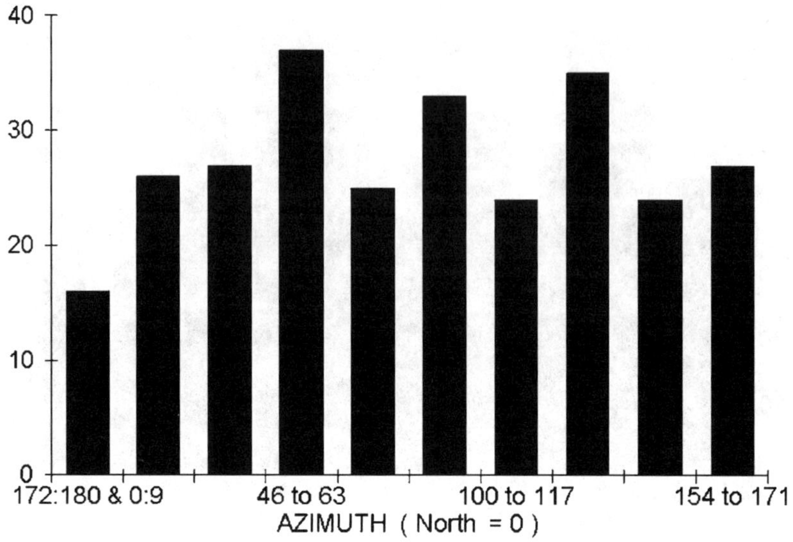

Figure 3. BIRD features associated with effigy mounds in Wisconsin (Total = 225).

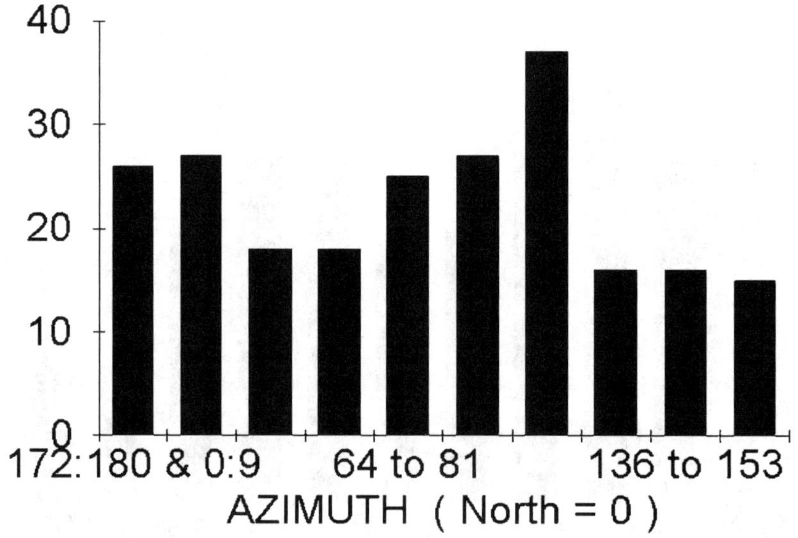

Figure 4. LINEAR MOUNDS features associated with effigy mounds in Wisconsin (Total = 704).

Table 2. Azimuth Values for Selected Astronomical Phenomena for Observers Near 43E North Latitude.

Phenomena		Declination (Angle above or below equator)	Azimuth (Position on horizon, North = 0)
North Celestial Pole		90°	0°
SUN	Summer Solstice	23° 27'	
	Sunrise		56°
	Sunset		304°
SUN	Equinox	0°	
	Sunrise		89°
	Sunset		270°
SUN	Winter Solstice	-23° 27'	
	Sunrise		122°
	Sunset		238°
MOON	North Extreme	28° 36'	
	Moonrise		50°
	Moonset		310°
MOON	South Extreme	-28° 36'	
	Moonrise		130°
	Moonset		230°
VENUS	North Extreme	26° 50'	
	Rise		51°
	Set		309°
VENUS	South Extreme	-26° 50'	
	Rise		127°
	Set		233°
SIRIUS		-16° 35'	
	Rise		110°
	Set		245°
PLEIADES		24° 04'	
	Rise		53°
	Set		303°

With this analysis, the largest number of features is found in the 100–117 degree azimuth range. This is due mostly to the excess of linear mounds, with this sector containing almost 60% more than average. This is between the east-west line (azimuth 90) and the southeast horizon location (azimuth 122–130) of southernmost rises for solar system objects: sun, moon and planets. These features could be associated with an event such as sunrise in late October–early November. However, ethnographic data related to a sunrise-related ritual at this time of the year would be needed to substantiate such a conclusion and this information is not available now. The only unique astronomical phenomenon that seems to occur in this sector is the rising of a star such as Sirius.

A large number of features is also found in the 10–27 region. More than an average number of features appear here for both animal mounds' features and linear mounds. This is difficult to explain for any astronomical purpose. The sun, moon or planets would not be found at these regions of the horizon. If the line is to represent the northeast horizon, it involves the rising of objects that are at declinations of about 40 degrees; the bright stars Deneb, Vega and Capella are close

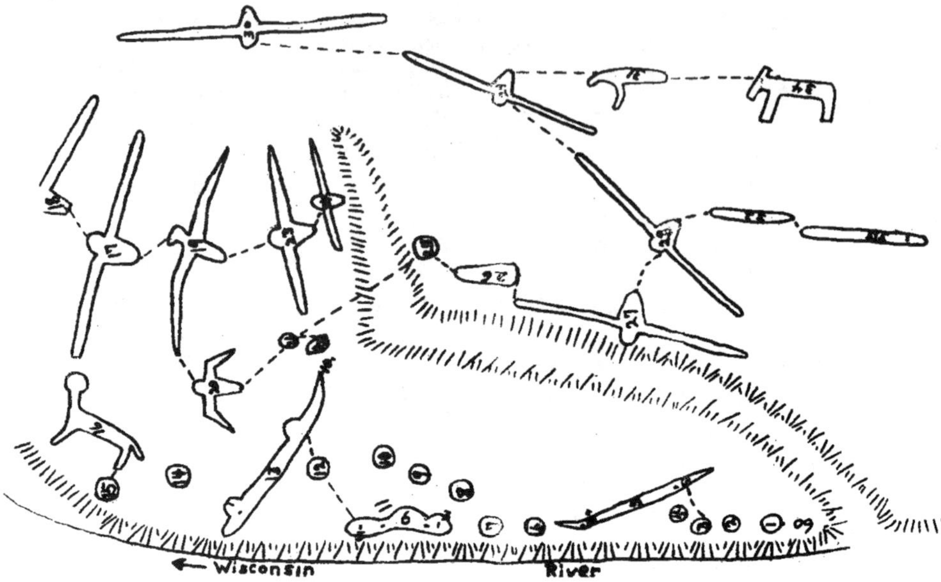

Figure 5. McClary Mound Group No. 4, Richland County, WI. Copy of sketch by T. H. Lewis in April, 1886. Northwestern Archaeological Survey, Notebook 25, page 39. With permission of Minnesota Historical Society.

to this. In the opposite direction to the southwest, this would be the setting of objects of almost -40 degrees declination; even Antares is higher than this, so viewing any astronomical phenomena along this region of the south-southwest horizon is unlikely.

Field surveys and measurements to check these calculated, predicted phenomena are essential. Observations were made at the December 1990 solstice to check if the extended wing bird mound at the Effigy Mound Park in Whitewater, Wisconsin (Stekel, Johns and Scherz, 1991), did mark the solstice sunset position. Actual observations show a much better alignment with a linear mound labeled #5 by C. E. Brown (1920) than with the bird wings. The alignment, if any, for this bird mound would more likely be with something further south than the sun, such as the south extreme for the moon or planets. This is especially speculative due to damage to the west wing.

Scherz (1990) has reported an extensive field survey at Lizard Mound County Park in Washington County, Wisconsin, near West Bend, Wisconsin. This site was not included in Lewis' survey. (Northwestern Archaeological Survey, 1895) Some ethnographic reports were used here to help relate astronomical phenomena and observations at this site. It may be possible to extend this procedure to other sites.

An April 1886 field survey of the McClary Mound Group No 4 in Richland County, Wisconsin, near Muscoda, Wisconsin, was reported by Lewis on pages 39–51 of notebook 25 (Northwestern Archaeological Survey, 1895). A transcribed copy of Lewis' field sketch is shown in Figure 5. We surveyed the mounds that remain of this site in July, 1990. Our map is included as Figure 7 (Stekel, 1991). Figure 6 is a similar map drawn from Lewis' survey data. There is an excellent correlation between the Figures 6 and 7.

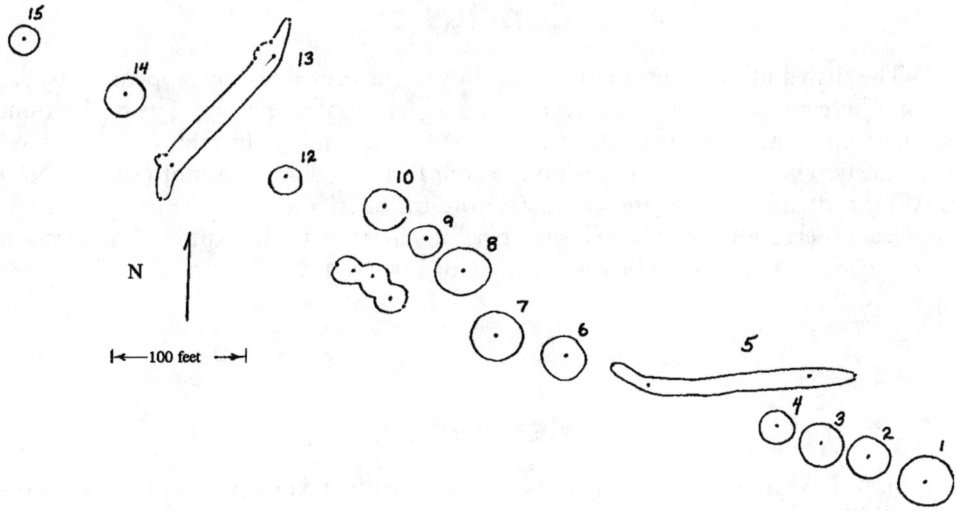

Figure 6. Scale Drawing based upon T. H. Lewis' Survey Data for McClary Mound Group No. 4, Richland County, WI.

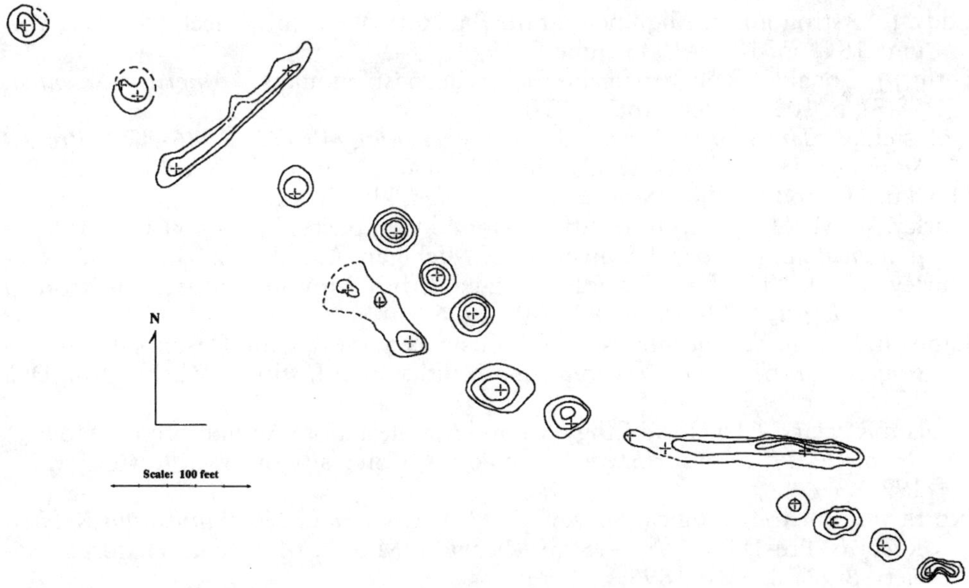

Figure 7. McClary Mound Group No. 4. R 1 W, T 9N SE 1/4 Section 34 Richland County, WI. Sketched and surveyed July 1990 by Frank Stekel, Frank Shadewald, and Shirley, Sharon and Sandra Stekel.

Conclusion

The distribution of effigy mounds' linear features does not appear to be random. There appears to be some recognition of cardinal directions. The bird mounds seem to show the most correlation with solstice phenomena, but this must be viewed cautiously. The abundance of mounds in the 10 to 27 degree azimuth sector (NNE-SSW) is difficult to associate with astronomical phenomena on the horizon. Astronomical observations may have been a part of determining effigy mound orientations, but they do not appear to be the only factor involved.

References

Aveni, A. F. *Skywatchers of Ancient Mexico*, Austin, TX: University of Texas Press, 1980.

Brown, C. E. Maps and notes dated September 21, 1920, Box 43 of the Charles E. Brown Manuscripts, *Wisconsin State Historical Society Archives*, Madison, WI, 1920.

Brown, C. E. "An Expert Describes the Maples Mounds," letter published in *The Whitewater Register*, Whitewater, WI, p. 1 and 3, September 30, 1920.

Carlson, John B. "America's Ancient Skywatchers," *National Geographic*, 177(3), p. 76–108, March, 1990.

Eddy, J. "Astronomical Alignment of the Big Horn Medicine Wheel," *Science*, Volume 184, pp. 1035–1043, June 7, 1974.

Fernie, J. Donald. "Stonehenge and the Archaeoastronomers," *American Scientist*, 78(3), p. 103, March–April, 1990.

Goldstein, L. *The Southeastern Wisconsin Archeology Project: 1986–87 & Project Summary*, University of Wisconsin-Milwaukee, Milwaukee, WI, July 15, 1987.

Hawkins, G. and J. White. *Stonehenge Decoded*, New York, NY: Delta Dell, 1965.

Hurley, W. M. "An Analysis of Effigy Mound Complexes in Wisconsin," *Anthropological Paper No. 59*, University of Michigan, Ann Arbor, MI, 1975.

Hurley, W. M. "The Late Woodland Stage: Effigy Mound Culture," *Wisconsin Archeologist*, 67(3–4), p. 283–301, 1986.

Lapham, I. A. "The Antiquities of Wisconsin as Surveyed and Described," *Smithsonian Contributions to Knowledge*, Smithsonian Institute, Washington, DC, 1855.

Mallam R. Clark. "The Iowa Effigy Mound Manifestation: An Interpretive Model," *Report 9, Office of the State Archeologist*, University of Iowa, Iowa City, IA, 1976.

Northwestern Archeological Survey. (T. H. Lewis') *Field Notebooks and Related Records*, Pre-1880, 1880–1895; Microfilm M 549, Minnesota Historical Society, St. Paul, MN, 1895.

Petersen, R. The Wisconsin Effigy Mounds Project, I, Luther College Archeological Research Center, Decorah, IA, October 15, 1979.

Radin, P. *The Winnebago Tribe*, Thirty-Seventh Annual Report of the Bureau of American Ethology, Smithsonian Institution, Washington, DC, 1923, Reprinted by University of Nebraska Press: First Bison Books, 1990.

Ritzenthaler, R. E. (Revised by L. Goldstein). *Prehistoric Indians of Wisconsin*, Milwaukee Public Museum, Milwaukee, WI 1965.

Scherz, J. Survey Report: Lizard Mound Park, Washington County, Wisconsin, *Ancient Earthworks Society, Inc.*, Box 1124, Madison, WI, 1990.
Stekel, F. "Indian Mounds Archaeoastronomy," *Report submitted to University of Wisconsin-Whitewater Physics Department*, UW-Whitewater, Whitewater, WI, 1990.
Stekel, F., L. Johns and J. Scherz. "Whitewater Effigy Mound Park: The Maples Mound Group," *Wisconsin Archeologist*, 72 (1–2), p. 118–126, March–June, 1991.
Stekel, F. "Effigy Mound Archaeoastronomy II," *Report submitted to University of Wisconsin-Whitewater Physics Department*, University of Wisconsin-Whitewater, Whitewater, WI, 1991.
Stout, A. B. and H. L. Skavlem. "The Archaeology of the Lake Koshkonong Region, *Wisconsin Archeologist* o.s. 7(2), pp. 47–102, April–June, 1908.
Weiss, Peter L. "Reflections on Refraction," *Science News*, Vol. 138, p. 236–237, October 13, 1990.
Williamson, R. A. *Living the Sky: The Cosmos of the American Indian*, Chapter 11, Boston, MA: Houghton-Mifflin, 1984.

Section 3

Central and South America

21

A Lunar Origin for the Mesoamerican Calendars of 20, 13, 9, and 7 Days

Martha J. Macri
Departments of Native American
Studies and Anthropology
University of California, Davis
Davis, CA

J.T. Goodman (1897) suggested that the "gods" of the numbers one to twenty matched the twenty Maya day glyphs. Eric Thompson (1950:88, 131–36) agreed that the first thirteen numbers do indeed correspond to 13 of the Maya day glyphs. Building on these observations, I propose that the Mesoamerican count of 13, the set of 20 day names, the Nine Lords of the Night, and the 7-day period of the glyph Y cycle, were at one time components of a synodic lunar count. The count of 13 originally commemorated the waxing of the moon from the first visible crescent to the full moon. Seven more days (added to 13 making 20) count from the full moon to the third quarter. The Nine Lords of the Night originally signified the moonless nights beginning with the third quarter moon, which rises at midnight, and continuing through the period of invisibility. The cycles, originally sequential, were, at some point, separated from the original schema and used as independent concurrent repeating cycles.

* * *

Calendars of Mesoamerica

The indigenous peoples of Mexico and Central America have for several millennia shared a calendrical system consisting of a year of 365 days, composed of 18 periods of 20 days plus 5 additional days, and a cycle of 260 days, composed of the combination of the numbers 1–13 with twenty day names. The unique matching of day designations in these cycles repeats every 52 years. The symbols (glyphs) for these days and months are known from several cultures, e.g. Maya, Epi-Olmec, Zapotec, Mixtec, Aztec, etc., and words for these days and months are known from virtually every linguistic group.

From the first century BCE to the Postclassic Period peoples of the Gulf Coast, Highland Guatemala, and the Maya also counted days from a starting point: September 8, 3113 BCE The date is given in single days, and multiples of 20 days, 18x2O days (360 days), 20x360 days, and 400x360 days up to 13 periods of 400x360 days at which point the count begins again. The Epi-Olmec (c. 200 BCE–250 CE) of the Gulf region and the Preclassic peoples of Highland Guatemala (100 BCE–100 CE) recorded this count with bar/dot numerals, with a placement of smallest periods (days) at the bottom, largest (400 years) on top. The lowland Maya, whose languages require the presence of numeral classifiers, nearly always recorded these dates with bar/dot numerals followed by period glyphs which functioned as numeral classifiers.

The Maya of the Classic Period (c. 250 C.E.–900 CE) also recorded the placement of a day in repeating cycles of 9 and 7 days. Classic Period monumental inscriptions included the position of the day in the current lunation, the position of this lunation in a repeating cycle of 6 lunations, the name of this lunation, and the length of the current month (29 or 30 days).[1] Table 1 summarizes these cycles, the methods by which they were recorded, and approximate earliest and latest attestations of the information.

Origin Theories for a 260 Day Count

The notion of a calendar of only 260 days, is, as far as I am aware, unique to ancient Mesoamerica. Scholars have long sought a motivation for such a time period. Some have suggested that it grows out of a fascination with the number 13 and the vigesimal system present throughout the region. This does not, however, explain why 13 might be important. One suggestion offered by Zelia Nuttal in 1928, and more recently by Malmström,[2] is that at the latitude of 14.8°, the latitude of Izapa, Guatemala and Copán, Honduras, 260 is the number of the days between the zenithal sun's southward and northward passage. Thompson (1950:98) raises several objections to this theory. The region for which the period is 260 days, is, in fact, quite limited when compared to the area in which the calendar was used. Further, he suggests that the 260 days ought to have been followed by the complementing 105 days rather than repeating continuously. Finally, Thompson objects to the assumption that the cycle originated in the region of Copán. Malmström's focus on Izapa is poorly chosen, since the site hosts numerous narrative sculptures but no hieroglyphic inscriptions.[3]

Others favor the explanation that 260 days approximates the length of human gestation (more precisely, the time from the first missed menstrual cycle to the day

1. For a detailed explanation of these calendars see Aveni (1980:133–173).
2. Malmström (1997).
3. Vail (1997).

Table 1. Mesoamerican Calendars

Cycle	# Days	Method of Recording	Earliest	Latest
365 day year	365	bar-dots/head variants + a glyph for each of 18 months and a 5-day period	Preclassic/Mesoam[1]	present/Mesoam
20 day names	20	a glyph for each day	Preclassic/Mesoam	present/Mesoam
trecena	13	bar-dots/head variants	Preclassic/Mesoam	present/Mesoam
long count	0–1,872,000	bar-dots/head variants with period glyphs	37BCE/Epi-Olmec	Postclassic Maya
9 day glyphs	9	a glyph for each day	Classic/Maya	present/Zapotec
7 day series	7	bar-dots/head variants with Glyph Y[2]	Classic/Maya	Classic/Maya
lunar count	29/30	glyphs for 20 with bar-dots/head variants	Classic Maya	Classic Maya
819 day sets	3276	trecena + day name given at 819 intervals (quarter points) 819 = 7 x 9 x 13	Classic/Maya	Classic/Maya
lunar sidereal count (27.3 x 3)	82	known from date intervals of multiples of 82	Classic/Maya[3]	present/Quiche[4]

1. Approximate periods for the Maya: before 250 C.E. is Preclassic; 250–900 C.E. is Classic; after 900 C.E. is Postclassic.
2. Yasugi and Saito (1991).
3. Dütting and Schramm (1988).
4. B. Tedlock (1992:190–197).

of birth). In highland Guatemala today, women measure the length of pregnancy by the moon, and this is offered as the motivation for the 260 day calendar.[4] Indirect support for this notion comes from the practice in several Mesoamerican cultures to name babies for the name and number of the day of their birth, the same name and number of the day they were conceived.

Several researchers have explicitly suggested a lunar origin for the 260 day calendar. Spinden[5] noted that two 260 day periods equal three eclipse seasons, Long[6] suggested that 260 days approximated nine lunar months. Jakeman (1947) offers as support for a lunar origin:

1. A year of months called *uob* 'moons' was recognized in the sixteenth century.
2. The presence of the lunar count in initial series dates (days from first crescent, month number in a cycle of 6 months, the name of the lunation, the length of the current month (alternating 29 and 30 days).
3. Frog *uo* used for the twenty day period, for moon *u*.
4. The glyph for moon also used for the number 20.

He proposed a chronology which begins with an original lunar calendar, then a count of lunar calendar years, a decimal system, the invention of writing, a ritual or market period of 7 days, month names, creation of the winal and the sacred year of 260 days, invention of the sacred almanac (divinatory calendar), invention of the long count, initial series dates (long count with lunar series and calendar round), record of the k'atuns, the 9 Lords of the Night, lunar count of the supplementary series, the 819 day cycle, the 18 month solar calendar, standard initial series dates, calendar round dating, stela cult, period ending dates, distance numbers, and ultimately, the adoption of the year-bearer count.

The Lunar Cycle

The thesis presented here is that the Mesoamerican calendrical cycles of 13 number, 20 day names, and repeating cycles of 9 and 7 are based on various segments of the lunar cycle. Several years ago while questioning why the number 13 was so important two different people suggested to me that it was because 13 is the average number of days from the first visible crescent to the full moon.[7] Beattie, a student of anthropology and archaeoastronomy, suggested that there might be a relationship between the lunar cycle and the counts of 13, the 20 day names, and the count of 9. I subsequently noted that several of the Maya head variant signs seem to have had a lunar significance (Macri 1982; 1984). At that time I mentioned, but did not explicate, the relationship between the head variants and 13 of the Maya day signs noted by Goodman and Thompson. What follows here is a brief examination of possible correlations between segments of the lunar cycle and the Mesoamerican day counts of 20, 13, 9, and 7.

4. Tedlock (1992:93).
5. Spinden (1924).
6. Quoted in Jakernan (1947).
7. I am grateful to Dee Beattie and Birch Moonwomon for this suggestion.

Over 2500 years ago, some Mesoamerican culture (perhaps several) observed and incorporated into their world view a lunar cycle that was composed of three segments: a 13 day waxing period, a 7 day waning period from the full moon to the third quarter moon, and a 9 day period from the third quarter through the period of invisibility. Some group—perhaps a single person, almost certainly speakers of a Mayan language, then took each of these three segments and created several independent repeating cycles. This community then had enough political power and cultural energy to set in motion the eventual spread of at least certain aspects of this new scheme to peoples throughout Mesoamerica.

The segments of the lunar cycle that they identified correspond to what we now understand as the trecena (the repeating count of 13 days), the twenty Mesoamerican day names (consisting of the trecena plus 7 added days), a count of 9 (known from the Aztecs as the Nine Lords of the Night), and a recently identified count of 7.

Thirteen and Twenty

The phenomenon of the waxing moon has held significance for peoples throughout the world. It is universally and directly observable; it counts forward to those nights in which the moon provides considerable light. Celebrations in numerous cultures have taken place on the night of the full moon. The average number of days from the first visible crescent to the full moon is 13.[8] There are two cycles used in Mesoamerica which testify to the importance of the period of the waxing moon: the count of 13 days, and the set of 20 day names (both continuous repeating cycles).

The Head Variants

A set of head variants, unique to the ancient Maya, represented the numbers one though 20 (Figure 1). The glyphs are distinct, however, only up to the number 13.

The number 13 sometimes has a unique sign, but it may, like the numbers 14–19, be composed of the head for the smaller number (3–9) with the addition of a skeletal jaw. The head variant for 10 is a skull, so these composite signs are actually based on a decimal system—as are the numbers 13–19 in Mayan languages. So the set of 13 unique head variants seems to have its origin in the count of 13, the set of repeating numbers combined with the set of 20 day names to form the 260 combinations of numbers and day names which comprise what is called the sacred calendar.

What do the head variants signify? Are they 13 gods? Is it possible to establish that the head variants have any lunar associations? Nearly all of the faces of the head variants represent approximations of the sounds of the words for the numbers in Mayan languages. In some cases, such as the numbers 1, 5 and 8 the match between phonetic elements on the head variants and the Maya word is clear. In others, such as 2, 5, 9 and 12, however, the sound signified by the head variant only approximates the number. Such approximations would not have been tolerated in the hieroglyphic script of the Classic Period. However, the origin of the head variants certainly predates the period of true writing. They may, in fact, represent one of the earliest attempts to represent words visually. The clearest correlations are shown in Table 2.

---

8. This may account for the sacredness of 13 to a number of cultures, including contemporary Western culture in which 13 is considered unlucky.

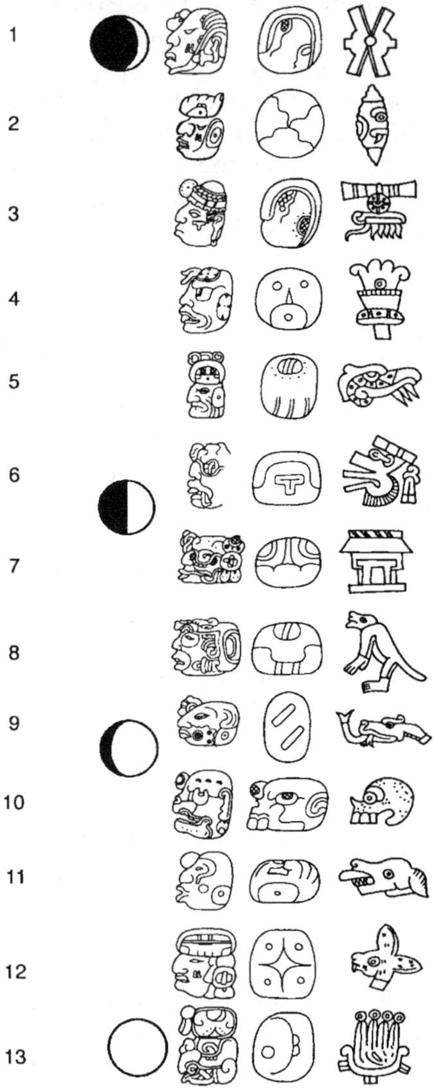

Figure 1. Lunar phases with the Maya head variants and the Maya and Mexican day signs, Kab'an/Movement through Muluk/Water (Maya head variants and day signs by Matthew Looper, courtesy of the Maya Hieroglyphic Database Project, Mexican signs after Caso (1967:6–7).

Table 2. Head Variants Phonetically Associated with Maya Numbers

#	Glyph	Yucatec Maya	English	Maya
1	woman's head	*na'* 'woman'	first	*nah*
2	fist above head	*k'ab'* 'hand'	two	*ka'*
5	man, year sign headdress	*haab'* 'year'	five	*ho'*
6	ax in eye	*wak'* 'chop'	six	*wak*
8	face with sprout scrolls	*wah* 'tortilla; corn'	eight	*waxak*
9	face with jaguar pelt	*b'alam* 'jaguar'	nine	*b'olon*
10	skull	*laj* 'finish; die'	ten	*lajun*
12	sky	*ka'an* 'sky'	twelve	*lajka'*

The fact that the origin of these so-called *gods* as rebus representations of the words for the numbers is based on words in Mayan languages suggests that Maya speakers invented the set of head variants. Clear lunar associations with these glyphs occur with the first, sixth, and thirteenth heads. In addition to signifying the word 'first', the woman's head used for 'one' is associated with the moon goddess of Maya tradition. In Classic Maya inscriptions this female head precedes women's names and represents *na* 'woman; mother'. As a head variant it simply stands for *nah* 'first'.

The ax in the eye of the head variant for 6 *wak* may suggest the first quarter moon which appears to be chopped (*wak'*) in half. The final consonant of these two words differ in glottalization, a distinction never ignored in the Classic script, but seemingly tolerated in this rebus-like system.

Table 3. Head Variants Not Phonetically Associated with Maya Numbers

#	Glyph	Yucatec Maya	English	Yucatec Maya
3	circle/plaited headdress		three	*ox*
4	sun'god'	*k'in* 'sun; day'	four	*kan*
11	Kab'an day sign		eleven	*b'uluk*
13	waterlily monster	13 Muwan	thirteen	*oxlahun*

The unique head variant for 13 (Table 3) consists of a beaked creature, sometimes called the "waterlily monster," who wears a headdress of foliage or the glyph for "winal" on his head, a glyph for 20 which itself has lunar associations. The word for full moon given in the Colonial Yucatec Motul dictionary is *oxlahun ka'an u* literally, '13 sky moon'. This confirms that even during the sixteenth century the number 13 was associated with the full moon. The Muan bird of Maya mythology is named 13 Sky in the Dresden Codex. The relationship between the full moon and Muan bird has never been fully investigated.

The significance of three and 7 is not clear. The sun 'god' for four may be semantically related by the four-part flower which signifies 'sun', and possibly indicating the four directions. The Kab'an day sign for 11 is discussed above.

The Head Variants and 13 Day Signs

It was observed first by Goodman (1897) that there was a relationship between the unique set of glyphs that personify the numbers 1–13 and the names/glyphs for 13 of the 20 day names (Table 4). Actually Goodman proposed a numerical value for all 20 of the day symbols, but it is only the first 13 of the day names that Thompson (1950:88, 131–36) confirmed as corresponding to the head variants of the numbers. Table 2 lists the numbers one to 13, the day names in, Yucatec beginning with Kab'an, the object depicted in the Maya glyph, the logographic value in Maya hieroglyphic writing with the English gloss, the phonetic value in the Maya script (when the sign appears without the cartouche), and the English gloss of the Mexican day name.

I propose that the head variants were created first as icons, as visible representations of the Maya numbers, and that the naming of the days came later in time. That sequence explains why the four-part flower symbol of the sun matches the day "flower," and why K'an which signifies corn *wah*, matches the "corn god" with the phonetic *wa* symbol on its head, and why Kimi 'death' matches the skull used for 10. The day signs were not named arbitrarily; at least some of them were derived directly from the already extant set of head variants.

A striking correspondence to the lunar cycle is the phonetic value of the thirteenth day sign Muluk, *uh*, the Maya word for 'moon'. This matches the beaked monster head for 13, which is possibly related to the Muan bird, whose name in the Maya Codices is *ox lahun ka'an*, the phrase which in sixteenth century Maya meant 'full moon'. In Mesoamerica both ancient and contemporary peoples see a rabbit on the face of the moon. The Aztec name for this day, 'rabbit' supports this association between this day and the full moon.

Seven Additional Days

Seven more day names (Table 5) were added to the 13 to make up the set of 20 (Figure 2). This takes the count from the first crescent, through the full moon, to the day of the third quarter moon. The significance to the ancient Maya of most of these day names is not clearly understood. Unlike the 13 day names previously discussed, most of these signs are never used in non-calendrical contexts.

Nine

Beginning with the third quarter, approximately 20 days from the first visible crescent, exactly 7 days from the day of fullness, the moon rises at midnight, and then about one hour later each night (Figure 3). So with the third quarter begins the period of moonless evenings. Beattie first proposed to me that the series of supernaturals called the Nine Lords of the Night by the ancient Aztecs may represent the 9 days during which the moon is not present in the night sky (Table 6). Thompson (1929) identified a sequence of 9 glyphs in the Classic Maya calendar which undoubtedly correspond to these supernaturals. The first glyph read *b'olon k'u*, literally '9 gods', corresponds to the *b'olon ti k'u* mentioned in the creation accounts below (see "Twenty," below). An extensive analysis of this cycle is currently in preparation.

Several features of the Nine Lords of the Night sequence are in keeping with the lunar hypothesis. The Maya glyphs for G6 and G7 both include female faces. They correspond to two female supernaturals found in the later Mexican sources—all of the Mexican *tonalamatls*, and all but one of the Borgia group of codices records the Nine Lords of the Night.[9] In the Borgia Codex they have lunar crescent symbols as nose ornaments. These days would represent the last two days the moon is

9. Quiñiones Keber (1995:158).

Table 4. 13 Day Names

#	Yucatec Name	Glyph	Logographic Value	English	CV Value	Mexican Nane (Eng)
1	Kab'an	?	*kab'*	earth	ka	earthquake
2	Etz'nab'	flint?	*tok'*	flint		flint
3	Kawak	?	*tun/haab'*	year	ku	rain
4	Ahaw	ruler	*nik*	flower	xo	flower
5	Imix	fist (5 fingers)	waterlily	ha		crocodile
6	Ik'	?	*nal*	wind		wind
7	Ak'b'al	?	night			house
8	Kan	?	*wah*	corn	wa	lizard
9	Chikchan	?	xa			snake
10	Kimi	skull	death			death
11	Manik'	thumb & finger	chi[1]			deer
12	Lamat	star	*ek'*?	star; moon?		rabbit
13	Muluk	?	*uh*	moon	u(h)	water

1. *chih* is 'deer' in Cholan languages, the Yucatec cognate is *keh*.

Table 5. 7 Day Names

#	Yucatec Name	Glyph	Logographic Value	English	CV Value	Mexican Nane (Eng)
14	Ok	dog	*ok?*	enter	dog	
15	Chuwen	?/monkey	se	monkey		
16	Eb'	jaw/skull?	grass			
17	B'en	?	*aj?*	reed		
18	Ix	jaguar	jaguar			
19	Men	eagle?	eagle			
20	Kib'	?	vulture			

Figure 2. Lunar phases with the Maya and Mexican day signs, Ok/Dog through Kib'/Eagle (Maya day signs by Matthew Looper, courtesy of the Maya Hieroglyphic Database Project, Mexican signs after Caso (1967:6–7)).

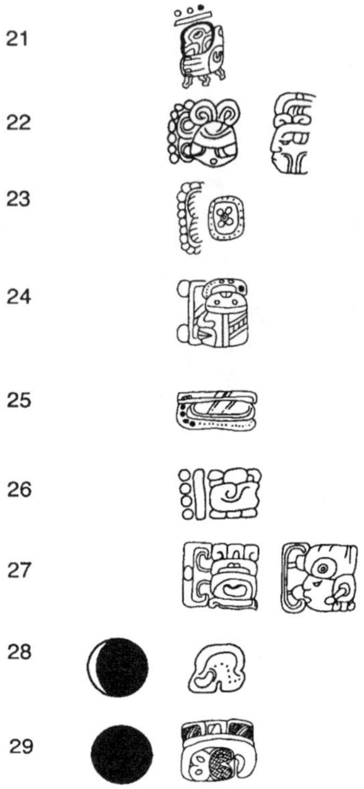

Figure 3. Lunar phases with the Maya glyphs G1–G9 (by author).

visible as a crescent in the morning sky at dawn. G8 is a bent feather *'o* when used as a phonetic symbol. By the twenty-eighth day from first visible crescent the moon would be invisible, or very nearly so. A bent feather is not unlike the slender crescent that might still be visible on this day.

G9, however, is the most compelling. It is the sun symbol, sometimes as a headdress, which is usually half darkened. On this day (and the one following in a 30-day month) the moon is not visible.[10] It is also, however, the time when a solar eclipse may occur, the only day of the lunar cycle when the moon has the possibility of blocking out the sun. The face of the Mexican god for this day is black.

10. The time of lunar invisibility—the time between the last sighting of the old crescent and the first sighting of the new crescent—can be as little as a day and a half (from morning to the evening of the following day) to as much as two and a half days. Because it is the brightness of the sun which makes the moon invisible, several factors are responsible for the variation, including the time of year (the orbits of the sun and moon are closer in the spring and fall), clearness of the sky, and attentiveness of the observer.

Table 6. Nine Day Cycle

#	G#	Maya Glyph	English	MexicanGod	Borgia
21	G1	Folon k'u	9 gods	fire god	
22	G2	(h)u ??	??	flint god	
23	G3	shield? flower?	young god		
24	G4	wuk ??	7 com/moon	corn god	corn
25	G5	ho' ??	5 mirror?	death god	skull
26	G6	na na?	two? woman	water goddess	U nose bead; skirt
27	G7	wuk?? na ??	woman	earth goddess	U nose bead; breast
28	G8	'o	feather (bent)	cave god	iguana? offering
29	G9	k'in	sun (darkened)	rain god	dark face

Twenty

Twenty is significant throughout Mesoamerica. Most ethnic groups base all large numbers on multiples of 20, most having fully vigesimal counting systems. The association between 20 and the number of digits on a person is explicit. In many Mayan languages 'one person' signifies '20'. The modern Lacandon Maya use 'all your fingers' for 4; 'one hand' for 5; 'two hands' for 10; 'one foot' for 15, and 'one person' for 20 (Bruce 1968:70). Several examples of glyphs and iconographic symbols from the Classic Period show two feet above a glyph for 20, confirming that 'one person; 20' is signified.

In addition to associating 20 with the 20 digits of the human body, the Classic Period Maya directly substituted glyphs, for the moon with the number 20. As mentioned above, the lunar significance of 20 is that it is the number of days from the first visible crescent to the third quarter of the lunar cycle. That is, 20 is the number of days during which the moon is visible in the evening sky. The 20 days of the moon are followed by 9 days in which there is no moon visible in the evening sky.

Confirmation from Ethnohistoric Document's

There are at least two Colonial Period sources which relate to the hypothesis that the counts of 20, 13, 9, and 7 have a lunar origin. The sixteenth century Motul Dictionary of Yucatec Maya, mentioned above, contains an interesting entry. In it *oxlahun ka'an uh*, literally '13 sky moon' is translated as 'luna llena; oposicion' [full moon, opposition].[11]

The following passages are from a colonial document written by Maya scribes in Yucatec, using Spanish orthography. The first tells of the creation of the world, the second of the creation of the winal, the 20 day period. *The Book of Chilam*

11. The Spanish translation of the phrase is transcribed incorrectly in the Cordernex (Barrera Vásquez 1980:612) as *luna flena o posición*.

Balam of Chumayel. Here the 13 days of the waxing moon fight with the 9 days of the moonless nights. Seven days are added to the 13. Also mentioned may be the two or the three days of darkness.

X THE CREATION OF THE WORLD
oxlahun-ti-ku tumenel bolon-ti-ku (Roys 1933:31)
thirteen gods were seized by **nine gods**
ualacito yahal cab, oxlahun pic dzac tu uuc u xocan yahal cab
during the creation **thirteen** infinite series <added> to the **seven** was the count of the creation of the world
catun ah cab tiob
dawned a new world for them
ci uba ca kin dzam, ox kin dzam (Roys 1933:32)
the **two day sinking** was declared, and the **three day sinking**

In the section telling of the creation of the winal, the 20 day count, it clearly states that it was formed from 13 and 7.

XIII THE CREATION OF THE WINAL
ca yalah tun bayla:
and <a voice> spoke as follows
oxlahun tuc uuc tuc, hun (Roys 1933:39)
thirteen entities and seven entities are one
lay zihci winal y uchci yahal cab,
created was the winal, created was the earth
tzolci caan y luum y cheob y tunich (Roys 1933:40)
set in order were the sky, earth, trees, rocks

These passages specifically mention the joining of 13 and 7 to create a period of 20 days, a conflict with a set of 9 days, and a sinking time (days of lunar invisibility?) of two or three days. This happened as the sky, earth, trees and rocks were set in order.

Political Implications

In Mesoamerica, the sacred numbers 20, 13, 9, and 7 drew their original significance as segments of the lunar cycle. The original 13-day count from the first crescent to the full moon was extended by 7 days to the third quarter. These 20 days make up the period during which the moon is visible in the evening sky. The count of 9 signified the 9 days during which the moon is no longer visible during the evening hours.

At an early point in Mesoamerican. prehistory, prior to any calendrical records, someone changed the relationship between the counts of 20, 13, 9, and 7 from consecutive cycles to simultaneously repeating cycles. As a result, the previously lunar calendar became divorced from the observable phenomena which engendered it. The cycles became arbitrary rather than observable, and the sequence of days could no longer be determined by direct observation, but had to be tracked by specialists.

Whatever the motives of these calendar creators, a strong case can be made for a lunar origin of the count of 13, and of the 20 day names. It would appear from the dictionary entry "13 sky moon" for 'full moon' that even in the sixteenth century the initiated were still aware of this lunar origin. The Nine Lords of the Night suggest a lunar association for a third day count. Still another Maya cycle, a count of 819 days, is arrived at by multiplying each of these segments of the lunar cycles: 13 × 7 × 9. Certainly, the ancient Mesoamericans enjoyed a veritable symphony of calendars.

References

Aveni, A., *Skywatchers of Ancient Mexico*. (Austin: University of Texas Press, 1988).

Barrera Vásquez, A., *Diccionario maya cordemex: maya-español español-maya*, (Mérida, Yucatán, México: Ediciones Cordemex, 1980).

Bruce, R.D., Gramática del Lacandón, (México, D.F.: Instituto Nacional de Antropologfa e Historia, 1968).

Caso, Alfonso. Los calendarios prehispinicos. (Mexico, D.F.: Universidad Nacional Antropología de México, 1967.

Diitting, D. and M. Schramm, "The Sidereal Period of the Moon in Maya Calendrical Astronomy," Tribus 37(1988):139–73.

Goodman, J.T., "The Archaic Maya Inscriptions," Appendix to *Biologia Centrali-Americana: Archaeology*, by A.P. Maudslay, (London 1897).

Jakeman, M. Wells, *The Ancient Middle-American Calendar System: Its Origin and Development*. Brigham Young University Publications in Archaeology and Early History, No. 1. (Provo, 1947).

Macri, M.J., *Phoneticism in Maya Head Variant Numerals*. Master's Thesis, University of California, Berkeley, (1982).

Macri, M.J., "The Numerical Head Variants and the Mayan Numbers," *Anthropological Linguistics* 27(1985):46–85.

Malmström, Vincent H., *Cycles of the Sun, Mysteries of the Moon*. (Austin: University of Texas Press, 1997).

Martínez Herández, J., *Diccionario de Motul maya-espafiol atribuido a fray Antonio de Ciudad Real y Arte de lengua maya por fray Juan Coronel*, (Merida, 1929).

Quiñones Keber, Eloise. *Codex Telleriano-Remensis: Ritual, Divination, and History in a Pictorial Aztec Manuscript*. (Austin: University of Texas Press, 1995).

Roys, R., *The Book of Chilam Balam of Chumayel*, (Washington, D.C.: Carnegie Institution of Washington, Publication no. 438, 1933).

Spinden, Herbert J. "'The Reduction of Mayan Dates." Peabody Museum of American Archaeology and Ethnology, Harvard University, *Papers*, Vol. VI, No. 4. Cambridge, 1924.

Tedlock, B., *Time and the Highland Maya,* (Albuquerque: University of New Mexico Press, 1992).

Thompson, J.E.S., "Maya Chronology: Glyph G of the Lunar Series," *American Anthropologist* 31(1929):223–31. 4.

Thompson, J.E.S., *Maya Hieroglyphic Writing*: An Introduction, (Washington, D.C.: Carnegie Institution of Washington, Publication no. 589, 1950).

Vail, Gabrielle, "Review of Cycles, of the Sun, Mysteries of Moon: The Calendar in Mesoamerican Civilization." *Australian Archaeology* 45(1997):72.

Yasugi, Y, and K. Saito, "Glyph Y of the Maya Supplementary Series," *Research Reports on Ancient Maya Writing*, 34, (Washington, D.C.: Center for Maya Research, 1991).

22

2000 Years of Continuity in Astronomical Symbols from Monte Albán to the Aztec Stone at Tenochtitlán

R. Robert Robbins
Department of Astronomy
University of Texas
Austin, TX

This paper presents new original discoveries at Monte Albán and the Plaza Mayor at Tenochtitlán that illustrate the persistence of cosmological belief systems in Mesoamerica over thousands of years. New observations suggest that the Danzantes Pyramid at Monte Albán is a Group E solar observatory and that a new alignment from Building J (the "Observatory") ultimately leads to the conclusion that the stars of Cassiopeia were an important grouping for the Zapotec. Several additional lines of evidence reinforcing this claim are presented. It is also suggested that the stars of Cassiopeia were important to the Aztec as well, some 2,000 years later, since these stars appear to be found on the perimeter of the Aztec "Calendar" Stone. There is also suggestive evidence that Tycho's Supernova 1572 can be found on the Stone. Finally, it is argued that it is easy to find a place for Cassiopeia in the current models of Mayan creation studies.

* * *

Introduction

Studied in isolation, the diversity of cultural groups in Mesoamerica can overwhelm students of the region with a body of detailed information so dense that it can make it hard to see the "forest for the trees." While the diversity is certainly fascinating, we might also reach different kinds of insights by devoting more effort towards the intercomparison of cultures—pursuing the commonalties in beliefs and customs that they exhibit over time. This paper presents astronomical discoveries spanning over 2000 years that illustrate the persistence of basic belief structures shared by the cultures of Mesoamerica.

The Danzantes Pyramid at Monte Albán as a Solar Observatory[1]

The magnificent mountain-top plaza called Monte Albán, just outside Oaxaca, Mexico, was constructed by the Zapotec and their predecessors from at least 400 BC. Figure 1 shows a site plan of the plaza, adapted from J. Acosta (1965). Various features that will be discussed throughout the paper are labeled on it.

The Danzantes is one of the most ancient features that can be found at Monte Albán, and it has been built over a number of times, starting intensively in the 3rd Century B.C (Period II). It is famous for the grotesque portraits of contorted "dancers" on its facades.

In October 1987, 10 days after the autumnal equinox, I visited Monte Albán, and my interest was captured by the physical resemblance between the Danzantes Pyramid and the well-known Group E structure at Uaxactún in Guatemala—a solar observatory that tracks the full seasonal range of the Sun's north-south motion along the horizon for a properly located observer. Figure 2 shows a photograph of Danzantes exhibiting the usual structure of such buildings. They have three symmetrically placed pedestals on a substructure, spaced so that the Sun rises over the northern pedestal at summer solstice and over the southern pedestal at winter solstice. And at equinox time, the Sun rises from behind the central temple, sending its rays through a window.

Since Danzantes is on the west side of the plaza, oriented approximately north-south at the southwestern corner, it would be used to follow sunset points, rather than sunrises, similar to the Mayan site of Dzibilchaltún.

I had only one day to observe the behavior of the Sun at Monte Albán, so it was very important to determine a correct viewing location quickly. Fortunately, an appropriate spot readily became apparent. Running roughly parallel to the north-south line of the temple is a loose stone wall (see Figure 1) that creates a courtyard in front of Danzantes. There is a tumble-down gate in the wall, the location of which is indicated in Figure 1 by a little gap. The gate shows clearly in aerial photographs (Ferguson and Rohn 1990), and falls closely along a line between the central pedestal of Danzantes and the doorway of the Palace across the Plaza (from which equinox sunset can also be seen). The gate is visible in the large mural that covers one entire wall in the Monte Albán room at the National Museum of Anthropology. The gate turned out to be precisely the location that makes the Danzantes function as a solar observatory.

I made observations of the altitude of the Sun as it marched toward sunset with a simple sextant and I also photographed it with a camera of known scale factor.

1. The observations in this section were first presented and discussed at the 1988 meeting of the AAS Historical Astronomy Division in Austin, Texas, but have not appeared in print yet.

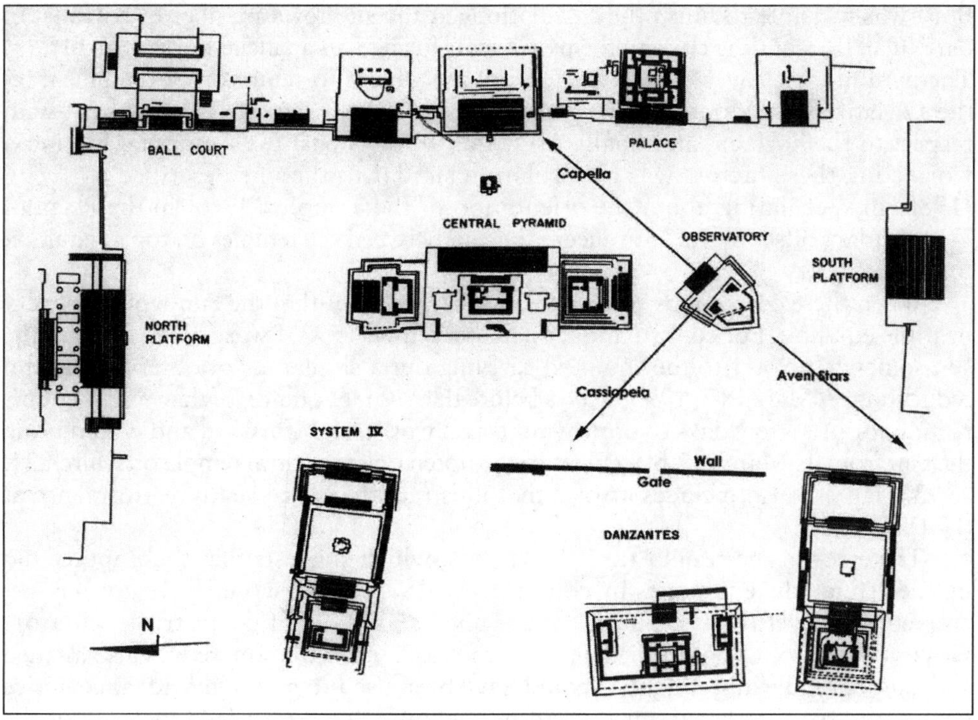

Figure 1. Map of Monte Albán Archaeological Zone, Oaxaca.

Figure 2. A Photograph of the Danzantes Pyramid with its function as a solar observatory indicated.

But I was not able to finish the calculations at the site for a number of reasons. In particular, (a) standing close to the pyramid, it formed an artificial horizon at 11°. (b) The pyramid itself was angled 3.25° from the general orientation of the plaza. (c) the 17° latitude of Monte Albán meant that the Sun was setting at angle of 73° with respect to the horizon, and finally (d) it was 10 days past the Autumnal Equinox. Correcting these factors was a calculation similar to the one described by Aveni (1980) in speculating about the orientation of the Templo Mayor in the Plaza of Tenochtitlán, offset by 7.5° to squeeze the Sun between the temples on top at equinox time.

When these corrections were made, it was apparent that the Sun would set very near the equinox. For confirmation, on a later trip in 1993, I was able to observe the sun somewhat closer to the equinox, carrying out a similar set of observations and reductions on March 17, 1993, 3 days before the Vernal Equinox of that year. The correction for just a few days of northward travel was straightforward and would bring the Sun from its March 17 placement to the notch on the central temple on March 21, 1993. Figure 2 also includes arrows that illustrate the entire yearly performance at the Danzantes.

The error in the results are 1 day or less, which suggests that the Zapotec did not determine the equinoxes by counting the days (which would have given a discrepancy of several days) but instead applied some sort of geometrical criterion. However, the front door of the Palace on the east side of the square was also aligned to admit equinox sunset light; it could have been used to give some advance notice of equinox, so that ceremonies could be prepared. The open-field location of the point of observation would make it possible for more people to appreciate the hierophonies than at Uaxactún.

Finally, on the back side of the Danzantes, one finds a number of small rooms facing out to the western sky. With the variegated skyline created by the rugged mountains in Oaxaca, it would be easy to maintain a horizon calendar from that vantage, giving a vital double-check on the function of the main pyramid itself. Especially near equinox, when the sun is moving rapidly, observers should be able to obtain a one-day accuracy.

Cassiopeia as an Important Asterism at Monte Albán

The early restoration at Monte Albán was carried out in the 1930s by Caso (1935) who suggested that Building J was an observatory because it contained many irregular chambers and compartments and some suggestive glyphs on its exterior. But this suggestion was not verified until the extensive observations of Aveni (1980) showed that the building called Mound J (the "Observatory) seemed to be the central feature from which the astronomical phenomena radiate.

Building J literally calls out for attention, resembling as it does an arrowhead or the "home plate" of a baseball game and also because of its unique orientation in relation to all the other buildings. Building J was apparently constructed in stages over a period of time and is highly asymmetric. The top and bottom sections of the building have different orientations, like a skewed layer-cake, and none of the external walls are parallel. The steps going up are not perpendicular to the front wall of the structure itself.

Aveni determined that the south-westerly direction indicated by the point of the "arrow" of Building J indicated a narrow 3° range of the southwestern horizon where 5 of the brightest 25 stars in the sky set one right after another—the stars of the Southern Cross combined with Alpha and Beta Centauri. They are indicated in Figure 1 as "Aveni Stars". Aveni also discussed (Aveni 1980) confirmation in the folklore suggesting that the Southern Cross was a celestial grouping of great importance to these cultures.

Aveni found additional alignments by considering lines of sight from Mound J to other buildings. In particular, he found a northeastward alignment that located a zenith tube in Pyramid P and another nearby line of sight indicating that the star Capella rose heliacally on the first calendar date when the Sun was at the zenith at that latitude.

A New Alignment and New Stars of Importance

Building J is honeycombed with sunken stairwells leading to hidden chambers and observing peepholes and while coming up from one of these, I found what appeared to be a previously unrecognized alignment again radiating from Building J. It consisted of a rectangular slot in the northwestern wall through which one can see the pyramids of System IV (see Figure 3). I took measurements between this direction and the western edges of the second-level pedestals on the main central pyramid (Building H) and correcting for the orientation of the central pyramid with respect to north, I ended up with a value of 43.5° west of celestial north, for an azimuth of 317.5°. This direction is indicated in the site plan of Figure 1, running from Building J towards System IV.

Precession calculations to 250 BC were carried out for a number of stars with nearby set azimuths, giving

	Azimuth (degrees and minutes)
Sightline to match	317 30
Alpha Cassiopeia	316 40
Beta Cassiopeia	319 35
Gamma Cassiopeia	320 51
Capella	311 24
Deneb	311 03
Vega	310 27

The stars of Cassiopeia center nicely on the 317.5° sightline, spreading about 4° along the horizon. (Aveni found a spread of 3° for his southern stars.) Not only are the other candidates 5° or more off, the stars of Cassiopeia will be shown below to have other additional properties that corroborate their importance.

Figure 3. The opening in the wall at Building J (the vertical slot at the lower right) forms a sightline towards System IV and Cassiopeia.

Further Arguments Supporting the Alignment

As many authors have rightfully discussed, one line of sight needs some corroboration to be convincing. In this case, happily, I believe that such evidence is available and that there are in fact a number of different arguments that support the reality of the proposed alignment.

One could begin with the fact that the stars of Cassiopeia are in fact fairly bright and they form a tight grouping that the eye would tend to discover. However, this kind of interpretation is always subject to challenge and there is much better evidence available. (I am indebted to a helpful reviewer who pointed out that Cassiopeia is a grouping of importance for the Navajo and other Northern American Tribal Groups.)

Next, I note that the stars of Cassiopeia exhibit a very interesting symmetrical relationship during the year with the five southwestern stars that Aveni has shown to be important. Figure 4 illustrates how the stars of Cassiopeia set in the NW just as the southern Aveni stars rise in the SE. Similarly, the stars of Cassiopeia rise in the NE just as the stars of the Southern Cross set in the SW. Further, both groups of stars are within a few degrees of the same angular distance from the meridian. But most dramatically, the two groups of stars are both in the Milky Way and connected by it to each other. As Cassiopeia rises in the NE, the Milky Way is passing through the zenith, just as it is when Cassiopeia rises. These behaviors (best visualized on a

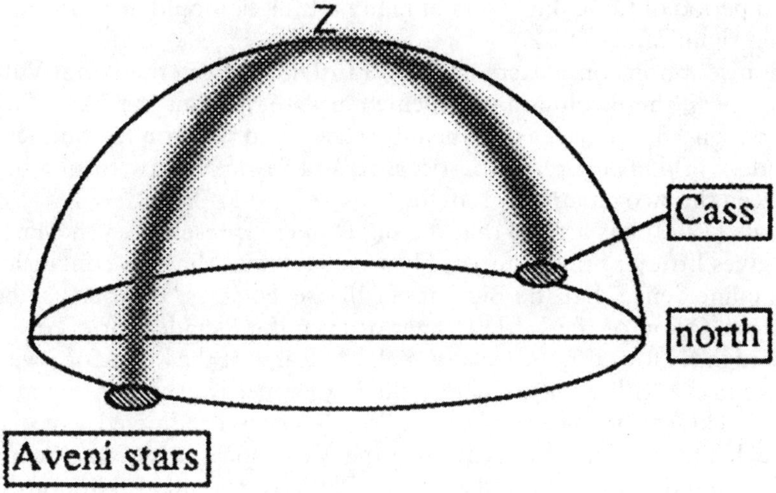

Figure 4. The symmetry between Cassiopeia and the Aveni stars.

celestial globe) are of course natural coincidences, but they could be assigned a function by a culture once their motions were understood.

Another method of building confidence in an alignment is to find some type of "function" the indicated stars could carry out; for example, rising heliacally at an appropriate time to indicate some important ritual or agricultural date. I was at first unable to detect any obvious function, but interesting results were soon obtained by seeking comparisons with folklore.

Cassiopeia and Mayan Beliefs

Recently published work in Mayan epigraphy have proposed a Mayan myth of creation that needs to be examined carefully by astronomers. Two principal figures in this effort have been Linda Schele (Maya Cosmos) and Dennis Tedlock, well-known for his translation of the Popul Vuh—an eighth century book of creation according to the Quiché Maya.

A critical moment in the myth comes on February 5, 3112 BC, when the World Tree (the Milky Way) is raised up. This marks the end of a dark time when the sky was laying face-down on the Earth. The rising of the Milky Way into the sky represents the start of creation. Landing in the tree and somehow empowering this event is a bird of magical powers who has been called the Great Celestial Bird.

Could that Great Bird be Cassiopeia? Many observers have noted a tendency of Cassiopeia to have the shape of a set of wings. Using Voyager II, I found that at sunrise on February 5, 250 BC, Beta Cassiopeia has just risen heliacally and Alpha is just rising. In other words, Cassiopeia rose heliacally on the day of creation! Since Mayan astronomers could not have guessed what the skies of 3112 BC looked alike, it would have been natural and necessary to start with the stars of their own epoch. Using 250 BC in this calculation is somewhat arbitrary; obviously, a myth develops over a period of time. But errors of many centuries should not change this basic conclusion significantly.

From inscriptions on pottery and the translations from the Popul Vuh by Tedlock and others, there comes a complementary story about the Hero Twins, who defeated the bird Seven Macaw—the false sun—and went on to conquer the gods of the underworld. The Great Celestial Bird and Seven Macaw form a bridge or a link between the two channels of myth.

Tedlock (1993) has argued that the Big Dipper represents Seven Macaw in the sky, but gives little argument for it. He seems to state that his conclusion comes from the coincidence that the Big Dipper has seven stars. But on that basis, one could invoke Orion or the Little Dipper or even the Pleiades (or even Cassiopeia on a dark night). However, the legends make it clear that the killing of Seven Macaw takes place in the Milky Way, whereas the Big Dipper is some 70 ° away from the Milky Way. This seems to be too large a discrepancy to be allowed, especially when there is such a reasonable alternative waiting. Cassiopeia is actually located IN the Milky Way, and the story takes place there. Perhaps it could also honor the important and beautiful Quetzal bird as well.

Cassiopeia Found on the Aztec "Calendar Stone"

I have already presented evidence and arguments from several different directions that the stars of Cassiopeia may indeed have been significant to the Mesoamerican cultures. At this point, however, I discovered that the Cassiopeia story was just beginning.

The famous Aztec "Calendar Stone" is of course not a calendar. It is, rather, a cosmological history of Mesoamerica and the cycles of creation and destruction that were thought to govern existence. Among the various aspects of the Stone still debated by scholars is the the question of whether the numerous shallow depressions pecked into the outside, apparently unfinished, perimeter of the Stone actually represent stars and constellations. (In this paper I will refer to these depressions as "holes.") Many of these pecked features have not been unambiguously identified with star groupings recognizable to us. High-quality photographic reproductions of the Stone can be found in Bernal (1980) and Paz (1990).

I would like to concentrate attention on only one portion of the Stone—the location on the lower-left outer portion where we find pecked holes connected by lines. No other group has this characteristic. I will argue that this figure can plausibly be identified as Cassiopeia. The argument is best visualized by examining Figures 5 and 6 as the discussion goes along.

In 5a, I have redrawn the figure at the edge of the Stone following the well-known drawing of Nuttall (1901) reproduced in (Aveni 1980). In 5b, I have removed the "Aztec" lines and redrawn the lines differently but still utilizing the same stars. What emerges is a decent resemblance to Cassiopeia, shown in Figure 5c. In the classical depiction of Cassiopeia, the faint star Kappa Cass is usually retained to make the grouping look more like a queen's throne. Covering Kappa gives the familiar "M" or "W" shape of the constellation's brightest stars. This reminds us that

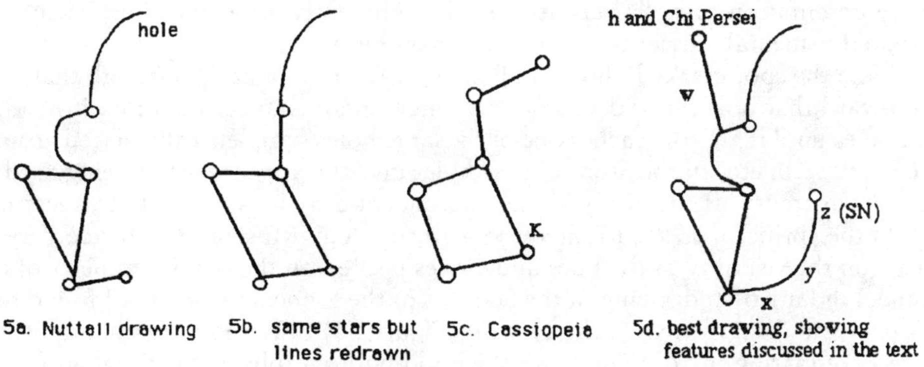

Figure 5. (a) Nuttall drawing of the edge of the Stone, (b) using the same stars but drawing the lines differently, (c) the stars of Cassiopeia for comparison, (d) A more accurate drawing reveals more information.

Figure 6. A detailed view of the pecked areas of the Stone.

the representation of stellar groupings with inherently arbitrary "lines" can be a formidable mental barrier to seeing other possibilities.

Note the spot marked "hole" in Figure 5a. There is a pecked trough that runs up toward that hole but it does not encounter a star at its termination. Instead, it continues until it almost reaches one of the large holes symmetrically placed around the outer perimeter of the Stone. These holes give the appearance of being used to support or transport the heavy Stone. Subsequent drawings will omit this feature.

In the spring of 1995, I visited the Anthropology Museum in Mexico City to check on the visibility of the lines and circles pecked on the outer perimeter of the Stone. I did my own drawing of the features of the Stone, labelled 5d. I found that my drawing disagreed with the Nuttall (Figure 5a) drawing in many important ways. Comparing my drawing with the good photographs of the Stone in Bernal (1980) and Paz (1990) should allow the reader to follow the arguments based on this newer drawing. I have added letters w, x, y, and z to clarify the discussion below.

A feature clearly visible on my drawing but missing on Nuttall's is the pecked trough w that runs from the semi-circle up to a previously unidentified hole. A quick consultation with a star atlas indicated that this new feature was actually the location of h and Chi Persei, a close pair of open clusters visible to the naked eye in dark environments. This was a complete surprise, and it further cements the identification of these stars as Cassiopeia. It could be argued that the Cassiopeia Hypothesis has passed one of the most stringent tests in scientific research—to be put "at risk" and *predict* something new and unforeseen. If there had been no object at that point, the argument would have been left with a dangling, unidentified object and the overall hypothesis would have been weakened.

Now I turn to discussion of another aspect of my drawing of the Stone—the feature that is indicated by **z**. In the Nuttall Figure 5a, there is a star indicated at y, but a close examination of this region on the Stone shows there is no star at position **y** on the Stone at all! What it does have is a "trough" of pecked incisions leading from **x** around **y** to **z** and a star very definitely pecked at **z**.

Upon further examination of star maps, it turns out that the location of **z** is very close to where astronomers find the remnant of Supernova 1572! At this point, doing a little speculating is irresistible In 1572, the region of Cassiopeia hosted a brilliant object which has come to be called Tycho Brahe's Supernova! Today astronomers can still detect emission from the remnant left behind as the radio object 3C10:G120.1+1.4 amidst the stars of Cassiopeia. The "z" drawn in represents fairly closely the position of the supernova with respect to the stars of Cassiopeia.

The new star was reported by Tycho on November 11, 1572, when he gave it an estimated brightness comparable to Venus at maximum. In November, Cassiopeia is high in the sky above the North Celestial Pole in the middle of the night, so it could not possibly have been missed. It continued a slow decline such that it was still about second-magnitude in May 1573. So an intriguing question is raised—is this a Mesoamerican record of this dramatic event?

It is impossible to discuss this question without being somewhat speculative but the possibilities deserve a hearing. 1572 is of course 51 years after the razing of Tenochtitlán by the Spanish, but the historical record tells us that the stone itself was toppled over and laid flat on the ground in the Plaza for a number of years. It was clearly treated with some reverence, possibly because the Spanish were aware of how important it was to the Indians. (In Cuzco, the Spanish executed all known possible pretenders to the throne of the Inka in the year 1572.)

If the Stone was exposed at all during occupation, it could have been further "worked," especially when such a spectacular object appeared. The fact that the Cassiopeia grouping has lines between the stars suggests a European influence, since no other known Mesoamerican constellation has such lines. At some date, the stone was ordered buried in the plaza by the church, but even then, the Spanish did not break it up; it was simply buried out of sight. It was finally rediscovered much later and accidentally, in 1790. I have been unable to find a record of the year the stone was finally buried, to determine directly if the Stone was still above ground in 1572. But if turns out to be true, this will be the second instance of a western hemisphere supernova record (see Robbins and Westmoreland 1991 for a discussion of the Crab Nebula).

Summary

The measurements and arguments in this paper are to my knowledge the first published evidence indicating that Cassiopeia was an important star grouping to the high cultures of Mesoamerica. To briefly summarize the arguments for this asterism:

- these stars have a visibility and unity in the sky. With no other bright stars nearby, the stars of Cassiopeia stand out to the eye.
- the alignment from Building J to the Cassiopeia azimuth set point.
- the symmetrical spatial relationships in the sky connecting Cassiopeia with the "Aveni" southern stars—in particular, their location in the Milky Way and the rise and set "dance" they execute.
- the discovery of the Cassiopeia grouping on the Aztec Stone.
- the intriguing possibility that the February 5 heliacal rise of Cassiopeia may represent the initiation of "Creation." in the character of the Great Celestial Bird.
- the "at-risk" prediction of the position of the star clusters h and Chi Persei.
- a possible discovery and location of the 1572 supernova.

I hope that researchers interpreting the codices and studying other aspects of the ethnographic record might also begin to find glyphs, figures, and references that could be identified with Cassiopeia . Once it is known that something exists or is important, it often becomes easier to find it and recognize it in other guises and places. These discoveries and their interpretation may be speculative, but they do hold together fairly well. We appear to be lucky in finding a grouping that was not too abstract to recognize. If a decent resemblance is not a good criterion, then what would be?

From Monte Albán to the Aztec Stone at Tenochtitlán, we have mentally transported Mesoamerican culture almost two millennia forward in time. And we find evidence that the same stars were important to cultures as "diverse" as the Zapotec, the Maya and the Aztec. It is apparent that while we discuss the abundant diversity of Mesoamerican cultures, we should at the same time continue to seek out the underlying and continuing fundamentals in their belief structures as revealed by their astronomy. It is apparent that central features of the myth structures are shared by cultures over many years of time. While the diversity of early cultures is fascinating, I believe that by making every effort to study cultures comparatively, we can often obtain a different perspective into the belief structures of our ancestors and other early groups, helping us place myths in space and time and trace the migrations of the creators of those myths and stories through space and time.

References

Aveni, A. 1980, Skywatchers of Ancient Mexico (Austin: University of Texas Press).

Acosta J. 1965, "Preclassic and Classic Architecture of Oaxaca", in Handbook of Middle American Indians, Volume 3, part 2 Editor: Gordon Willey, Austin: University of Texas Press, p. 815.

Bernal, I. 1980, A History of Mexican Archaeology (London: Thames and Hudson) p. 83.

Caso, A. 1935, Los Exploraciones en Monte Albán, Temporada 1934–35. (Mexico City: Instituto Panamerican de Geografía e Historia), 18.

Nuttall, Z. 1901, "The Fundamental Principles of Old and New World Civilizations" Peabody Museum of American Archaeology and Ethnology papers, 2:1–602. Cambridge, Mass.

Paz, Octavio. 1990, Mexico. Splendors of Thirty Centuries. (New York: Metropolitan Museum of Art).

Peterson, W. & Rohn, A. 1990, Mesoamerica's Ancient Cities (Boulder: University Press of Colorado) p. 65–70.

Robbins, R. Robert & Westmoreland, R. 1991, "Astronomical Imagery and Numbers in Mimbres Pottery" The Astronomy Quarterly 79, pp. 65–88.

Schele, L., Freidel, D.,and Parker, J. 1993, Maya Cosmos (Morrow 1993).

Tedlock, Dennis & Barbara, "A Mayan Reading of the Story of the Stars", Archaeology, December 1993.

23

Jupiter in Classic and Postclassic Maya Art

Susan Milbrath
Florida Museum of Natural History
Gainesville, FL

Jupiter, the powerhouse of planets in Old World Astronomy, has yet to be identified in Maya iconography. Study of the patterning of dates on monuments at several Maya sites suggests a link between the planet Jupiter and one of the principal gods of Maya rulers, known as GII in the Palenque inscriptions. GII is probably named Tahil or K'awil in the Classic Maya period, and he seems to be the counterpart of God K in the Postclassic period. As the youngest brother of the Palenque triad, GII is a natural counterpart for the youngest of a triad of mythological brothers known from modern Mopan Maya folklore. In this legend, the older brothers are Venus and the Sun, the same roles the elder brothers take at Palenque. The youngest brother in the Mopan legend is Mars or Jupiter. Monuments with GII often bear dates that correlate with Jupiter's retrograde at Palenque. The Sarcophagus Lid at Palenque depicts the deceased king Pacal in the guise of God K. This transformation or apotheosis took place in A.D. 683 on a date coinciding with the first stationary point of Jupiter. There is a high correlation between monuments representing this god at Yaxchilán and dates coinciding with Jupiter's period of retrograde motion during the Classic period. In the Postclassic period, a similar association occurs in the Dresden Codex Venus pages, where God K is associated with a date marking the rise of the Morning Star at a time that Jupiter was in retrograde motion.

Introduction

The deity that scholars designate as God K clearly has a celestial nature in Maya imagery. A winged God K is found in architectural contexts that suggest an association with the sky (Taube 1992:69). Postclassic mural painting from Tulum depicts a skyband with a winged God K probably dating around A.D. 1400 (Figure 1). A skyband arches over a winged God K on a painted capstone from the Temple of the Owls at Chichen Itza, probably dating to the terminal Classic period, between A.D. 800–900 (Figure 2). As we will see, God K is associated with Katun cycles involving Jupiter and Saturn, especially in Classic Maya times (A.D. 250–900). There seems to be a link between God K, Katun ending dates, and Jupiter's position in

Figure 1. God K's celestial aspect as winged deity on sky band in Late Postclassic mural from Tulum (after Taube 1992:fig. 34c).

Figure 2. Winged God K in jaws of coiled serpent with sky band arching overhead on painted capstone from Terminal Classic Temple of Owls at Chichén Itzá (Structure 5C7; after Tozzer 1957:fig. 384).

retrograde. God K may be the embodiment of the planet Jupiter in a number of representations, especially those linked with kingship during the Classic period.

God K in the Postclassic and Colonial Period

God K is well represented in Maya Postclassic codices, dating between A.D. 900–1521. Studying the gods in the codices at the turn of the century, Schellhas

Figure 3. God K's Postclassic portrait head glyph with elongated snout bearing smoking mirror (after Taube 1992:fig. 32a).

Figure 4. Postclassic God K has long snout with branching element inset with mirror on Dresden Codex page 12a (after Taube 1992:fig. 32e).

(1904) recognized 15 deites and named them with the letters A through O (excluding the letter J), because their glyphic names could not be interpreted at that time. Subsequent research suggests that images interpreted as God F should be reclassified, and scholars have identified five additional gods, known by the letters A', CH, Q, R, and S (Taube 1992:11). The fact that there are 19 such gods is intriguing in terms of astronomy, because the Dresden Codex Venus pages records a list of 20 gods that includes the sun (God G), moon (Goddess I), and God K (Kelley 1976:fig. 28). Thompson (1972:65, 67) proposed that these gods are companions of Venus, perhaps representing constellations of the zodiac in conjunction with Venus. Since, the sun and moon are among those named, the list clearly goes beyond the constellations along the sun's path. A zodiac-like sequence appears in the Paris Codex 23–24, picturing 13 constellations located on or near the ecliptic (Milbrath 1999). Thus the list of 20 astronomical gods in the Dresden Codex could represent 13 constellations, plus the sun, moon, and five naked-eye planets. Since God K does not resemble any of the zodiac figures in the Paris Codex, it seems reasonable to propose that he takes the role of a planet in the list of gods.

God K's Postclassic name glyph shows his head with an elongated snout bearing a smoking mirror (Figure 3). Postclassic figures of God K have a long snout bearing branching elements (smoke?), sometimes incorporating a mirror element, as in the Dresden Codex 12a (Figure 4). God K's smoking mirror can be traced back to Classic period times. The mirror itself may be an astronomical image, for mirrors are prominent in expressions referring to Venus positions in the Dresden Codex Venus pages (46–50).

God K can be identified as the sky god called Bolon Dzacab in Colonial period sources. This identification, first proposed Seler at the turn of the century, is now widely accepted (Seler 1960–1961, I:376–377; Taube 1992:73). Thompson (1970:281) notes that a passage from the *Chilam Balam of Chumayel* describes Bolon Dzacab taking the seeds of edible plants to the thirteenth layer of heaven at the time of the great flood, when nine gods defeated thirteen gods in a cosmic battle. The Motul dictionary glosses Bolon Dzacab as "perpetual thing," meaning eternal, however Thompson (1972:90) translates God K's name as "nine generations." He proposes that this name may have something to do with the nine lords of the underworld who ruled over a repeating sequence of nine nights recorded in Classic period inscriptions (Thompson 1970:227, 280). When combined with other Classic period calendar cycles, the nine lords cycle forms larger cycles of time that may have been used in astronomical calculations.

Around 1566, Landa described Bolon Zacab (Bolon Dzacab) as a god worshipped at the new year (Tozzer 1941). This may be why God K is prominent in ceremonies associated with years named Ben and Etznab in the Dresden Codex. On Dresden Codex 25b, God K is linked with a column of Ben year-bearer glyphs symbolizing the eastern direction (Figure 5; Thompson 1972:90–93). Associated glyphs say that cacao is God K's food. Cacao, the chocolate bean, was the principal currency for Mesoamerica, hence God K is a deity associated with great wealth. This may be why God K is surrounded by cacao pods on a capstone from Chichen Itza (Figure 2). In the years named Etznab (years that followed Ben), depicted on page 26c of the Dresden Codex, God K offers a beheaded turkey (Figure 6). God K clearly takes a role involved in some way with the annual new year ceremonies. If God K can be identified as the planet Jupiter, his association with the annual cycle can be explained by the fact that the planet is visible in the sky for about one year (mean interval 367 days) during each synodic period.

Page 46 of the Dresden Codex Venus table shows God K killed by a projectile hurled by the Morning Star god at heliacal rise. This role suggests that he is a planetary god other than the Morning Star. Study of the layout of the Venus pages indicates that Venus spearing a victim symbolizes opposition, rather than conjunction (Milbrath 1995, 1999). For this reason, the Venus god uses an *atlatl*—a long distance projectile weapon. The heliacal rise date on page 46 is 5 Kan 7 Xul, corresponding to 1221 January 26 O.S. in the early Postclassic period, when Jupiter was at the opposite side of the sky from the emerging Morning Star. This date also coincides with the retrograde period of Jupiter. As we will see, Classic period monuments representing God K often bear dates that correlate with Jupiter's position in retrograde.

A late Postclassic mural from Mound 1 at Santa Rita, Belize, shows God K with his hands tied behind his back, a prisoner or a victim destined for sacrifice, a role like that of God K in the Venus pages (Figure 7). Despite certain Central Mexican style details, the Santa Rita murals are essentially Maya (Taube 1992:4). A Maya 1 Ahau date appears with God K, an interesting detail in light of the pattern linking Jupiter events to Katuns ending 1 Ahau during the Classic period (Thompson 1960:158, 228). The 1 Ahau date recurs at a Katun end approximately every 20 years (20 × 360 days). The whole Katun series represents a period of approximately 256.25 years (13 × 20 × 360 days).

Figure 5. Ben new year with God K in temple of east (Dresden Codex 25b; after Villacorta & Villacorta 1977).

Figure 6. Etz'nab new year with God K offering beheaded turkey and drops of blood (Dresden Codex page 26c; Villacorta & Villacorta 1977).

God K plays a prominent role in Katun ceremonies in the Paris Codex 2–12. Owing to the structure of the Katun cycle, each of the 13 Katun lords was associated with a different Ahau date designating the Katun end. Each Katun lord ruled a period of 19.71 years (20 × 360 days). Kelley (1985:238) notes that the interval between successive conjunctions of Jupiter and Saturn is between 19 and 20 years (the approximate length of a Katun). The conjunction of Jupiter and Saturn approximately every 20 years cannot be expected to align with a specific date in the Katun cycle over long periods of time, but those instances when there was a conjunction at the Katun end certainly would have been notable. In the conjunctions between Jupiter and Saturn, Jupiter is the dynamic actor in the scenario, essentially running laps around the slower moving Saturn. In this light, the God K head carried to the enthroned Katun lord in the Paris Codex may symbolize Jupiter transported across the sky to a new encounter with Saturn. The interval of the Katun approximates the period between successive conjunctions of Jupiter and Saturn; and every three Katuns (approximately every 59 years) the conjunction event would be expected to occur in the same part of the sky (Kelley 1985:238).

Many scholars conclude that God K is related to the Postclassic Mexican god Tezcatlipoca of Central Mexico, also associated with a smoking mirror (Milbrath 1999). Tezcatlipoca is a god with multiple astronomical aspects and associations. One

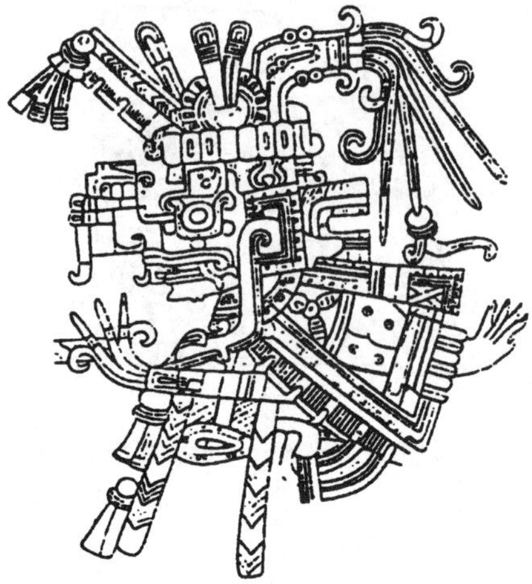

Figure 7. Late Postclassic God K bound like prisoner in murals of Mound 1, Santa Rita, Belize (after Taube 1992:fig. 32f).

Figure 8. Classic period relief of God K at Sayil has both serpent leg and smoking mirror of Central Mexican Tezcatlipoca (after Taube 1992:fig. 34a).

of Tezcatlipoca's aspects is Mixcoatl, a god linked with Venus; another is Huitzilopochtli, a solar god (Nicholson 1971:412, 426). A few representations fuse Tezcatlipoca with the lunar god Metztli, according to Nicholson. He also points out that some accounts refer to Tezcatlipoca as "another Jupiter," whereas others link him with Ursa Major. Among the many guises of Tezcatlipoca, the black Tezcatlipoca seems to be closest to God K. The Codex Borgia (54) shows the black Tezcatlipoca as the victim of the Morning Star in the south (Seler 1963, II:121). And it seems significant that the Venus pages of the Dresden Codex name God K as a victim of the Morning Star, also linking him with the south (Kelly 1976:fig. 28).

Tezcatlipoca and God K both share an association with mirrors and serpents. A mirror brow is one of the diagnostic traits of Tezcatlipoca in images from the Codex

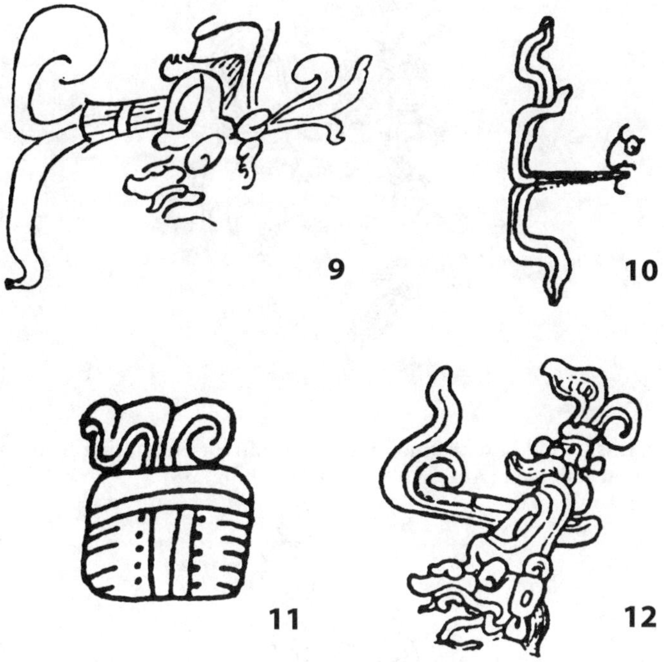

Figures 9–12. Classic images of God K's smoking forehead mirror resemble smoking cigars and symbols for fire emanating smoke (after Schele & Miller 1983:figs. 3k to n).

Borgia (14, 69), a manuscript probably dating to the Late Postclassic period (Milbrath 1989:117). Tezcatlipoca has a smoking mirror on his head and a serpent foot in the Codex Vaticanus A (3v, 44v), a Colonial period Aztec codex. Although a serpent foot is not seen in Postclassic images of God K, his ophidian aspect is apparent in the Late Postclassic Codex Madrid (31b). A Classic period relief of God K at Sayil has both the serpent leg and smoking mirror of Tezcatlipoca (Figure 8). The manifestation of these traits in the Maya area appears to be earlier than in Central Mexico, indicating that some aspects of the cult first developed among the Maya.

The Classic Period God K and his Counterpart GII

There is clear continuity tracing the imagery of God K back through time, however, some stylistic modifications are evident. Most notably, the smoking mirror changes position between the Postclassic and Classic periods (compare Figures 4 and 5 with 8 and 9). In the Classic period, God K carries a mirror on his brow, rather than on his snout. Sometimes a torch emanates from the mirror, its smoke resembling cigar smoke or the smoke from fires (compare Figures 8–11). In other depictions, a smoking celt or axe emanates from God K's mirror (Figures 12, 13, 14). God K's nocturnal nature is evident in images from the Palenque Palace that show the mirror-browed God K wearing a star (Figures 15, 16). One of the earliest known

Figure 13. Late Classic God K with smoking celt or mirror emanating from mirror on brow (Pier C of House A at Palenque; after Robertson 1985b:fig. 38).

Figure 14. Early Classic God K with smoking celt emanating from Akbal mirror brow, Tikal Stela 31 (after A. Miller 1986:pl. 19).

images of God K has a smoking celt emerging from a mirror with markings like the glyph for night known as Akbal (Figure 14).

In the Classic period, God K's head with a mirrored brow or a smoking mirror alone can name God K (Figure 17). A number of glyphs show God K's name associated with the number nine, providing a link with the Colonial period Bolon Dzacab (Thompson 1970:fig. 7). Schele (1976:25–26) first recognized that "God K in hand" (T1030e) is a glyph usually denoting the presentation of God K at the time of heir-designation or accession. This is just one of many Classic Maya contexts that indicate God K was important as a god of lineage and rulership.

Thompson (1970:226, 289) emphasizes God K's association with vegetation, and he proposes that the Colonial period God K was known as Kauil, meaning "surplus of our daily bread [maize]." Stuart's (1987:15) phonetic reading of God K's name as *k'awil* (*kauil*) provides support for linking God K with maize (Taube 1992:78). Copán Stela 11 depicts the ruler Yax Pac in the guise of God K with maize foliation on his brow, suggesting an aspect of God K that may be linked with *k'awil* (Fash 1991:fig. 108). Nonetheless, more recent interpretations of *k'awil* suggest that the term refers to stone sculptures, flint, and stone axeheads associated with lighting (Freidel et al. 1993:194–200). Schele (1992:120) notes that on Copán Stela

Figure 15. Mirror-browed God K wearing T510f star (West Court of Palenque Palace; after Robertson 1985b:fig. 358).

Figure 16. God K's head emerges from jaws of double-headed serpent flanking portrait of ruler on North Palace Substructure, Palenque (after Robertson 1985b:fig. 339b).

11, an 819-day passage refers to the serpent foot as the *way* or alter-ego of K'awinal (God K).

Although K'awil is now more commonly used in translating God K's name, it should be noted that his name has also been read as Tahil, based on linguistic connections with the word for obsidian (*tah* or *toh*), suggesting a connection with Tohil, the Quiché god of lightning and storms (Taube 1992:75–76; Tedlock 1985:365). Tohil was lord of a Quiché lineage at the time of the conquest. Indeed, God K is similarly connected with lineage and lightning. As Taube points out, God K's headdress elements—mirrors, fire, burning axes, torches, and cigars—all may allude to lightning. In my opinion, God K is a planet linked with meteorological phenomena, especially storms and lightning.

God K appears as GII of the Palenque Triad in mythological texts recording the birth of three brothers. The Cross Group gives GII's birth date as 1.18.5.4.0 1 Ahau 13 Mac (2697 B.C.). GII glyphs are like those naming God K, but they show the full figure in a reclining posture (Figure 18). Berlin (1963:93) named this god GII because he always the second listed in texts naming all three brothers of the Triad, however, his mythic birthday makes him the youngest of the brothers, with the sun as the middle brother and Venus as the eldest brother (Milbrath 1999; Schele

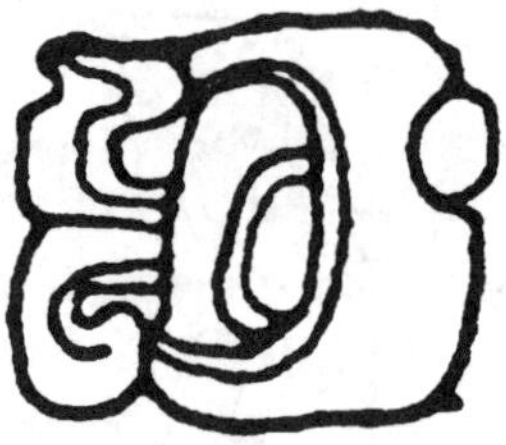

Figure 17. Classic glyph for God K representing smoking mirror (after Schele & Miller 1983:fig. 3i).

Figure 18. Palenque triad god known as GII, shown full figure in reclining posture, representing counterpart of God K (Temple of Foliated Cross at D2; after Berlin 1963).

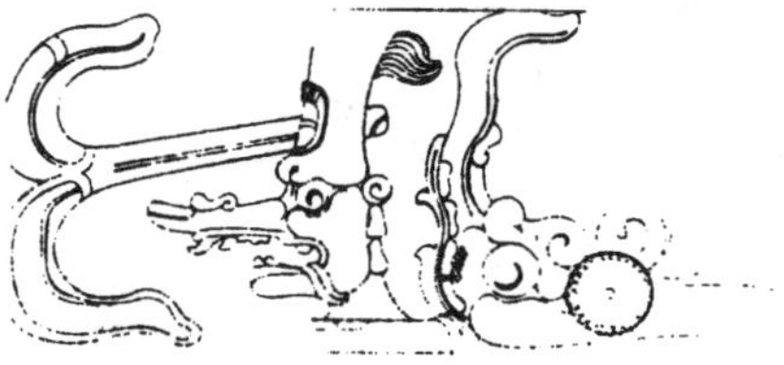

Figure 19. Mirror-browed God K wearing T510b star on altar in South Palace Substructure, Palenque (after Robertson 1985b:fig. 409a).

1976:10). This evokes connections with a modern Mopan tale in which Venus and the sun are the elder brothers, and the youngest brother is Mars or Jupiter. As will be seen, the pattern of Classic period dates provide strong evidence linking GII/God K with Jupiter. Furthermore, God K is specifically named in an 819-day cycle known to refer to Jupiter.

The 819-day cycle featuring God K has a high incidence of Jupiter and Saturn events in the Classic period inscriptions (Justeson 1989:103). The 819-day phrases often name God K in the fifth position (Kelley 1976:57–58, fig. 17). A companion glyph (T739) in the fourth position also is used in Glyph Y of the Supplementary Series where it is governed by a seven-day cycle that may be a 'planetary week' (Yasugi and Saito 1991).

In light of the link between God K and rulership, it is interesting to note that a number of Maya rulers incorporate God K in their personal names or titles. God K is seen in the headdress of Stormy Sky (K'awil Chaan) on Tikal Stela 31, a fifth century monument that records his accession to the throne (fig. 14; A. Miller 1986:54,

82). Here God K's torso appears on a sky glyph. This is Stormy Sky's personal name worn as a costume accessory (Coggins 1990:84).

On Maya lintels and stelae, God K frequently appears in the jaws of a double-headed serpent held by rulers as a royal insignia (Cohodas 1982:113; Stuart 1984:19). On architectural reliefs from Palenque, God K's head emerges from the jaws of a double-headed serpent flanking the portrait of a ruler (Figure 19). Whereas God K's serpent foot often seems to represent a naturalistic snake, the double-headed serpent with an upturned bulbous nose seems to be linked with the front head of the Cosmic Monster (Milbrath 1999).

Ceramic vessels depict God K with other astronomical deities. God K appears with the Moon Goddess in scenes featuring a large boa (Robicsek & Hales 1981:16–19). Typically, God N emerges from the serpent's mouth while its body wraps around the Moon Goddess, and the serpent's tail bears an image of God K (Milbrath 1999). Occasionally the snake is represented as an extension of God K's leg, recalling God K's serpent foot on the monument from Sayil (Figure 8). A number of relief-carved pottery vessels show God K in a specific relationship with God L, a god of the Morning Star in the Dresden Codex (Milbrath 1995, 1999). One vessel depicts God L offering the head of God K (Coe 1973:116, no. 56). 'Poison bottles' represent God K seated facing God L with an offering between the two (Robicsek 1978: fig. 210, pl. 238). Some vessels depict God L with God K's head on his cape (Robicsek 1978:173, figs. 188–189).

God K has multiple aspects most clearly seen in ceramic representations. A painted pot depicts two aspects of God K, one with the smoking mirror headdress, the other with a flower positioned on his brow (Coe 1982:47, no. 19). Another pot shows one figure holding a glyph compound referring to Bolon Dzacab and another holding the head of God K with a smoking mirror (Coe 1982:118, no. 62). God K is even shown as quadripartite in some representations (Coe 1982:54).

Jupiter Events and God K on Classic Maya Monuments

Late Classic monuments with God K images often bear dates that coincide with Jupiter's retrograde motion. When Jupiter is not in a significant position, a Saturn event may be substituted. Often the dates are associated with dynastic rituals or rituals relating to Katun events. As noted above, the Katun cycle itself reflects the pattern of intervals between successive conjunctions or oppositions of Jupiter and Saturn. In calculating these Classic period dates, I employ the 584,285 correlation preferred by Thompson (1960) in his later work, one supported by a number of scholars today (Milbrath 1999).

At Naj Tunich, God K's portrait-head glyph appears with the date 13 Ahau, referring to the Katun ending 9.17.0.0.0 13 Ahau 18 Cumku (771 January 18; Stone 1995:142, fig. 6-27). This is especially intriguing because God K's head is represented in the Katun-ending ceremonies of the Postclassic. The 13 Ahau date falls less than three weeks after the first stationary point of Jupiter (771 January 1; Meeus n.d.). Saturn was also in retrograde motion. Such correlations probably led the Maya to focus on observations of these two planets at the Katun end.

By studying a pattern of dates in the life of Chan Bahlum, Lounsbury (1989:table 19.1) correlates Palenque monuments with Jupiter's retrograde, although he does not identify specific imagery that refers to Jupiter. At six years of age, the young ruler was designated as heir to the throne on 9 Akbal 6 Xul (9.10.8.9.3; 641 June 12), a date that correlates with Jupiter's departure from second stationary point (Lounsbury 1989:table 19.1). Some 30 years later, the ruler began his reign under the patronage of Jupiter. At the time of Chan Bahlum's accession on 8 Oc 3 Kayab (684 January 5), Jupiter had just passed its second stationary point and was crossing the meridian at dusk (second stationary point 683 December 23; Meeus n.d.). Both his heir designation and his accession date were timed by Jupiter's departure from second stationary point (Lounsbury 1989).

The small figure on the central panels of temples in Cross Group, previously interpreted as Pacal (Schele 1976:13–14), is now recognized as Chan Bahlum at six or seven years of age, when he was designated as heir to the throne (Figures 20–22; Robertson 1991:20). The date of heir-designation event is positioned near the young Chan Bahlum, whereas the accession date appears with the taller figure representing Chan Bahlum as an adult (Figure 21). All three temples show different aspects of God K, indicating there may be at least three manifestations of God K at Palenque. The Temple of the Sun shows Chan Bahlum at the time of his accession, holding a manikin of God K with a smoking brow (Figure 22). The Temple of the Cross depicts the young Chan Bahlum standing on a skeletal head with an elongated snout and the number nine (*bolon*) alluding to the Bolon Dzacab aspect of God K (Figure 21; compare Thompson 1970: figs. 9p to q). This image is captioned by the heir-designation event above the smaller figure of Chan Bahlum. The Temple of the Foliated Cross depicts God K emerging from a spiral shell with a maize plant in hand (Figure 20; Dütting 1984:23; Robicsek 1979:115; Stuart 1978:167). The young Chan Bahlum stands on God K's shell, indicating a direct link with the maize aspect of God K.

Temple XIV at Palenque shows that after his death Chan Bahlum continued his relationship with God K and the planet Jupiter (Figure 23). The text opens on the left with a 9 Ik 10 Mol date followed by a statement that God K was displayed under the auspices of the Moon Goddess; then the date is repeated on the right after an interval of almost a 100,000 years (Schele & Miller 1986:272, fig. VII.2). The display of God K is mentioned again, followed by a reference to 9 Ahau 3 Kankin (9.13.13.15.0; 705 October 31), a date that falls about three years and nine months after the death date (3 × 365 days + 260 days). Lounsbury (1989:250, fig. 19.5) links this posthumous date with Jupiter's departure from second stationary point (second stationary point on 705 October 21; Meeus n.d.). Saturn was also in retrograde at this time, and the moon passed Jupiter a few hours before dusk. The lunar event is especially significant because Chan Bahlum's deceased mother is compared to the Moon Goddess in the texts. Apparently Lady Ahpo Hel takes the role of the moon on the panel, for she holds the God K manikin, just as the moon holds Jupiter in conjunction.

Dumbarton Oaks Relief Panel 1, another posthumous portrait of a ruler holding a God K ceremonial bar, is also associated with the time of retrograde motion for both Jupiter and Saturn. Fox and Justeson (1978:55–56) note that the ruler's date of death on 6 Akbal 11 Pax (9.15.1.6.3; 732 December 11) is a few days from Saturn's second stationary point (732 December 15; Meeus n.d.), and an interval at the end of the inscription counts 378 days (the length of Saturn's synodic period) to 9.15.2.7.1 when Saturn was once again at its second stationary point (733 December 29; Meeus

Figure 20. At time of his heir designation, youthful Chan Bahlum stands on shell housing God K, who holds sprouting maize that embodies young ruler; subsequently Chan Bahlum appears fully mature at time of accession; Temple of Foliated Cross at Palenque (Robertson 1991:fig. 153).

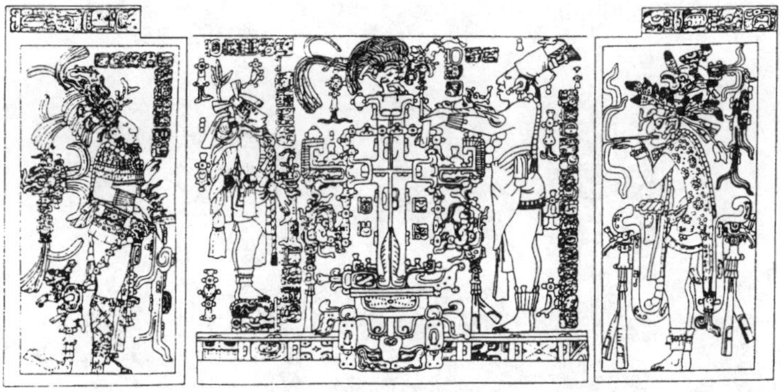

Figure 21. Young Chan Bahlum standing on Bolon Dzacab aspect of God K at time of heir designation in A.D. 641, and mature Chan Bahlum at 40 years of age, when he was crowned king in A.D. 684; Temple of Cross at Palenque (Schele & Miller 1986:fig. II.6).

n.d.). More recently, Justeson (1989:109) points out that the death date is also connected with the first stationary point of Mars (732 December 18; Meeus n.d.). It should noted that the date also coincides approximately with the midpoint of Jupiter's retrograde motion (732 October 10 to 733 February 15; Meeus n.d.). Fox and Justeson suggest that astrology played a role because astronomical events were of significance when they coincided with historical events. On the other hand, Marcus (1992:440) maintains that the historical events sometimes were manipulated to fit the astronomical events. In my opinion, the death event may actually refer to the ruler's apotheosis, for such an event could be adjusted to fit the proper astrological conditions.

Figure 22. Chan Bahlum holding God K manikin on right of a war shield, from Late Classic Temple of the Sun, Palenque (see 3-6b; after Robertson 1991:fig. 95).

Figure 23. Apotheosis of Chan Bahlum on October 31, 705 (Julian) correlates with time when Jupiter departed from second stationary point; Lady Ahpo Hel holds manikin of God K, just as moon held Jupiter in conjunction, from Temple XIV at Palenque (Schele & Miller 1986:fig. VII.2).

The most important image of apotheosis involving Jupiter is seen on the Sarcophagus Lid of the Temple of the Inscriptions at Palenque, a pyramid that houses the tomb of Chan Bahlum's father, Pacal (Figures 24, 25). God K and the Jester God

Figure 24. Pacal's apotheosis as God K took place when Jupiter reached its first stationary point in A.D. 683 (detail of Sarcophagus Lid of Temple of Inscriptions at Palenque; after Robertson 1985a:fig. 73).

Figure 25. Pacal's apotheosis as God K took place when Jupiter was at its first stationary point in August of 683 (Sarcophagus Lid of Temple of Inscriptions at Palenque; Robertson 1985a:fig. 73).

emerge from either end of a bicephalic serpent bar, and Pacal appears in the guise of God K at the base of a cosmic tree (Schele 1976:17; Robertson 1991:18). This scene has an astronomical context, for he is surrounded by a sky band. On the edge of the lid, his death or some other event related to his death is recorded on the Calendar Round 6 Etz'nab 11 Yax (9.12.11.5.18; 683 August 26). David Stuart has deciphered the event glyph as "he entered the road" (*och bih*; Schele 1992:133). Heinrich Berlin (1977:137) notes that this event (T100:585) refers to the end of a reign, and it may signal death or apotheosis. This expression probably alludes to Pacal's apotheosis, rather than his death. Coggins (1988:74–75) suggests that Pacal is shown dancing out of his tomb on the Sarcophagus Lid, indicating an image of apotheosis. The apotheosis date is only two days short of Jupiter's first stationary point (683 August 28; Meeus n.d.). It is especially noteworthy that Pacal's son, Chan Bahlum, was not inaugurated as the new ruler until Jupiter had begun to move forward again four months later.

Jupiter also appears to be involved with the posthumous portrait of Yax Pac of Copán. On Stela 11, Yax Pac appears a guise of God K, one of only two known portraits of a deceased ruler in the guise of God K (Fash 1991:177, fig. 108). Like Pacal's image of apotheosis, the Copán monument is associated with the retrograde period of Jupiter. The stela bears the date 8 Ahau, inferred to be that of the Lahuntun ending on 9.19.10.0.0 8 Ahau 8 Xul, marking the halfway point of the Katun cycle. On this date (820 April 30), Jupiter was on the eastern horizon at dusk, approximately at the midpoint of it retrograde motion (February 18 to June 21; Meeus n.d.). Furthermore, Jupiter was in conjunction with the full moon on that night.

On Bonampak Stela 1, God K appears on the headdress of an ancestor figure emerging from a cleft in a Cauac (Witz) monster on the base of the monument (Figure 26). The stela marks the first Lahuntun ending of Chaan Muan's reign, recorded in a fragmentary Initial Series date of 9.17.10.0.0 (780 November 26; Mathews 1980:64; Proskouriakoff 1993:163–164). At this time, Jupiter had been in retrograde motion for three weeks (first stationary point 780 November 5; Meeus n.d.). Jupiter, visible high above the eastern horizon at dawn, was especially brilliant for it was precisely in conjunction with Uranus, a planet that was bright enough at this time to be seen with the naked eye. Meanwhile, Saturn was invisible in conjunction with the sun.

A lintel from Bonampak Structure 6 depicts a ruler holding a double-headed God K serpent bar (Schele & Grube 1994:109, fig. 180). Both images of God K have mirrored brows with axe inserts, and one has Akbal markings on the mirror, suggesting an opposition of light and dark. The only date on the monument, 7 Chuen 4 Zotz (9.8.9.15.11; 603 May 11), does not correspond to Jupiter's retrograde, but Saturn was in retrograde motion. Saturn may be significant in the imagery of God K because God K encodes the relationship between Jupiter and Saturn in his role as a Katun god.

Dos Pilas Stela 14 shows a ruler displaying a God K manikin scepter (formerly Stela 25; Houston 1993:tables 3-1, 4-1, fig. 3-24). Two passages have a glyph compound mentioning God K. The monument bears an Initial Series date of 9.14.0.0.0 6 Ahau 13 Muan (711 November 29) and a Calendar Round of 8 Ix 2 Cumku (9.14.5.3.14; 717 January 15) associated with a 'shell star' compound. The Initial Series date corresponds to the full moon. It also marks a time when the Evening Star was about to reemerge. Neither date correlates with Jupiter's retrograde, but the Katun-ending date marks a time that Jupiter was overhead at dawn and it coincides precisely with Saturn's first stationary point (711 November 29). The second date also approximates Saturn's first stationary point (717 January 29; Meeus n.d.). Here Saturn may substitute for Jupiter in the context of monuments depicting God K imagery.

Dos Pilas Stela 15 also shows the display of a God K scepter (Houston 1993:fig. 3-25). Among several dates, it bears a Calendar Round date of 7 Ahau 3 Kayab

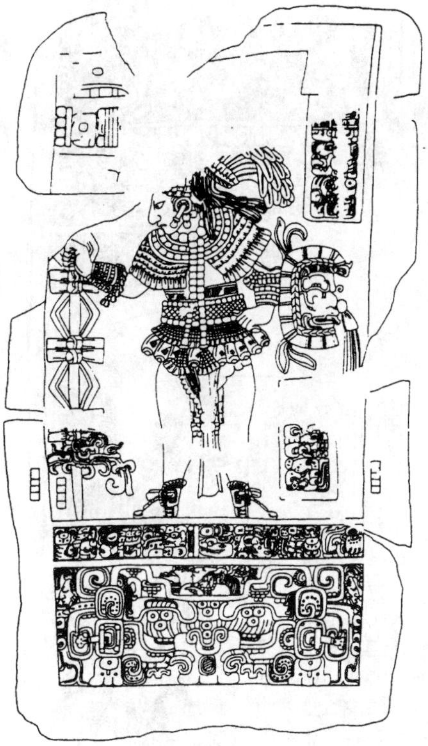

Figure 26. Bonampak Stela 1, with an ancestor bearing God K insignia emerging from Witz monster (Mathews 1980:fig. 3).

(9.14.10.4.0; 721 December 26) that coincides with Jupiter's retrograde motion (first stationary point 721 November 15; Meeus n.d.). Another date (9.14.9.10.13 1 Ben 16 Tzec; 721 May 13) coincides with Saturn's retrograde motion.

Dates on some Dos Pilas monuments depicting God K cannot be linked to the retrograde of either Jupiter or Saturn, but they do record subdivisions of the Katun that could be related to observations of Jupiter and Saturn. Among such examples, we find Dos Pilas Stela 11 depicting the display of the God K scepter on a Calendar Round date of 12 Ahau 8 Kankin (9.14.5.0.0; 716 November 2; formerly Stela 22; Houston 1993:table 3-1; fig. 3-27).

Monuments from Yaxchilán clearly link God K and Jupiter events. On Stela 1, God K emerges from open jaws of a double-headed serpent represented with a sky band body and deer attributes (Figure 27). The Initial Series date on the monument is 9.16.10.0.0 1 Ahau 3 Zip (761 March 11; Proskouriakoff 1993:112). This date coincides with the new moon falling about two weeks after Jupiter's first stationary point (761 February 27; Meeus n.d.). A fragmentary Calendar Round date, reconstructed as 1 Oc 18 Pop (9.16.8.16.10; 760 February 14), records an 819-day event that is also a little over two weeks after Jupiter's first stationary point (760 January 27; Meeus n.d.). The two dates may be purposefully paired to show Jupiter's retrograde in two sequential cycles.

Yaxchilán Lintel 30 names God K in an 819-day event on 1 Ben 1 Chen (708 July 20; 9.13.16.10.13), a date reached by counting back 397 days from the birth

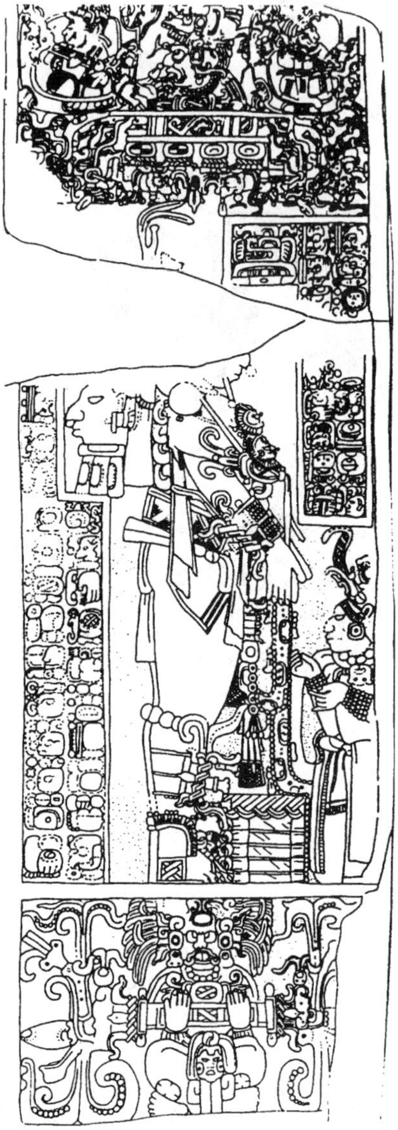

Figure 27. Sun God holding skeletal serpent at base; overhead double-headed sky serpent with God K in its jaws, from Late Classic Yaxchilán Stela 1 (Tate 1992:fig. 124).

of Bird Jaguar IV recorded on 8 Oc 13 Yax (709 August 21; 9.13.17.12.10; Tate 1992:appendices 2–3). Although neither date pertains to the period of Jupiter's retrograde, the distance number is a close approximation of the synodic period of Jupiter (398.9 days). The 819-day event on 1 Ben 1 Chen refers to the east and red before mentioning God K at A5. At this time, Jupiter and Saturn were in Gemini in the east at dawn, and Saturn was in conjunction with the waning moon, while Venus was on the eastern horizon in Cancer about to disappear as the Morning Star. One synodic period later, Jupiter was once again in the east at dawn at the time of Bird Jaguar's birth in A.D. 709. Now Mars was approximately at the midpoint of its retrograde motion, above the eastern horizon in Leo (second stationary point 709 September 10; Meeus n.d.). At this time, Jupiter and Saturn were precisely in conjunction in Cancer, and Venus was only a few degrees away, now also in Cancer. This massing of three planets above the eastern horizon at dawn may have been considered an auspicious omen for the ruler's birth, and the count backwards could relate to a symmetry seen in the previous synodic period with another celestial triad formed by Jupiter, Saturn, and the moon in Gemini at dawn.

A ruler holds the God K manikin on the back of Yaxchilán Stela 11, which bears a 1 Imix 19 Xul date (9.15.19.1.1; 750 May 29) falling at the midpoint of Jupiter's period of retrograde motion (Tate 1992:fig. 136). Another date (9 Ahau 18 Xul), exactly four years earlier, is positioned in a like fashion just above the ruler on the front side of the stela; this date coincides with the time that the moon passed by as Jupiter and Mars were precisely in conjunction. Both dates correlate with times that Saturn was in retrograde motion.

Table 1. Yaxchilán Depictions of God K Associated with Jupiter Retrograde Stationary points from Meeus n.d.; Monument dates from Tate 1992

Stela 11:		9.15.19.1.1 (5/29/750+)	1st st. pt. 3/28/750 2nd st. pt. 7/28/750
Lintel 42: God K event		9.16.1.2.0 (6/6/752*+)	1st st. pt. 6/8/752
Lintels 6 and 43:		9.16.1.8.6 (10/10/752*+)	2nd st. pt. 10/5/752
Lintel 7: God K event		9.16.1.8.8 (10/12/752*)	2nd st. pt. 10/5/752
Lintel 40:		9.16.7.0.0 (3/27/758*)	2nd st. pt. 3/29/758
Stela 1: 819-day event		9.16.8.16.10 (2/14/760)	1st st. pt. 1/27/760 2nd st. pt. 5/30/760
Stela 1:	IS	9.16.10.0.0 (3/11/761)	1st st. pt. 2/27/761 2nd st. pt. 6/30/761
Lintel 38:		9.16.12.5.14 (6/23/763)	1st st. pt. 5/7/763 2nd st. pt. 9/5/763

*Date within 7 days of a stationary point
+Date also coincides with Saturn's retrograde period

Surveying all the dated Yaxchilán monuments provides a test for the link between God K imagery and Jupiter events. At Yaxchilán, 8 of the 15 monuments (53%) that have God K images bear dates that coincide with Jupiter's retrograde, allowing a seven-day window on either side of retrograde (Table 1). This percentage seems significant in terms of statistical analysis. Even allowing a generous 30-day window on either end of the period, Jupiter's mean period of retrograde motion represents less than half of the planet's synodic period (45%; Aveni & Hotaling 1994:S40). And the percentage of Jupiter's mean period of retrograde using a seven-day window is only 33% of its synodic period (120 + 7 + 7 or 134 days out of 399 days total). In a random sample of dates using a seven-day window, you would expect only around 33% of the dates to coincide with retrograde, but the God K monuments at Yaxchilán show a relatively high percentage of such dates. The correlation is strengthened by the fact that the sample is not subject to the problems inherent in selecting the date that works best, because all the Yaxchilán monuments in Table 1 bear only a single date, except Stela 1 which has two dates, both of which coincide with Jupiter's retrograde motion. The first date in Table 1 accurately pegs the

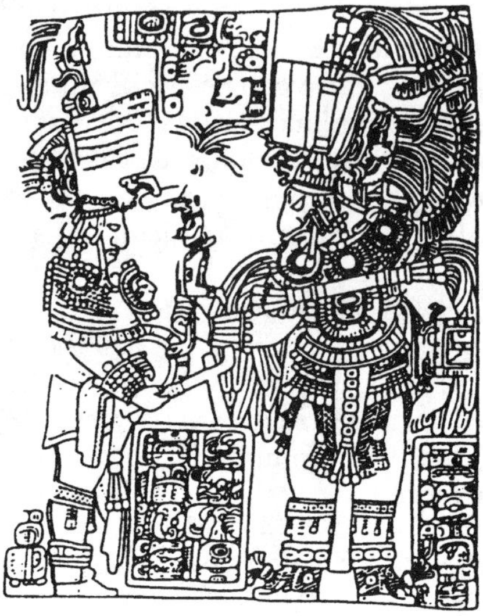

Figure 28. Yaxchilán Lintel 42 showing God K manikin (Tate 1992:fig. 148).

mid-point of Jupiter's retrograde, and the next four fall within seven days of the first or second stationary points of Jupiter.

Three other Yaxchilán lintels depicting God K bear dates approximating Jupiter's first stationary points (within 21 days). Lintel 39's Calendar Round date of 4 Imix 4 Mol (9.15.10.0.1; 741 June 25), one day after the Lahuntun ending, approximates Jupiter's first stationary point (741 July 10; Meeus n.d.). Lintels 32 and 53 record a Calendar Round date of 6 Ben 16 Mac (9.13.17.15.13; 709 October 23) that Tate (1992:appendix 2) correlates with the position of Saturn and Jupiter aligned together at stationary points. The fact that Jupiter and Saturn were close together (within 2°) is certainly significant. This date is less than three weeks before Jupiter's first stationary point on 709 November 10, and is even closer to Saturn's first stationary point on 709 November 2 (Meeus n.d.). Since the superior planets slow down as they approach retrograde, it seems that an approximation of the Jupiter event is acceptable.

God K events are noted in texts on four Yaxchilán monuments, all of which depict a ruler holding a God K manikin (Lintels 3, 7, 42, 52; Tate 1992:appendix 2). God K events on Lintels 7 and 42 correlate with a stationary point of Jupiter, and the God K event on Lintel 42 also relates to a time that Saturn was also at its first stationary point (Figure 28; Table 1). The God K event on Lintel 52 coincides with Saturn's second stationary point (766 February 3, Meeus n.d.). Lintel 3 records a Hotun ending (9.16.5.0.0; 756 April 6), and the same Hotun ending appears on Lintel 54, also depicting the ruler holding a God K manikin. Neither Jupiter nor Saturn were in retrograde on the Hotun date recorded on these two monuments, but the date is a subdivision of the Katun that could be used to track the position of Jupiter in relation to Saturn.

In dynastic affairs, Saturn's stationary point may have been considered a suitable substitute, if Jupiter was not at the appropriate position for a royal ritual involving God K. Lintel 1, showing Bird Jaguar IV with the God K manikin, has a date that approximates Saturn's first stationary point (752 April 2; Meeus n.d.). The recorded event on this date is his accession on 11 Ahau 8 Zec (9.16.1.0.0; 752 April 27). Mars was in retrograde on this date (second stationary point 752 May 5; Meeus n.d.).

Allowing a 21-day window on either side of retrograde indicates that 13 of 15 monuments with God K imagery have dates relating to the retrograde periods of Jupiter or Saturn, and six of these correspond to the retrograde periods of both planets (Lintels 6, 32, 42, 43, 53, Stela 11). Five bear dates linked only with the retrograde of Jupiter (Lintels 7, 38, 39, 40, Stela 1). And two other God K monuments bear dates that relate only to Saturn's retrograde (Lintels 1, 52).

Those Yaxchilán monuments with dates coinciding with Jupiter's retrograde that do not depict God K generally show events related to the ballgame, warfare and blood offerings. Step VII from Structure 33 links a ballgame event on 9.15.13.6.9 to Jupiter's stationary point (Tate 1992:appendix 2, fig. 111). On Lintel 10, a capture event falls within the period of Jupiter's retrograde motion (9.18.17.12.6; 3/14/808; Tate 1992:appendix 2). A similar correlation is evident on Stela 18, depiciting Shield Jaguar with a captive, where the text records a capture event on 3 Eb 14 Mol (9.14.17.15.12; 729 July 11), followed by a reference to God K as a sky god (C3) and a bloodletting event at C6 (Tate 1992:appendix 3, fig. 145).

Looking at the entire sample of dates Tate (1992) published for Yaxchilán, 26% of the total (29 of the 109 dates) fall in Jupiter's retrograde period or seven days on either side of retrograde. This figure falls below the expected random frequency of 33%, using a seven-day window, but when we narrow our focus to those monuments depicting God K, the percentage bearing dates relating to Jupiter's retrograde relatively high at 53%. Thus Yaxchilán provides strong evidence that God K is associated with Jupiter's retrograde motion.

Turning to the site of Tikal, we find another set of monuments to test for a possible pattern linking God K images to Jupiter's retrograde period. Although I will not analyze the monuments statistically, a pattern similar to Yaxchilán is notable, especially on monuments that have only one date.

Tikal Stela 20 shows Yaxkin Chaan K'awil (Ruler B) wearing a headdress crowned by God K; his fret-ended mouth mask (possibly a stylized serpent jaw) is decorated with a star glyph (Figure 29). The inscribed date records the first Katun ending of his reign on 2 Ahau 13 Zec (9.16.0.0.0; 751 May 3). This stela is paired with Altar 8 in Twin-pyramid Complex P (Group 3D-2), one of the many Twin-pyramid complexes apparently devoted to Katun ceremonies. The Katun ending 9.16.0.0.0 provides a very good correlation with Jupiter's first stationary point (751 May 2; Meeus n.d.). Saturn was also in retrograde motion, and Mars rose at dawn, while Venus was visible near its maximum altitude. This was an auspicious interlocking of astronomical events with the calendric cycles marking a period ending. This Katun ending was surely the focus of major ceremonies involving Jupiter, for the planet stood still to honor Yaxkin Chaan K'awil on the first Katun completed in his reign.

A lintel from Structure 5D–52 shows Yaxkin Chaan K'awil holding a God K manikin scepter; an eroded date probably refers to the Lahuntun ending on 3 Ahau 3 Mol (9.15.10.0.0, 741 June 24; Jones 1977:52, fig. 17). On this period-ending date, Jupiter was about to enter retrograde (first stationary point on 741 July 10; Meeus n.d.). This Lahuntun ending approximates Jupiter's first stationary point, and it marks a halfway point leading up to the exceptional Katun 9.16.0.0.0, when Jupiter reached its first stationary point at the end of the Katun.

The same 3 Ahau 3 Mol Lahuntun date opens the inscription on Lintel 3 of Temple IV, which features God K in the jaws of a Cosmic Monster arching over the

Figure 29. Yaxkin Chaan K'awil (Ruler B) wearing star mask and headdress crowned by God K at time of Jupiter's first stationary point (Tikal Stela 20; Jones 1977:fig. 15).

ruler's head (Figures 30, 31). Of the four Calendar Round dates on Lintel 3, only the Lahuntun ending on 3 Ahau 3 Mol shows a relationship with Jupiter's retrograde, approximating the planet's first stationary point. The lintel depicts God K displaying a mirror glyph in his hand, positioned like the flat-hand mirror compound used to mark Venus positions in the Venus Table of the Dresden Codex.

The Lahuntun ending 3 Ahau 3 Mol, opening the inscription on Lintel 2 from Temple IV, similarly coincides with Jupiter's first stationary point. Here Yaxkin Chaan K'awil (Ruler B) holds his God K manikin scepter as a Jaguar War God protector looms over him (Jones 1977:fig. 12). The latest date on Lintel 2 (9.15.15.14.0 3 Ahau 13 Uo; 747 March 5) falls within retrograde periods of both Jupiter and Saturn. Schele and Grube (1994:187, 1995:40–46) note that a solar eclipse on 744 January 19 was a prelude to the star war event recorded on the lintel. According to Lounsbury (1989:255), the star war event (B8) was timed by Jupiter's departure from second stationary point. This war event dates to 7 Ben 1 Pop (744 February 2; 9.15.12.11.13), coinciding approximately with the midpoint of Saturn's retrograde period, and only two weeks after Jupiter's second stationary point (744 January 17; Meeus n.d.). Justeson (1989:table 8.8) points out that a number of war events coincide with the retrograde periods of at least one of the superior planets.

On Lintel 3 of Temple I, a Water-lily Jaguar looms over another Tikal ruler carrying the God K scepter as an insignia of lineage and authority (Jones 1977:fig. 1). Originally designated as Ruler A, he is variously called Ah Cacao Caan Chac and Ah

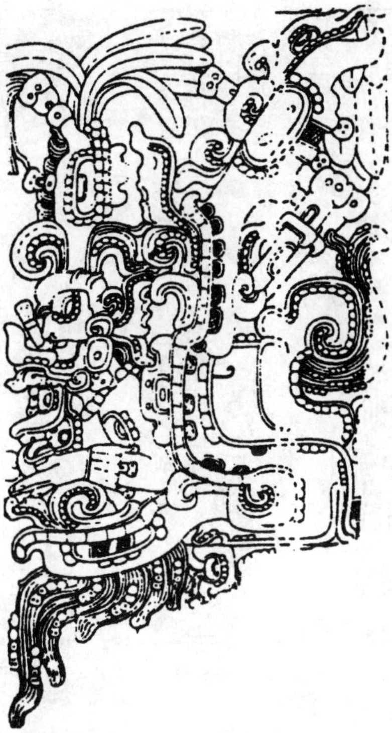

Figure 30. God K in jaws of Cosmic Monster arching over ruler's head (detail of Tikal Lintel 3, Temple IV; after Jones 1977:fig. 11).

Cacaw, or more recently as Hasaw Chaan K'awil. The date 11 Etz'nab 11 Chen (9.13.3.7.18; 695 August 3) at A3–B3 falls in Saturn's period of retrograde motion. On 12 Etz'nab 11 Zac (9.13.3.9.18; 695 September 12), forty days later, a bloodletting event coincides approximately with Jupiter's first stationary point and Saturn's second stationary point (Aveni & Hotaling 1994:table 1; Schele & Freidel 1990:445).

Stela 5 from Tikal probably also features God K, although here the manikin scepter is almost completely effaced (Jones 1977:table 1, fig. 13). The text begins with a fragmentary Calendar Round date, reconstructed as 9.15.3.6.8 3 Lamat 6 Pax (734 December 6), less than two weeks before Jupiter's first stationary point (734 December 19; Meeus n.d.). The date also coincides with Saturn's retrograde motion, and it approximates the time Venus was most brilliant as the Evening Star. This date appears with a distance number (9.11.12) that counts forward to 4 Ahau 8 Yaxkin (9.15.13.0.0; 744 June 8), a period ending that is the dedicatory date of the monument. Since 9.11.12 is a lunar interval, the moon was once again at the last quarter. At this time, Jupiter was in conjunction with the sun and Saturn was departing from its second stationary point (744 May 17; Meeus n.d.). These two dates seem to correlate with events involving the moon, Jupiter, and Saturn.

As early as the mid fifth century, Tikal dynastic monuments link God K with Jupiter's retrograde motion. On the Early Classic Tikal Stela 1, a ruler holds a serpent bar bearing a God K hatchet manikin with a snake leg, the prototype for the serpent-legged God K-manikin scepter of the Late Classic period (Proskouriakoff 1993:16; Taube 1992:fig. 37c). The monument is believed to date to the reign of Stormy Sky

Figure 31. God K carried in jaws of front head of Cosmic Monster arching over Yaxkin Chaan K'awil (Ruler B) of Tikal; inscription opens with Lahuntun ending date (9.15.10.0.0 3 Ahau 3 Mol) in A.D. 741 that approximates Jupiter's first stationary point (Lintel 3 of Temple IV at Tikal; Jones 1977:fig. 11).

(K'awil Chaan) or shortly thereafter. The fragmentary dates remain debatable, but one reconstruction suggests a date of 9.0.15.11.0 (451 April 28; Freidel et al. 1993:424, n. 58). At this time, Saturn was at the midpoint of retrograde (451 January 18 to 451 June 7) and Jupiter was approaching its second stationary point (451 May 9; Meeus n.d.). Jupiter, Saturn, and Mars were all clustered together in Virgo to the east, while the Evening Star was making its first appearance in Taurus to the west.

Stela 31 from Tikal is one of the earliest known images associating God K with a living ruler from Tikal (Coggins 1990:85). The ruler Stormy Sky wears a headdress with God K, here representing his personal name (Figure 14). Perhaps Jupiter was Stormy Sky's patron planet in the Early Classic period, just as Jupiter guided events in the lives of Late Classic rulers, especially at Palenque and Yaxchilán. The Initial Series date recorded on Stela 31 (9.0.10.0.0 7 Ahau 3 Yax; 445 October 16) corresponds to a time when Jupiter was approaching its second stationary point (445 November 23; Meeus n.d.). This date also approximates the heliacal set of Mars. Other coordinating events include the new moon and Venus approaching its maximum brilliance as the Morning Star (Milbrath 1999). The ruler himself seems to be compared to Hun Ahau, the underworld sun and Morning Star, while his father seems to be apotheosized as the Sun God overhead (Milbrath 1999).

One of the earliest known images of God K appears on the Leyden Plaque, probably also from Tikal. Here the ruler Moon Zero Bird holds a double-headed serpent that bears God K and the Sun God in its two jaws (Schele & Miller 1986:121, pl. 33b). The recorded date, 8.14.3.1.12 1 Eb 0 Yaxkin (320 September 14), corresponds approximately to Jupiter's heliacal rise just prior to the fall equinox. Jupiter's dawn rise so close to the fall equinox coordinates the cycles of the sun and Jupiter.

Alternate Identifications of God K

Schlak (1989) correlates dates on monuments depicting God K manikin figures with periods when Mercury was visible, using maximum elongation as a key to determining visibility. His sample is relatively small, and he concentrates on manikin scepter imagery, excluding many of the God K images discussed here. When these monuments are included, a link with Jupiter events emerges. Among the 20 dates that Schlak correlates with the visibility of Mercury, eight coincide with Jupiter's retrograde motion, allowing a seven-day window. Of those that do not, a significant number come from monuments with multiple dates. If we are free to select another date, we often find one that correlates with Jupiter's retrograde motion. Since Mercury is visible 65% of the time (Aveni & Hotaling 1994:S39), a statistical study should be done with a complete corpus of God K monuments from several sites to determine whether there is an unusually high percentage of God K monuments that correspond to times when Mercury was visible.

In a more recent study, Schlak (1996:table 4) takes his original 20 dates and adds another variable to his analysis (altazimuth), reflecting Mercury's height above the horizon. He also adds 12 mores dates linked with God K glyphs (Schlak 1996:table 3). Even with the expanded list, Schlak has omitted a number with God K images incorporated in my study. He makes no claim that any of the dates are precisely maximum altitude or maximum elongation, only that they coincide with Mercury's visibility. Of the 12 new dates, one-third are within five days of one of Jupiter's stationary points (Dates 3, 9, 10, 12). All but three dates can be linked with times that either Jupiter or Saturn were in retrograde.

Dennis Tedlock (1992:250–258, 266–267) proposes that Mars is represented by GII in the Triad and Hun Hunahpu in the Popol Vuh. He reconstructs Hun Hunahpu's birthday as -8.5.0 before the current Maya epoch, concluding that this god shares GII's 1 Ahau birthday, a date he suggests is specifically associated with the heliacal rise of Mars, although he provides no convincing proof for this assertion. Tedlock notes that the birth dates of both gods give intervals that are evenly divisible by 780, the mean synodic period of Mars. He also finds a link between Vucub Hunahpu and Jupiter, but here his evidence seems even more tenuous. Furthermore, it should be pointed out that these interpretations contradict his earlier identification of a falcon character in the Popol Vuh as Jupiter and the monkeys brothers as Mars (Tedlock 1985).

Other planetary identities proposed for God K include Clemency Coggins's (1990:84–85, fig. 5.1a) suggestion that God K represents Venus as a god of lightning connecting the earth with the sky. She believes that Stormy Sky's personal name is a pun for K'uk' Ka'an ("bird-serpent"), referring to Kukulcan as Venus. Although a link with lightning seems likely, the connection between God K and the feathered serpent seems tenuous, especially since Coggins makes no attempt to link God K to dates recording significant Venus positions.

God K and Jupiter in Cultural Context

Scholars clearly agree that God K has a planetary identity, but they are not in accord about which is God K's planet. The data presented here indicates that God K embodies the planet Jupiter and the Katun cycle involving Jupiter and Saturn in the Classic period. Jupiter events were apparently important in the lives of Classic Maya rulers, and the rulers often appear with the insignia of God K. Jupiter is implicated in the dates associated with images of dead rulers in the guise of God K, suggesting that God K is involved in the apotheosis of Maya rulers. At Yaxchilán, Palenque and Tikal, monuments representing God K usually bear dates coinciding with Jupiter's retrograde motion. God K might have more than one planetary aspect, for when a Jupiter event is not apparent, Saturn seems to act as a substitute. Between Katun 9.15.10.0.0 and 9.17.0.0.0, there seems to be intense interest in recording dates that coincide with Jupiter's and Saturn's retrograde, probably because the Katun 9.16.0.0.0 links the retrograde periods of both Jupiter and Saturn, and precisely marks the first stationary point of Jupiter. God K is also part of the 819-day inscriptions that seem to relate to Jupiter and Saturn events, pointing to a likely association with one of these two planets.

The Palenque Triad places Venus and Jupiter (GI and GII) as the older and younger brother of the sun, as they are in a 20th-century Mopan Maya myth. GII is clearly linked with Postclassic God K. Postclassic evidence also points to an association between God K and Jupiter, based on God K's role in the Katun cycle and his representation in the Venus pages of the Dresden Codex at a time that Jupiter was in retrograde motion, when the planet was positioned opposite to the rising Morning Star.

It seems to be clear that God K is linked with monuments recording dates that coincide with Jupiter and Saturn's retrograde motion. In the Yaxchilán sample, 13 of the 15 God K monuments bear dates that coincide with the retrograde of Jupiter or Saturn, allowing a 21-day window on either side of retrograde. A similar pattern seems to be evident at Tikal, but here the picture is somewhat more complex because most of the monuments bear multiple dates.

Classic period representations of God K connect him with lineage, rulership, lightning and thunder. His imagery often involves celestial associations, including sky bands, sky glyphs, and stars. His mirror may also be a celestial symbol. He appears in scenes with the Moon Goddess and the Morning Star, images that may show celestial conjunction.

As a god of thunder and lightning, God K shares the same role as Jupiter in Ancient Greece and Rome. Perhaps this notion has an explanation in nature, one that is connected with folklore linking Jupiter to storms. God K's role as the god of Maya kings also presents an overlap with ancient European traditions identifying Jupiter as the planet of kings. There is a natural basis for the association with kings. Jupiter seems to dominate the night sky. Jupiter is rarely absent from the sky (mean disappearance interval 32 days), unlike Venus which frequently visits the underworld and is never seen overhead. Jupiter is second only to Venus in brilliance. Steady, strong, and ever-present, Jupiter makes a perfect celestial ruler, the counterpart to the good ruler on earth.

References

Aveni, A. F. and L. D. Hotaling. 1994, "Monumental Inscriptions and the Observational Basis of Maya Planetary Astronomy" Journal for the History of Astronomy XXV, p. S21

Berlin, H. 1963, "The Palenque Triad" Journale des Société des Americanistes 52, p. 91

Coe, M. D. 1973, The Maya Scribe and His World (New York: Grolier Club).

Coe, M. D. 1982, Old Gods and Young Heros: The Pearlman Collection of Maya Ceramics (Jerusalem: The Israel Museum).

Coggins, C. C. 1988, "Classic Maya metaphors of death and life" RES: Anthropology and Aesthetics 16, p. 66

Coggins, C. C. 1990, "The Birth of the Baktun at Tikal and Seibal", in Vision and Revision in Maya Studies, Ed. F. S. Clancy & P. D. Harrison, p. 79 (Albuquerque: University of New Mexico Press).

Cohodas, M. 1982, "The Bicephalic Monster in Classic Maya Sculpture" Antropologica XXIV, p. 105

Dütting, D. 1984, "Venus, the Moon, and the Gods of the Palenque Triad" Zeitschrift für Ethnologie 109, p.7

Fash, W. F. 1991, Scribes, Warriors and Kings (London: Thames and Hudson).

Fox, J. A. & J. S. Justeson. 1978, "A Mayan Planetary Observation" Contributions of the University of California Archaeological Research Facility 36, p. 55

Freidel, D., L. Schele & J. Parker. 1993, Maya Cosmos: Three Thousand Years on the Shaman's Path (New York: William Morrow and Company).

Houston, S. D. 1993, Hieroglyphs and History at Dos Pilas: Dynastic Politics of the Classic Maya (Austin: University of Texas Press).

Jones, C. 1977, "Inauguration Dates of Three Late Classic Maya Rulers of Tikal, Guatemala." American Antiquity 42, p. 28.

Justeson, J. S. 1989, "The Ancient Maya ethnoastronomy: an overview of hieroglyphic sources", in World Archaeoastronomy. Selected Papers from the 2nd Oxford International Conference on Archaeoastronomy, Ed. A. F. Aveni, p. 76 (Cambridge: Cambridge University Press).

Kelley, D. H. 1976, Deciphering the Maya Script (Austin: University of Texas Press).

Kelley, D. H. 1985, "The Lords of Palenque and the Lords of Heaven", in Fifth Palenque Round Table, 1983, vol. VII, Ed. M. G. Robertson & V. M. Fields, p. 235 (San Francisco: Pre-Columbian Art Research Institute).

Lounsbury, Floyd G. 1989, "A Palenque king and the planet Jupiter", in World Archaeoastronomy: Selected Papers from the 2nd Oxford International Conference on Archaeoastronomy held at Merida, Yucatan, Mexico, 13–17 January 1986, Ed. A. F. Aveni, p. 246 (Cambridge: Cambridge University Press).

Mathews, P. 1980, "Notes on the Dynastic Sequence of Bonampak, Part I", in Third Palenque Round Table, 1978, part II, Ed. M. G. Robertson, p. 60 (Austin: University of Texas Press).

Meeus, J. n.d., "Stations of Jupiter, Saturn, Venus, and Mars", ms in possession of the author.

Milbrath, S. 1989, "A Seasonal Calendar with Venus Periods in Borgia 29–46", in The Imagination of Matter: Religion and Ecology in Mesoamerican Traditions, edited by Davíd Carrasco, p. 103. BAR International Series 515 (Oxford: BAR).

Milbrath, S. 1995, "A New Interpretation of the Dresden Codex Venus Pages", in Cantos de Mesoamérica: Metologías científicas en la búsqueda del conocimiento prehispánico, Ed. D. Flores, p. 257 (Mexico City: Instituto de Astronomía, Facultad de Ciencias, University Nacional Autónoma de México).

Milbrath, S. 1999, Star Gods of the Maya: Astronomy in Art, Folklore, and Calendars (Austin: University of Texas Press) in press.

Miller, A. G. 1986, Maya Rulers of Time: A Study of Architectural Sculpture at Tikal, Guatemala (Philadelphia: The University Museum, University of Pennsylvania).

Nicholson, H. B. 1971, "Religion in Pre-Hispanic Central Mexico", in Handbook of Middle American Indians 10, Ed. R. Wauchope, Archaeology of Northern Mesoamerica, part 1, Ed. G. F. Ekholm & I. Bernal, p. 395 (Austin: University of Texas Press).

Proskouriakoff, T. 1993, Maya History, Ed. R. A. Joyce. (Austin: University of Texas Press).

Robertson, M. G. 1985a, The Sculpture of Palenque vol. II, The Early Buildings of the Palace (Princeton: Princeton University Press).

Robertson, M. G. 1985b, The Sculpture of Palenque vol. III, The Late Buildings of the Palace (Princeton: Princeton University Press).

Robertson, M. G. 1991, The Sculpture of Palenque vol. IV, The Cross Group, the North Group, the Olvidado, and Other Pieces. (Princeton: Princeton University Press).

Robicsek, F. 1978, The Smoking Gods: Tobacco in Maya Art, History and Religion (Norman: University of Oklahoma Press).

Robicsek, F. 1979, The Mythical Identity of God K. In Tercera Mesa Redonda de Palenque vol. IV, Ed. M. G. Robertson & D. C. Jeffers, p. 111 (Chiapas, Mexico: Pre-Columbian Art Research Center).

Robicsek, F. & D. M. Hales. 1981, The Maya Book of the Dead: The Ceramic Codex (Charlotte: University of North Carolina).

Schele, L. 1976, "Accession Iconography of Chan-Bahlum in the Group of the Cross at Palenque", in Proceedings of the Segunda Mesa Redonda of Palenque, Ed. by M. G. Robertson, The Art, Iconography & Dynastic History of Palenque, part III, p. 9 (Pebble Beach: Robert Louis Stevenson School).

Schele, L. 1992, Workbook for the XVI Maya Hieroglyphic Workshop at Texas (Austin: University of Texas, Department of Art History and the Institute of Latin American Studies).

Schele, L. & D. Freidel. 1990, A Forest of Kings: The Untold Story of the Ancient Maya (New York: William Morrow).

Schele, L. & N. Grube, 1994. The Proceedings of the Maya Hieroglyphic Workshop: Tlaloc-Venus Warfare, March 12–13, 1995 at the University of Texas, Ed. P. Wanyerka.

Schele, L. & J. H. Miller, 1983. The Mirror, the Rabbit, and the Bundle: "Accession" Expressions from the Classic Maya Inscriptions. Studies in Pre-Columbian Art and Archaeology, no. 25 (Washington, D.C.: Dumbarton Oaks).

Schele, L. and M. Miller, 1986. The Blood of Kings: Dynasty and Ritual in Maya Art (Fort Worth: Kimbell Art Museum).

Schellhas, P. 1904, Representation of Deities of the Maya Manuscripts. Peabody Museum of Archaeology and Ethnology Papers IV, p. 3

Schlak, A. 1989, "Jaguar and Serpent Foot: Iconography as Astronomy", in Word and Image in Maya Culture, Ed. W. F. Hanks & D. S. Rice, p. 260 (Salt Lake City: University of Utah Press).

Schlak, A. 1996, "Venus, Mercury, and the Sun: GI, GII, GIII of the Palenque Triad" RES: Anthropology and Aesthetics 29/30, p. 181
Seler, E. 1960–61, Gesammelte Abhandlungen zur Amerikanischen Sprach-und Altertumskunde, 5 vols. (Austria: Graz).
Seler, E. 1963, Comentarios al Códice Borgia, 2 vols and facsimile (México, D. F.: Fondo de Cultura Económica).
Stone, A. 1995, Images from the Maya Underworld: Naj Tunich and the tradition of Maya Cave Painting (Austin: University of Texas Press).
Stuart, D. 1978, "Some Thoughts on Certain Occurrences of the T565 Glyph Element at Palenque", In Tercera Mesa Redonda de Palenque, vol. IV, Ed. M. G. Robertson & D. C. Jeffers, p. 167 (San Francisco: Pre-Columbian Art Research Institute).
Stuart, D. 1984, "Royal auto-sacrifice among the Maya" RES: Anthropology and Aesthetics 7/8, p. 7
Stuart, D. 1987, "Ten Phonetic Syllables" Research Reports on Ancient Maya Writing, no. 14. (Washington, D.C.: Center for Maya Research).
Tate, C. E. 1992, Yaxchilán: The Design of a Maya Ceremonial City (Austin: University of Texas Press).
Taube, K. A. 1992, The Major Gods of Ancient Yucatan. Studies in Pre-Columbian Art and Archaeology, no. 32 (Washington, D.C.: Dumbarton Oaks).
Tedlock, D. 1985, Popol Vuh. Simon and Schuster, New York.
Tedlock, D. 1992, "Myth, Math, and the Problem of Correlation in Mayan Books", in the Sky and Mayan Literature, Ed. A. F. Aveni, p. 247 (Oxford: Oxford University Press).
Thompson, J. E. S. 1960, Maya Hieroglyphic Writing: An Introduction (Norman: University of Oklahoma Press).
Thompson, J. E. S. 1970, Maya History and Religion (Norman: University of Oklahoma Press).
Thompson, J. E. S. 1972, A Commentary on the Dresden Codex: A Maya Hieroglyphic Book (Philadelphia: American Philosophical Society).
Tozzer, A. M. 1941, Landa's Relación de las Cosas de Yucatan: A translation. Peabody Museum of Archaeology and Ethnology Papers XVIII.
Tozzer, A. M. 1957, Chichen Itza and its Cenote of Sacrifice. Memoirs of the Peabody Museum, Harvard University, vols. 11, 12.
Villacorta, J. A. & C. A. Villacorta. 1977, Codices Mayas. Second edition, Tipografía Nacional, Guatemala.

24

Analysis of the Nazca Spirals

Phyllis Burton Pitluga
Adler Planetarium
Chicago, IL

Building upon the information presented by this author at the previous Oxford Conference on Archaeoastronomy in Bulgaria (that Peruvian plant and animal figures and their related lines, triangles and quadrangles were associated with Milky Way dark spots as guided by Andean ethnoastronomy and represented two significant cultural times of year), this paper expands the survey to the spirals, truncated spirals and their associated lines and triangles. The analysis of their desert distribution, azimuths, sky correlations and ethnographic connections will be examined.

* * *

Introduction

Along the south coast of Peru there are gigantic ground drawings on a barren rocky desert. These drawings have been studied on and off for some seventy years to understand who put them there, how and why. Archaeologists have presented convincing evidence that these features were made by people living down in Ingenio Valley, just beside the great concentration of drawings. The dating of habitation is from about 400 B.C. The designs on the ceramics found both in the valley and with the desert drawings indicate that the construction extended to about 600 A.D. This time span covers the Late Paracas through Middle Nasca time period (Silverman & Browne 1991).

The markings adjacent to Ingenio Valley include twenty-eight huge plant and animal contour-line figures, hundreds of lines extending to the desert horizon and eight immense plazas (Pitluga 1993). In addition, there are four spirals with triangles along the valley edge, and three truncated spirals (flattened on one edge) a couple of kilometers in from the border. Previous investigators have not analyzed the spirals and truncated spirals as comparable duplicates (Hawkins 1969; Aveni 1990; Reiche 1993; Reinhard 1988) correlated with the polar regions of the sky.

Methodology

The selection criteria for sight lines were that each is a manmade line, beginning at a mound, extending in a straight unbroken line to the desert horizon and is physically connected with a spiral or a truncated spiral.

A Kern A-1 astronomical theodolite was used in determining the azimuths of the lines. Three sun sightings were taken and corrected utilizing standard references and calculations to know the orientation of the theodolite. For each line, there is a record sheet with the altitude of the horizon and three ground azimuths. The three measurements were averaged (most varied by +/- 3') and then were corrected with each of the three sun sightings to obtain an average azimuth. A computer search of the LodeStar Plus astronomical data base revealed which celestial objects third magnitude or brighter appeared above these lines during the date range of 400 B.C. to 600 A.D.

In the analysis, line azimuths and bright stars are considered aligned if within +/- 20 arc minutes in azimuth. The acceptable altitude range for a star to be considered at the horizon is based upon atmospheric refraction and an extinction coefficient of 0.25, +/- 20 arc minutes (Schaefer 1986).

Spiral and Truncated Spiral Data

Along the northern border of the desert there are four spirals at the wide ends of triangles. Three truncated spirals are located about two kilometers in from the border. All three have their flattened sides toward the north and their accompanying lines pointing southward. The spiral/triangle azimuths concentrate toward the north whereas the truncated-spiral lines concentrate toward the south, clearly in non-random patterns.

The azimuths of the four triangles with the spirals have a similar span on the horizon of 343° to 354°. A careful examination of the 100-meter spiral and triangle that runs perpendicular to the lines and figure of the spider (Figure 1) reveals a manner in which the spiral/triangles may have been developed. The spider is argued to have aligned with the setting stars of Orion with lines constructed at three different times (389 B.C., 129 B.C. and 201 A.D.) (Pitluga 1993). Because there is a smaller older spiral underneath the bigger spiral beyond the eastern side of the triangle that edge may provide the first alignment with the sky. The eastern edge of the triangle (azimuth 353°20', altitude 3°08') would have aligned with the setting of the star Alpha Ursae Majoris (magnitude +1.8) in approximately 100 A.D., at the same moment that the stars of Orion were setting beyond the spider (Figure 2).

By 200 A.D. precession had shifted the stars of Orion northward and Alpha Ursae Majoris westward. By about 450 A.D. the western border of the triangle (azimuth 349°15', altitude 3°08') would have aligned with Alpha Ursae Majoris and with the outer edge of the newer bigger spiral as the stars of Orion set (Figure 3). In a similar style, the other spiral/triangles could be related to the motions of the northern stars.

Looking at the truncated spirals, their associated lines aim southward. Truncated spiral #1 (Figure 4) has a 60 centimeter-wide line running to its west side that

Figure 1. Spiral and triangles.

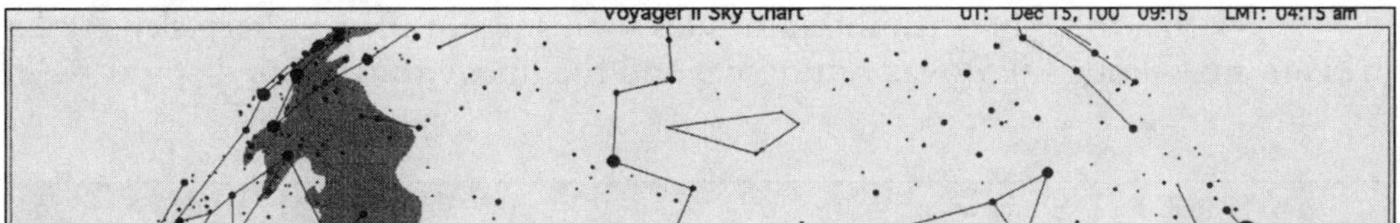

Figure 2. Sky showing Alpha Ursae Majoris setting near 'N" of "North" and Orion setting near "West."

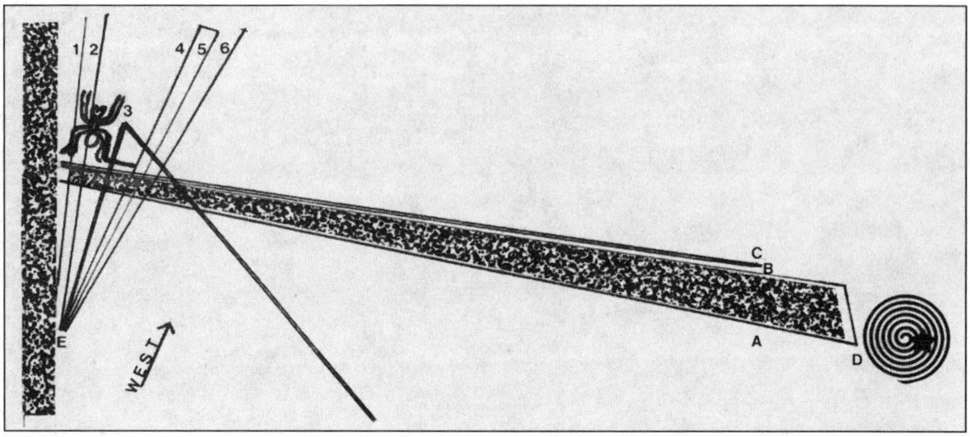

Figure 3. Diagram of spider, triangle, spiral relationship (Lines 1, 2, 4 and 5 associated with stars of Orion; A, B, C align with Alpha Ursae Majoris; D is the spiral with the small star indicating where the smaller spiral is located underneath the big spiral; E origin of spider sight lines.

extends to the horizon (azimuth 194°14, altitude 1°). Truncated spiral #2 (Figure 5) has two lines that extend to the horizon (azimuth 165°01', altitude 42.5'; and azimuth 170°27', altitude 37') Truncated spiral #3 (Figure 6) has one wide line cutting through it and extending to the horizon (azimuth 196°50', altitude 1°56').

Correlating these line azimuth/altitudes with stars of the southern sky, Alpha Eridani (magnitude +0.5) was setting beyond the line with truncated spiral #1 in

Figure 4. Truncated spiral #1 (line to west not in photo).

Figure 5. Truncated spiral #2 with lines.

399 B.C. (Figure 7). At truncated spiral #2, Alpha Eridani rose beyond line 165°01' at the same moment that Alpha Hydrae (magnitude +2.86) rose beyond line 170°27'

Figure 6. Truncated spiral #4 (car tracks mostly within line borders).

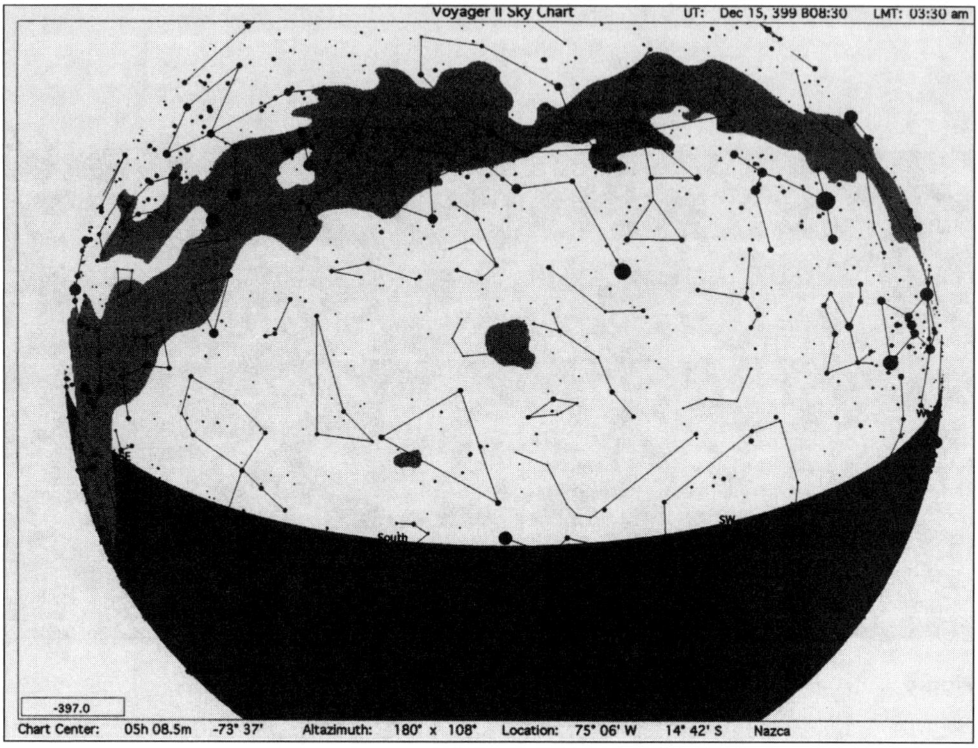

Figure 7. Alpha Eridani setting at the center horizon, as seen from truncated spirals #1 and #3.

in 389 B.C. (Figure 8). Alpha Eridani set beyond the truncated spiral #3 in line with its one line extending to the horizon in 130 B.C.

The pole star during this time span was the star Beta Hydrae (magnitude +2.8). The star Gamma Hydrae (magnitude 3.24) was just barely circumpolar, just grazing the horizon at lower culmination. The effect of refraction by the atmosphere on Gamma Hydrae would have been to lift this star by about half a degree as it approached lower culmination, giving the impression of a flattened or truncated circle as it moved around the pole star. At the latitude of this site, 14°42' South, the circumpolar region would extend down from the South Celestial Pole 14°42' on a flat airless terrain. Interestingly, from the beginnings of the lines, the angular diameter of each truncated spiral is, respectively, 11°42', 15°25', and 12°12'.

Spiral and Truncated Spiral Interpretation

The Earth is considered to be the spindle whorl of an immense spindle which the Kogi universal mother works incessantly to produce yarn for weaving. The cosmic spindle is right-spun or clockwise, which is the common spinning technique of most South American Indians in most circumstances (Urton 1981).

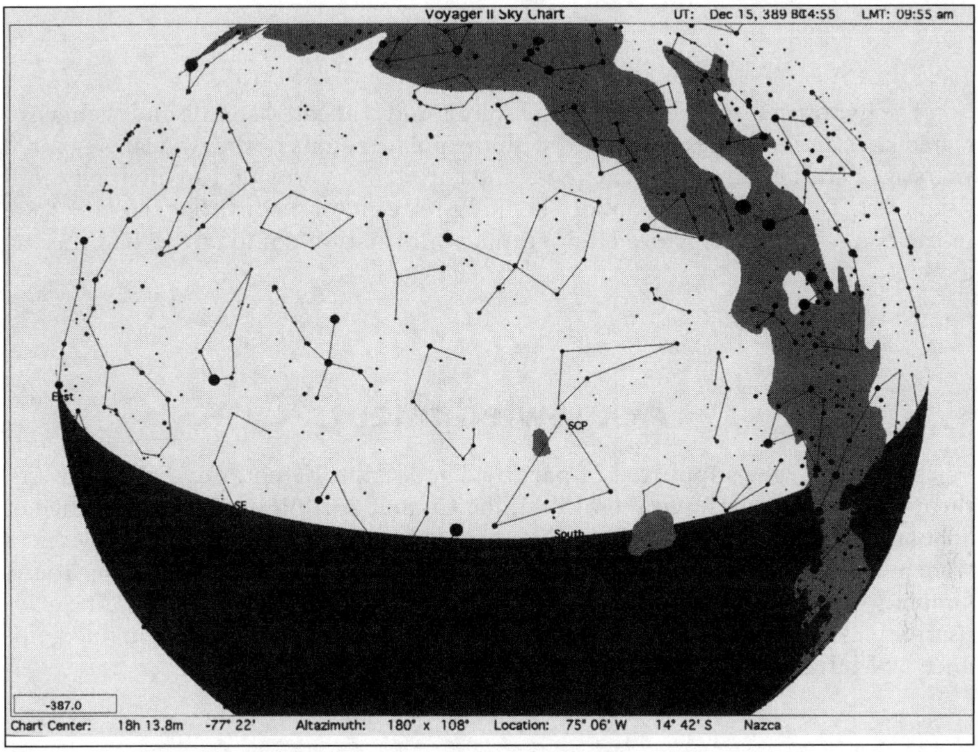

Figure 8. Alpha Eridani and Alpha Hydrae rising as seen from truncated spiral #2 (note Beta Hydrae near SCP—South Celestial Pole—and nearly circumpolar Gamma Hydrae to the upper right of "South."

To the present-day descendants of the indigenous Peruvians the spinning of yarn and the spindle whorl have directional significance. On the island of Taquile in Lake Titicaca, the people make a clear distinction of the direction of spinning because of the magical properties attributed to counterclockwise spun yarn. From Inca times, it was reported that thread was usually spun clockwise (right spun) except in manufacturing articles to be used in sorcery. Also, the religious lore employing left-spun yarn is rich, diverse and strong and present in offerings to Pachacamac (Mother Earth) (Prochaska 1988).

Relating this to the polar regions of the sky, stars to the south move clockwise (the direction of spun yarn for common items). Stars to the north move counterclockwise (the direction of spun yarn for ritual use). In the Nazca valley there is a spiral and triangle commonly called the needle and thread. On the hillside there are two figures etched into the hillside, a llama and a cotton flower gone-to-seed (with the cotton puffed out from the sepal). This may be iconographically relating the spirals to the theme of spinning of cotton and alpaca wool.

More ethnographic information is needed to support the hypothesis that the spirals and truncated spirals are related to the poles of the sky. The drawings on the ground represent both directions of motion. The big spiral is a double spiral so that one can move outward from the center in a clockwise and also in a counterclockwise direction. The truncated spirals sweep back and forth from their centers outward in alternating clockwise then counterclockwise directions.

Conclusion

The hypothesis tested was that the spirals and truncated spirals and their associated lines and triangles correlated with the polar regions of the Andean sky 2000 years ago. The measurements, analyses and correlations provided an internally consistent test. This same interpretation should be relevant for features like this elsewhere in the Nazca drainage. More ethnographic information could strengthen this argument.

Acknowledgments

This work was supported in part by The Wenner-Gren Foundation for Anthropological Research (Grant #4598), the Council for International Exchange of Scholars (Senior Fulbright Scholarship, serial grant), and Renate Reiche. Services were provided by EnturPeru, Aerocondor Airlines, Faucett Airlines, and Lima Tours. Graphics were supported by The Adler Planetarium & Astronomy Museum. The manuscript was improved by constructive reviewer's comments. The invaluable guidance by Maria Reiche is paramount.

References

Aveni, A. 1990, "An Assessment of Previous Studies of the Nazca Geoglyphs" in Lines of Nazca, Ed. A. Aveni. Philadelphia: American Philosophical Society.

Hawkins, G. 1969, Ancient Lines in the Peruvian Desert: Final Scientific Report for the National Geographic Society Expedition (Washington DC: National Geographic Society).

Pitluga, P. 1993, "New Methodology in the Astronomical Analysis of the Nazca Figures," Proceedings of the 4th International Oxford Conference on Archaeoastronomy, "submitted".

Prochaska, R. 1988, Taquile: Weavers of a Magic World (Lima: Arius).

Reiche, M. 1993, Contribuciones a la Geometria y Astronomia en el Antiguo Peru (Lima: Asociacion Maria Reiche Para Las Lineas de Nazca).

Reinhard, J. 1988, The Nazca Lines: a New Perspective on their Origin and Meaning (Lima: Editorial Los Pinos).

Schaefer, B. 1986, "Atmospheric Extinction Effects Archaeoastronomy" no. 10 (supplement to Journal for the History of Astronomy) 17, S32.

Silverman, H. & Browne, D. 1991, "New Evidence for the Date of the Nazca Lines" Antiquity 65, PP. 208–220.

Urton, G. 1981, "The Use of Native Cosmologies in Archaeoastronomical Studies: The View from South America", in Archaeoastronomy in the Americas, Ed. R. Williamson. College Park, MD: University of Maryland Center for Archaeoastronomy p. 292.

25

Inti, the Condor and the Underworld: The Archaeoastronomical Implications of the Newly Discovered Caves at Machu Picchu, Peru

James S. Westerman
Chicago, IL

Could archaeoastronomy help to explain the enigmatic archaeologic site of Machu Picchu? The recent discovery of subterranean caves in an eastern sector of the ruins coupled with the observation of an architecturally directed sunrise beam which only penetrates into the caves on historically important days in the Incan calendar, seems to answer this question with a qualified yes! In this paper I will address the importance of caves in Incan origin mythology, the presence of sun position measurement sites in Machu Picchu and specifically the Condor Group, the historical and theoretical significance of the times of the year that the restricted sunlight penetrates into the newly discovered caves, and finally the overall architecture of the site in and around the caves. Based on this analysis a case will be offered explaining how this particular site may be a key for explaining why the city of Machu Picchu was built at its spectacular location in the Andes mountains north-west of Cuzco, Peru.

* * *

Caves and the Incan Creation Myth

After the capture and subsequent execution of the Inca Atahualpa in 1532, the Spaniards began systematically gathering information about the culture of the conquered Incas. In general the gatherers of this information were called the chroniclers as they recorded their information in chronicles or books which were then sent to Spain for collection and analysis. Only the very early chroniclers such as Juan de Betanzos and Pedro Cieza de Leon actually *witnessed* Incan rituals untainted for the most part by the presence of the Spanish conquerors. "Cieza and Betanzos were the first to record continuous renderings of myths of creation and origin as told in Cuzco. In Andean eyes, the two themes — the creation of the world and the origin of the Incas, or the origin of human beings in a given place went hand in hand."[1] These two chroniclers both recorded that the Incas believed that they were created by the creator god Viracocha on the Island of the Sun in Lake Titicaca and were sent underground to travel to their homeland where they emerged from a cave at a place they described as Pacariqtambo, or the inn of the dawn. Sarmiento de Gamboa, the official historian of the Viceroy Francisco de Toledo who governed Peru from 1569 to 1581, claimed in his official history that there were three caves or windows located at the place of emergence and that one of the Inca brothers who emerged from the caves was later enticed to re-enter one of them where he was walled up and abandoned by his other siblings.[2]

There was no specific place mentioned to Betanzos or Cieza as to where the site of Pacariqtambo existed.[3] The history of how a specific site south of Cuzco became associated with this legend is well told in Gary Urton's *The History of a Myth*. I do not have sufficient space in this paper to elaborate all the nuances as to what was involved in concretizing the mythical origin point of emergence of the original Incas from the ground except to say that there were apparently some very self-serving motivations by a specific individual, Rodrigo Sutiq Callapina, to have this point be located in his homeland. Also, after having personally visited the sites presently identified with Pacariqtambo, in my opinion, the physical characteristics of the sites are not in congruence with the myth. In spite of this, because of the early mention of caves in both Betanzos' and Cieza's versions of the origin myth, there is probably some validity to the idea that the Incas truly did think they emerged from the ground at a place called Pacariqtambo. What is in question to me is where this place was located for the Incas. Because the Spanish had repeatedly shown the Incas that they were contemptuous of their belief system and their sacred places, there could have existed strong reasons for the Incas to try to keep their sacred origin place secret from the Spanish.

Astronomical Observation Points at Machu Picchu

There are many locations in the ruins of Machu Picchu where solar sightings seem to have been made by the Incas. In general sunrise solar observations were done by observing beams of sunlight cast through openings. Vertical solar measurements were done using gnomons like the Intihuatana stone. According to the chroniclers we can infer that the ancient Andean peoples had official observers whose duty it

1. MacCormack, page 100.
2. Urton 1990, page 27.
3. Ibid., page 29.

was to position themselves in a particular place and wait for the sun to rise or set.[4] When the sun appeared through a strategically placed window or rose or set over a specified point on the horizon as viewed from the observation point, the official observer would know that a specific calendar date had arrived.

It appears that the *Torreón* acts as a sunrise observation point for the June and December solar sunrise measurements.[5] The *Intimachay* cave seems to have been used for December solstice sunrise sightings.[6] The *Intihuatana* seems to have been used for solar zenith measurements on October 30th and February 13th (Figure 1). Other sites in Machu Picchu which may also have been observatories are a large terrace rock near the gatekeepers hut on the terrace overlooking the ruins. This worked rock has on its sides man made holes which possibly held some kind of device which could have been used for solar or stellar sightings.[7] There is a point on the western side near the quarry area of Machu Picchu where a semi-circular basin has been carved in a stone and which may have been used for sunset observations.[8] In the sacred rocks area in the northeastern area of the ruins there is a etched stone with a compass rose design which may have been some sort of sundial.[9] (Figure 2)

The Condor Group

In June, 1995 James Westerman and Alfredo Valencia Zegarra discovered two new caves served by a single stairway in the southeastern urban sector of Machu Picchu. These caves were found beneath a huge boulder which is suspended above the ground in an area of Machu Picchu commonly called the Condor Group. The architecture of the buildings to the east of the caves is such that it restricts the sun's shining into the new caves to a few specific days each year. In Machu Picchu the range of sun horizon traverse is from approximately 64 degrees on June 20th to 113 degrees on December 20th, a range of 49 degrees. The sun does not appear to move across the sky for approximately eight days at each of the solstice points. The true azimuth from the foot of the stairs in the cave to the gap between the buildings through which the dawn sun beams into the Condor Group is approximately 74 degrees. The distance from the stairs to the gap is 14.5 meters. Allowing for the thickness of the thatched roofs which would have existed in antiquity on the two buildings (Figure 3), the effective width as viewed from the cave for the window opening at the level of the eastern horizon would have been approximately .5 meters. Using the equation circumference = pi × diameter where diameter = 29 meters gives a circumference of 91.11 meters. An arc of 0.5 meters for a circle of 91.11 meters in circumference calculates to approximately 2.0 degrees [(.5/91.11) × 360°]. Assuming the sun has a diameter on the horizon of .5 degrees, according to astronomical calculations the entire disk of

4. Bauer & Dearborn, page 55.
5. Dearborn & White, 1983.
6. Dearborn & White & Schreiber, 1987.
7. Comments by Zuidema, R. T., 1996, on "The Mystery of Machu Picchu" television program, The New Explorers, Kurtis Productions, Chicago, IL.
8. Alfredo Valencia and Fernando Astete, 1995, "private communication."
9. Ibid.

Figure 1. Intihuatana stone at noon October 30, 1986 showing it casts no shadow at the Zenith. All photographs by author except as noted.

Figure 2. Compass and etched rock in sacred rock area of Machu Picchu.

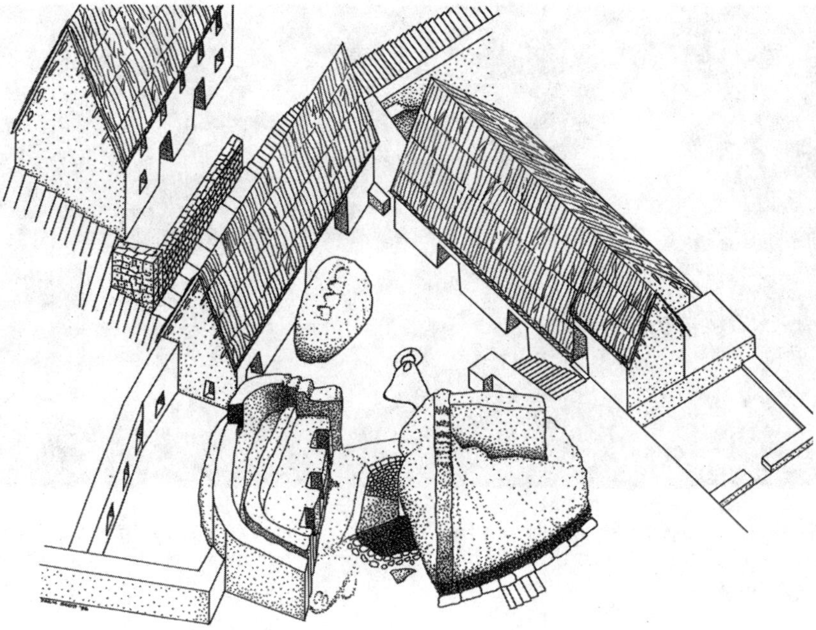

Figure 3. Artist's reconstruction of the Condor Group. (Drawing commisioned by author from Jack Scott).

the sun would be visible at the foot of the stairs in the cave for five days twice a year.[10] Due to the overhang caused by the suspended boulder over the cave the sun is only visible in the cave for the first few minutes after it clears the eastern horizon (Figure 4). The days of the year when this occurs are April 20 to 24 (Figure 5) and August 16 to 20 (Figure 6). Due to the roof lines which existed during Inca times on the two buildings which form the gap, and the overhang of the suspended boulder above the cave, these would have been the only times of the year when the sun would have been visible in the caves.

Historical and Theoretical Significance

We know from eyewitness accounts recorded in the chronicles by Betanzos, Cieza, and the unknown chronicler that the Incas held religious ceremonies at sunrise in Cuzco during these times of the year.[11] Also according to the aforementioned chroniclers, mummies of the deceased Incas were present and participated in these ceremonies. During the August period planting rituals took place and during April harvest ceremonies were observed.

---

10. Phyllis Pitluga of the Adler Planetarium, Chicago, IL, using Voyager II Version 2 computer program.
11. MacCormack, pages 74–5, 103.

Figure 4. View of eastern horizon from foot of stairs in the center cave. In antiquity only the section of the horizon between the two buildings would have been visible due to the roofs of the two buildings.

Tom Zuidema has theorized that the Incas took measurements when the sun was directly overhead at noon in Cuzco (Lat.13°30'S). a position he calls the zenith.[12] The shaped Intihuatana stone (Lat.13°9'S) on the hill on the western side of Machu Picchu, which effectively is situated at almost the same latitude as Cuzco, casts no shadow at noon on October 30 and February 13, the dates of the zenith. Dr. Zuidema has also theorized that the Incas were aware when the sun was at a position of 13°30' north latitude in Cuzco a position he calls the anti-zenith. The times of the year when this occurs in Cuzco and also Machu Picchu are April 23 and August 18. These are the same days when the sunrise rays shine through the gap in the Condor Group into the newly discovered caves. If the Incas did design the architecture of the Condor Group to effect the sun only entering into the caves on these dates and used this area as a solar observatory, then this is the first known confirmation of Zuidema's theory about the anti-zenith measurement.

The Architecture of the Condor Group

One of the severe misconceptions about this area of Machu Picchu is how it ever received the name "The Prison Group". In various books and by many authors this designation is attributed to the modern discoverer of Machu Picchu, namely Hiram Bingham. I have researched all of Bingham's published works and have had personal discussions with Dr. Richard Burger, the Director of the Peabody Museum of Natural History at Yale University and keeper of Bingham's personal

12. Zuidema 1981.

Figure 5. Dawn sun beam April 21, 1994.

notes. Dr. Burger claims he has never seen anything in writing by Bingham designating this area as a prison. Bingham in his works called this area the "Unusual Niches Group."[13] He did not know that the three caves existed in the area, but he was impressed by all the assorted niches that surround the area which we now know contains the caves. He speculated that these niches could have been used to hold the mummies of the deceased Incas. The architecture of this grouping of buildings seems to have been for ceremonial activities probably connected to the dates the sun shines into the caves (Figure 7). The building that forms the south or right side of the sun gap as viewed from the caves was a two story building with a kitchen and area for the keeping of guinea pigs on its lower level. Guinea pigs were used in antiquity, and are still used by the Andean people to this day, for ceremonial banquets. According to Bingham's floor plan, the entry into the Condor Group was restricted in antiquity to the doorway leading off the long stair-

13. Bingham, page 84.

Figure 6. Dawn sun beam August 19, 1995.

way which runs along the east side of the group. The wall of the building that served as a vestibule to this entrance comprises one side of the sun gap. On the floor of the area there is a large stone with a carved head of a raptorial bird which looks somewhat like the head and neck ruff of a male condor. On the days when the sun shines directly into the cave, the beam cast by the opening is centered on the carved bird's head. We excavated around this rock, but aside from the carved area around the eastern corner it appears to be completely natural. There was no subterranean drainage in the area of the semi-circular carving around the beak which many tourist guides claim was a basin with a drainage hole to collect sacrificial liquids. There is a large boulder suspended by three points over the entrance to the center cave. When seen in profile the eastern most face of this boulder gives somewhat the appearance of a human lower jaw and head struggling to raise upward (Figure 8). To the north of the carved bird's head there is a large rock with carved indentations in it which could have been used as a seat (Figure 9). It is ideally situated to view the sunbeam passing into the cave. Beneath the suspended boulder there are two caves (Figure 10). The cave to the south or left

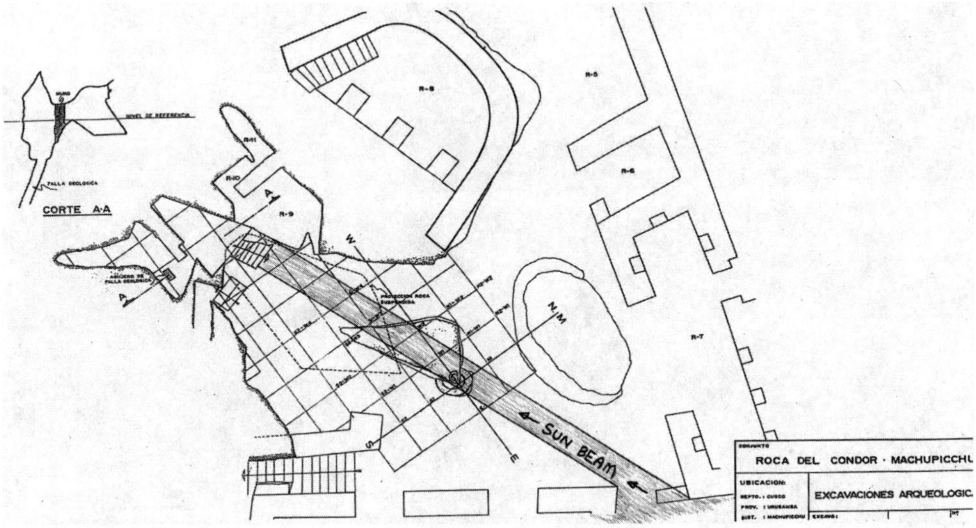

Figure 7. Floor plan of the Condor Group. (From Valencia Z., A. Informe Excavaciones Arqueologicas).

was walled off in antiquity and has been sealed off again since our discovery of it. Inside this left cave there was discovered a deep fissure going down approximately four meters into the ground. On top of the fissure the Incas had placed rich black earth and on top of this there were stacked rocks up to the top of the cave. The middle cave has stairs and Incan walls in it. It ends at a fissure hole which serves as a drainage point for the rainwater which enters the cave and keeps the water from entering into the left cave. In this cave there was discovered a crystal as well as Incan pottery fragments and stone hand tools. The upper or third cave has three small rooms and was discovered by Alfredo Valencia in 1968. There is an open pit in the area of this upper cave and the three niches on the curved walled rock to the north face into this pit (Figure 11). From these niches a section of the beam passing into the lower center cave can be observed. The other large niches to the rear of the suspended boulder are situated so that only four of the nine can see into the pit.

Analysis of the Findings

It appears that the suspended rock and the two lower caves are natural. The structure of the third upper cave looks like it was roofed over by the Incas. Traces of plaster were also found inside the upper cave. The stairs, walls and black dirt and other stone work in the two lower caves were clearly placed there by the Incas. When you combine all the elements discussed you have a natural cave setting which was modified by the Incas to allow the sun to only beam into the center cave entrance at certain specific times of the year, times which were known to be ritually significant, and which also are coincident with what Tom Zuidema calls the anti-zenith. From the point of view of the Andean cosmovision, the times of the year that the sun

Figure 8. Side view of eastern side of the suspended rock.

Figure 9. Dawn light into Condor Group showing rock with "seats" carved into its top.

was directly in a vertical position, both from above (the zenith) and from below (the anti-zentih) in relation to their conception of the three parallel planes of the

Figure 10. Entrance to the caves. The lower opening is to two caves. The upper opening is to a cave discovered by Alfredo Valencia in 1968.

Figure 11. View of pit on north side of the suspended rock.

Figure 12. Plan of Machu Picchu. (From: Hemming J. & Ranney, E. Monuments of the Incas).

universe (*Alaj Pacha*—the world above, *Aka Pacha*—the world we live in, *Manka Pacha*—the world below) probably was very significant. Symbolically, at the dawn of the anti-zenith, the beam of light from the sun, or god Inti as the Incas thought of it, casts its light over the image of a condor's head, the sacred bird of the Andean people, and then goes into the caves or entrance to the underworld. The enormous boulder was raised up apparently by natural forces and stairs were provided for people to emerge out of the caves. One cave was walled off. To me all these elements point to a reenactment of the emergence of the Incas out of the ground as portrayed in their creation myth. Due to the overhang of the suspended rock over the caves the sun can only be seen in the caves at the first few minutes of the sunrise. This would truly make this an "inn of the dawn" or Pacariqtambo. It just may be that all of Machu Picchu was constructed based on the existence of this natural area. If you study the floor plan of the ruins you will note that most of the buildings are laid out in straight lines except for this area (Figure 12). If this area was the Incas sacred point of origin then it is conceivable they would have come back to their "homeland" on ceremonial occasions. There are eight roads which lead into Machu Picchu from all corners of the Incan empire[14] indicating that it was a focal point for some reason. The reason the Spaniards were never told about the place could have been because the Incas were determined to maintain a code of silence about this their sacred homeland.

Of course any idea about the reality of Machu Picchu to the Incas is only based on conjecture and studied hypothesis. We can never know the truth with certainty. But nevertheless, the combination of archaeoastronomy and architecture in the Condor Area presents one of the strongest and most impressive cases that Machu Picchu may have been the true Pacariqtambo of the Incas.

14. Frost and Bartle, page 11.

References

Aveni, A. F. 1981, "Horizon Astronomy in Incaic Cuzco", in Archaeoastronomy in the Americas Ed. R. A. Williamson, Los Altos, CA: Ballena Press, p. 305

Bauer, B. 1991, "Pacariqtambo and the Mythical Origins of the Inca" Latin American Antiquity, 2(1), p. 7.

Bauer, B. & Dearborn, D. S. P. 1995, Astronomy and Empire in the Ancient Andes (Austin: University of Texas Press).

Betanzos, J. translated and edited by Hamilton R. 1996, Narrative of the Incas (Austin: University of Texas Press).

Bingham, H. 1930, Machu Picchu A Citadel of the Incas (New Haven, CT: Yale University Press).

Bingham, H. 1986, Lost City of the Incas (New York: Atheneum).

Dearborn, D. S. P. & White, R.E. 1982 "Archaeoastronomy at Machu Picchu" Annals of the New York Academy of Sciences Vol. 385, p. 249.

Dearborn, D. S. P. & White, R. 1983, "The 'Torreon' of Machu Picchu as an Observatory", Archaeoastronomy No. 5 (JHA, xiv), p. S37.

Dearborn, D. S. P. & White, R. 1989, "Inca Observatories: Their Relation to the Calendar and Ritual", in Proceedings of the Second International Oxford Conference Merida, Yucatan, Mexico, Ed. A.F. Aveni, Cambridge, UK: Cambridge University Press, p. 462.

Dearborn, D. S. P. & White, R. & Schreiber, K. J. 1987, "Intimachay: A December Solstice Observatory at Machu Picchu, Peru", American Antiquity, 52(2), p. 346.

Frost, P. & Bartle, J. 1995, Machu Picchu Historical Sanctuary (Lima, Peru: Nuevas Imagenes S.A.).

Hemming, J. & Ranney, E. 1982, Monuments of the Incas (Boston: Little Brown and Company).

MacCormack, S. 1991, Religion in the Andes (Princeton, NJ: Princeton Univ. Press).

Urton, G. 1981, "The Use of Native Cosmologies in Archaeoastronomical Studies: The View From South America", in Archaeoastronomy in the Americas, Ed. R. A. Williamson, Los Altos, CA: Ballena Press, p. 285.

Urton, G. 1990, The History of a Myth (Austin: University of Texas Press)

Valencia Z., A. 1995, "Informe Excavaciones Arqueologicas, "private communication."

Valencia Z., A. and Gibaja O., A. 1992, Machu Picchu, La Investigación y Conservación del Monumento Arqueológico despues de Hiram Bingham (Cuzco, Peru: Municipalidad del Cuzco).

Zuidema, R. T. 1981, "Inca Observations of the Solar and Lunar Passages Through Zenith and Anti-Zenith at Cuzco", in Archaeoastronomy in the Americas, Ed. R. A. Williamson, Los Altos, CA: Ballena Press.

26

The Astronomical Significance of a Procession, a Pilgrimage and a Race in the Calendar of Cuzco

R. Tom Zuidema
Department of Anthropology
University of Illinois
Urbana, IL

Four landscape features, defined as *sucanca*, played a central role in the system of astronomical observation of Cuzco, the Inca capital. One indicated the December solstice sunrise and another June solstice sunset, but for the remaining two less information is available. In addition, all four were used in a highly integrated system of ritual visits on dates different from their astronomical uses. On the basis of systematic contrasts and relationships between the two systems, it is argued that of the remaining sucancas, one indicated sunrise for the beginning of the wet season, some two months before the December solstice, and the other sunset for the beginning of the dry season, half a year apart. These dates would have coincided, or have been close to, the respective dates of the sun's first passage through zenith and its first antizenith set.

* * *

Introduction

In 1979, Aveni and I gave two papers on horizon astronomy in Cuzco, capital of the Inca empire until 1534. We based our research on some Spanish reports of Inca astronomy as these began to appear after 1550.

A major part of Aveni's (1981) analysis was dedicated to a set of four pillars on the western horizon, 2 km from the plaza in the city. From the data of a well-informed chronicler (Anonymous 1908) on the location of the pillars and their ob-

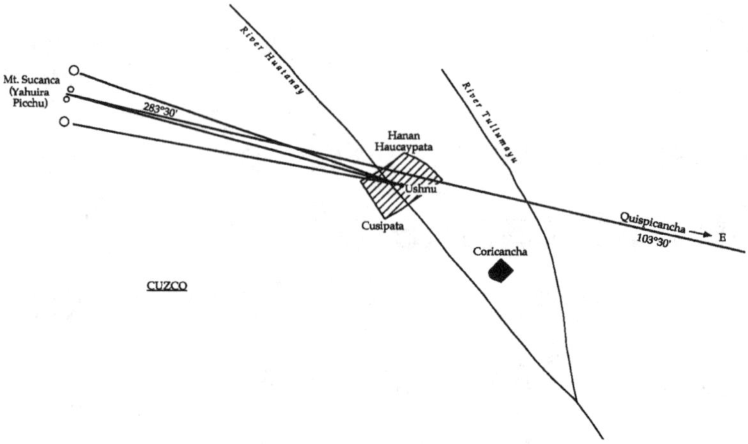

Figure 1. The two uses of Mt. Sucanca for observing sunset.

servation from the plaza, Aveni suggested a date for sunset between the inner pillars around August 18.[1] The chronicler also specified the distance between the outer pillars and thus I estimated that the sun would have passed all pillars in about a month (Figure 1. After Zuidema 1981a).[2]

My paper was concerned with a related problem. I wondered about the significance of a landscape feature said to be used for astronomical observation at the far eastern end of the Cuzco valley (Zuidema 1981a). Observation from the pillar mountain (103½° azimuth) would have coincided with, or been close to, a sunrise on the days when, at the latitude of Cuzco (-13½°), the sun goes through zenith (Oct.13 and Feb. 13). A sunset in the reverse direction (283½°), toward the pillar mountain, would indicate dates close to half a year apart from the first ones (Apr. 26 and Aug. 18). The second of these dates would have been appropriate for the August observation from the nearby plaza of sunset between the inner pillars. This reverse direction would have referred to an "antizenith" sunset, being immaterial whether the Incas knew that then the sun went through nadir.

In a series of later articles I argued—based on information about a procession at the time of harvest in April—that, in fact, an observation of sunset was made in the antizenith direction focused on the pillar mountain (Zuidema 1993, 1996, 1997). But now the observation was made from a third observation point just outside Cuzco. Thus, we remain in doubt about the astronomical value of the distant feature. In this paper, I will not return to the problem as initially proposed by Aveni and me in 1981 but accept as essentially correct the astronomical conclusions then drawn. Instead, I will concentrate on the calendrical use of ritual visits to four landscape features, including the two mentioned, said to be used for astronomical ob-

1. The anonymous chronicler mentions only one date for the astronomical use of the pillars. I will keep to this date, contrary to the suggestion of Bauer & Dearborn (1995 pp. 73–4) that the pillars might have been used for various dates as observed from different locations.

2. I correct a wrong calculation of Bauer & Dearborn here. They came to a period of 35 days for the sun's passage through the outer pillars adding this to the time of passage through the inner pillars. They do not make this error in the third diagram of their figure 4 (p. 71). Here they accept the distances as proposed by me.

servation. Remarkably enough, most visits were not carried out on days as reconstructed for the astronomical use of those locations but on quite different ones. The contrast will help me to arrive at more definite answers to questions about the Incaic dates of observation.

The Sources of Information

The four places were all described by Molina [1574] for their use in a system of visits. Although other landscape features were also used for observing a sunrise or sunset, they were not included in the same system and I do not need to discuss them here. Molina also knew about the system of *ceques*—41 directions organizing the distribution of *huacas*, sacred places, in or close to the Cuzco valley—which included the four places as *huacas*. The system had been discovered previously by Polo de Ondegardo [around 1559] (1981, 1990) (included in Cobo [1653] pp. 169–186) and thus I consider also a more general observation made by him on astronomical markers. Cobo elaborated on this observation later, probably with access to more extensive information of Polo from 1559 and now lost. Their observations were not made, however, in the context of the description of the ceques but elsewhere.

Polo (1981 pp. 463–4) and Cobo (1956, pp. 141–4) refer in their general descriptions of Inca astronomy also to a calendrical use of the *ceque* system. Sacrifices were brought to *huacas*, following the *ceques* from the inside out. The remains of those sacrifices were then brought to the places of astronomical use; Polo and Cobo specify two places by name. The first, *Pucuy Sucanca*, was the *sucanca* indicating a sunrise or set for the beginning of the wet season (*pucuy*). The second, *Chirao Sucanca,* indicated a sunrise or—set for the beginning of the dry season (*chirao*). The *sucancas* were not, however, as Polo's text might lead us to think, outside the ceque system but included into it as *huacas*. In other words, a *sucanca* could be located as *huaca* on a ceque because of its astronomical use while the ceque itself represented a quite different calendrical value. Polo and Cobo's descriptions lead to the conclusion that the *ceque* system, next to its role in organizing space, also regulated astronomical and calendrical time. They evoke an integration of space and time only vaguely being aware how that system worked. Nevertheless, their idea is fundamental for reconstructing the Inca calendar.[3]

The Four *Sucancas*

Let me discuss first the four places, three identified as *sucanca* by either Molina or the ceque system. The fourth was not labeled *sucanca*, but as its ritual and as-

3. From the information of Polo and Cobo it is clear that the Incas dealt with a system of fixed periods in the tropical year that these authors recognized as "months." Another issue is, how in this context the Incas made ritual use of certain synodic lunar periods. While I do not need to deal with the whole problem here, the argument of this article will lead to the identification of two lunar periods in the context of the solar calendar.

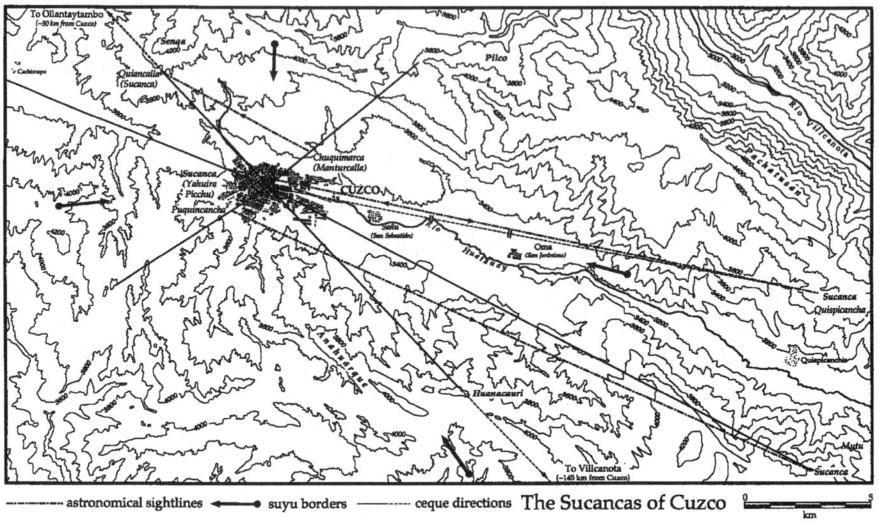

Figure 2. The four *sucancas* and their suggested use for observation.

tronomical uses were integrated with those of the other three, I will deal with it as such (Figure 2).[4] After a review of the *sucancas* and the documented astronomical use of two, I turn to the visits.

Two *sucancas* were located on the far eastern side of the valley (Molina pp. 68–9). The first was near another *huaca*, the mountain Mutu. Modern maps identify various places as Mutu in an area at a distance of + 25 km from Cuzco. The *sucanca* should have been located in or near this area and possibly closer to Cuzco but Molina's information does not allow a more precise location. We know, however, the approximate direction from Cuzco.

The second *sucanca* was near the mountain Quispicancha, *huaca* of an Incaic town of the same name. Together they still represent an impressive archaeological site. The modern village now called Quispicanchis is a colonial resettlement further down at the foot of the mountain. While we can locate Mt. Quispicancha precisely, the *sucanca* might have been located closer to Cuzco.

The other two *sucancas* were much closer to Cuzco at the western end of the valley. With respect to the first, Molina (1989, pp. 106–7) knows the mountain of the pillars as Yahuira—one of Cuzco's six sacred mountains (Sarmiento [1572] ch. 31)—and mentions it for a visit just before the December solstice. He does not refer to the August observation of the four pillars, although he (p. 74) records the importance of the structure in the plaza, the *Ushnu*, from where they were observed.

4. All four places can be located well, irrespective from the inclusion of three as huacas in the Ceque system. Molina (pp. 126–7) refers to the use of the ceques as straight directions. In the map, I have assumed that the sucanca locations also defined their ceque directions. There is no need to enter into the important question how closely other huacas were located to the direction of their common ceque (Bauer 1998). For two reasons, the sucanca near Mt. Quispicancha presents a special problem. Its name, as given by Molina, does not occur in the *ceque* system. Probably, the *sucanca* can be identified with the last *huaca* of a group of three *ceques* that I have indicated by a dotted line as parts of one general direction. Only in this case, the second *ceque* began beyond the space represented by the first and the third beyond that of the second.

Near Yahuira was a place called Picchu which name today has replaced the old one of the mountain. The Anonymous chronicler (pp. 151, 158) does not mention the mountain. But as he specifies the distance of the *Ushnu* to the pillars and the date of their observation, he enables us to identify Yahuira as their location. However, the ceque system calls the mountain, not the pillars, Mt. Sucanca.

The last place was called Quiancalla and formed part of another sacred mountain (of the six mentioned by Sarmiento), Senqa, northwest of Cuzco.[5] While it was not called *sucanca*, it had two "markers (Spanish *mojones*) or pillars" probably similar to the ones on Mt. Sucanca. Irrespective of its use for astronomical observation (at a distance of seven km as the crow flies), the pillars would have identified the place for use in ritual visits.

Comparing the four *sucancas* to each other, the distant ones had no pillars; they were both called *sucanca* by Molina. The nearby ones had pillars and the Anonymous author comments that the group of four pillars of Mt. Sucanca was used for viewing in August at a distance of around 2 km. But in addition he makes the intriguing statement that they also could be seen from a distance of about 12 to 18 km. No pillars can be distinguished from that far and we may wonder if the author really meant what is claimed. I suggest that, because of the distant view, only the mountain itself was called Sucanca. It served a double purpose of observation: as a *sucanca* from a distance and for its pillars from nearby.

Coming now to the observations, we have documentation for three *sucancas*. Easiest to solve is the problem of Quiancalla mentioned for sunset during the June solstice (*ceque* system in Cobo p. 172). Around this time of the year the king went to a temple of the sun north of Cuzco, Chuquimarca, built on the hill of Manturcalla (Molina pp. 69–70; Cobo pp. 176, 215), which places can be well located. Chuquimarca is the only possible candidate for making the June solstice sunset observation behind Quiancalla.

The Incas applied an identical method observing the December solstice sunrise at the end of the month of *Capac raymi*, the "royal feast" that included the initiation of noble boys. People brought a statue of the Sun to the temple of Puquincancha south of Cuzco, together with solar paraphernalia as brought to Chuquimarca. Only the *sucanca* near Mutu could have served then an observation from the temple. It was to be one of sunrise.[6]

While we have reasonably good empirical evidence on the astronomical use of Quiancalla as a *sucanca* and of Mutu sucanca, such kind of information seems to be missing for the other two *sucancas*.[7] Our only documented evidence on Mt. Sucanca is for the observation of its pillars. We also have no direct information on the

5. We can locate Quiancalla more precisely than Bauer & Dearborn (pp. 80–9) realize (Zuidema 1981b). Nearby were said to be a spring of water of the still existing village of Huarhuaylla, a road still in use entering the valley from the northwest, and a structure, identified as a "fortress," later replaced by a Spanish aqueduct but still called "fortress" by local people.

6. Aveni and I identified Puquincancha in 1979 and Bauer & Dearborn (1995, pp. 90–1) confirm the location. They do not mention the *sucanca* and do not put on their map the nearby area of Mt. Mutu. Contrary to their judgment, the observation of the December solstice sunrise from Puquincancha to the *sucanca* near Mutu is well supported. I should remark one significant difference between the two solstice observations. The June one was made in the middle of its solar month but the December solstice at its end.

7. I will call the eastern *sucancas* after their nearby mountains, although these did not necessarily define the exact directions from Cuzco of the *sucancas*.

astronomical use of Quispicancha sucanca. Especially here, the ritual data will help to reconstruct that use.

Before turning now to the visits honoring the four *sucancas*, let me make a methodological remark concerning the observation of the last two. As viewed from opposite sides of the valley, Quispicancha sucanca and Mt. Sucanca are below their more distant horizons although they surely could have helped to identify features of interest there. Thus, Bauer and Dearborn (p. 92–8) declared them of no interest studying Incaic techniques of astronomical observation. Such a judgment is, of course, putting the cart before the horse. The Incas called these places *sucanca*; it is to us to find out the reasons why. The former splendor of Mt. Sucanca (Yahuira) as sacred mountain within the limits of Cuzco, and the similar significance of Mt. Quispicancha to the town of Quispicancha, demonstrate that both mountains were of paramount political importance in their common valley. A sightline between them and extended in both directions to the distant horizons could well have served legitimate purposes including astronomy.[8]

I will support the claim that one of the four *sucancas* was used as a *sucanca* for the beginning of the wet season (*Pucuy*) and another of the dry season (*Chirao*). Polo does not suggest any specific dates, and Molina and the *ceque* system fail to mention the terms *Pucuy* and *Chirao*. First I thought that the two seasons could have started, respectively, in February, when two Inca months occurred with the name *Pucuy*, and in August. In a recent article (Zuidema 1992), I studied extensive documentation from the XVIIth century in Central Peru indicating that the *Pucuy* season started at the beginning of November with the coming of rains, and the one of *Carhua* (a synonym for *Chirao*) near the end of April, after harvest (Figure 3). Modern ethnographic information supports similar dates all over southern Peru. In line with these dates, Betanzos [1551] (p. 71), our first chronicler to report on the Incaic solar months, adds *Pucuy* to the name of the month preceding the December solstice. Best candidates for indicating the dates in November and April will be Quispicancha sucanca and Mt. Sucanca as viewed from each other.

The Ritual Visits

The *sucancas* were visited in three periods of the Incaic calendar:

- A procession and a pilgrimage by priests of the Sun in the month of the June solstice;
- a commemoration of those two visits in the second month before the December solstice;
- a race by young men ending in the second month after the December solstice.

Both June rituals, one marking space within the Cuzco valley as delimited by the horizon and the other outside the valley, occurred on the same days when the king

8. The early Spaniards admired Mt. Yahuira because of its beautiful terraces but dismantled these after the earthquake of 1650, using the stones for the reconstruction of the city. Mt. Quispicancha is surrounded by a huge pre-Inca wall. On top of it are petroglyphs and an old police report mentions graves containing golden objects. With no other huaca in the valley was such care expended on its ceremonial importance.

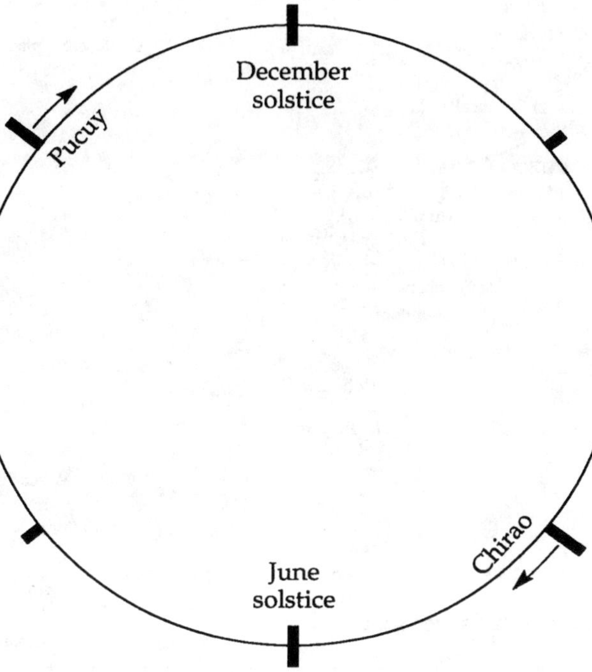

Figure 3. The six suggested dates of observation.

stayed in Chuquimarca. His activities included the (probably daily repeated) observation of sunset determining the exact day of the June solstice. In the procession, priests went, before dawn and on each day of the month, to mountain Huanacauri, SE of Cuzco and foremost among its six sacred mountains (Figure 2). Here they sacrificed a llama at sunrise. Back in Cuzco at noon, they sacrificed a llama in the central temple of the Sun. A third llama was sacrificed at sunset on mountain Alpitán because "the sun sets behind it" (Molina pp. 67–8). This name does not allow any interpretation or identification. But as the mountain was visited at the same time when Quiancalla served a sunset observation from Chuquimarca, I suggest that Alpitán was another name for Quiancalla. Its direction from Cuzco is exactly opposite that of Huanacauri and at comparable distance. On the same days when the king saw sunset approach, reach, and leave Quiancalla, the priests arrived there.

During the time that the king was in Chuquimarca and the procession was repeated, other priests went to a distant temple of the Sun in the same direction as Mt. Huanacauri from the temple in town (Molina pp. 68–9). Their goal was the temple of Villcanota, said to be dedicated to the December solstice (Figure 4).[9] The

9. This conclusion can be drawn combining four definitions given in the Aymara dictionary of Ludovico Bertonio [1612] using the word *villca*.

Villca: The sun as they said in ancient times, and now they say *inti*.

Villca cuti: The solstice when (the sun) begins to return from the tropic of Capricorn to (that of) Cancer.

Villca: Place of worship ("adoratorio") dedicated to the sun or other idols.

Villcanuta: Very famous place of worship between Sicuana and Chungara. It means house of the sun, according to the barbaric indians.

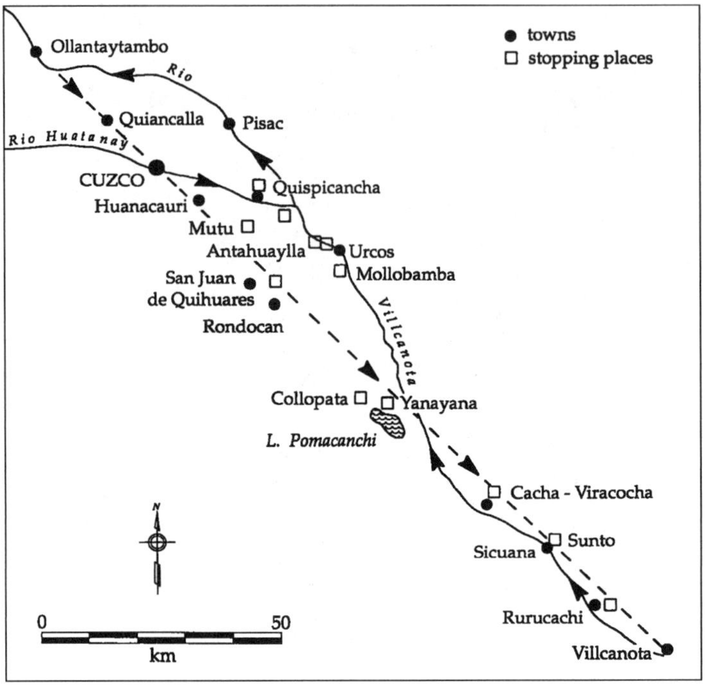

Figure 4. The roads of the procession, the pilgrimage and the race.

priests left Mutu sucanca some ten days before the day of the June solstice, and returned to Quispicancha sucanca some ten days after the event.[10] Villcanota was clearly visited for a purpose in conjunction with the processions from Huanacauri to Quiancalla. But why did the priests include the eastern sucancas in the pilgrimage? The visits in the other two periods of the year will contribute to an answer. Some observations can be made now.

The comparison to another procession—not considered here (Zuidema 1993, 1996, 1997)—shows that in Inca opinion the procession from Huanacauri to Quiancalla reflected the path of the Sun through the sky from dawn to dusk. Moreover, it probably was meant to replicate the sun's path from the December solstice sunrise to the June solstice sunset. Both intentions were, however, primarily symbolic as the direction from Huanacauri to Quiancalla, to be called the "solstitial axis," was turned some 20° clockwise from the axis that could have made it useful for practical purposes of Inca astronomy. By starting the pilgrimage from Mutu sucanca, then turning to the direction from Cuzco to Villcanota, the Incas tied a place of astronomical observation to a ritual (the pilgrimage) of calendrical intent. The pilgrimage to Villcanota and back projected the royal observation and the priestly procession to Quiancalla onto a wider landscape and cosmological scale.

The procession and pilgrimage are referred to again, in the second month before the December solstice, for a daily succession of sacrifices to all the same *hua-*

10. For an analysis of the names and locations of the 21 *huacas* as other "temples of the Sun" that the priests daily passed on their pilgrimage to Villcanota and back, see Zuidema 1982 figure 16.5 pp. 441–2.

cas and *sucancas* (Molina p. 98). We may question whether the sacrifices were actually brought to all those places or rather addressed to them from Cuzco, as a close reading of the text seems to indicate. Still, we have to ask why again.[11]

In June, the rituals had kept in mind the "solstitial axis." Why was the repeat not done for such a solstitial reason? An answer can be found questioning the use of the two different roads, from and to the Cuzco valley. In June, the pilgrimage had departed through the mountains and returned by the river Villcanota, leaving the river only when approaching Quispicancha. Then, in the dry season, the waters of the river are low. But in November, with heavy rains coming, they are swelling. Urton (1981 p. 69, 71, 172) has demonstrated how much native cosmology still takes into account this seasonal change of the Villcanota river. The commemoration of the pilgrimage in November could well reverse the two directions. The first sacrifice would have been addressed to Quispicancha sucanca and the last to Mutu sucanca. Then approached the time of December solstice, Mutu sucanca being observed from Puquincancha. Such use of the pilgrimage would imply that Quispicancha sucanca was observed at the beginning of the second month before that solstice. Commemoration stressed the astronomical use of both *sucancas* and worship helped to "carry back" the rising waters to their source against the current of the river.

The involvement of the river waters becomes even more explicit, and useful for solving the calendrical problem, if now we look at the race in the third period of the year (Figure 4). It was known under two names: *Mayucati* "following (the waters of) the river" and *Pura upiay* "drinking (the waters of the river?) (during) full and waning moon."[12] According to Cobo (p. 212–3)—his example is the only one to make explicit the differential use of solar and lunar periods in the Inca calendar—, the *Mayucati* ritual followed a soli-lunar schedule. With a first new moon after the beginning of its solar month, it began six days after full moon.[13] Molina (p. 146–7) discusses the feast for the first month after the December solstice, *Camay quilla*. Other chroniclers assign the other name, *Pura upiay*, either to the feast in relation to this month (Cabello Valboa [1586] p. 350), or, as a month name by itself, to the next month (Fernandez [1571] vol. 2 p. 86; Gutierrez de Santa Clara [1600] vol. 3 p. 256). We realize that the lunar ritual would have fallen more frequently in this second, less important month. When late, it even could have coincided with the end of the month itself.

The sequence of movements to and from Villcanota *before* the December solstice may have reversed the one in June. The *Mayucati* or *Pura upiay* ritual executed *after* the December solstice clearly meant to express a complementary movement. Its professed intention was to accompany the ashes of the year's sacrifices cast into the river Huatanay. First, runners followed the ashes down this river from Cuzco to the Villcanota river. The confluence is between Quispicancha sucanca and the place where in June the pilgrims had returned from the Villcanota river. The runners then followed the ashes in this river down to the Inca town of Ollantay-

11. Molina, in using Spanish month names for Incaic months, normally is one month ahead of all other chroniclers. Thus his Incaic month corresponding to "October" is understood as "November."

12. For the term *pura*, see also Urton 1981 p. 82–5.

13. Cobo explains the detail about the New Moon after the initiation rituals of the noble boys. Although their last ritual had occurred on the day of the December solstice, they maintained their fast till the next new moon. Only if new moon arrived early would they sacrifice a llama in her honor but not if late.

tambo. Here the river was considered strong enough to carry the ashes to the Ocean without help. Two days later, the men returned to Cuzco in a competitive race (Zuidema in press). The river did not play a role anymore and we can assume that the runners followed a road through the mountains as straight as possible. No names of places en route are mentioned and we don't know if the men would have visited Quiancalla. As it was, however, a custom to worship Cuzco on gaining first view,[14] the visit would have been likely. Quiancalla and Ollantaytambo are located in the same direction from Cuzco.

The direction from Ollantaytambo over Quiancalla to Cuzco and the direction from Cuzco over Huanacauri to Villcanota constitute each other's, almost precise, prolongations. There emerges for analysis a year-long ritual in which the combined use of the direction from Ollantaytambo to Villcanota is counteracted by an equally combined use of the river from Villcanota to Ollantaytambo. We can consider the direction from Ollantaytambo to Villcanota also as an extension and reversal of the "solstitial axis" between Huanacauri and Alpitán (Quiancalla). Its principal significance (given the inclusion of the temple of the Sun in Villcanota) would also have been solstitial. Then, why was the commemoration of the pilgrimage and the race timed so outside a solstitial use? An answer is suggested comparing the movements in the three periods to each other.

Comparing the Ritual Visits

Clearest is the timing of the *Mayucati* with the race back to Cuzco. It was part of rituals that began with the December solstice and ended within 53 (30 + 15 + 8) days later. The second limit is the fixed date of 12 February in the Gregorian calendar. Only after this day in the Inca calendar, tributes arriving from the whole empire were allowed to enter Cuzco (Polo [1571] 1990 pp. 46, 56, 80, 83, 84). It was the day of the sun's second passage through zenith. The race from Ollantaytambo, with its use of the solstitial axis and the visit to Quiancalla sucanca "dragged" the influence of the December solstice toward the day of this event (13/2).

Having in mind the calendrical importance of the Mayucati, we can consider in a similar way the commemoration of the June rituals. Molina refers to them in the context of other rituals in the two months before the December solstice. His description can be combined with that of other chroniclers, so let me address the problem briefly. The commemoration took place at the same time as two other rituals: the initiation of the noble boys—when they first carried out exercises of endurance outside Cuzco—and the Itu rituals inside town—whose execution were the prerogative of king and high nobility. While Polo (1981 p. 461) and Cobo (p. 220) discuss the Itu for the month of November, Ramos Gavilán [1621] p. 157) claims (again, probably, based on Polo's lost description) that they followed a lunar schedule including a joint manifestation in the plaza of older men and initiandi. Clearly, the manifestation is the same act as the one that Molina describes for the 15th to 20th days of *Capac raymi* suggesting that these days were celebrated from full moon

14. Chroniclers refer to this custom and I observed a survival of it in 1954 entering Cuzco by bus. When the city appeared in view, all the men took off their hats.

on.[15] Some time after these lunar days came two days around the December solstice, one for earpiercing the boys. We can apply to the lunar rituals before this solstice the same type of calculation as to those after. Each sequence fitted within a 53 day period. At their earliest, the Itu and initiation rituals began with a new moon after the sun's first zenith passage. At their latest, they ended just before the December solstice.

I suggested that the commemoration of the pilgrimage began with an attention to Quispicancha sucanca and ended with one to Mutu sucanca. We can conclude that prior to this lunar version of the pilgrimage, observation was made of sunrise behind Quispicancha sucanca and some time after, observation of sunrise behind Mutu sucanca. I assume that the same observations also delimited the lunar celebrations after the December solstice.

With these remarks I return to the June rituals and suggest why the pilgrimage ended at Quispicancha sucanca. By moving to and from the temple of Villcanota, the priests helped the sun turn around the solstice, strengthening it with prayers, offers and sacrifices (Guaman Poma ff. 246(248), 248(250)). Beginning their pilgrimage at Mutu sucanca underlined the fact that this place had served the observation of the December solstice sunrise half a year apart. The visit was advanced by ten days, not only because of logistic necessity of travel to Villcanota. Such a period, before and after the solstice, was also needed observing how the sun slowly moved along the horizon, coming to a standstill in the middle of its twenty days. But the pilgrimage did more. The fact that Quispicancha sucanca was observed two months before the December solstice, suggests that the priests should be present there half a year later. By visiting the *sucanca*, the priests would have made explicit the importance of defining the period around the June solstice opposite, and equal to, that from first to second observation of Quispicancha sucanca around the December solstice. The priests could have stayed till August. They would have seen sunset as the exact reversal of the second zenith sunrise half a year apart.

The Role of Mt. Sucanca (Yahuira, Picchu) in Observing Zenith Sunrise

So far, I refrained from making the claim that Mt. Sucanca should have been the location from where zenith sunrise behind Quispicancha sucanca was observed. No chronicler seems to mention that role. Given the importance of ritual visits for understanding the role of other sucancas in astronomical observation, I finish the discussion considering the value of two pieces of information making a corresponding claim for Mt. Sucanca.

First is a statement by Cobo which, probably again, derives from Polo's lost information. But writing almost a hundred years later than Polo, he adds his own

15. Molina does not give a hint that the Capac raymi rituals might have been lunar, although he outlines the lunar character of the rituals in the next month, *Camay quilla*, and in *Coya raymi* (September). The present discussion makes clear, however, the lunar sequence in the first part of *Capac raymi*.

interpretation, trying to bring his data more in line with the Christian calendar. Re-integrating them into the Incaic system of *sucancas*, let me point out first the changes.

Cobo (p. 142), in his chapter on astronomy and the calendar, claims that there were two markers ('padrones') or pillars east of Cuzco, for the December solstice sunrise, and two west, for the June solstice sunset. He calls the eastern ones *Pucuy sucanca*—"for the beginning of winter where the year was divided in half"—and the western ones *Chirao sucanca*—"for the beginning of summer"—, adding these names to the horizon features and not to the places from where they were observed. (Polo (1981) only had said that *Pucuy sucanca* was for the beginning of winter and *Chirao sucanca* of summer.) Cobo confuses the beginning of the wet season (*pucuy*) with the western concept of the beginning of winter (December solstice) and of the dry season with that of summer (June solstice). He betrays his error when, in describing the Incaic year, he begins with the December solstice and the month of *Capac raymi*, the "Royal month." (He even says that the second month began on January 20). Polo (1981 p. 464) had explicitly claimed that the Inca year began with *Capac raymi* as the month before the December solstice. Betanzos had already made the same claim, adding *Pucuy* to the month's name. From their descriptions we can be sure that the wet season (*pucuy*) also began ahead of this solstice. Even so, they did not claim that this season began with *Capac raymi*. From the argument developed here, we discovered that the *Pucuy* season began two months before the December solstice.

But Cobo also makes a highly intriguing statement which now serves us well. *Pucuy sucanca* "was observed from the city itself," for the beginning of the wet season. Only Quispicancha sucanca could possibly have been identified with *Pucuy sucanca*. One would not have been able to see this *sucanca* from downtown Cuzco because of landscape features in between. As the direction of observation from Mt. Sucanca (Yahuira, Picchu), at the border of the Incaic city, is the same as from its plaza, we can now safely suggest that, in fact, Mt. Sucanca was used for making the *Pucuy* observation "from the city" toward Quispicancha sucanca. But its name Sucanca was not derived from this use. Clearly, Mt. Sucanca itself was observed from the other end of the valley when the *Chirao* (dry) season opened, that is around the time of first antizenith sunset (April 26). I now suspect that the *sucanca* and the mountain of Quispicancha were one and the same, as Mt. Sucanca and Mt. Yahuira as the mountain of Cuzco also were. Both mountains received their identities as *sucanca* because of the distant view that they received from each other within the system of four *sucancas*.

We can now appreciate Molina's description of Mt. Yahuira in the context of the ritual visits. He mentions how its top was visited by the noble initiates when they received there from the king their golden earspools and new loincloths of grown-up men. This act initiated their six days of manifestations with the older men in the plaza. The spools were to be worn with a thread of wool from their ears until the day of the December solstice when their ears were actually pierced. Following the lunar schedule, Mt. Yahuira could have been visited as early as 15 days after the first observation of zenith sunrise from here and as late as six days before the December solstice. Similar to the use of Quiancalla, the importance of Mt. Yahuira (Sucanca) for the zenith sunrise observation was "dragged" toward the December solstice. We can imagine that the Inca king, in a position of "Axis Mundi," was compared to his father the Sun first of all when the latter passed the zenith position in the sky. Being united with the initiandi on top of Mt. Yahuira, the king imparted on them the afterglow of his solar connection. It seems as if the relay visits to the western *sucancas* allowed the moon to carry the sun from his first to his second zenith passage when the sun was most needed for sending his rains.[16]

Conclusions
The Wet Season – Dry Season Axis

In this paper I considered only the highly integrated system of four sucancas, combining Polo and Cobo's descriptions of observing *Pucuy* and *Chirao sucancas* with Molina's descriptions of visiting *sucancas*. Molina paid much attention to Mt. Yahuira but did not seem to integrate its role into that of the other *sucancas*. However, the observation for the beginning of the Pucuy season was made from here and the mountain came to the fore as the actual linchpin of the system. Located at the confines of the city, it shared a central role with Coricancha and with the *Ushnu* from where the close observation of the pillars was made. In contrast, the other three *sucancas* served concerns with places on the distant horizon and beyond.

The chroniclers reported on only one astronomical observation fully, that of the passage of sunset through the pillars on Mt. Sucanca. Perhaps it was their only occasion of actually witnessing an observation. But even here they could not record information on the astronomical purpose or the precise date. All other conclusions about the astronomical system in Cuzco had to be reached with the help of ritual data. These data proved, however, to be sufficient and helped to arrive at precise dates of interest to the Inca calendar. One of the most difficult problems in the study of the prehispanic calendar as a register of Cuzco's socio-political, economic and ritual practices, is to decide which of the available pieces of evidence give us best control over their trustworthiness. At the moment, I consider Molina's exact description integrating a procession and a pilgrimage with a race into one year long ceremony as lending itself best to such a goal. On that basis I propose that the combination of two solstice dates, two dates of zenith sunrise and two dates of anti-zenith sunset formed the hard core of the Inca calendar (Figure 3).

References

Anonymous. 1908, "Discurso de la sucesion y gobierno de los Yngas." in Juicio de Limites entre el Peru y Bolivia, vol 8, Chunchos., ed. V. Maurtua, 149–65. Lima.

Aveni, A. F. 1981, "Horizon Astronomy in Incaic Cuzco." in Archaeoastronomy in the Americas, ed. R. A. Williamson, 285–304. Los Altos, Cal./College Park, Md: Ballena Press/Center for Archaeoastronomy.

Bauer, B. S., & D. S. P. Dearborn. 1995, "Astronomy and Empire in the Ancient Andes: The Cultural Origins of Inca Sky Watching." (Austin: University of Texas Press).

16. This suggestion reminds of a finding of Urton (pp. 71–7) in the modern village of Misminay, some 50 km from Cuzco. People observe how the sun coming from the north passes through the "center" of the sky, but consider as uninteresting when the sun is south of zenith. Thus "when the crops are inside the female earth, they are in the domain of the moon,..." and "...during the period of the sun's movement from the zenith to the south and back, the principal celestial body is the moon, not the sun."

Bauer, B. S. 1998, "The Sacred Landscape of the Inca. The Cusco Ceque System." (Austin: University of Texas Press).
Bertonio, L. 1984, "Vocabulario de la Lengua Aymara." Ed. X. Albo, and F. Layme, Cochabamba: CERES.
Betanzos, J. de. 1987, "Suma y Narración de los Incas." Ed. M. Martín Rubio, Madrid: Atlas.
Cabello Valboa, M. 1951, "Miscelanea Antártica. Una historia del Perú antiguo." (Lima: Universidad Nacional Mayor de San Marcos. Instituto de Etnología).
Cobo, B. 1956, "Historia del Nuevo Mundo." Vol. 2. Ed. F. Mateos (Madrid: Atlas).
Fernandez, D. 1963, "Primera y Segunda Parte de la Historia del Perú." in Crónicas del Peru vols 1, 2. ed. J. Pérez de Tudela (Madrid: Atlas).
Guaman Poma de Ayala, F. 1987, "Nueva Crónica y Buen Gobierno." Ed. J. V.Murra, R. Adorno and J.L. Urioste (Madrid: Historia 16).
Gutiérrez de Santa Clara, P. 1963, "Quinquenarios o Historia de las Guerras Civiles del Peru." in Crónicas del Peru vols 2, 3. Ed. J. Pérez de Tudela (Madrid: Atlas).
Molina, C. 1989, "Fábulas y Mitos de los Incas." Ed. H. Urbano (Madrid: Historia 16.
Polo de Ondegardo, J. 1981, "Los errores y supersticiones de los Indios sacadas del Tratado y averiguación que hizo el Licenciado Polo." in El Catecismo del III Concilio Provincial de Lima y sus Complementos Pastorales (1584–1585), Ed. J.G.Durán, pp. 457–78. (Buenos Aires: Universidad Católica Argentina).
———. 1990. "El mundo de los Incas." Ed. L. González & A. Alonso. (Madrid: Historia 16).
Ramos Gavilán, Al. 1988, "Historia del Santuario de Nuestra Señora de Copacabana." Ed. I. Prado Pastor. (Lima: Ignacio Prado).
Sarmiento de Gamboa, P. 1943, "Historia de los Incas." Ed. A.Rosenblatt. (Buenos Aires: Emecé editores, S.A.).
Urton, G. 1981, "At the Crossroads of the Earth and the Sky. An Andean Cosmology." (Austin: University of Texas Press).
Zuidema, R. T. 1981a, "Inca observations of the solar and lunar passages through zenith and anti-zenith at Cuzco." in Archaeoastronomy in the Americas, Ed. R. Williamson, p. 319, Los Altos, Cal./College Park, Md: Ballena Press/The Center for Archaeoastronomy.
———. 1981b, "Comment on J. H. Rowe: "Archaeoastronomy in Mesoamerica and Peru" Latin American Research Review 16 (3), p.167.
———. 1982, "Bureaucracy and Systematic Knowledge in Andean Civilization." in The Inca and Aztec States, 1400–1800, Eds. Collier, Rosaldo & Wirth, p.419, New York: Academic Press.
———. 1992, "El Encuentro de los Calendarios Andino y Español." in Los Conquistados. 1492 y la Poblacion Indigena de las Americas, Ed. Heraclio Bonilla, p. 287. Quito: Tercer Mundo editores, FLACSO.
———. 1993, "De la Tarasca a Mama Huaco. La Historia de un Mito y Rito Cuzqueño." in Réligions des Andes et Langues Indigènes Equateur—Perou—Bolivie avant et après la Conquête Espagnole. Actes du Colloque III d'Etudes Andines., Ed. P. Duviols, Aix-en-Provence: Publications de l'Université de Provence.
———. 1996, "Fête-dieu et Fête de l'Inca. Châtiment et Sacrifice Humain comme Rites de Communion." in Le Corps de Dieu en Fétes. Ed. A. Molinié. Paris: Les Editions du Cerf, pp. 175–222.
———. 1997, "Cosmovisión Inca y Astronomía en el Cuzco: Nuevo Año Agrícola y Sucesión Real." in Pensar América. Cosmovisión Mesoamericana y Andina, Ed. A. Garrido Aranda. Córdoba, Spain: Obra Social y Cultural Cajasur, Ayuntamiento de Montilla. pp. 249–270.

———. in press. "Relaciones Rituales y de Intercambio entre el Cuzco y el Bajo Río Urubamba." in Intercambio y Comercio en los Andes: La Interacción Tierras Altas—Tierras Bajas. Eds. F. Cárdenas & T. Bray. 49 Congreso Internacional de Americas, Quito 1997.

Section 4

Europe, Asia Minor, and Egypt

27

Elements of Greek Astronomy and Religion in Minoan Crete

Goren Henriksson and Mary Blomberg
Astronomical Observatory
Department of Classical Archaeology
and Ancient History
Uppsala University
Uppsala, Sweden

Archaeoastronomical investigations of important Minoan monuments show orientations at two peak sanctuaries to the heliacal rising and setting, and the acronychal rising and cosmical setting of Arcturus ca. 1830; all dates are B. C. E. unless otherwise stated. The heliacal rising occurred one lunar month before the autumn equinox and this event relative to the phases of the moon would have made it possible to regulate a lunisolar calendar like that of the later Mycenaeans and Greeks. In addition the summer solstice and the equinoxes are clearly marked on one of the peaks by alignments to sunrise and sunset behind distant foresights.

The corridor in the Central Sanctuary Area of the palace at Knossos was oriented towards sunrise at the equinoxes and a device was arranged such that the exact day of the equinoxes and also the eleventh day preceding the spring and following the autumn equinox were clearly shown. The result was a simple method for keeping an accurate lunisolar calendar. A similarly placed corridor in the palace at Zakros was oriented towards the moon at its southernmost risings in the Middle Bronze Age (ca. 2000–1600) This and the recovery of many figurines with moon-like head-dress from the nearby peak sanctuaries together with the orientation of the chamber graves in the cemetery at Armenoi within the major standstills indicate a religious role for the moon in Minoan Crete, perhaps a forerunner of the Greek goddess Artemis. Our findings indicate a long tradition of astronomical observations on the part of the Minoans and some transmission of their accumulated knowledge to the later Mycenaeans and Greeks. The implications of these findings for the history of Greek astronomy and religion are discussed.

* * *

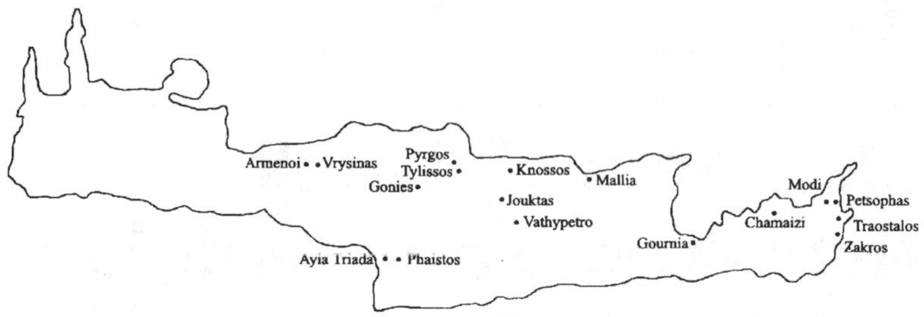

Figure 1. Map of Crete with location of Minoan sites in the Uppsala project. Not to scale.

Introduction

Minoan astronomical observations and orientations occurred in a context of similar activity in contemporary Europe, Mesopotamia, and Egypt (North 1994). We have copious written evidence from both Mesopotamia and Egypt bearing on the early astronomical activity in these neighbouring regions to Crete (Clagett 1995; Cohen 1993; Neugebauer 1969, 1975; Parker 1978; van der Waerden 1974). We have little precise evidence of astronomy in Greece before the 4th century (Dicks 1970). However Martin P. Nilsson (1920), through a careful analysis of the surviving material, recognised the existence on the Greek mainland perhaps as early as the 8th century of the eight-year intercalary cycle for maintaining a lunisolar calendar. This early use of the cycle was an enigma to Nilsson, as he knew of no adequate evidence for the requisite astronomical knowledge among the Greeks. Neither the civilisation of Mesopotamia nor of Egypt had a lunisolar calendar with regular intercalations and, in general, the astronomical interests of the two areas were directed towards quite different aims than those of the Greeks (Clagett 1995; Parker 1978; Cohen 1993). Our knowledge of early European astronomy is still too vague to say much of comparative use concerning the aims motivating it.

In a paper delivered at the Uppsala symposium on religion and power in the ancient Greek world (Blomberg & Henriksson 1996), we argued for a Minoan origin of the Greek lunisolar calendar. Much of the evidence which we used was not known when Nilsson made his study. It consisted primarily of the indications in the Linear B texts for a Mycenaean lunisolar calendar and of the orientations at the Minoan peak sanctuary on Petsophas, which would have served very well towards maintaining a lunisolar calendar. We have continued with our investigations of Minoan orientations and present here the final results from two peak sanctuaries and results to date from the palaces at Knossos and Zakros (Figure 1). The evidence corroborates our view of systematic astronomical observations on Crete in the Middle Bronze age (ca. 2000 to 1600), the period of one of the two high points of the Minoan civilisation. We follow up the presentation of the evidence with our conclusions as to the significance of the particular Minoan astronomical interest for the history of Greek astronomy and religion. The Gregorian calendar has been used for all datings.

Petsophas

At Oxford IV we gave the preliminary results from our investigations of the peak sanctuary on Petsophas (Blomberg & Henriksson 1993). The sanctuary lay above the important Minoan harbour town near modem Palaikastro (Figure 2). The site seems to have been selected because the sun could have been observed to set at the equinoxes behind the isolated conical peak of Modi which lies about seven kilometers away to the west and on which there was a similar sanctuary. A shift in position north or south of the structure on Petsophas of only 30 meters results in a change in the observations by one day. Also the moon would have been observed to set behind Modi sometimes at the equinoxes (Figure 3). Another important astronomical fact is that when the the sanctuary was in use the sun rose at the summer solstice just south of the highest peak on Karpathos, Kali Limni, 90 kilometers to the northeast (Figure 4), in the same direction as the main axis of the sanctuary building (Figure 2).

In the final evaluation of our data, we became aware that the walls AA' and AB, considering the mean errors, were oriented in the same direction as the heliacal rising and setting, and also the acronychal rising and cosmical setting of Arcturus in the years around 1830 (Table 1). The horizon can be seen in both directions from the comer at A. The orientations of the walls have been calculated by the orthogonal regression method from 10 measurements for AA' and 6 measurements for AB. The visibility of Arcturus was calculated according to the parameters in Tables 1 and 2.

Traostalos

In view of the results from Petsophas we decided to measure the orientations at the contemporary peak sanctuary on the highest part of the nearby mountain massif of Traostalos. The site lies nine kilometers due south of Petsophas and is visible from there. Modi is also clearly visible as well as the entire eastern coast of Crete (Figure 1).

The Minoan structure is on the northeastern slope about 35 meters from the highest point and only a few meters from a broad plateau. There are two walls which correspond in orientation to AA' and AB on Petsophas (Figure 5). That is, considering the margins of error, they could have been oriented originally in the same direction as the heliacal rising and setting and the acronychal rising and cosmical setting of Arcturus in the years around 1830. On Traostalos, however, these walls seem to have been part of the structure itself, whereas on Petsophas they lie adjacent to the main building. The wall AB is in poor condition and some of its stones have been moved; others are in situ. We judged that further points measured here would not have contributed to a more accurate determination of the orientation. The wall AA, on the other hand, is partly carved from the bedrock and is well defined. We determined the orientation of the walls at both sites on two different occasions at intervals of about a year each between the measurements and we got very similar results on each occasion (Table 1). This provides some reassurance that our choice of measuring points for determining the original orientations is not unduly subjective.

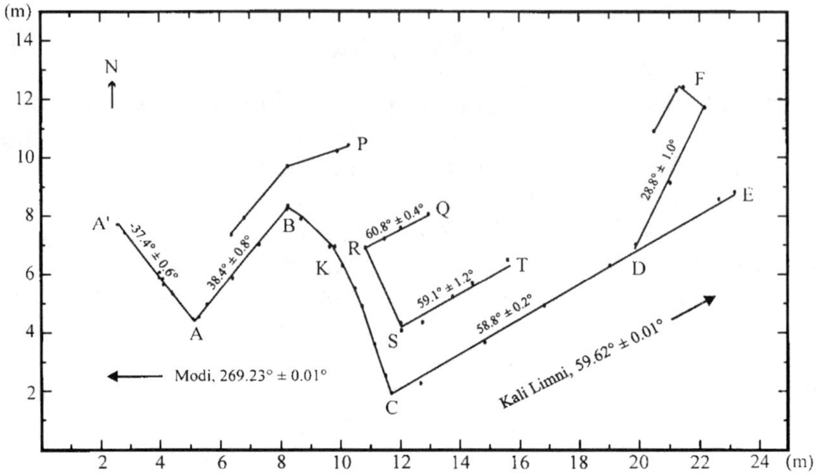

Figure 2. Petsophas. Orientation of the walls.

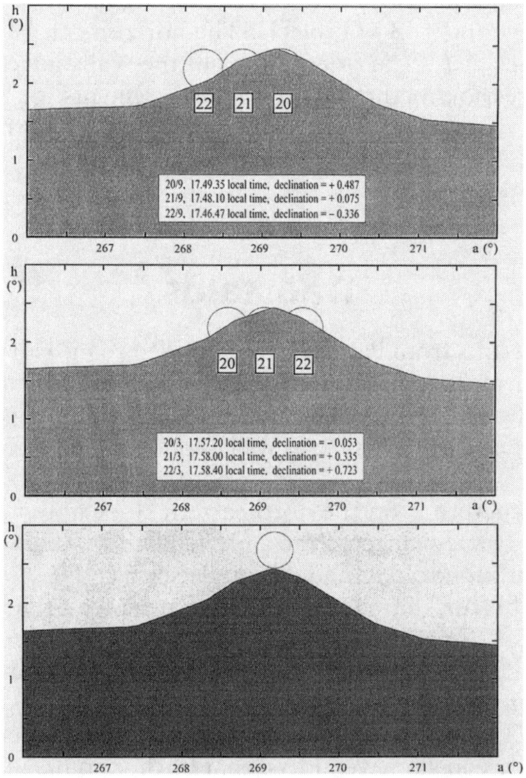

Figure 3. Top and middle: sunsets behind Modi as observed from Petsophas, 1997 September 20, 21 & 22 and March 20, 21 & 22 B. C. E. (refraction for t = + 10' Q. The relationships are valid for many centuries. Bottom: full moon setting behind Modi as observed from Petsophas at the autumn equinox, 2121 September 21, 05.13.45 local mean solar time (refraction for t = + 10' Q.

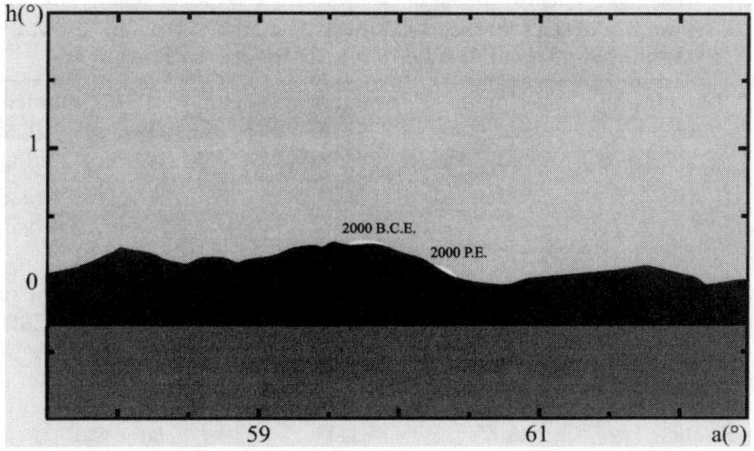

Figure 4. Theoretically calculated position of sunrise at the summer solstice above Kali Limni, Karpathos' highest peak (1215 m), 2000 June 23 B.C.E., 04.39.3 5 local mean solar time (sun on the left), as it would have been observed from the peak sanctuary on Petsophas (H. 255 m). The sun on the right shows sunrise 2000 June 21 present era, 05.01.38 Greek standard time. Refraction calculated for + 20' C and atmospheric pressure 760 mm. Hg.

Modi lies close to the direction of AA' and Arcturus will have appeared directly above that peak at its heliacal setting in the years around 1730 (Figure 6). Another useful orientation on Traostalos is that of the wall BY which is the same as the extreme western limit for the circumpolar stars at that latitude (Figure 7).

We evaluated our measurements of the walls on Petsophas and Traostalos using Student's T-test with pooled variance. Table I provides a summary of the results. AA' and AB on Petsophas do not differ significantly in their deviation from the north-south direction. This indicates that they may have intentionally been built symmetrically with respect to north. The same is true for Traostalos, but the result here is not entirely convincing due to the poor condition of the wall AB. We have also analysed the correlation in the directions of the two pairs of walls. There is 90% probability that they were intentionally oriented in the same directions (Figure 7). In view of the margin of errors we cannot exclude the possibility that the eastern wall of the pair was used to observe both the heliacal and acronychal risings and the western wall was used to observe both the heliacal and cosmical settings. We find no more likely explanation for this correlation than observation of the same celestial object. Arcturus is a very likely candidate as its heliacal rising and setting and acronychal rising and cosmical setting occurred in the directions of the walls for the period when the sanctuaries were in use. Since it can be difficult under unfavourable weather conditions to see the star in the early morning and late evening light, the orientation of some device such as walls would have improved the chances of observing its first appearance.

Figure 8 is a theoretical presentation of the change in azimuth for the heliacal rising and setting and acronychal rising and cosmical setting for Arcturus from 2200–1500 at the latitude of Petsophas. The measured orientation of the wall AB at both sites can in principle correspond to either the heliacal or the acronychal rising, and the corresponding optimal construction dates can be read from the horizontal axis. The walls AN can in similar fashion correspond to the heliacal or cosmical setting and their optimal construction dates read in the same way.

Table 1. Summary of the Measured Azimuths and Mean Errors for the Walls on Petsophas and Traostalos. All Dates Are Before the Present Era.

Site	Orientation	Azimuth	Mean error	No.	Comment
Petsophas	AA'	322.6° (-37.4°)	±0.6°	10	Arcturus' heliacal and cosmical setting ca. 1830
	AB	38.4°	±0.9°	6	Arcturus' heliacal and acronychal rising ca. 1830
	CE	58.8°	±0.2°	8	sunrise at the summer solstice ca. 2000
	ST	59.1°	±1.2°	4	"
	RQ	60.8°	±0.4°	8	"
	DF	28.8°	±1.0°	9	later wall
	to Modi	269.23° (–90.77°)	±0.01°		Sunset at the spring and autumn equinoxes ca. 2000
	to Traostalos	179.31°	±0.01°		true south 180.00°
Traostalos	AA'	320.6° (-39.4°)	±0.8°	8	Arcturus' heliacal and cosmical setting ca. 1830
	AB	38.9°	±1.8°	4	Arcturus' heliacal and achronychal rising ca. 1830
	BY	325.0° (-35.0°)	±0.8°	8	western limit for the circumpolar stars at Traostalos' altitude (35.15°)
	to Modi	320.28° (-39.72°)	±0.01°		Arcturus' heliacal setting ca. 1730
	to Petsophas	359.31° (-0.69°)	±0.01°		true north 0.00°

As there is 90% probability that the pairs of walls at the two sites were intended to have the same orientations, we determined the weighted mean of their deviation from north, which is 38.3' ± 0.4'. From this mean value we get the optimal construction date, 1827 ± 77 years. This gives us an independently determined date for the sanctuaries which agrees very well with the archaeologically determined date of 1900–1800 (MacGillivray and Driessen 1990; Manning 1995).

The risings and settings of Arcturus at the two peak sanctuaries occurred in the Middle Minoan Period at four times of the year that would have been useful for calendaric purposes in Crete. An important astronomical fact which must be kept in mind here is that the positions of the stars change slowly due to precession. This means that particular risings and settings of Arcturus today occur nearly two months later than they did in the early Middle Bronze Age. In the case of Arcturus it is also important to take into account the proper motion.

At the time when the walls were built on the two mountains, the heliacal rising of Arcturus occurred on the 24th of August. This is one lunar month before the autumn equinox. We have independent evidence from the Hellenistic Period that the

Table 2. Parameters Used in Calculating the Visibility of Arcturus at Its Heliacal Rising and Setting, Acronychal Rising and Cosmical Setting.

Parameters	Sources
Visual extinction coefficient = 0.13	Calibrated according to the observations at
This means that the extinction coefficient in Vmagnitude = 0.17	Athens of Schmidt 1865. The advantage of using Schmidt's observations is that they were made before modern air pollution.
Extinction Air Mass	Computed using the tables from Bemporad 1904
Visibility of faint stars, taking into account contrast at faint levels of light.	Siedentopf 1941
Brightness of the sky at twilight	Ljunghall 1949
Refraction for air temperature = + 10° C and air pressure = 760 mm Hg	Reasonable estimates
Arcus visionis, heliacal rising and setting = 9.3° Altitude for Arcturus = 2.9°	Our own calculations based on the above sources
Arcus visionis, acronychal rising and cosmical setting = 8.2° Altitude for Arcturus = 2.0°	Our own calculations based on the above sources

year in Crete began at the first crescent moon following the autumn equinox (Guarducci 1945). If this seems like a long time for the continued use of a calendar, we can compare it with the longer periods of the Egyptian use of their calendar and of our use of the Julian calendar. The only change in the latter has been the slight adjustment in the length of the year made by Pope Gregory XIII in 1582 of the present era. Crete was an exception to most other areas of Greece where the year began at the first sight of the crescent moon after either the spring equinox or the summer solstice (Samuel 1972). Thus the heliacal rising of Arcturus may have served to herald the equinox and also the beginning of the new year. The orientation of the palace at Knossos seems also to have had the autumn equinox as its object (see below). Arcturus could have had an important role in the regulation of a Minoan lunisolar calendar, as the star's heliacal rising relative to the phases of the moon would have indicated when it was time to intercalate a month, intercalations being necessary to compensate for the incommensurate motions of the moon and the sun, and thereby to keep the months in the correct seasons. A simple rule could have been used, for example: 'whenever the new moon appears in the evening sky on one of the I I days following the morning rising of Arcturus, it signals the beginning of an intercalary month'. Similar rules were used in Egypt and the Near East (Clagett 1995; Cohen 1993).

The acronychal rising would have occurred on the day in the Middle Bronze Age corresponding to our 2nd of February, which was midway between the winter solstice and the spring equinox. This may indicate that there was also a division of the solar year into eight equal parts. Such a year would have been practical for farmers, whereas the lunisolar calendar would have been necessary for seasonal religious observances.

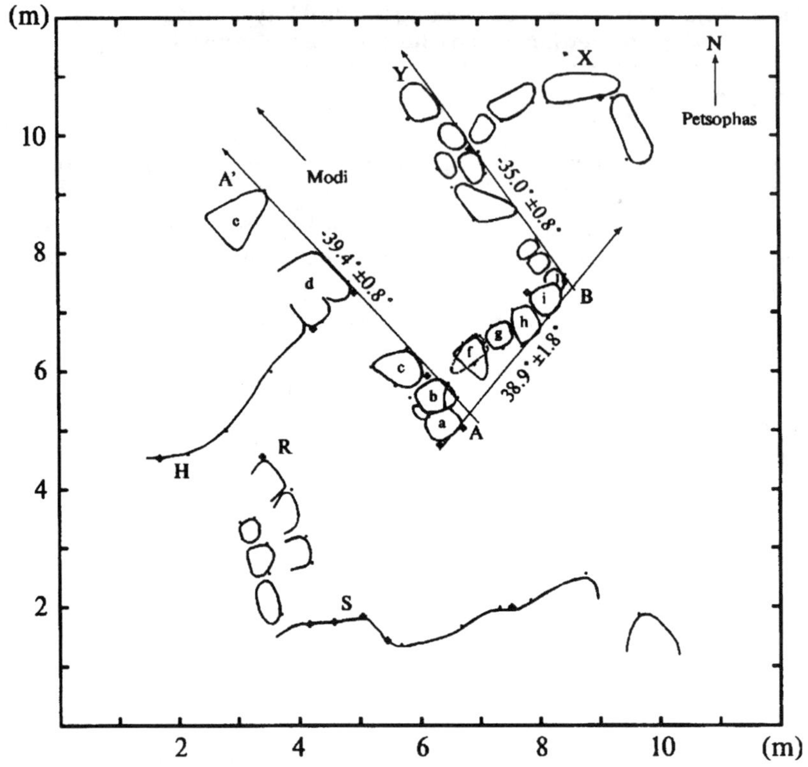

Figure 5. Traostalos. Orientation of the walls.

The cosmical setting on the I I th of May and the heliacal setting on the 13th of October would have been practical for indicating the limits of the sailing season in Crete. The peak sanctuaries were close to the two important Minoan ports in eastern Crete, those at Palaikastro and Zakros. The beginning of May and October have been the traditional dates for the sailing season in the Aegean until recent times. We find this information in a number of ancient and modem sources (Lambrou-Phillipson 1991; Casson 1986). The Pleiades were used to tell the limits in the Archaic period (ca. 700–500). By that time, due to precession, the settings of Arcturus were no longer appropriate; we find them, nevertheless, associated with storms at sea down into the first century (e. g. Aratos, 3 d cen.; Geminos, 1st cen.) This may have been a superstitious survival from the Bronze Age when the settings and risings were signals for the beginning and end of stormy weather and, consequently, the limits of the sailing season. There are a number of survivals in later sources of references to stellar positions which do not conform to the time of their writing, but which would have been correct in an earlier period (Blomberg 1992). Similar effects of conservative forces can be detected in the Egyptian astronomical texts (Neugebauer 1969; van der Waerden 1974).

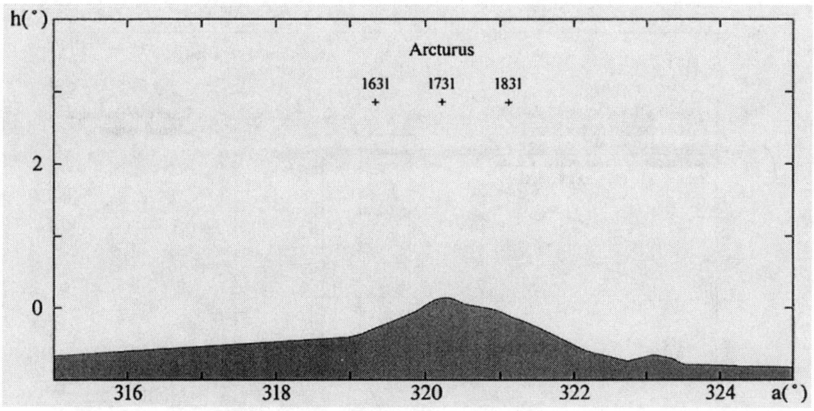

Figure 6. Heliacal setting of Arcturus above Modi as it would have been observed from Traostalos in the years 1631, 1731, and 1831 B.C.E.

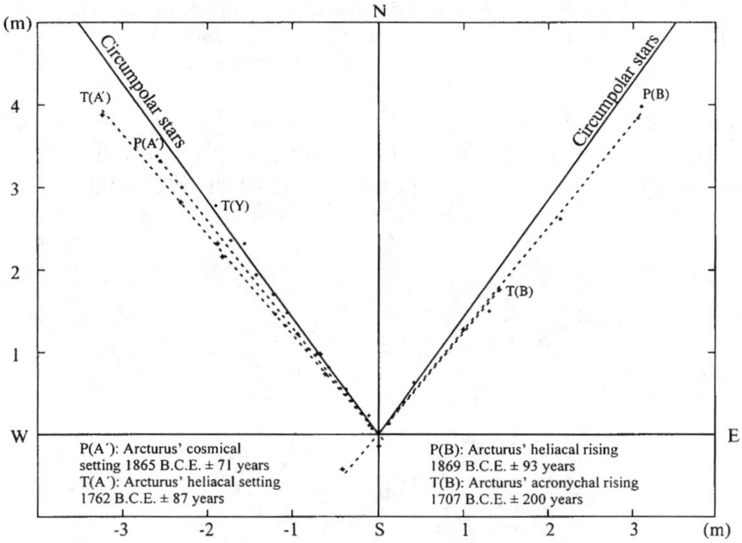

Figure 7. Orientation of the walls AA' and AB on Petsophas and AA', AB, and BY on Traostalos. The wall BY on Traostalos is oriented towards the extreme western limit for the circumpolar stars at that latitude.

Knossos

It was early observed that the main axes of the Minoan palaces are neither north-south nor eastwest and that this could be due to orientations towards the heavenly bodies (Marinatos 1934; Shaw 1977). However no detailed study of the orientations and their objects has been presented. We have found that in the palace at Knossos (Figures 9 & 10), the so-called Corridor of the House Tablets—named after the Linear B tablets with ideogram of a house which were found there—is oriented to sunrise behind the Ailias ridge on the morning of the equinoxes. The

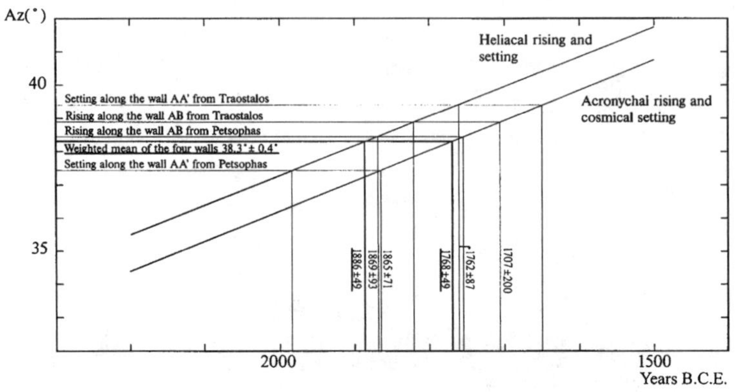

Figure 8. The measured azimuths of the walls on the Minoan peak sanctuaries at Petsophas and Traostalos can be found on the vertical axis. The corresponding optimal years for observations of the heliacal rising and setting and the acronychal rising and cosmical setting of Arcturus can be found on the horizontal axis from the computed lines. For every wall there are two possible optimal years, but only the identifications which give the smallest deviations are accepted and the corresponding years have been presented with their mean errors.

distance to the ridge is about 775 meters. The palace lies on a hill which had been the site of a village from about 8000. The village grew continually and about 2000 the entire summit was levelled for the construction of the first palace. There is nothing in the terrain itself which constrains the present orientation of the building.

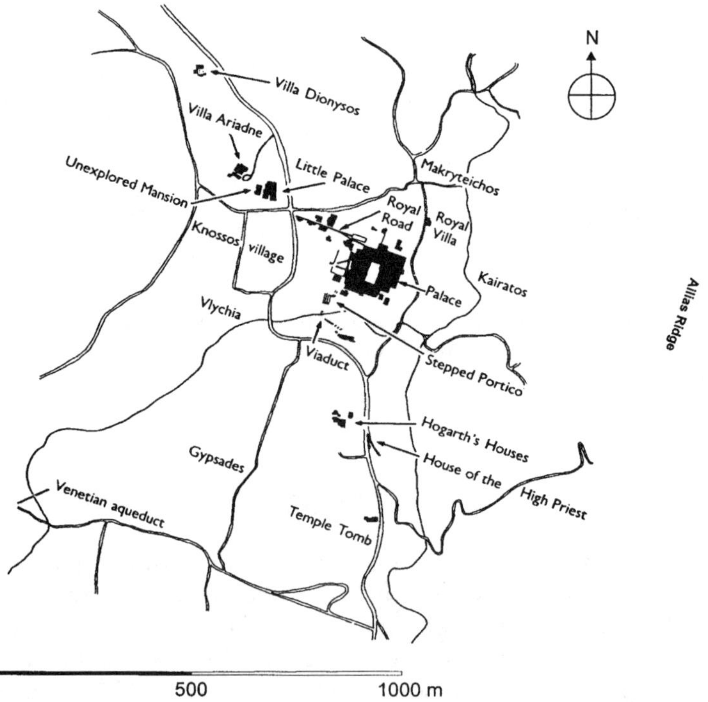

Figure 9. Knossos Area. Courtesy authors of The Aerial Atlas of Ancient Crete.

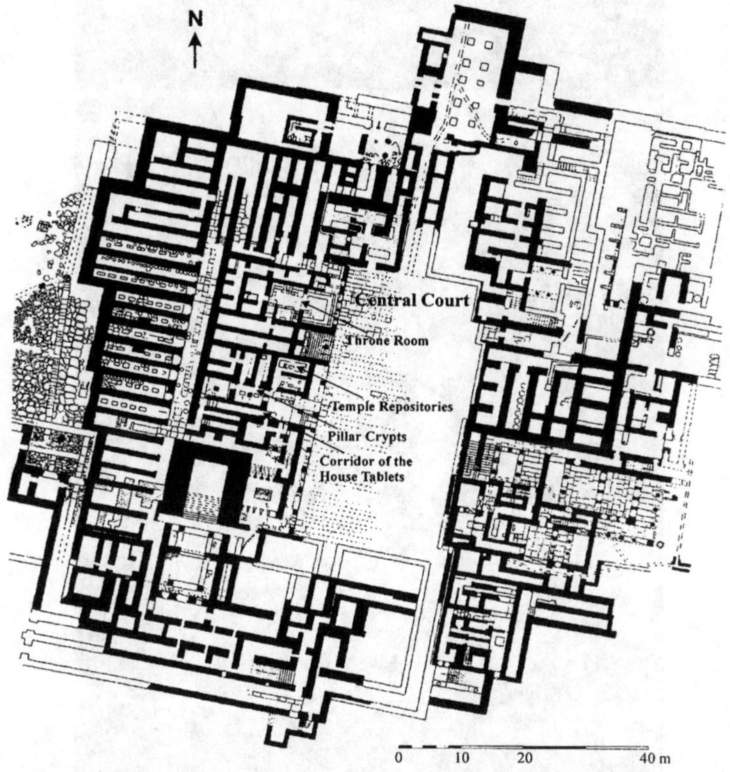

Figure 10. The palace at Knossos. Courtesy authors of *The Aerial Atlas of Ancient Crete*.

The longer axis of the Central Court is 10.4' east of north, for the western facade (Figure 10). In the west wing lay the most important cult rooms—the throne room, the pillar crypts, and the temple repositories—with doors opening towards the east. The floor level here was retained throughout the 700-year life of the palace and has resulted in a difference of about a meter up to the Central Court of the last phase of the palace (Figure 11).

In the Central Sanctuary Area of the palace lies the Corridor of the House Tablets from which we see the top of the Ailias ridge (Figure 11). The orientation of the Corridor to the rising sun, as it appeared above the Ailias ridge on the morning of the equinoxes from a point near the middle of an unusual concave stone just to the north of the door in the southern wall of the Corridor (Figures 12 & 13), is valid if four conditions obtained:

1. the floor level of the Corridor was about the same as it is today,
2. the wing opposite had no more than two stories,
3. there was a structure such as a doorjamb or pillar at point A (Figure 13),
4. the height of the Corridor's eastern door was not much less than 2.0 meters (about the equivalent of six Minoan feet, according to the different estimations which have been made). There is no archaeological evidence against these conditions.

On the morning of the autumn equinox, the rays of the sun would have illuminated the southern half of the concave stone. The stone is in situ and the question is whether or not it is the remnant of an installation of some kind which existed in

Figure 11. Eastern doorway of the Corridor of the House Tablets with the Ailias ridge in the background.

the Minoan period to heighten the effect of the sun's rays. The sun never reaches further into the Corridor. If liquid is placed in the stone, a small distinct reflection is cast in the southwest comer of the Corridor at the moment of sunrise on the equinoxes (Figure 14). At sunrise close to the equinoxes the upper limit of the shadow on the southern wall will touches the top of the middle double axe in the southern wall, however this may be fortuitous (Figures 14 & 15).

The sun illuminates the concave stone for the last time eleven days following the autumn equinox (Figure 13). Since this is the difference between the solar year and twelve synodic months, the phase of the moon eleven days following the autumn equinox will be same as at the next autumn equinox, when the sun will reappear at the upper north-eastem comer of the door, as seen from the stone. Using a similar rule of thumb as described above for Petsophas, the Knossians would have known when to intercalate a month: 'when the new moon appears in the evening on any of the I I days following the autumn equinox, it signals the beginning of the intercalary month'. These appearances of the sun will be similar but reversed at the spring equinox and not useful for predicting the phase of the moon at the following spring equinox.

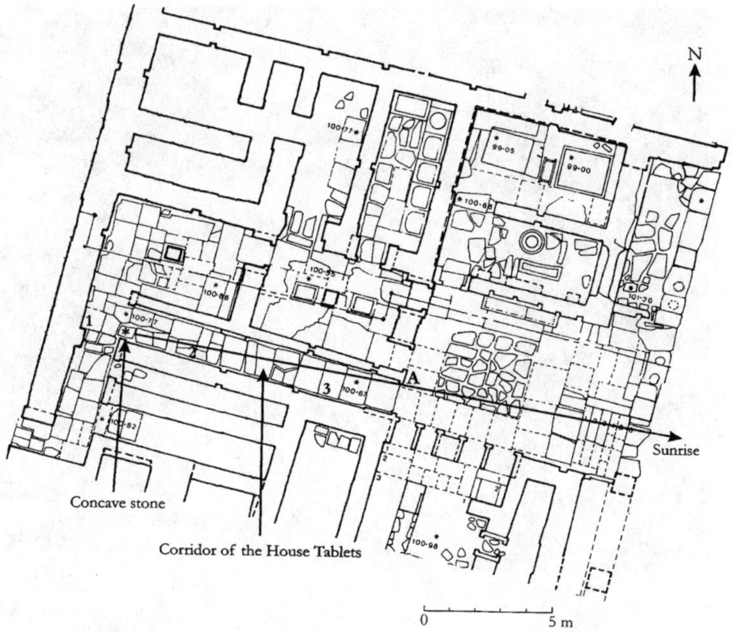

Figure 12. Central Sanctuary Area of the palace at Knossos. A: frame to the east doorway of the Corridor of the House Tablets; *: concave stone; 1-3: positions of the double axes engraved in the south wall. The line to sunrise indicates the orientation from the stone to the sun on the morning of the equinoxes. Orientation of the south wall of the Corridor = 102.6'. Courtesy S. Hood and W. Taylor, reproduced with permission of the British School at Athens.

We think it likely, therefore, that the autumn equinox was the object for the orientation of the Corridor and we conclude that the general orientation of the palace was the result of the need for a reliable method for determining the day of the autumn equinox, showing when it was time to intercalate a month, and predicting the phase of the moon for calendaric and religious purposes. This implies that the Minoans were using a lunisolar calendar, as did their successors the Mycenaeans and the Greeks, and that their year began in the Bronze Age at the first crescent moon following the autumn equinox, as it did in Crete also in the Hellenistic period (Guarducci 1945).

Zakros

At Zakros the western side of the palace's central court is oriented 37.6' east of north (Figures 16 & 17). It was earlier observed that the direction at right angles to the major axis of the court was close to the direction of moonrise at the southern major standstill (Shaw 1977). We have independently reached the same conclusion, but specifically with respect to the northernmost corridor of the west wing and the cliff opposite (Figures 17 & 18). The corridor adjoins the religious areas of the palace, as was the case with the Corridor of the House Tablets at Knossos. The distance to the cliff is about 710 meters. The first palace at Zakros, like the one at

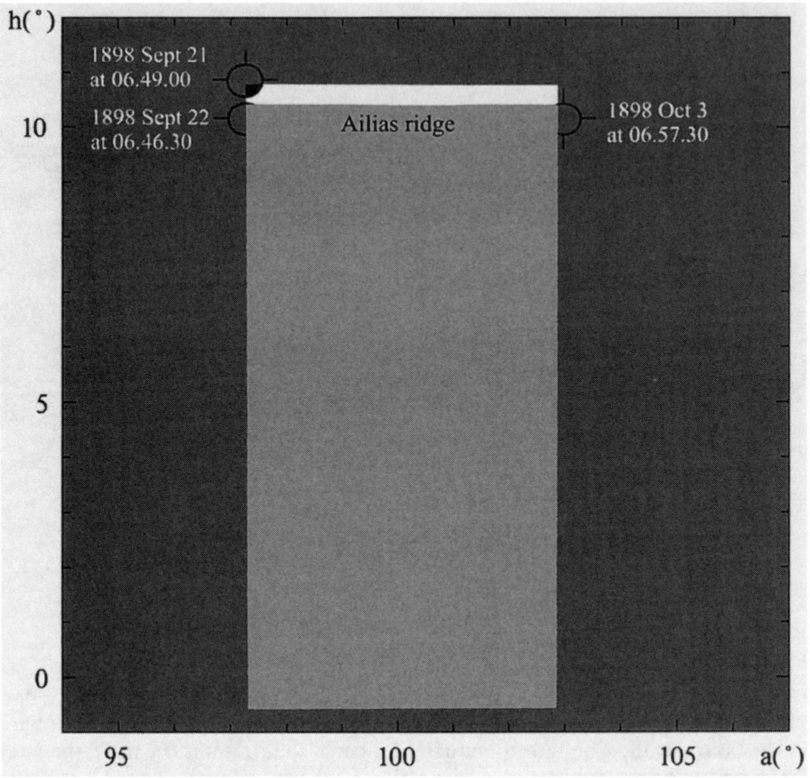

Figure 13. At the upper left the sun as it appeared from the concave stone in the Corridor of the House Tablets at the autumn equinox 1898; all dates B.C.E. At the upper right the sun eleven days after the autumn equinox. Based on measurements made with the digital laser theodolite SOKKIA SET 4C, which gives an error of only a few millimetres. The orientation error is less than 1/100 of a degree. Such exact measurements reduce the likelihood that we introduce errors into the results.

Knossos, was built early in the Middle Bronze Age. Also here, as at Knossos, the location does not inhibit the orientation of the palace.

From the corridor the cliff can have been used as a foresight for observation of moonrise at the southern major standstill. This observation would have been especially impressive in the case of the full moon at the summer solstice when the moon would have followed the contour of the hill as it climbed into the evening sky (Figure 19). The traces of an older foundation course along the southern wall of the corridor may be evidence that this was the exterior wall of the building at some earlier period. At a later rebuilding, the view to the cliff may have remained visible through the door to the open court opposite (Figure 20).

We have found that the orientation of the corridor (126.5' east of north) is very close to the azimuth for the northern edge of the cliff, which is very near the azimuth to the southernmost point for the upper limb of the full moon at the summer solstice at the beginning of the Middle Bronze Age (126.27). The moon no longer rises so far south due to the change in the inclination of the earth's axis.

Figure 14. The rays of the sun reach the concave stone in the Corridor of the House Tablets at the moment of sunrise on the autumn equinox. The shadow touches the top edge of the double axe inscribed in the southern wall of the Corridor.

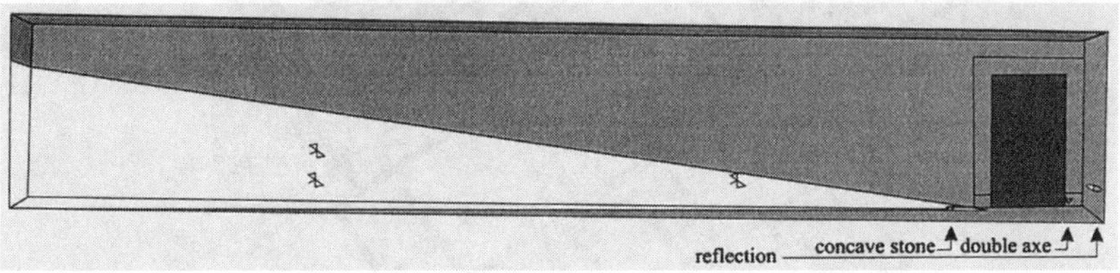

Figure 15. Drawing of the effects of the sun at sunrise in the Corridor of the House Tablets at the equinoxes. By Allan Klynne.

We propose that the orientation of the palace and its position on the plain was the result of the wish to know when the moon had reached its southern-most point, most probably for religious rather than calendaric reasons.

Conclusions

Minoan astronomical observations were nearly contemporary with those of other Bronze Age cultures in the lands around the Mediterranean-Europe, the Near East, and Egypt. It is likely that there was exchange of astronomical ideas and information among these cultures just as there is documented exchange of raw materials and manufactured products. Nevertheless all four area maintained their own cultural

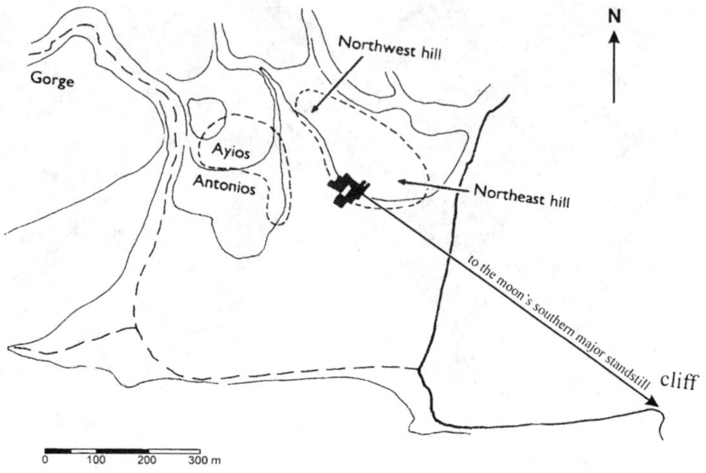

Figure 16. The area of the palace at Zakros. Courtesy authors of The Aerial Atlas of Ancient Crete. The original has been modified to include the cliff.

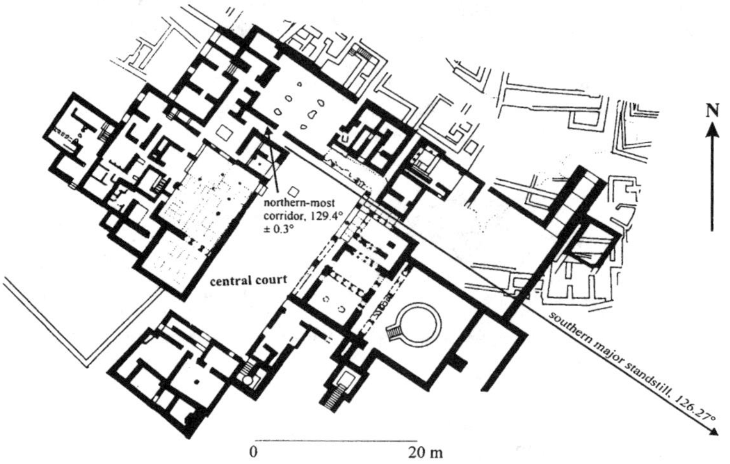

Figure 17. The palace at Zakros. Courtesy authors of The Aerial Atlas of Ancient Crete.

identities and these are reflected in all aspects of their respective societies—their art, architecture, religion, leadership and, not least, their astronomy. There is no good support at present to justify the view that any one of these cultures was the innovator or leader in astronomy. Astronomical orientation in the British Isles may be older than those elsewhere (North 1994), but this is likely to be a logical consequence of local incentives playing on natural human curiosity and intelligence and, therefore, not at all unlikely to occur independently in other areas and periods. The evidence we have indicates centuries-long development of astronomy for specific purposes in different cultural areas. We support a model of astronomical development in early societies being based primarily on culture-specific factors in the societies. This does not rule out the possibility of influence coming in as a result of exchange with other peoples.

Figure 18. Zakros. The ridge opposite the northem-most corridor of the west wing.

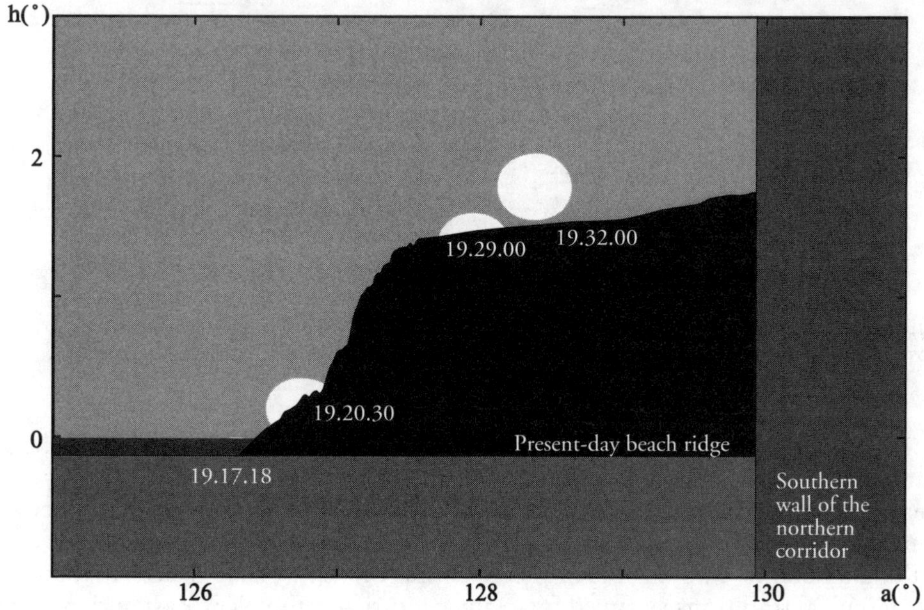

Figure 19. The extreme southern rising of the full moon at the summer solstice 2034 June 23 B.C.E. from the northem-most corridor of the west wing of the palace at Zakros. The time is mean solar time.

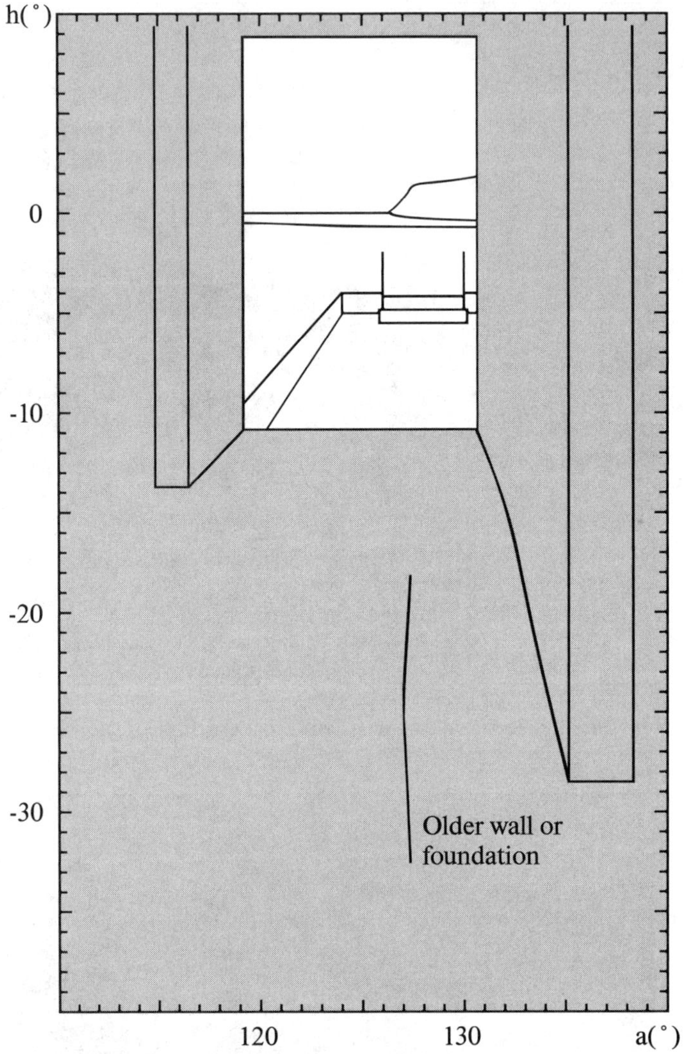

Figure 20. The view from a position 20 cm from the south-westem comer of the northem-most corridor of the palace at Zakros. Eye level is 160 cm above the present-day floor.

We conclude that, at least in part, later Greek astronomy was based on the accumulated astronomical knowledge of the Minoans which had been conveyed to them via the Mycenaeans. This is made the more feasible by the fact that other traditions and achievements from Bronze Age Crete survived in later Greece. There is nothing inherently implausible in such a course of events, as it is clear that ancient astronomical traditions survived for many centuries in the Near East and in Egypt.

The long tradition of Greek didactic poetry (e.g. Hesiod, ca. 7 th cen.; Aratos, 3rd cen.) has been argued to be the later stages of an even longer oral tradition which facilitated the preservation of practical knowledge. We think it likely that the conservative character of tradition, likely to have been especially strong where the sacred heavenly bodies were concerned, would have been a strong contributing factor in the preservation of Minoan astronomical knowledge. We submit that our findings have significance for the history of both Greek astronomy and religion.

As evidence in the case of astronomy we argue that the Minoans and the later Greeks were using observations of celestial bodies for two specific and similar purposes:

1. The risings and settings of the bright star Arcturus were being used to indicate dates of practical importance, e. g. to mark the limits of the sailing season and the time to intercalate a month in a lunisolar calendar for the Minoans and the time of the year for pruning the grapevines and for recording historical events for the Greeks. The risings and settings of Arcturus continued to be important for farmers, historians, and probably the general population, from Hesiod (ca. 7th cen.) down into Roman times. Thucydides (5th cen.), for example, placed historical events in the year by relating them to the positions of Arcturus. This implies that his readers would have understood the references. There are many instances in Greek literature of a similar use of other stars.

2. The major positions of the sun and the moon were used to maintain a lunisolar calendar. There is evidence that such a calendar was also being used by the Mycenaeans, suggesting that they were the intermediaries through whom the astronomical knowledge of the Minoans was conveyed to the Greeks (Blomberg & Henriksson 1996). It is well-known that the Greeks of the historical period used a lunisolar calendar and that their religious observances occurred on specific days of specific months. We know also that it was of great importance that these celebrations occur in the proper seasons (Geminos, 0 cen.). Regular use of both the eight-year cycle and the nineteen-year cycle for the sun and moon were used at earlier dates in Greece than elsewhere in the Mediterranean area, although the Babylonians were using the nineteen-year cycle shortly after the traditional date of its introduction at Athens (Geminos, 0 cen.). There is no evidence of a lunisolar calendar of the same type as that of the Greeks in any other contemporary area.

These similar practical uses of the heavenly bodies in the same geographical area by cultures which succeeded one another, and each of which clearly admired and borrowed from its predecessor, implies a continuous astronomical tradition. In other words the history of Greek astronomy can be extended to include a chapter on local origins in the Bronze Age Aegean, notably in Minoan Crete.

In the case of religion, we add the results from Zakros to our earlier evidence for a Minoan goddess in some way associated with the moon (Blomberg & Henriksson 1996). The orientation to the southern major standstill strengthens the arguments for Mycenaean and Minoan religious interest in the moon on Crete. Such interest in the moon has also been indicated by the results obtained in the study of the more than 220 Late Bronze Age graves at Armenoi in central Crete, as the orientations of all but possibly one lie within the major standstills (Papathanassiou, Hoskin, & Papadopoulou 1992, 1993; Hoskin & Papathanassiou 1996).

Both the origins of Greek religion and Greek belief in the divinity of the sun, moon and stars are relevant for the question of continuity from the earlier periods to historical times, a question of great current interest. Many articles have been written on the Bronze Age origins of Greek religion and books on the subject regularly have a chapter on this aspect (e. g. Nilsson 1950; Burkert 1985). Most scholars recognise that the celestial bodies were in some sense divine for the Greeks, but the origin and character of this divinity are not well understood.

The gaps in our knowledge of early Greek religion are largely due to the fact that the sources for its study from the Mycenaean and Minoan periods are mainly symbols of debatable meaning on small objects of various kinds and therefore not easily interpreted. The symbols in Minoan art which seem more or less clearly to refer to the heavenly bodies have been collected by Goodison (1989), who interprets them largely as referring to the sun. Many of them could refer, instead, to a star or to the moon. We know very little of the Minoan conventions for representation in some branches of art. Therefore any additions to our knowledge of religion in the Bronze Age Aegean are very valuable. The evidence for a Minoan goddess associated with the moon, for example, when compared with the later associations of the Greek Artemis with the moon and the fact that Artemis may be one of the divinities mentioned in the Linear B tablets (Ventris & Chadwick 1973), strengthen the arguments for continuity in Greek religion from the Bronze Age.

Acknowledgments

We would like to thank the members of the boards of the following foundations for making our research possible: the Swedish Council for Research in the Humanities and Social Sciences, the Gunvor & Josef Andr Foundation, the Axel & Margaret Ax:son Johnson Foundation, the Magn. Bergvall Foundation and the Helge Ax:son Johnson Foundation. We are also grateful to the

Greek Archaeological Service for permission to study the sites. We would like to thank as well C. Davaras, former ephor, and N. Papadakis, present ephor of antiquities at Ayios Nikolaos, A. Karetsou, ephor of antiquities at Iraklion, Berit Wells and Bodil Nordstr6m of the Swedish Institute, all of whom have been helpful in furthering our work. We thank also Lennart Bondesson, professor in mathematical statistics at Uppsala University, who helped us with the statistical evaluation of our measurements. We are grateful as well to the authors of the Aerial atlas of Ancient Crete (Myres, Myres, & Cadogan 1992) for allowing us to use their figures 17.1 & 17.3 (Knossos), 44.1 & 44.3 (Zakros).

We owe a special debt of gratitude to Peter E. Blomberg for his help with the measuring and his contributions to our discussions.

References

Aratos, 3rd cen., *Loeb Classical Library* 129, *Aratus*, ed. G. R. Mair. 1955, reprinted 1989, Cambridge MA: Harvard Univ. Press.

Bemporad, A. 1904, "Zur Theorie der Extinktion des Lichtes", *Mitteilungen Grossh. Sternwarte zu Heidelberg* 4.

Blomberg, M. 1992, "The Meaning of /G F in Hesiod", *Opuscula Atheniensia* 19, p. 49.

Blomberg, M. & Henriksson, G. 1996, "*Minos Enneoros*. Archaeoastronomical. Light on the Priestly Role of the King in Crete", *BOREAS, Uppsala Studies in Ancient Mediterranean and Near Eastern Civilizations* 24, *Religion and Power*

in the Ancient Greek World. Proceedings of the Uppsala symposium 1993, ed. P. Hellström & B. Alroth, Uppsala: Acta Universitatis Upsaliensis, p. 27.
Blomberg, M. & Henriksson, G. 1993, in *Proceedings of the Fourth 'Oxford' International Conference on Archaeoastronomy, Stara Zagora, Bulgaria, August 1993*, in press.
Burkert, W. 1985, *Greek Religion*, (Cambridge MA: Harvard University Press)
Casson, L. 1986, *Ships and Seamanship in the Ancient World*, 2nd ed. (Princeton: Princeton Univ. Press).
Clagett, M. 1995, *Memoirs of the American Philosophical Society* 214, *Ancient Egyptian Science 2, Calendars, Clocks, and Astronomy*, Philadelphia: American Philosophical Society.
Cohen, M. E. 1993, *The Cultic Calendars of the Ancient Near East* (Bethesda MD: CDL Press).
Dicks, D. R. 1970, *Early Greek astronomy to Aristotle* (London: Thames and Hudson) Geminos, I" cen., Parapegma, ed. G. Aujac. 1976, Paris: Collection des Universitds de France Goodison, L. 1985, Bulletin of the Institute of Classical Studies, supplement 53, Death, Women and the Sun, London: Univ. of London.
Guarducci, M. 1945, "Note sul Calendario Cretese", *Epigraphica* 7, p. 72 Hesiod, ca. 7 th cen., Loeb Classical Library 57, *Hesiod*, ed. H. G. Evelyn-White 1920, reprinted 1977, Cambridge MA: Harvard Univ. Press.
Lambrou-Phillipson, C. 1991, "Seafaring in the Bronze Age Mediterranean: the Parameters Involved in MaritimeTravel" *Aegaeum 7, Thalassa LEgee Prehistorique et la Mer*, ed. R.
Laffineur & L. Basch, Li&ge: Univ. de Li~ge, p. I I.
Ljunghall, A. 1949, "The Intensity of Twilight and its connection with the Density of the Atmosphere", *Meddelanden från Lunds Astronomiska Observatorium* 13, ser. 2, No. 125.
MacGillivray, J. A. & Driessen, J. 1990, "Minoan settlement at Palaikastro", *Bulletin de Correspondance Hellénique*, supplément 19, *L 'habitat Égéen préhistorique*, ed. P. Darque & R. Treuil, Athens: École française d'Athènes, p. 395.
Manning, S. W. 1995, *Monographs in Mediterranean Archaeology 1, The Absolute Chronology of the Aegean Early Bronze Age*, Sheffield: Sheffield Academic Press.
Marinatos, S. 1934, in *Proceedings of the First International Congress of Prehistoric and Protohistoric Sciences, London August 1–6, 1932* (London: Oxford Univ. Press), p. 197.
Myres, J. W., Myers, E. E., & Cadogan, G. 1992, *The Aerial Atlas of Ancient Crete* (London: Thames and Hudson).
Neugebauer, 0. 1969, *The Exact Sciences in Antiquity*, 2nd ed. (New York: Dover)
Neugebauer, 0. 1975, *History of Ancient Mathematical Astronomy* (New York: Springer-Verlag).
Nilsson, M. P. 1920, *Skrifter utgivna av Humanistiska Vetenskapssamfundet i Lund* 1, *Primitive time-reckoning*, Lund: C. W. K. Gleerup.
Nilsson, M. P. 1950, *The Minoan-Mycenaean Religion and its Survival in Greek Religion* 2nd & (Lund: C. W. K. Gleerup).
North, J. 1994, *Fontana History of Astronomy and Cosmology* (London: Fontana Press).
Papathanassiou, M., Hoskin, M., & Papadopoulou, H. 1992, "Orientations of Tombs in the LateMinoan Cemetery at Armenoi, Crete", *Archaeoastronomy* 17, p. S43.
Papathanassiou, M., Hoskin, M., & Papadopoulou, H. 1993, "Orientations of Tombs at Armenoi, Crete: Addenda," *Archaeoastronomy* 17, p. S54.
Papathanassiou, M. & Hoskin, M, 1996, "The Late-Minoan Cemetery at Armenoi: a Reappraisal", *Journal of the History of Astronomy* 27, p. 53.

Parker, R. A., 1978, "Egyptian Astronomy, Astrology, and Calendrical Reckoning", *Dictionary of Scientific Biography* 15, supplement 1, ed. R. Adams & L. Zejszner, New York: Charles Scribner's sons, p. 706.

Samuel, A. E. 1972, *Handbuch der Altertumswissenschaft* 1:7, *Greek and Roman Chronology*, Munich: Beck.

Schmidt, I F. 1865, "Über die Dämmerung", *Astronomische Nachrichten* 63, No. 1495.

Shaw, J. W. 1977, "The Orientation of the Minoan Palaces", in *Antichità Cretesi: Studi in Onore di Doro Levi* 1, Catania, p. 47.

Siedentopf, H. 1941, "Neue Messungen der visuellen Kontrastschwelle", *Astronomische Nachrichten* 271, 193.

ThucyCatas, 5th cen., *Loeb Classical Library* 108, *History of the Peloponnesian War*, ed. G. P. Goold 1928, reprinted 1991, Cambridge MA: Harvard Univ. Press.

Ventris, M. & Chadwick, J. 1973, *Documents in Mycenaean Greek*, 2nd ed. (Cambridge: Cambridge Univ. Press)

van der Waerden, B. L. 1974, *Science Awakening 2, The Birth of Astronomy*, Leyden: Noordhoff International Publishing.

28

Scratching the Surface of Astronomy in the Land of the Bible: Archaeology, Texts, and Astronomy

Sara L. Gardner
University of Arizona
Tucson, AZ

The use of astronomy by the people in ancient Israel is a subject that has yet to be explored in biblical or archaeological scholarship with the exception of J. Glen Taylor's monograph *Yahweh and the Sun* (1993). Taylor's approach is innovative because he uses texts, artifacts and archaeology, and approaches a new subject for this field. Cult complexes were oriented to the eastern horizon as early as the Neolithic period (5500–4500 BCE). Celestial images appeared in paintings and on artifacts in the Chalcolithic period (4500–3300 BCE). Uniquely from the Middle Bronze Age in northern Israel a solstice site was found in the Golan Heights that will be discussed as well as the elements of high places (*bamot*), a particular type of cult installation. In the Early Bronze Age (3300–2000), Middle Bronze Age (2000–1550 BCE), Late Bronze Age (1550–1200 BCE) and in the Iron Age (1200–587/6 BCE), temples oriented in the general direction of the east. Biblical and extra-biblical texts recorded the importance of astronomy by specifically condemning the places and people who "worshiped the Host of Heaven." This survey will cover approximately 4,000 years, and its primary focus is to apprise the reader of the plethora of evidence relating to astronomy and to establish the importance of investigating the use of astronomy by the people of ancient Canaan and Israel.

* * *

Introduction

The Old Testament lands of Phoenicia, Israel, Judah, Amurru, Ammon, Moab and Edom-Canaan in ancient times, and Syria-Palestine in modern times-extended from modern day Syria through Lebanon to the Sinai in the south, and west of the Jordan River. The ancient occupants shared similarities with the Mesopotamians and Egyptians; nations they traded with and/or were governed by. These similarities were reflected in their pottery, customs, social structures, and subsistence patterns throughout their history, from the Neolithic period through the Iron Age. Therefore, it is not surprising that the religions of the biblical lands embraced a belief system or religion that was similar to their neighbors.

Mesopotamian and Egyptian religions deified the sun, moon, and stars, and their astronomer-priests observed the movements of their gods and aligned them to seasonal changes (Whitefield 1995: 12–18). Early observations of the celestial sky laid a foundation for mathematical astronomy, but modern scholarship focuses on the written calculations that recorded astronomical observations with minimal studies on temple orientation except for the Pyramids of Egypt (Krupp 1984: 289–320), In Syria-Palestine temple orientation is commented on in generalities, and only one solstice site has been identified (see below). Scholarship ignores the astronomy of Syria-Palestine, when such studies would provide data for a critical understanding of the calendars and religions of the Canaanites, pre-Israelites, and Israelites. The following statement attests to modern scholarship:

> One may assume that the ancestors of Israel and the early Israelites themselves followed some sort of calendar (or calendars), but the extant sources do not permit one to determine what its (their) nature may have been (Vanderkam 1992: 814).

The statement implies that there is no way to discover what system the Israelites and Canaanites (pre-Israelites) used to organize their calendars and schedule their sacred events, when we have enough evidence to speculated on the type of calendar that they used. Prior to the Babylonian exile (587/6 BCE) the Israelites commenced their year in autumn with the "feast of ingathering (harvest)" which was at the "turn" or at the end of the year[1] (Ex 34:22; Ex 23: 16; Vanderkam 1992: 817). An extra-biblical calendar text from Gezer (ca. 925 BCE) listed twelve periods that were divided according to the agricultural seasons:

Two months of gathering,
Two months of sowing,
Two months are late planting;
One month of gathering (cutting) flax,
One month of gathering (cutting) barlies,
One month of gathering (and) all;
Two months of pruning,
One month of summer.

(Borowski 1987:88)

Agrarian societies based their calendars on the movement of the sun, moon, and stars, and used that movement to announce a new month or season. For the

1. Both "turn of the year" and "end of the year" are used in the Hebrew Bible to describe the end of the year that followed the harvest.

Israelites, that importance of the heavens is conspicuous; El, the god of Abraham, Isaac and Jacob, created "... lights in the expanse of the heavens to separate the day from the night, and let them be for signs and for seasons, and for days and years..." (Genesis 1: 14),[2] and then, created the sun and moon (Genesis 1: 16). The "lights in the expanse of the heavens" as "signs and seasons" described stars that marked the movement of the celestial sky in rhythm to seasons of the earth such as the "feast of ingathering," and emphasized their role in the marking of "days and years." The calendar text from Gezer divided the year into lunar units identified by seasonal activity: harvest, planting, late planting, hoeing, feasting, vine tending, and summer fruit. The above demonstrates that biblical and extra-biblical texts verify that the people of Canaan were cognizant of the movement of the sun, moon, and stars.

I believe that the archaeological evidence exists to support an argument for astronomical observation and that astronomy was important to the people of Syria-Palestine. The following will survey cult sites and artifacts from the Neolithic period to the Iron Age cult sites and artifacts. Biblical texts will be used to describe high places and establish them as candidates for observation sites; and temple design will help to determine the importance of orientation.

The Late Neolithic Period–Early Bronze Age (5500–2250 BCE)

During the Neolithic period, agriculture, animal domestication, and new social structures were introduced, and in the Early Bronze Age, metallurgical technology advanced (Mazar 1992a: 35–150). These innovations presented major changes in the way humans subsisted, organized their society, and interacted with the larger world around them.

A Neolithic Period Cult Site

The majority of sites found in the Neolithic period (5500–4500 BCE) were agricultural villages in the fertile zones and seasonal camps in the deserts; pottery and long-distance trade were introduced at this time (Mazar 1992a: 49–58). Another innovation was the delineation of sacred space or cult sites.[3] A large cult complex was found in the 'Uvda Valley near Timna in the Negev Desert of southern Israel (Yogev 1993: 1493–95). The complex (11 × 12 m) was oriented in general N/S/E/W directions with a cult area situated in the westernmost corner[4] (Figure 1). The corner room had a group of sixteen pointed standing stones,[5] and east of the complex is a row of carnivores and

---

2. The New Oxford Annotated Bible with the Apocrypha (1991) will be used unless otherwise specified.

3. Open-air sanctuaries such as the one in the 'Uvda Valley are cult sites that use courtyards or other delineated unroofed areas to conduct religious ceremonies (Mazar 1992b: 162).

4. The N/S/E/W directions used in following discussion are based on the directions used in the published excavation reports or subsequent articles. The reports generally do not state if the directions are based on True North or Magnetic North nor are any astronomical data available at this time. Therefore, at best the following discussion is subjective and speculative.

5. These small upright stones ranging from 20–30 cm appear to be symbolic rather than functional (Yogev 1993: 1494).

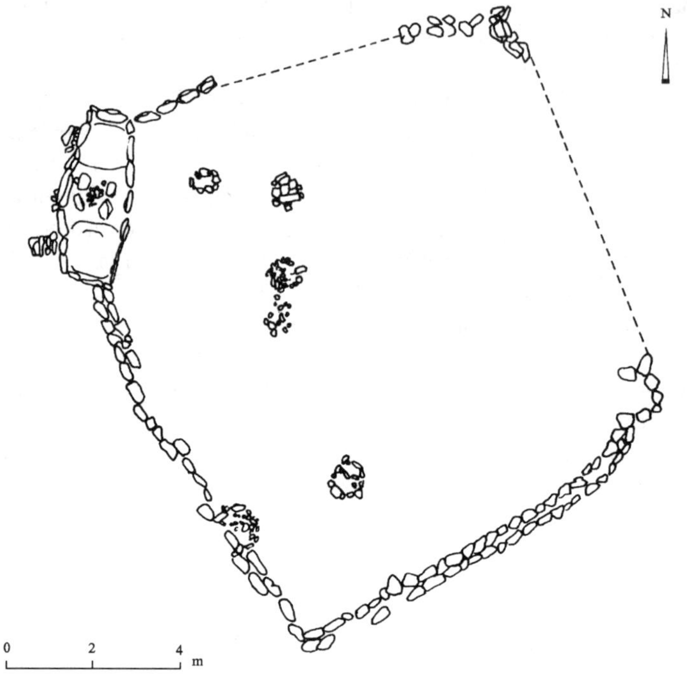

Figure 1. Courtyard of ʻUvda Valley Neolithic cult complex.

antelopes whose shapes are outlined with flat upright stones (Figures 2 and 3). They faced in an eastern direction to low lying hills. Radiocarbon dates place this complex in use from 4610 ± 90 and/or 4450 ± 200 BCE. Orienting a cult room, temple and/or temple complex to the eastern horizon became a main design feature in the following Chalcolithic period, Early Bronze I, and Early Bronze II (4500–2700 BCE), a period that spanned almost 2000 years.

The Chalcolithic Period and Early Bronze Age Cult Sites

Chalcolithic and Early Bronze temples were found at Megiddo in northern Israel, at Teleilat Ghassul near the Jordan River in Jordan, and at Gilat in southern Israel. Their cult buildings were designed over a broadhouse plan; that is, a long rectangular room with the entrance in one of the long sides and a cult niche where the cult object was placed in the west wall (Figure 4). According to A. Kempinski (1992a: 56) the object caught the rays of the rising sun through the opposing east entrance. This design continued until the latter part of Early Bronze II (3000–2700 BCE), when the temple design changed to a rectangular building with an entrance on the short side and was divided into two rooms, a hall and an inner sanctuary. The change in the temple plan was accompanied with the addition of two columns in front of the entrance; at Megiddo the temples were no longer oriented to the east (Kempinski 1992a: 58).

At Megiddo a round open-air platform or "high place" appeared next to the temple in Early Bronze III (2700–2300 BCE; Figure 5). It was on the southeast side

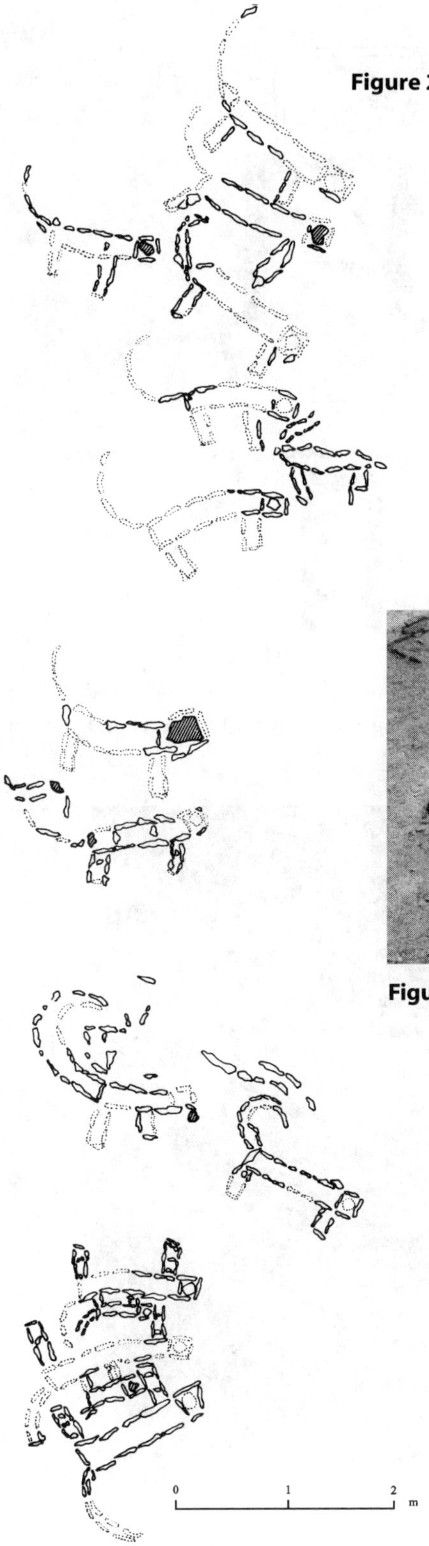

Figure 2. (left) Stone outlines of animals NE of courtyard.

Figure 3. (above) Closeup of stone outline of a scorpion.

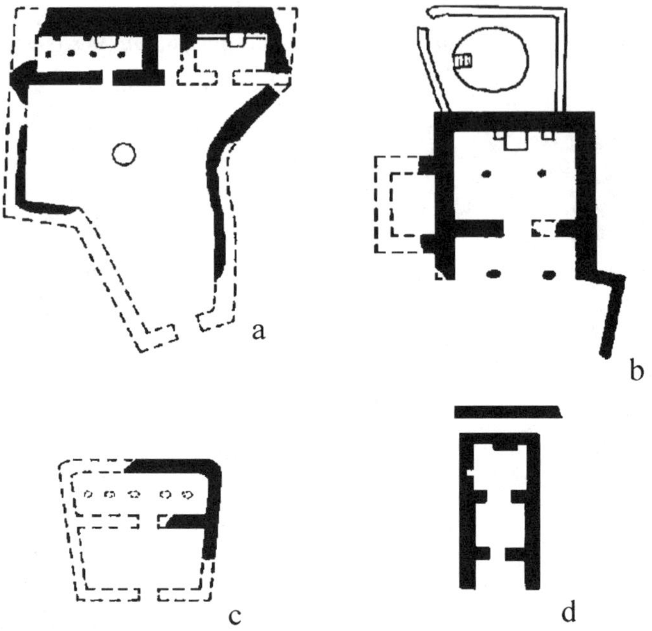

Figure 4. Chalcolithic period and Early Bronze Age temple plans found in Israel and Jordan.

Figure 5. Megiddo *bamah* (2700-2300 BCE).

of the Early Bronze II temples. The excavated remains were 8 m in diameter and 1.4 m high; the original height was unknown. It was surrounded by a wall with an entrance in the southeast corner, and was accessed by stairs on the east. During the same period it was enlarged, and its courtyard was attached to the temple (Kempinski 1992c: 174). Today, the horizon as seen from the platform is punctuated with two mountains, Mount Tabor and Mount Moreh. Mount Tabor is the place where the sun rises at the June solstice today (personal observation). Although Megiddo may have been used for astronomical observation, the site of Rujim el-Hiri in the Golan Heights northeast of Megiddo was definitely a solstice site during periods contemporary with Megiddo.

Rujim el-Hiri is a megalithic Bronze Age complex located near the Sea of Galilee, and has been identified as a solstice site by Y. Mizrachi (1992). The site was comprised of a massive central stepped stone tumulus (ca. 20 m in diameter and 4.5 m high) that contained a burial chamber. The tumulus was located near the center of the structure and was surrounded by four concentric stone walls. The shape of the complex was a slightly flatten circle which was 155 m in diameter north to south, and 144 m in diameter east to west (Figure 6). The outer wall had entrances on the northeast and southeast. In the center of each entrance was a megalith with two additional megaliths east of the complex. With the help of A. Aveni, Y. Mizrachi determined that "if one was to stand at the center of the complex during the June solstice sunrise of 3000 BC, the first gleam of sunrise would appear at the center of the Northeast entryway in Wall 1. Given the same scenario for the same observer

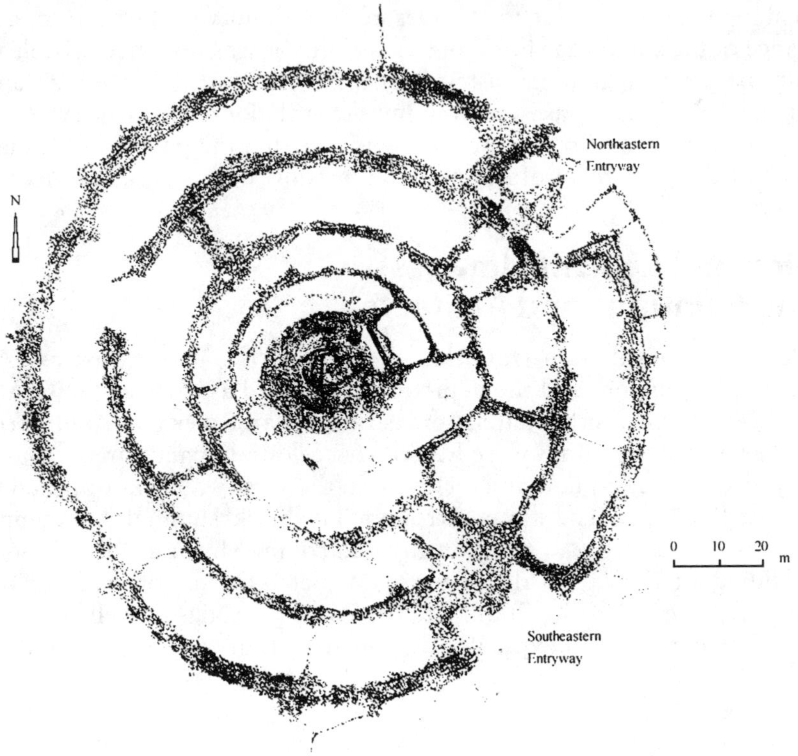

Figure 6. Rujim el-Hiri complex (Middle Bronze Age).

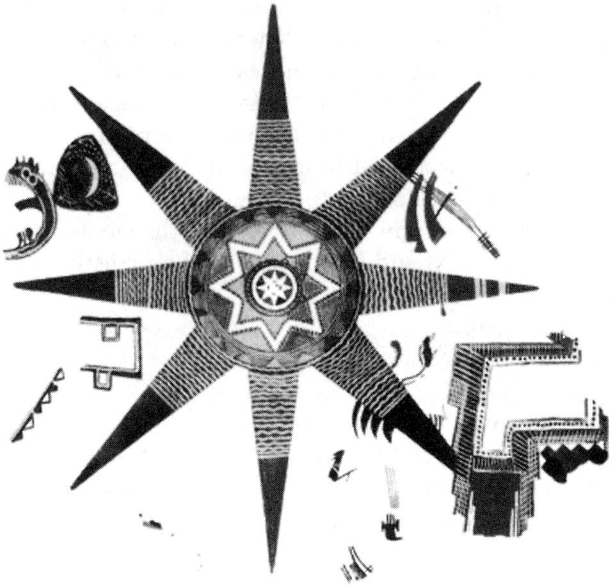

Figure 7. Teleilat Ghassul star (Chalcolithic period).

in a post 2000 BC date, the first gleam of sunrise would appear way-off from the center of the same entryway"[6] (Mizrachi 1992: 114–15). The apparent movement of the sun along the horizon due to changes in the obliquity of the ecliptic caused the appearance of the sun on the horizon to shift until it was no longer visible through the northeast entryway and the eastern megaliths. The author suggested that further studies might reveal more precise astronomical correlations (Mizrachi 1992: 112–18). Nonetheless, Rujim el-Hiri may be vital evidence that the people of Canaan knew how to chart the movement of the sun, and possibly other celestial objects.

Religion and Celestial Images: Walling Paintings and Artifacts

Celestial images demonstrate that the people of Canaan were aware of astronomical phenomena, and the depictions appeared as early as the Chalcolithic period at Teleilat Ghassul situated across the Jordan River from southern Israel. Several plaster wall paintings were found with celestial images. In one building an eight-rayed star (1.84 m in diameter) was on the west wall that opposed an eastern entrance (Figure 7). The star is surrounded by masks, animals, and temples and testify to the cultic nature of the painting and room (Mazar 1992a: 75–6). In another building, on the west wall, a painting depicted the sun or moon rising above mountains (Figure 8; North 1960: Plate II). These paintings as well as other paintings from Teleilat Ghassul explicitly demonstrate that celestial phenomena were

---

6. The map directions for Rujim el-Hiri were based on True North, but data on azimuths was not included in the 1992 dissertation by Y. Mizrachi. The dissertation referred to alignments, celestial bodies/phenomena without specific data (Mizrachi 1992: 112–118).

Figure 8. Teleilat Ghassul rising luminary (Chalcolithic period).

observed and recorded, but other evidence exists as well. Suns and stars were represented in eight-, seven-, and five-rayed figures on incised pebbles, pendants, and ossuaries as well as permanent features such has the paved area around an altar at Megiddo from Early Bronze I (3300–3000 BCE; Elliot 1977: 12).

A large paved open area, which sloped to the east towards Mount Tabor and Mount Moreh (see above), was the cult center at Megiddo, and at its center was a stone altar that contained a Neolithic female figurine (Loud 1970: 54). Thirty-six of the paving stones around the altar were incised with a variety of human, animal figures, and geometric shapes; three had astral symbols. First, a figure with raised arms and wearing a decorated belt was shown holding a lyre in the left hand, and a crescent moon rose to his right (Loud 1970: Pl. 273: Drawing 5; Figure 9). Second, a star appeared to the right of three figures (Loud 1970: Pl. 273,: Drawing 6;

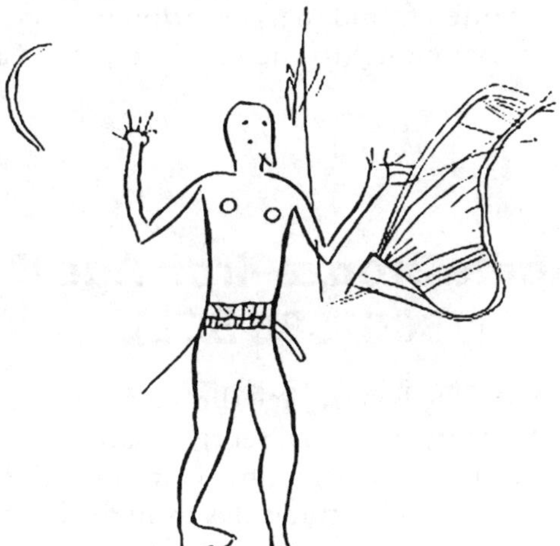

Figure 9. Drawing from the incised pavement stones at Megiddo; figure with harp and crescent moon (3300 BCE).

Figure 10. Drawing from the incised pavement stones at Megiddo; three figures around a star (3300 BCE).

Figure 10), and third, a star was incised with another geometric shape, an irregular rectangle (Loud 1970: Pl. 282: Drawing 35; Figure 11). Unfortunately, the images from the paving stones were poorly drawn and need a closer examination to see if more symbols or other indicators are present. Nonetheless, the Teleilat Ghassul paintings, paving stones and other artifacts contain important representations of astral images, and most often found in association with cult centers which indicates the importance of astronomy to the religion of the Canaanites and pre-Israelites.

Middle Bronze–Iron Age II (2300–586 BCE)

In Syria-Palestine the Middle Bronze Age through Iron Age II were periods of increased population, settlement, and social reorganization under a centralized authority. There was a significant degree of cultural continuity during this period, and many of the religious and technological traditions from the Bronze periods persisted (Dever 1990). Early in the Iron Age the Israelites emerged as a nation under Saul, David, and Solomon. After the death of Solomon, the nation split into Judah in the south, and Israel in the north. Conflict arose over issues of sovereignty, governmental policies, and religious beliefs. In the 10th and 9th centuries BCE written ac-

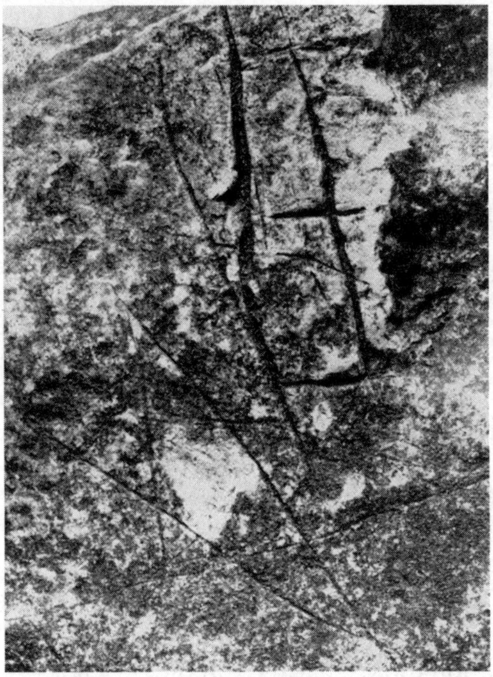

Figure 11. Drawing from the incised pavement stones at Megiddo; star and geometric shape (3300 BCE).

counts began to record political and religious conflicts and from these biblical and extra-biblical accounts the foundations of Israelite institutions are extrapolated.

Biblical References and Their Implications

The biblical texts described two types of cult complexes: "high places" such as the one at Megiddo from the Early Bronze period, and temples. A high place or *bamah* was one element of a complex whose major cultic elements also included standing stones (*massabot*), and sacred poles or pillars (*asherim/Asherah*). Literary examples testified to a concurrent use of these elements. "...They set up for themselves standing stones (*massabot*) and sacred poles (*asherim*) on every high hill and under every green tree; there they made offerings on all the high places (*bamot*)..." (II Kings 17: 10–11). "...They also built for themselves high places (*bamot*), standing stones (*massabot*), and sacred pillars (*asherim*)..." (I Kings 14: 23). Biblical texts indicated that the sacred poles were trees or were made of wood (Deut 16:21, II Kings 18:4, 23:14–15; Day 1992: 486). References to high places, standing stones, and sacred poles occurred repeatedly throughout the two histories of ancient Israel in the Old Testament: I & II Kings and I & II Chronicles.

The "high place" was an open-air cultic installation that consisted of a raised platform on which various cultic activities were performed (Barrick 1992: 196–200; Mazar 1992a: 350). In the early Iron Age the Israelites sacrificed on the high places (I Kings 3:3), which were situated on mountains (I Kings 11:71; Jer 17:3), hills (I Kings 14:23; II Kings 16:4), in valleys (Jer 7:31; 32–:35), at the top of the street (Ez 16:

24–25, 311), and in all the towns (I Kings 17: 9). They were found throughout Judah and Israel at Gibeon, Dan, and Bethel. Furthermore, Micah asked "What is the high place of Judah? Is it not Jerusalem?" (Micah 1:5). Ahaz, the king of Judah (8th century BCE), built high places and erected altars to Baal and Asherah and "... worshiped all the host of heaven and served them" (II Chr 33:3). His son Hezekiah (715–687 BCE) sought to reform the Yahwistic religion of Judah and centralize the cult in Jerusalem. As part of his reform he destroyed the high places, shattered the standing stones, and cut down the sacred poles (II Kings 18: 4). During his reign Josiah of Judah (640–609 BCE) also initiated a religious reform in the name of Yahweh. In the periods prior to Josiah's reform the images of the sun were used in temple worship, because the biblical text informed the reader that Josiah "... removed the horses that the kings of Judah had dedicated to the sun at the entrance of Yahweh... then he burned the chariots of the sun with fire..." (II Kings 43:11). He killed the priests of the high places in Judah and the area surrounding Jerusalem and "those who burned incense to Baal, to the sun and to the moon and to the constellations and to all the host of heaven" (II Kings 23: 5). Furthermore, biblical texts attested to knowledge of a solstice, and lunar standstill:

> Sun, stand still at Gibeon,
> and Moon, in the Valley of Aijalon.
> And the sun stood still, and the moon stopped,
> Until the nation took vengeance on their enemies
> (Jos 10: 12b–13a)[7]

Gibeon was a biblical "high place"; the site has been identified by modern day archaeology, and the Valley of Aijalon has not been firmly identified, in other words, today archaeologists are not in agreement if the designated location of the valley is correct. High places were referred to approximately 100 times in the Old Testament, and 80% of the references link them to cultic acts (Barrick 1992: 197).

High Places, Standing Stones and Sacred Poles/Pillars

"High places" with standing stones were prominent in the Middle Bronze Age throughout southern Canaan and found in rural and urban contexts. The round altar at Megiddo from the Early Bronze Age was rebuilt in Middle Bronze II (2250–1750 BCE). Numerous standing stones were added (Mazar 1992a: 178–180). At Nahariya a similar high place was attached to a temple and a small platform (5.5 × 5 m) was built to the north and a row of standing stones were added east of the platform (Kempinski 1992b: 174). Another high place was found on the summit of Mount Ebal, where a large platform with a ramp approach faced the eastern horizon and to the mountains (Mazar 1992c: 294).

7. The first two lines of Joshua 10: 12b have been identified as the original poem. The next two lines were a later interpretation of the first two lines by a second writer or editor. The final lines that refer to the astronomical event are: "Is this not written in the Book of Jashar? The sun stopped in midheaven, and did not hurry the whole day" (Jos 10: 13b) were added still later by a third writer or editor in an attempt to explain the original and second set of lines (Miller and Tucker 1974: 84). From an astronomical perspective the first two writers were aware of the solstice and lunar standstill, and the third writer was not.

Figure 12. Standing stone at Shechem (1450-1200 BCE).

Open cult sites from Iron Age I were found at Hazor in northern Israel, Arad in southern Israel, and in the Samarian hills of central Israel. The cult shrine found in the Samarian hills was 20 m in diameter with a standing stone to the east of the paved area (Mazar 1992a: 350–51; 1992b: 174; 1992c: 292–93). In the 10th century BCE Jeroboam, the new king of Israel, built a cult center at Tel Dan in northern Israel (I Kings 12: 3 1). South of the main structure was a walled open-air platform; the walls opened to the east and south, and the platform had the bases for two columns to the north for sacred poles or pillars (Mazar 1992b: 184–85).

"High places" with standing stones were found at Nahariya, Megiddo, and Rujim el-Hiri; standing stones were found in pairs or, in the case of Gezer, a group of ten. A single standing stone was also used in a cult complex at Shechem, which dated to 1450–1200 BCE (at the end of the Late Bronze period). It was about 1.5 m width, and a little over 1 m thick (Figure 12; [Campbell 1993: 13–52]). Another single stone was found at an open-air shrine at Tell el-Farah (N), that is biblical Tirzah (Mirochedji 1993: 438). The group of ten standing stones from Middle Bronze II (2250–1550 BCE) found at Gezer ranged in height from ca. 1.25–4.0 m high (Figure 13). They appeared to be aligned in north/south direction with an altar on the western side of the alignment, which suggests an eastern face (Dever 1992: 1000). To date no viable function of these stones has been suggested, although, no one doubts that they had a cultic function.

Celestial Images: Standing Stones, Sacred Poles and Stars

Celestial images found on decorated stele, altars, and plastic artifacts from Hazor suggest how standing stones, or perhaps, sacred poles/pillars (*asherim*) or both, functioned. On the southwest side of the tell at Hazor a temple yielded two statues and ten stelae (ca. 33–101 cm high; Mazar 1992a: 254). The statue on the right was a seated figure with a crescent moon engraved on his chest identifying him as a moon god

Figure 13. The Standing Stones at Gezer (1600 BCE).

(Gonen 1992: 253). The statue on the left was a lion, which was one of the symbols of Asherah, the goddess of Israel and Judah. Nine of the ten stelae were undecorated, but the central stele depicted a disc cradled in a upward turned crescent with two knobs underneath. The emblem rose above two upright arms and hands, which represented the consort of the moon god according to Y. Yadin (Figure 14; Yadin 1970: 222, Gonen 1992: 232). A temple on the north end of the mound contained a basalt statue of a god with the star emblem on the chest and an altar with two upright stones or poles which have the same emblem rising above them (Figure 15; Gonen 1992: 228). There is a striking similarity between the upright hands from the central stele above and the upright poles on the altar: each were postured stiffly, and have a celestial emblem rising above (Figure 16; Mazar 1992a: 286). The depictions demonstrate that standing stones and/or sacred poles may have been used to mark the path of the rising celestial body, the sun, moon, or star.

Figure 14. Israeli archaeologist, Yigael Yadin, with the stelae from Hazor (Late Bronze Age: 1550 BCE).

Figure 15. (left) Altar from Hazor (Late Bronze Age).
Figure 16. (right) Hazor Stele (Late Bronze Age).

Temple Orientation

One practical method that can be use to determine whether or not an investigation of astronomy was a technology that the Canaanites and Israelites practiced is to begin by surveying the orientation of cult buildings, the placement of entrances, and/or cult niches or statues. An examination of the interaction that the niches or statues or standing stones may have had with celestial phenomena such as the rays of the sun entering an entrance or striking sacred objects would enable the scholar to determine if further studies were viable or if they would be productive. Yet, such studies are lacking, and the relationship between direction and astronomical phenomena is limited to an occasional comment with the exception of J. Glen Taylor's monograph *Yahweh and the Sun: Biblical and Archaeological Evidence for Sun Worship in Ancient Israel* (1993). He focused on temples from the Iron Age and Hellenistic period, and used textual and art historical evidence to establish that the Israelites were worshiping Yahweh as a sun god. Unfortunately, his architectural data for orientation was drawn from excavation reports and was therefore flawed (see note 5). Excavation reports are not uniform in the recording of map directions; sometimes the direction used is True North, Magnetic North, or plotted randomly from the confines of the dig house. J. Glen Taylor assumed that the direction on site plans is either True North and Magnetic North and adjusted accordingly: "...Calculations are based on the assumption that the compass point in the published report...

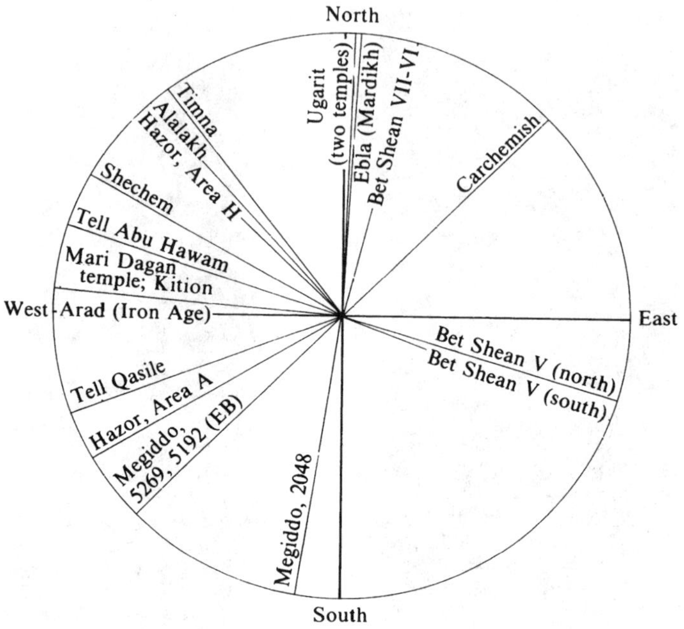

Figure 17. Compass drawings of temple orientations.

points to True North. If the compass point reflects magnetic north, the angle of the temple would be..." (Taylor 1993: 266, 268, 270, 272, see Footnotes). In his summation on solar orientation, he concluded: "...while orientation to the sun is not always characteristic of sun temples, the lack of clear orientation to the sun on the part of the Yahwistic cultic structure attested provides no firm indication of the presence of solar elements within the cult..." (Taylor 1993: 86).

G. R. H. Wright called for studies on the orientation of temples because he found this information sadly lacking (Wright 1985: 274). One study that surveyed the orientation of selected temples from Mari in Syria in the north to Timna in the Negev Desert in the south to Kition on Cyprus in the Mediterranean, through the Middle Bronze, Late Bronze and Iron Ages, diagramed data on temple orientation from excavation reports (Figure 17; Mazar 1992b: 178), A. Mazar states "...the diagram indicates that the orientation of temples in Israel follows no consistent pattern. The majority of the temples...point to the western part of compass rose... (but) no set rule can be established...." (1992: 187). The above statement either was erroneous or confused, when it stated that the direction of the temples was to the west, their entrances generally point on the eastern horizon.

Cult Sites from the Chalcolithic Period to Late Bronze Age

Temples were designed over the broadhouse plan in the Chalcolithic and Early Bronze Age; in the latter part of the Early Bronze Age a shift in temple design occurred and the longhouse plan was used (see above [Kempinski 1992: 54–58]). Temples continued to be designed over the longhouse plan with variations through the Late

Bronze Age. The open air sanctuaries were found throughout, and "high places" were used most prominently during the Iron Age, although they are mentioned in the Hebrew Bible in Middle Bronze Age context and we have one example at Megiddo (Mazar 1992b). Unfortunately, the number of excavated Iron Age temples are limited, and the primary reliance for their description comes from the biblical description of the Temple of Solomon. The detailed directions given for the placement of elements in relationship to north, south, east, and west point to the importance of orientation for this religious, national monument (II Kings 7: 15–22; I Kings 7: 23–39; I Kings 8: 64; Meyers 1992: 357–59). Ultimately, the archaeological reports provide orientation evidence for open-air sanctuaries, "high places", and temples, albeit flawed. Taking excavation reports or subsequent articles that include a direction arrow with the site or temple plan for well-known sites, the following statistics emerge (the directions should be in all cases considered to be general at best):

1. Neolithic to the Early Bronze Age
   5 cult sites were selected, and all entrances or important features were oriented east.
   100% of the selected sites were oriented east.
2. Middle Bronze Age
   11 cult sites were selected, and 9 entrances or important features were oriented east, and 2 entrances oriented southeast.
   82% of the sites were oriented east.
   18% were oriented southeast.
3. Late Bronze Age
   4 cult sites were selected: 2 entrances or important features were oriented east, 1 entrance opened southeast, and 1 entrance opened north.
   50% of the sites were oriented east, 25% southeast, and 25% north.
4. Iron Age
   3 archaeological cult sites and 1 biblical description (Solomon's Temple) were selected; 2 archaeological cult sites and the biblical cult site opened east; 1 site could not be determined.
   75% of the sites orient east, and 25% unknown.

In summation, 18 of the 24 selected sites oriented east, and 3 southeast; so 86% of the sites oriented east, 12.5% southeast, 8% north and 4% unknown. In response to A. Mazar's statement that "...the orientation of temples in Israel follows no consistent pattern...," the above statistics suggest that one could surmise that the direction east was important to the people of the southern Levant and the preference of the architects of Canaanite and Israelite temples. Consistency was present in the planning of cult sites, which were commonly oriented to the eastern horizon. Data need to be gathered to determine if a relationship between the cult sites and astronomical phenomena exists. Such data, which spans a period of about 5000 years, needs to be appraised critically in relationship to social, cultural, and technological changes.

Conclusion

I believe that astronomy was central to the religious beliefs of Canaan, and began no later than the Neolithic period, and by the Chalcolithic period is firmly entrenched technologically and religiously. The orientation of cult sites throughout the long history of Syria-Palestine suggests an interest in astronomical phenomena on the eastern horizon. The wall paintings at Teleilat Ghassul, designs on pebbles, pendants and ossuaries, and the stone pavement from Megiddo indicate that astronomy was important to the people and had a natural place in the religious belief system. Standing stones can be indicative of astronomical knowledge, and could have been used for data collection (see Ruggles 1988).

Scholarship on archaeoastronomy in Syria-Palestine is lacking, and when research is done it begins with mathematical astronomy (see Mizrachi 1992), or is based on inaccurate data (see Taylor 1993). The researcher is faced with another problem, the destruction of significant data by Hezekiah and Josiah and other Judean kings who later occupied the Levant (not mentioned) that would normally be found at "high places" with their standing stones and sacred poles. For example, the standing stone at the entrance of the temple at Shechem was broken, and its original height is unknowable (see Figure 12). Different questions need to be asked: foremost among them, did the Canaanites, pre-Israelites, and Israelites track the movement of the sun, moon and stars? Furthermore, did they mark the important events of the sun-the solstices and equinoxes as well as the lunar phases? If so, where and how did they take observations? What were the instruments of their observation system, and how were they used? Were the "high places," standing stones and sacred poles instruments used to collect data? Were the data then communicated to the populace through the temple and its rituals by telling or celebrating the time to plant, harvest, begin the new year, etc.?

"High places," open-air sanctuaries, temples and astronomy in the southern Levant can be better understood only through better research and the recognition that the ancient Canaanites and Israelites may have understood astronomy as did their neighbors in Mesopotamia and Egypt.

This is dedicated to Dr. Raymond White, University of Arizona, my mentor and greatest supporter, and without whom I have never written a dissertation on archaeoastronomy. Thank you. SLG

References

Barkay, G., "The Iron Age II–III," in *The Archaeology of Ancient Israel* (ed. A. Ben-Tor; New Haven: Yale University Press, 1992), 302–73.

Barrick, W. B., "High Place," in *The Anchor Bible Dictionary* (ed. D. N. Freedman; New York: Doubleday, 1992), vol. 3, 196–200.

Borowski, O., *Every Living Thing: Daily Use of Animals in Ancient Israel.* (Walnut Creek: AltaMira Press, 1987).

Campbell, E., "'Shechem," in *The New Encyclopedia of Archaeological Excavations in the Holy Land* (ed. E. Stern; Jerusalem: Israel Exploration Society & Carta, 1993), vol. 4, 1345–54.

Day, J., "Asherah," in *The Anchor Bible Dictionary* (ed. D. N. Freedman; New York: Doubleday, 1992), vol. 1, 483–86.

Dever, W. G., "Gezer," in *The Anchor Bible Dictionary* (ed. D. N. Freedman; New York: Doubleday, 1992), vol. 3, 998–1003.

———. *Recent Archaeological Discoveries and Biblical Research* (Seattle: University of Washington Press, 1990).

Elliot, C., "The Religious Beliefs of the Ghassulians ca. 4000–3100 B.C.," *Palestine Exploration Quarterly* 109 (1977), 3–25.

Gonnen, R., "The Late Bronze Age," in *The Archaeology of Ancient Israel* (ed. A. Ben-Tor; New Haven: Yale University Press, 1992), 211–57.

Kempinski, A., "Chalcolithic and Early Bronze Age Temples," in *The Architecture of Ancient Israel* (ed. A. Kempinski; Jerusalem: Israel Exploration Society, 1992a), 53–59.

———. "Fortifications, Public Buildings, and Town Planning," in *The Architecture of Ancient Israel* (ed. A. Kempinski; Jerusalem: Israel Exploration Society, 1992b), 68–80.

———. "The Middle Bronze," in *The Archaeology of Ancient Israel* (ed. A. Ben-Tor; New Haven: Yale University Press, 1992c), 159–210.

Krupp, E. C. (ed.), *Archaeoastronomy and the Roots of Science* (Boulder: Westview, Inc., 1984).

Loud, G., *Megiddo II: Seasons of 1935–39* (Chicago: University of Chicago Press, 1970).

Mazar, A., *Archaeology of the Land of the Bible* (New York: Doubleday, 1992a).

———. "Temples of the Middle and Late Bronze Ages and the Iron Age," in *The Architecture of Ancient Israel* (ed. A. Kempinski; Jerusalem: Israel Exploration Society, 1992b), 161–87.

———. "The Iron Age I," in *The Archaeology of Ancient Israel* (ed. A. Ben-Tor, 258–301, Yale University Press, 1992c).

Meyers, C., "Temple, Jerusalem," in *The Anchor Bible Dictionary* (ed. D. N. Freedman; New York: Doubleday, 1992), vol. 6, 351–69.

Miller, J. M. & G. M. Tucker, *The Book of Joshua* (Cambridge: Cambridge University, 1974).

Miroschedji, P. de, "Farah, Tel el-," in *The New Encyclopedia of Archaeological Excavations in the Holy Land* (ed. E. Stern; Jerusalem: Israel Exploration Society & Carta, 1993), vol. 2, 432–38.

Mizrachi, Y., *Rujm el-Hiri: Toward an understanding of a Bronze Age megalithic monument in the Levant* (Ph.D. diss., Harvard University; Ann Arbor, Michigan: UMI Dissertation Services, 1993).

The New Oxford Annotated Bible with Apocrypha, New Revised Standard Version (eds.B. M. Metzter, and R.E. Murphy; New York: Oxford University Press, 1991).

North, R., *Ghassul 1960* (Excavation Report, Rome: Pontificio Instituto, 1961).

Ruggles, C. L. N. (ed.), *Records in Stone: Papers in Memory of Alexander Thom* (Cambridge: Cambridge University Press, 1988)

Taylor, J. G., "Yahweh and the Sun: Biblical and Archaeological Evidence for Sun Worship," in *Ancient Israel* (Sheffield: *Journal for the Study of the Old Testament, 1993*).

Vanderkam, J. C., "Calendars: Ancient Israelite and Early Jewish," in *The Anchor Bible Dictionary* (ed. D. N. Freedman; New York: Doubleday, 1992), vol. 1, 814–820.

Whitefield, P., *The Mapping of the Heavens* (London: The British Library, 1995).

Wright, G. R. H., *Ancient Building in South Syria and Palestine* (Leiden-Köln, 1985).

Yadin, Y. "Symbols of Deities at Zinjirli, Carthage and Hazor," in *Near Eastern Archaeology in the Twentieth Century* (Garden City: Doubleday & Company, Inc., 1970) pp.199–231.

Yogev, O. "'Uvda Valley," in *The Encyclopedia of Archaeological Excavations in the Holy Land* (ed. E. Stern, vol. 4, 1493–95, Jerusalem: Israel Exploration Society & Carta, 1993).

29

Bedroom Politics and Celestial Sovereignty

E.C. Krupp
Griffith Observatory
Los Angeles, CA

Celebrated at the vernal equinox, the ritual intimacy of the Hittite king and queen was a seasonal exercise in bronze age fertility magic in the thirteenth century B.C. The ceremony was associated with an open-air shrine at Yazilikaya, in central Turkey. It is located on a hillside just two miles northeast of Hattusha, the Hittite capital. The entire assembly of gods depicted in reliefs on the sanctuary's rocky outcrops is there to witness the nuptials of Teshub, the Weather God of Heaven, and Hepat, the Mother Goddess and Sun Goddess of Arinna. On earth, the Hittite royal couple performed their own version of the marriage rites, but the function of the ceremony went beyond revitalizing the productivity of the land. Although the imagery at Yazilikaya is dominated by the themes of divine sacred marriage and the renewal of fertility, the fundamental purpose of the sanctuary is ratification of the king's franchise. Yazilikaya is essentially a rock art site, the kind of place usually associated with less centralized societies. Yazilikaya exploits the natural setting, however, and its use by an architecturally sophisticated civilization that erected palaces, temples, and libraries in the same neighborhood invites further inquiry. Accordingly, the site's potential as the symbolic center of the Hittite universe is explored.

* * *

Yazilikaya, a well-known rock-relief shrine (see Figure 1) near Hattusha, the Hittite capital, is associated with a seasonal ceremony of renewal that originated in Sumer at about 3500 B.C. Later known in Babylon as the *akitu*, the ceremony was an equinox New Year festival (Pallis), and it continued to be part of Mesopotamian royal ritual for at least 1200 years. Analysis of the content of the reliefs at Yazilikaya supports the presence of a similar tradition in ancient Anatolia among the Hittites in the thirteenth century B.C. (Bittel 1970, 107).

As experts in power politics and military expansion, the Hittites established a presence in Anatolia in 1900 B.C. By the seventeenth century B.C., they had consolidated their power in an imperial dynasty, and a century later even Babylon was invaded by them (Gurney 1980). As competitors, Egypt and Assyria defended their frontiers against the Hittites after 1450 B.C. The famous battle Ramesses II fought

Figure 1. Yazilikaya, a suburban shrine on the slopes northeast of the Hattusha, the Hittite capital, integrates carved reliefs and ancillary structures with the high natural pinnacles of the rocky hillside outcrop. Here, the view is northwest from the southeast zone of the site. The open space in the center of the picture is the south end of Chamber A. (Photograph E.C. Krupp).

at Qadesh, on the Orontes River in what is now Syria, is commemorated on the carved walls of the Ramesseum and Luxor Temple. Ramesses II built this mortuary temple near Thebes, the capital of Egypt during most of the New Kingdom era (1570–1070 B.C.).

Hattusha, the Hittite capital, and its ceremonial satellite Yazilikaya, are close to the village of Bogazköy, about 124 miles east of Ankara, in central Turkey. Fortifying their political center in a natural stronghold on a tributary of the Halys River, in 1520 B.C. the Hittite capital had massive ramparts (see Figure 2) that enclosed temples, palaces, storerooms, living quarters, and a library of 3000 inscribed clay tablets. The imposing walls and ornamented gates of Hattusha advertised political power and military strength.

Yazilikaya, however, is outside the city walls, about two miles to the northeast. On leaving the city for the isolated hillside sanctuary, it was first necessary to pass

Figure 2. Hattusha was protected by massive ramparts, and the gates into the city were ornamented with symbols of power. The Lion Gate, shown here, is on the west side of the city. It and much of what remains of the "upper town" is attributed to King Hattushili III.

through the hallowed burial grounds that flanked the route. Yazilikaya was reached by a road that actually began at the great ruined temple (Temple I) judged to have been dedicated to the Weather God of Hatti. The slope above the city is punctuated with a massive outcrop of exposed rock. Its high pinnacles are miniature mountains that enclose a complex of natural chambers and passages dominated by the two galleries where the reliefs were carved onto the rocky walls (see Figure 3). These sculptures inspired its current name, which in Turkish means "inscribed rock" (Bittel 1970, 91). In the main chamber, the sculpted surfaces depict a procession of figures that approach the central panel from both sides as the two halves of the parade wind around the rocks (see Figure 4). Hieroglyphic Luvian inscriptions accompanying these figures identify them by name as gods, and although the site is unmistakably a national Hittite shrine, the names of the gods are Hurrian. Linguistically unrelated to the Hittites, the Hurrians came from the kingdom of Mitanni in southeast Turkey, eastern Syria, and northern Iraq. They exercised a strong cultural influence across the Anatolian plateau, and despite the absence of an ethnic connection with the Semitic peoples of Mesopotamia, the Hurrians embraced some Mesopotamian traditions and transmitted them to their neighbors (Grayson).

The Hurrian names for the gods at Yazilikaya are apparently the product of direct Hurrian involvement in Hittite dynastic affairs. Although sponsorship of the reliefs is attributed to the Hittite king Hattushili 111 (1275–1250 B.C.), his queen, Puduhepa, was the daughter of a Hurrian priest. According to Hittite documents, she carried tablets listing the names of the gods to her new husband. Their son and Hattushili III's successor, Tudhaliya IV (1250–1220 B.C.), also commissioned Yazilikaya's sculpture, and he is depicted in the reliefs.

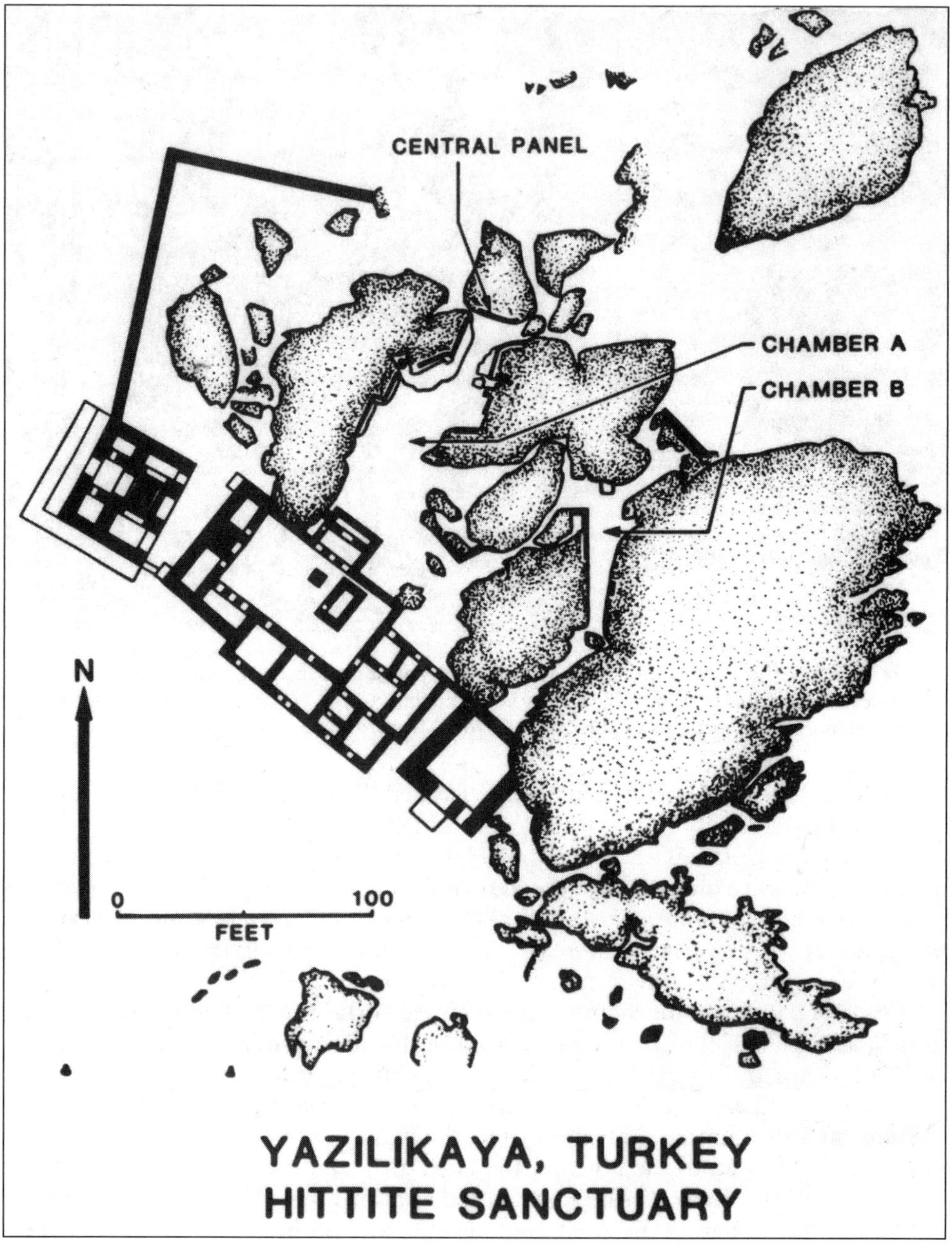

Figure 3. Access to Yazilikaya's natural enclosures was restricted by the construction of temples and walls. Most of the construction is south of the dominant crags of rock. The central panel of carved reliefs is located at the north end of Chamber A, which is dominated by a double procession of gods assembling to witness the seasonal nuptials of the Weather God and the Mother Goddess. The underworld connotations of Chamber B associate it with royal funerary traditions. (Griffith Observatory, Joseph Bieniasz).

Detailed analysis of other Hittite sources permit the identification of the pantheon on parade at Yazilikaya. The central panel is dominated by the weather and

Figure 4. Half of the divine procession in Chamber A winds around the natural walls of rock on the left and ends at the central panel, part of which is visible in the center of this view, on the lower face of a natural pyramid of rock. (Photograph E.C. Krupp).

storm god (see Figure 5), a personification of the fertilizing power of sky we know as Teshub, his Hurrian name. He is the Weather God of Heaven. The Babylonians called him Marduk and recognized his supremacy among the gods. In Assyria, he bellowed in the tempest as Adad. For Canaan, the lord of lightning and cloudburst was Baal, and in other parts of Asia Minor, thunder's hammer was held by Tarhun. Throughout the ancient Near East, he mated with Mother Earth and fathered the new life she seasonally brought to term.

The Hittites never met a god they didn't like, and Hittite state religion found places in the divine hierarchy for local variants of the universal gods. For that reason, the ranks of heaven did not include the Weather God of Heaven but local sponsors of the power of the storm like the Weather God of Hattusha, whose cult was linked specifically to the capital city.

The earth's capacity for seasonal rebirth is embodied in the mother goddesses of antiquity, and to the Hurrians, the mother of all earth was Hepat. As Teshub's bride, she coupled with him to deliver and parent the world's cyclic renewal. She faces Teshub at Yazilikaya as his queen and partner in the ceremony of Sacred Marriage (see Figure 6). All of the other gods have convened to witness and ratify a divine ritual that reinstates the sovereignty of the highest gods and the hierarchy of divine lineage through the renewal of the matrimonial bond between the divine rulers of heaven and earth. The syncretic character of Hittite culture enlarged the legion of gods into a huge family (Eliade 1978, 140) that was headed by the divine and cosmic couple, who partnered the fertility of the sky with the fecundity of the earth to perpetuate the world. These heads of the cosmic household are joined by Sharruma, their son and heir. He attends his mother and leads the retinue of female deities curving

Figure 5. The meeting of the Weather God of Heaven (left) and Hepat, the Mother Goddess and Sun Goddess of Arinna (right), is the focus of the central panel. Erosion is overtaking the reliefs in the wake of an ill-advised effort to create a replica of the carvings, a process which compromised the protective natural patina of the rock. (Photograph E.C. Krupp).

around the walls on the east side of the gallery. Sharruma and Hepat each stand upon the back of a lion, a symbol of sovereignty, and the first two goddesses that follow are supported by a doubled-headed eagle, a Hittite dynastic emblem. Teshub makes a pair of pedestals out of two deified mountains, Nanni and Hazzi, which emphasize his command of the summits where the clouds brew and the storms stampede. Bulls bellow like the thunder and were symbolically affiliated with the ancient storm gods. One of Teshub's two sky bulls also appear in the sacred marriage scene, and behind him the local incarnation of the rainstorm, the Weather God of Hattusha, seems to be performing the duties of best man. All along the west walls, male gods are lined up for the wedding. The only female in the troops is Shaushga, the planet Venus (see Figure 7). She is, however, a warrior goddess and is respected in the ranks of men. Other soldiers in the divine cortege include the grain god, the god of the waters, the moon god, the Hittite king deified with the name of the sun, two bull-men who stand on an earth symbol and hold up an emblem of the sky, and many more.

YAZILIKAYA, TURKEY

Figure 6. Both high rulers of the Hittite gods enjoy an elevated perspective. The Weather God of Heaven stands upon the bowed backs of a pair of deified mountains. As the sponsor of storms, the Weather God was a sky sovereign. His consort, Hepat, was no less influential and divine, and she is supported by a lion, an emblem of royal power, and the lion balances each paw on a separate peak. Smaller in stature than either partner in the ruling couple, the local divine agent of atmospheric power also occupies mountaintops on the far left. Sharruma, the son of the Weather God of Heaven and Hepat, stands on his own lion behind his mother, and he embodies the principle of royal succession. Two more goddesses follow him, and the double-headed eagle below their feet is a symbol of dynastic power. A pair of celestial bulls stampeded in the Weather God's tempests, and they both put in an appearance here. (Griffith Observatory, Joseph Bieniasz).

Conferred with solar privilege, Hepat, the Great Mother Goddess, was also regarded as the Sun Goddess of Arinna, one of the cities in the Hittite empire (Akurgal undated, 78). Why the Hittites solarized her remains unknown, but the title connects her with celestial power, seasonal transformation, and cyclical time. It put her on a heavenly footing with her husband. Although the real meaning of her solar aspect is lost, it remains consistent with the protocol of vegetation growth, the growth of flocks and herds, and the cultivation of orchards and fields. Partnered with the rain, the sun midwifes the pregnant earth.

The Sacred Marriage of the gods makes a cosmic matrimonial altar out of Yazilikaya, where the nuptials of the procreative power of the sky and the reproductive capacity of the earth are staged in the main gallery with an audience of gods. This principle of divine renewal was mimicked by the king and his consort in the Babylonian *akitu* ceremony at the New Year to commemorate the annual revival of the world. Performed at the vernal equinox, the ritual inaugurated the year with the imagery of fertilization and procreative renewal (Krupp 1991, 141). The ceremony actually originated in Sumer and was performed as seasonal ceremony of renewal by 3500 B.C. In any case, the king ritually mated with his partner in rites intended to revive time and the life of the land. As understudies for the storm god and the mother goddess, the king and queen reestablished their status as the energizers of the world.

Figure 7. The gods on parade at Yazilikaya include celestial spirits like Shaushga, the Hittite/Hurrian personification of the planet Venus. (Photograph E.C. Krupp).

We would be misled, however, to see in Yazilikaya nothing more than a royal shrine to vegetative fertility. The gods portrayed there and the ceremonies performed there were not just coaxing a little fecundity back into the world. Yazilikaya's rock-reliefs are regarded as the highwater mark of Hittite ideological art. Archaeologists judge it to be unique and call it "the most impressive of all Hittite religious structures" (Macqueen 123). Although other reliefs on living rock are known, nothing is as elaborate as Yazilikaya. The reliefs, for example, a Gâvur Kalesi, about 37 miles southwest of Ankara, also spotlight the divine royal family (see Figure 8). They are monumental, but they are not complex. Yazilikaya, on the other hand, is the Sistine Chapel of Hittite religious art, and like the *akitu*, it has a political dimension. In a bedroom demonstration of divine sovereignty, it is also a ratification of royal power. Among the gods, the Sacred Marriage culminates in a conjugal union that defines the hierarchy of cosmic power through divine lineage. The gods personified the revivifying forces of nature, and their pairing renewed world order. Hepat is the agent of world order and rhythmic renewal, and her acceptance of

Figure 8. The nuclear royal family of gods is also carved in heroic scale on the exposed rockface at Gâvur Kalesi, a Hittite fortified shrine and royal tomb. In purpose it may have paralleled Chamber B, at Yazilikaya, but the gods do not convene here to legitimize dynastic mandate through attendance at a Sacred Marriage. The two figures shown here probably represent the Weather God of Heaven and Sharruma, his designated heir. (Photograph E.C. Krupp).

Teshub's embrace advertises his license to her lap and his dominion over the gods. Hittite society, like the community of the gods, was feudal and rooted in rank and privilege. When the king and queen ceremonially coupled, they renewed far more than the fruit of the fields. Their copulation was political. Impersonating the gods, the king and his paramour renewed sovereignty through sacramental sex. Her caress of the king demonstrates the legitimacy of his line and endorses his bond with the ruler of heaven. In heaven and on earth, bedroom politics promoted celestial sovereignty. Engineering its status through analogy, Hittite royalty identified itself with the highest gods.

In conjugal embrace, the king and queen confirm they comprise a family, and royal family values—in heaven and on earth—establish the privilege of power.

There is more at stake in the bedroom than the procreative power of dynasties. Leadership is the real issue. Ideology validates leadership.

Additional affinities with kingship emerge from other reliefs at Yazilikaya. Closer to the entrance of the main chamber, a singular portrait of Tudhaliya IV balances him on mountaintops like the Weather God. By name, he is equated with the deified mountain, the sacred summit that brews the energizing fluid of life where earth touches heaven (Akurgal undated, 77). He is dressed for ritual like a priest (Alexander 97), and his outstretched arm supports a cartouche that symbolizes royal sovereignty. It is framed on the left and right by a symbol that looks a little like an Ionic column, and this double-column hieroglyph meant "Great King" (Akurgal 1973, 314). These "columns" support a winged pair of "sun" disks. Both are usually equated with the sun (Alexander 100) without explanation for the doubling. It is possible, however, that one of these celestial symbols represents Venus, and if so, the entire composition may share thematic elements with a symbolic tradition that reaches back to the Akkadian period of Mesopotamia (2360–2180 B.C.). A well-known Akkadian cylinder seal in the British Museum (Seal of Adda) includes the sun god, the goddess of Venus, the god of thundershowers, and the god of the waters with a mountain horizon profile in a scene thought to represent the world-renewing vernal equinox New Year sunrise and the celestial circumstances of the *akitu* ceremony (Krupp 1983, 204). The hieroglyph that normally appeared in the middle of this emblem was replaced with an image of the Mountain God (the Weather God of Hattusha, who is the local avatar of Teshub, the Weather God of Heaven) (Alexander 98; Akurgal undated, 77).

In fact, the purpose of the image of Tudhaliya IV in the main gallery may be documentation of his transformation into a god (Alexander 97). Because the Hittites regarded the king's death as his induction to the company of the gods, the depiction of Tudhaliya IV may incorporate a funereal allusion.

Divine patronage of the king is restated in the smaller natural chamber to the east of Yazilikaya's main gallery. Although the cast carved into the stone is smaller, part of the script is more explicit. Sharruma, the designated heir of the Teshub and Hepat, is an unambiguous personification of royal succession, and he is shown protectively embracing King Tudhaliya IV. The king, in fact, was the primary agent of the gods' will on earth (Macqueen 115). A cartouche is engraved above Sharruma's shoulder, and King Tudhaliya's cartouche is cut into the wall where the natural chamber meets a niche. Bittel (1970, 111) spotlighted the cartouche as significant location that deserved "special attention" in future excavations.

A peculiar relief to the left of Tudhaliya in Sharruma's protective custody and almost opposite a panel of 12 marching armed gods (see Figure 9) is known as the Sword God (see Figure 10), and it introduces additional references to the underworld and the dead in this area of Yazilikaya (Bittel 1972, 52). Resembling a giant upright sword with its point in the ground, the lion-ornamented hilt of the Sword God sculpture ends in the head of a god. Other Near Eastern texts associate sword offerings with the god of the underworld, and a Hittite text reinforces this connection with the declaration that the Weather God of Heaven deported eight unwelcome gods to the underworld by turning them into swords and sticking them in the ground. There is also other Near Eastern textual evidence linking "12 gods of the cross-roads" with the god of the underworld. Access to the smaller sanctuary is provided by a narrow tunnel of a crack that conveys the impression of entering the earth, and it is guarded by reliefs of two fierce, lion-headed, winged demons cut at the entrance. Human skeletal remains were recovered from the small gallery

Figure 9. A cavalcade of a dozen gods equipped with swords commands a comer at the main passage of Chamber B. These may represent the 12 gods of the crossroads and may refer as well to the underworld. (Photograph E.C. Krupp).

along with the bones of birds, at least one of which seemed to be part of a sacrificial offering. Integrating all of these clues, Bittel judged the smaller chamber of Yazilikaya to be "a temple of the dead, say the funeral temple for a king..." (Bittel 1970, 110.)

Divine legitimation of Hattushili III's usurpation of the Hittite throne from his nephew (Alexander 141) partly explains the intent of Yazilikaya's program of political art. The project was also congruent with the agenda of Puduhepa, Hattushili's Hurrian queen. She temporarily ruled the Hittite empire prior to the accession of her son Tudhaliya IV, and the formalization of the congress of gods with Hurrian names appears to reflect her influence. Finally, it was also in Tudhaliya IV's interest to put the ideological power of the shrine to work with a mandate to rule from the gods of creation and renewal (see Figure 11).

Cosmic approval of the king's authority is documented in a palace dedication prayer that reflects divine investment in the ruler.

Figure 10. Yazilikaya's "Sword God" relief also suggests an underworld and funerary aspect of Chamber B. The point of the blade punctures the earth. The hilt is a pair of lions topped by the upper body of the god. (Photograph E.C. Krupp).

> "To me, the King, have the gods-the Sun God and the Storm God-entrusted my land and my house...(The gods) have taken care of the kings. They have renewed his strength and set no limits to his years."

Political power is guaranteed, in part, by the productivity of the land. State religion magically activated the divine agents of prosperity in a concrete demonstration of the procreative power of the king. The Sacred Marriages of gods and rulers renewed the economic footing of kingship and so reinforced the political basis of kingship. Through sacramental sex and seasonal renewal, this ideology revitalized the economy and sustained sovereignty.

Recognizing the theme of divine Sacred Marriage in the content of Yazilikaya's reliefs, the German archaeologist Kurt Bittel, who directed excavations at Bogazköy, echoed an earlier suggestion by H. Zimmern (1918), who entertained the use of Yazilikaya as the place where the New Year Festival, sacred nuptials and all, was

Figure 11. In Chamber B, Sharruma, the sponsor of royal succession, protectively embraces Tudhaliya IV. Although Tudhaliya, as the son of King Hattushili III and Queen Puduhepa, held the option on the Hittite throne, his claim was enhanced by the ambitious program of art and ritual Puduhepa engineered at Yazilikaya. In witnessing the marriage of the Weather God of Heaven and Hepat, the gods were actually sanctioning the royal status of Hattushili and Puduhepa. Divine approval of the rules of royal succession was a natural consequence. (Photograph E.C. Krupp).

staged, an idea later endorsed by Gurney (1958, 108). Further research by Heinrich Otten (1956) exposed an ancient account of the Hittite New Year ceremony.

> "In. honor of the Weather God at (?) the beginning of the New Year a great festival of heaven and earth was celebrated. All the gods assembled and entered into the house of the Weather God. Whichever god harbors anger (?) in his soul shall chase the evil anger (?) from his soul. (Now) eat at this feast, drink! Satisfy your hunger and quench your thirst. The King's and the Queen's life hail! (Hea)ven's and the earth's (life) (?) hail! the grain's life (hail!)" (Bittel, 108)

The mountain-like character of Yazilikaya's rocky pinnacles allow it to function as the skyscraping "house of the Weather God." We also know the small temples that fronted the sanctuary and restricted access to it were not designed for continuous occupation, and that would argue for intermittent, seasonally significant use. A nearby stream may have been regarded as the home of the fertility goddess who dallied with heaven as the year turned. The evidence, then, leans in favor of Yazilikaya as place of fertile energy and marital alliance. It is a center of creation where ritual activity mirrored divine and cosmic renewal with the imagery of a sacred bedding, a mini-mountain love nest where transcendental nuptials were performed above the Hittite capital with the blessing of the gods. The timing of these ceremonies with the New Year—and vernal equinox—reenacts the beginning of time with royal pillow talk. Yazilikaya—populated by the stand-ins for the sovereign gods—functions as the place of Creation.

Although we have a reasonable familiarity with the home team of Hittite gods that played ball with nature on behalf of kingship, the development of their most important sanctuary among the rocks at Yazilikaya throws a curve at our attempt to score some understanding of the operation of Hittite ideology. It is curious that an imperial civilization at the height of its power and stability elected to house a national shrine on a hillside above the capital in a small complex of natural enclosures open to the sky. The Hittites were perfectly capable of engineering massive public works, yet despite Hattusha's construction of great temples, palaces, and walls, the empire invested ideologically in what would be regarded as rock art in a less complex society. The choice was deliberate and is no doubt partly explained by the desire to operate ritually in an environment that echoes the mountaintop territory of the high gods of creation and renewal. That recognition, however, fails to tie a bow on the divinity packaged within Yazilikaya's rocky wrapper.

As a locus for the creation and ritual renewal of power, Yazilikaya was a cosmic modem that put the king on-line with the gods, an earth navel that nourished the ruler with divine vigor. It was a junction of worlds where the realm of the dead, heaven, and earth met in burials, mortuary imagery, celestial gods, and portraits of the Hittite ruler. Although the mortuary content of the small gallery seems to conflict with the rituals of renewal implied in the main gallery, the presence of both themes is consistent with the cosmological character of the sanctuary. It has, in fact, the properties of a vestibule on the world axis, at the center of space and the source of time. Mircea Eliade (1954, 12) summarized the properties and fundamental meaning of such places. They operate as the Center of the World, the prime focus of sacred power and the prestigious point of contact with spirits and gods. This Center is often equated with a sacred cosmic mountain that permits heaven to consort with earth. The mountain often symbolizes the world axis, which is frequently identified with the polar axis and the hub of the sky. The summit of the Central World Mountain is the navel of the earth, where world order is nourished by the flow of celestial power. It is the point of Creation and Paradise, the eternal realm of divinity (Eliade 1954, 16). The cosmic marriage bed is made there by heaven and earth to precipitate the Creation of the World (Eliade 1954, 24). Capital cities and primary temples occupy the Center of the World and replicate the structure of the cosmos (Eliade 1957, 39). Ceremonies performed there recapitulate the beginning of time, and the ruler, a terrestrial officer of cosmic order, is a participant in the rituals of renewal. His place is the center of the center, and his performance is a natural consequence of the centralization of power. At the world's sacred center, with approval and assistance from the gods, he makes the world habitable by establishing its order and energizing its life in harmony with the structure and rhythm of the universe.

The cosmic mountain, the fertility of the land, the procreative power of nature, and the gods all converge at Yazilikaya, where the bedroom politics of royal ritual reignited dynastic authority in the New Year's cyclical return to the primordial Creation and the world-ordering era of cosmic gods. Charged with mystery and majesty, it made theatric use of the natural environment—open sky and fertile earth, where the gods move at will. In erotic imitation of sovereign gods, the Hittite king and queen made Yazilikaya the center of the Hittite universe and refertilized the ideology of kingship at the heart of the Hittite empire (Krupp 1996, 143–152).

References

Akurgal, E. no date, The Art of the Hittites (Harry N. Abrams, Inc., Publishers, New York).

Akurgal, E. 1973, Ancient Civilizations and Ruins of Turkey (Haset Kitabevi, Istanbul).

Alexander, R. 1986, The Sculpture and Sculptors of Yazilikaya (Associated University Presses, London).

Bittel, K. 1970, Hattusha: The Capital of the Hittites (Oxford University Press, New York).

Bittel, K. 1972, Guide to Bogazk6y (Bahqelievler, Ankara).

Eliade, M. 1954, Cosmos and History: The Myth of the Eternal Return (Harper and Row, Publishers, Incorporated (1959 reprint), New York).

Eliade, M. 1957, The Sacred & the Profane: The Nature of Religion (Harcourt Brace Jovanovich (1959 reprint), New York).

Eliade, M. 1978, A History of Religious Ideas, Volume 1—From the Stone Age to the Eleusinian Mysteries (The University of Chicago Press, Chicago).

Grayson, A. 1980, "Mitanni," in The Encyclopedia of Ancient Civilizations, Ed. A. Cotterell, New York: Mayflower Books.

Gurney, O. 1958, "Hittite Kingship," in Myth, Ritual, and Kingship, Ed. S. Hooke, London: Oxford University Press.

Gurney, O. 1980, "The Hittites," in The Encyclopedia of Ancient Civilizations, Ed. A. Cotterell, New York: Mayflower Books.

Krupp, E. 1983, Echoes of the Ancient Skies (Oxford University Press (1994 updated reprint), New York).

Krupp, E. 1991, Beyond the Blue Horizon-Myths and Legends of the Sun, Moon, Stars, and Planets (Oxford University Press (1992 reprint), New York).

Krupp, E. 1997, Skywatchers, Shamans, & Kings: Astronomy and the Archaeology of Power (John Wiley & Sons, Inc., New York).

Macqueen, J. 1986, The Hittites and Their Contemporaries in Asia Minor (Thames and Hudson Ltd, London).

Pallis, S. 1926–27, Historisk-Filologiske Meddelelser, Volume 12, The Babylonian Akitu Festival (Andr. Fred. Host & Son, Copenhagen).

30

The Decans of Ancient Egypt: Timekeepers for Worship, or Worshiped Beyond Time?

Kurt Locher
Guest Investigator
Astronomical Institute
Berne University
Berne, Switzerland

The annual rhythm of the decanal phenomena as recorded on ancient Egyptian coffin lids is the world's oldest preserved evidence of a division of nighttime into hours. That is why main investigators, O. Neugebauer, M. Clagett, and C. Leitz called these representations 'clocks' indicating to the hour-priests the times of their nightly rituals. A different opinion, currently shared by most Egyptologists, is that the belt of the 36 decanal constellations had initially been created as a scale of successive harbingers of the rising of Sothis, the heliacal one in particular being a symbol of both the annual revival nature and the entrance into eternal life after death and the days' sojourn of the dead body in the Embalming House. In this view the hour-priests' nightly rituals were to fill the waiting-time before that greater phenomenon, comparable to Christian Advent rituals aimed at bridging time rather than registering it.

* * *

Historical Background

It was almost two centuries ago when, soon after the decoding of the hieroglyphs, ancient Egyptian textual sources were found to contain a considerable amount of astronomical data. At the same time their connection with the calendar, especially with regard to the synodic lunar cycle and the heliacal rising of Sirius, was recognized.

For the discovery of the calendrical role of Sirius it helpful that this star had already occurred in a classical text by Plutarch (Nachstädt et al, 1971), mentioning

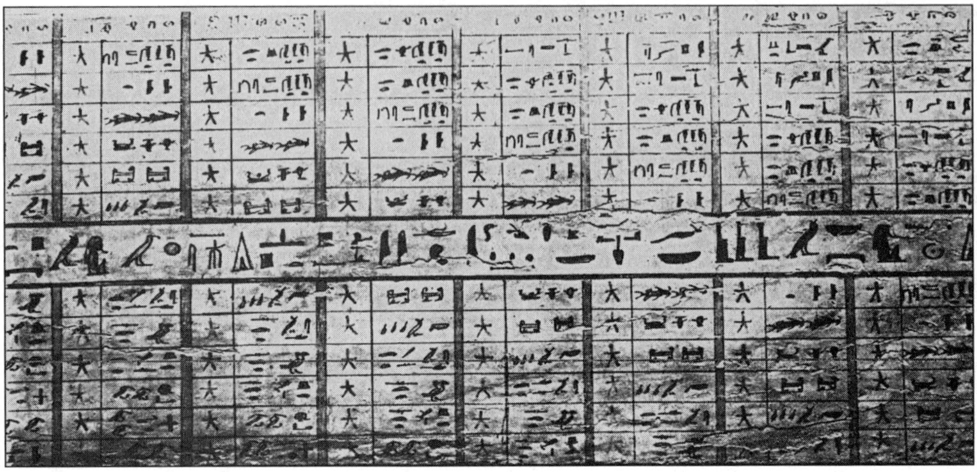

Figure 1. The 12 × 36 matrix on the loffin lids representing the visibility of each of the decans at one of the 12 nightly hours.

its relation to the goddess Isis and to the annual rising of the Nile as well as giving its name Sothis, which has since turned out to be a Greek mispronunciation of Egyptian Sopdit.

A reference to Orion and Ursa Major, which appears in the same Plutarchian text, might have been similarly helpful, but the place of Orion in the system of the 36 decans could not be established prior to the relatively late discovery of the first specimens of the 17 presently known Middle Kingdom astronomical coffin lids nor before the still later understanding of their signification.

We owe this discovery to Daressy (1900) who at the turn to the twentieth century also brought about a turn in our knowledge of historical roots of the 36 decanal constellations, dating them back from the New to the Middle Kingdom. In this same publication he also gave a sufficient summary of the previous knowledge so as to spare the student the task of reconsulting nineteenth century publications on the subject. However, Daressy lacked sufficient astronomical insight to recognize that the 12 × 36 matrices on the coffin lids represented visibility of each of the decans at one of the 12 nightly hours in one of the 36 calendrical decades. Instead of that visibility he was thinking of some kind of astrological influence changing according to hours and seasons. He was even mislead into believing that the stellar proper motion was so large as to be able to disrupt constellations during a few millennia.

It was left to Pogo (1932) to show that each entry of a decan's name in the 12 × 36 matrix meant its actual visibility phase at respective hour and decade, and, with the exception of Leitz (1995) every scholar involved in the debate since has agreed that this phase meant the rising. Figure 1 shows a typical section of such a 12 × 36 matrix, in which each decan's name is repeated along a diagonal line, each time ceding the nightly hour (the matrix line) to its neighbor in the sky in the following decade (the matrix column).

Beginning in 1960, and seconded by Egyptologist R.A. Parker, O. Neugebauer (1960–1969) finally engaged in compiling the entire corpus of Egyptian astronomical texts, after having before realised a comparable task with regard to the Hellenistic, Byzantine, Ethiopian and Assyrian material. Under the influence of the high mathematical standard of the latter, he was on the one hand inclined to overestimate the mathematical content of the other cultures' astronomy but on the other hand also to belittle their astronomical achievements because of their mathematical standards still being below the Assyrian one.

Although there are no time-reckoning devices known from the epoch of the earliest astronomical coffin lids, sundials and water clocks emerged many centuries later. Neugebauer saw in these lids instruments invented to indicate the nightly hours with the purpose of telling the hour-priests the times for their rituals, an idea shared till today mainly by Clagett (1995) and Leitz (1995).

The Alternative View

Inherent in ancient Egyptian thinking was the belief that deities were manifest. Thus profession of faith or apostasy as in Judeo-Christian religion were unthinkable. This conception is widely held in current Egyptology and best put into words by Assmann (1990). If deities are also present in asterisms, they are therefore doubly manifest.

The rising of a decan in the sight of the hour-priest is thus an epiphany in which the decanal deity manifestly participates. In this view there is no reason to ascribe to it the role of measuring time. Measurable time is too abstract and too modern a concept at least with regard to the Middle Kingdom, in whose texts, accordingly, the term "hour-priest" does not yet occur.

The Decans as Harbingers

There is wide textual evidence that the heliacal rising of Sirius held its calendrical and symbolical role already in the Old Kingdom when the notion of decans was still absent from textual sources. On the other hand, the Old Kingdom Pyramid Texts give Orion an outstanding mythological role besides, in connection with Sirius, which can be best interpreted as that of a harbinger of the latter, considering that Orion's rising precedes that of Sirius both seasonally in the heliacal rhythm and later in the year nightly.

A harbinger enters the scene through the same door as the deity which he announces. This is well realized in the sky if we postulate Orion's main figurative feature as having the same declination and thus the same rising place on the horizon as Sirius, which can easily be done by seeing Orion's crown as composed of the stars that form the girdle and sword of the Greek's and our own constellation, a view strikingly evident in the form of his crowning hieroglyph S3h, as is shown in Figure 2 reproduced from Locher (1993). Krauss (in press) has recently adopted my view of this in his comprehensive work on the Pyramid Texts.

This declinational coincidence of Orion's crown with Sirius holds best for the precessional epoch of about 2500 B.C., i.e. the beginning of the Old Kingdom, and is probably the celestial cause for the subsequent creation of the entire decanal belt, since the rising of conspicuous Orion may itself have given rise to the positing of another harbinger constellation, thus starting a chain of correlated constellations

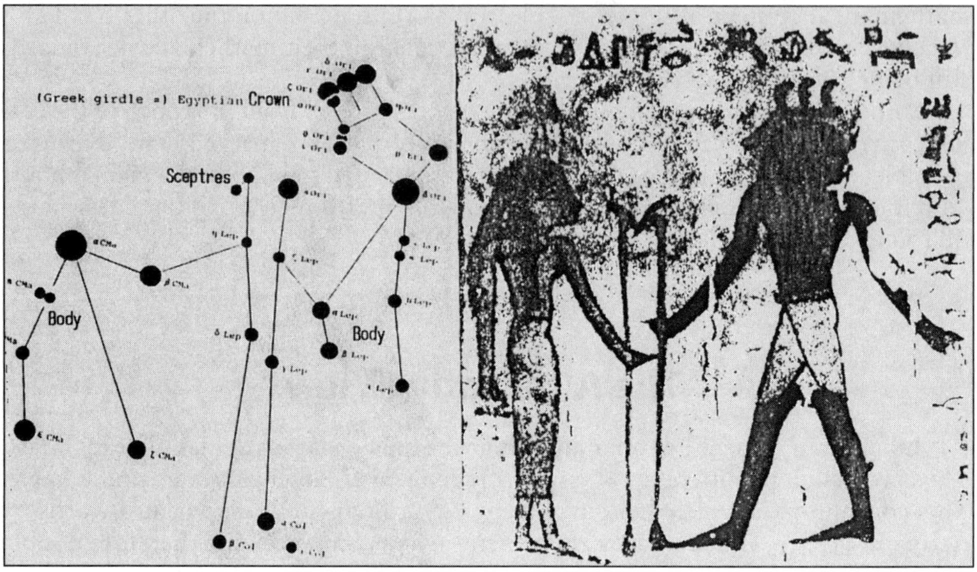

Figure 2. The Greek view of the Orion belt forms a crown in this image.

running all around the sky westwards with decreasing right ascension. This results in a view of the decans as successive harbingers to the main event of the heliacal rising of Sirius, as such bridging the waiting time for it in a role comparable to that of the Christian advent rituals.

The Celestial Identification

The 70 days of invisibility of Sirius between its heliacal setting and rising was of great symbolic significance in certain epochs. It found ritual expression in the 70 days' sojourn of the body in the embalming house between death and burial. Both the state of embalmment itself and the prescribed amount of time symbolized the transitional stage on the way toward another life period, corresponding to the new season of Sirius' visibility. The entire corpus of ancient Egyptian texts contains one single reference (Lange and Neugebauer, 1940) which ascribes a 70 days' invisibility to *all* the decans. Despite its late date, roughly a millennium after the epoch of the decans' genesis, Neugebauer (1960–1969) relied on this reference and was thus lead to postulate a course of the decanal belt roughly parallel to the ecliptic, which we are contradicting here on the grounds of our harbinger concept.

Except for their intersection in the Scorpius-Capricornus region, the two concepts must lead to different celestial identifications. While Neugebauer refrained from any identifications beyond those made by Plutarch (Nachstädt, 1971), attempts at identifying the decans in the sky were made by Böker (1984), Davies (1985), and Leitz (1995) who all followed a similar concept of a belt not parallel to the equator for different reasons.

By contrast, our equator-parallel idea was successful in identifying the most conspicuous decanal features such as Orion's crown, the "Ovoid" (Neugebauer's

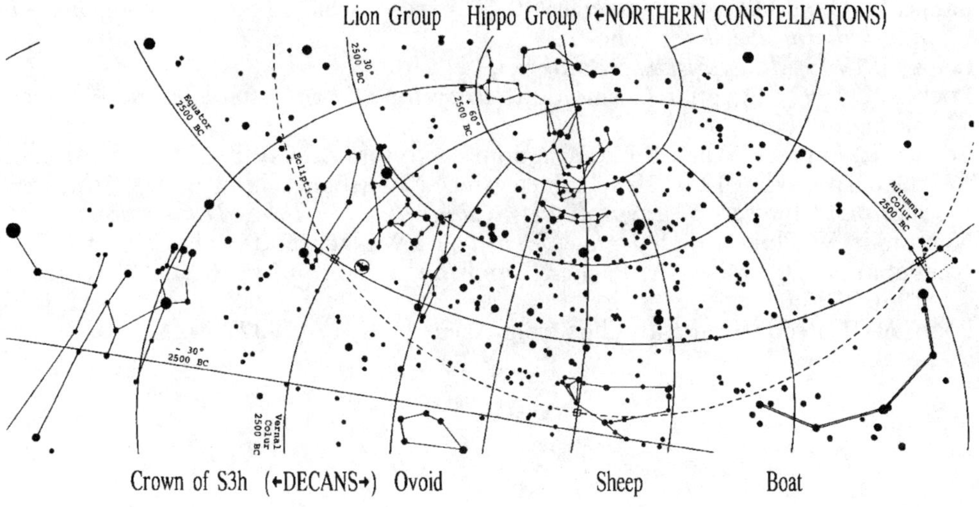

Figure 3. The most conspicuous decanal features seen in the sky.

term for a philologically controversial name), the Sheep, and the Boat, as is shown here in Figure 3.

The Increased Textual Basis

Since the outstanding Neugebauer and Parker ((1960–1969) work, it has been possible to extend the number of known decanal coffin lids from 12 to 17 (Locher, in press), and new criteria developed by Kahl (1993) have lead to a better stemmatic classification of these lids.

References

Assmann, J. 1990 *Ma'at: Gerechtigkeit und Unsterblichkeit im Alten Aegypten*, München.

Böker, R. 1984 "Ueber Namen und Identifizierung der ägyptischen Dekane," *Centaurus*, 27, 189–217.

Clagett, M. 1995 *Ancient Egyptian Science II: Calendars, Clocks, and Astronomy*, Philadelphia.

Daressy, M.G. 1900 "Une ancienne liste des décans Egyptiens," *Annales du Service des Antiquités de L 'Egypte*, 1, 79–90.

Davies, V.L. 1985 "Identifying Ancient Egyptian Constellations"; *Archaeastronomy* 9, S102–S104.

Kahl, J. 1993 "Textkritische Bemerkungen zu den Diagonalsternuhren des Mittleren Reiches," *Studien zur Altägyptischen Kultur*, 20, 95–107.

Krauss, R. *Astronomische Konzepte und Jenseitsvorstellungen in den Pyramidentexten*, Wiesbaden, in press.

Lange, H. O. and O. Neugebauer, 1940 "Papyrus Carlsberg No. 1," *Kongelige Videnskabernes Selskab*, 1, no. 2.

Leitz, C. 1995 *Altägyptische Sternuhren*, Leuven.

Locher, K. 1993 "Egyptian Cosmology," *Encyclopedia of Cosmology*, ed. N.S.Hetherington, New York.

Locher, K. in press "The Middle Kingdom Astronomical Coffin Lids, extension of the corpus from 12 to 17 specimens since Neugebauer and Parker," *7th International Congress of Egyptologists, Cambridge UK 1995, Transactions*.

Nachstädt, W. Plutarch, *De Iside et Osirlde*, ed.W.Nachstädt et al., Leipzig 1971.

Neugebauer, O. and R. A. Parker, *Egyptian Astronomical Texts*, Providence 1960–1969.

Pogo, A. "Calendars on coffin lids from Asyut" *Isis*, 17 (1932) 6–24.

31

Metal Artifacts As a Mirror of Ancient Georgian Astronomical World View

Irakli Simonia and Tsitsino Simonia
Abastumani Astrophysical Observatory
Institute of Architecture and Building
Tbilisi, Republic of Georgia

We describe three early metal objects from Georgia: a plate from the 16th–14th century BC, a spoon from the 4th–3rd centuries BC, and a cup from the 3rd century BC, and suggest that the decoration of each is based on astronomical symbolism.

* * *

Background

Ruins of ancient observatories, fragments of instruments, numerous neglected manuscripts and books, so far unclaimed discoveries, forgotten names—all these make up the ancient Georgian astronomy until now unknown to the world. The ancient state of Georgia, where this science developed, is situated in the Caucasus, along the very border between Europe and Asia.

The Black Sea, the Caucasian mountains and thick forests created an exceptionally beautiful landscape and contributed to a healthy, temperate climate. From time immemorial the land had been inhabited by Georgian tribes. The state of Georgia has existed in this territory for over 25 centuries. Through all those years full of hardships, the Georgians were creating and keeping up their own language, culture, and world outlook. Numerous masterpieces of literature, art and architecture are surviving evidence of the originality of the culture of this people. The oldest Georgian states—the West Georgian Kingdom of Colchis and the East Georgian Kingdom of Kartli—were established in the 6th century BC. It was the time when civil societies and special governmental institutions began taking shape in those two states. In the 3rd century BC the eastern and western Georgian states were united into one country. Before the 4th century AD the Georgian people had been following different religions, such as fire worship, Zoroastrianism, and others. In the early 4th century AD Orthodox Christianity took root in Georgia and within a century became the state religion. This process was enhanced by the strong political and cultural influence of the Roman Empire's eastern provinces which maintained close relationships with Georgia at that time.

Celestial phenomena had always been a subject of particular interest for Georgians. The old Georgian folk tales and legends frequently refer to celestial luminaries and various celestial phenomena. Those legends laid special emphasis upon the "domination of the celestial laws," and "inevitability of punishment by the powerful celestial forces." Perceiving the grandness of the occurring events, the Georgians ascribed some mysterious nature to the sky and celestial phenomena. Examining such artifacts as old Georgian decorations, tools, weapons, household utensils made of brass, bronze, silver, or gold, one can discern celestial luminaries—the Sun, the Moon, the stars—reproduced in different shapes and sizes.

In those times the religious system of Zoroastrianism played an important part in the life of the Georgian people. Their principal deity was the Moon, a symbol of a man-warrior. The sacred animal of the Moon was the Bull. A bull was frequently offered as a sacrifice. The shape of the Bull's horns seemed to remind the Georgians of the Moon. The walls of ritual edifices and dwelling houses frequently bore various images of the Bull and its horns. Different figures and figurines of the Bull and other sacred animals were widely common in the country.

The period from the 16th century BC to the 4th century AD in Georgia was the time of the accumulation of the most elementary, primary knowledge of the sky and celestial bodies; the application of the stored up knowledge in the culture, folklore and art; and the practical use of the knowledge, namely creation of chronology and calendars. This period witnessed the development of the ancient Georgians from mere observers of the sky to skilled astronomers and astrologers at the Royal Court. Old Georgian manuscripts present some evidence that a lunar calendar was used in this country as early as the 2nd century BC and remained in use up to the end of the 3rd century AD.

Because this period is largely unknown outside of Georgia, we describe here three artifacts from this time—a bronze disk (16th–14th centuries BC), a silver spoon (4th–3rd centuries BC), and a silver cup (3rd century BC). We propose astronomical interpretations of the designs on these three objects. Although these interpretations are not unique, and other possibilities exist, we hope that our suggestions will lead to further study of this period.

Bronze Disks ca. 16th–14th Century BC

Disk-shaped bronze plates were found by a Georgian archeologist, O. Gambashidze (1986), in women's burials at Bornigele and Zadengora in Georgia. In all over 30 such plates were recovered and were dated to the 16th–14th centuries BC. The plates are massive, some tens of centimeters in diameter (Figure 1), with curious relief designs on the surface. The plates have previously been studied from the

Figure 1. Bronze plate found in Georgia dating from 16th–14th century BC, a possible "cosmogram".

points of view of archeology, history and art. Ours is an attempt to analyze and interpret the surface designs from a purely astronomical viewpoint.

All the recovered plates bear similar designs containing the following basic common elements: a hole of a particular diameter through the center of the plate; surrounding the hole, a pattern involving raised hemispherical images and crescentic holes; further out a set of radial slits; beyond them an embossed rim. The plate illustrated here (Figure 1) has a tab seemingly designed to suspend it. The entire pattern is concentric and symmetric.

The whole sky and the strict regularity of celestial phenomena would have impressed and attracted those who watched them: Georgian tribes practiced land farming and, therefore, needed some kind of a calendar. We thus suggest the following astronomical interpretation of these bronze plates. We argue that they can represent the world view of the Georgian tribes of that time. The designs contain perforated and raised single or repeating components. Supposedly, the people who made these plates tried to represent celestial luminaries that they saw in the sky. The hole through the center could have been meant to represent the Sun. The repeating crescent-shaped holes could represent the Moon. The repeating spherical projections stood for the planets, stars, etc. The Sun is in the sky during the day's light hours; the Moon and the stars and planets are in the selfsame sky at night. The daytime and the nocturnal luminaries are thus represented together on the same plate, with all the celestial bodies being depicted on its surface which was supposed to represent the sky. Thus, the plates with relief designs could have served as a kind of flat model of the Universe.

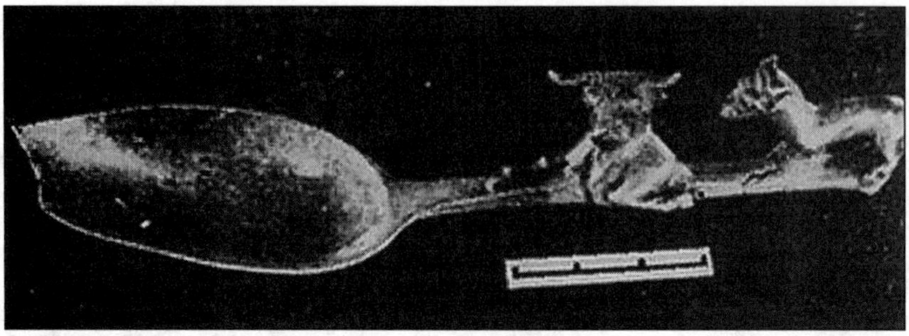

Figure 2. Silver spoon found in Western Georgia dating from 4th–3rd century BC, with figures of bulls on the handle.

Why are the images of the Moon and the planets arranged in a repeated pattern? We think that by repeating the images the ancient authors tried to show the permanent visible motion of the luminaries. That attempt to show motion suggests that the plates' makers used to watched the celestial phenomena for a certain period of time (a day, a month, a year) before representing what they had seen on the plate. So we suggest the term "cosmogram" for the plates with these patterns, each cosmogram thus holding a certain body of information.

Fire, light and heat were naturally important components of the life of Georgian tribes. A particularly high place of pride was given to the Sun as a source of heat and light. The image of the Sun would thus be put in the center of a cosmogram, all the other luminaries being distributed around it. A planet could be represented in the form of repeating triangular holes. Such an arrangement of the celestial bodies and such a world outlook are quite curious. It is by no means heliocentrism, since it is hardly possible that the ancient people could possess so much knowledge. Nevertheless, there ought to be some kind of definition. And we suggest the term "symbolic heliocentrism." The phenomenon of "symbolic heliocentrism" seems to be free of any limits or national features. It is not unlikely that some other land farming tribes in other parts of the world created artifacts bearing reflections of their possibly heliocentric world outlook.

Other Metal Objects ca. 4th–3rd century BC

The methods of metal design in Georgia underwent some changes during subsequent centuries. The primitive dot-and-line pattern covering different articles—decorations, weapons, tools—were replaced by raised, three-dimensional images of sacred animals possibly symbolizing celestial bodies. Silver, and occasionally gold, figurines of the Bull, the Deer, and the Lion became common. Excavations at Vani, an ancient town in western Georgia, revealed a remarkable number of finds in the form of sacred animals (O. Lordkipanidze, 1984). One of the most interesting finds was a silver spoon with its handle decorated with two figures of the Bull (Figure 2). That spoon dated back to the 4th–3rd century BC. Representation of that sacred animal on such a common household article suggests a possible ritual (or religious)

Figure 3. Silver cup found in Western Georgia dating from 3rd century BC, with possible calendric pattern.

meaning: the spoon may have been used during special ceremonies. That the spoon showed the images of the Bull offers one possible interpretation: that the role of the Moon was still important in the world view of the Georgians at this time.

O. Lordkipanidze (1986) reported other metal finds from Vani. They were articles with more elaborate, raised, three-dimensional designs. One of them was a silver cup dating from the 3rd century BC (Figure 3). A curious relief design decorates the outside of the bottom of the cup. A raised, hemispherical central detail of the design is suggestive of an astronomical object. A closed line made in relief and running from the center outward toward the edge resembles an orbit. Close to that orbit are four (possibly sacred) animals, facing four directions. A starlike design is made out near the orbit. Examining the photograph in a clockwise direction, one can notice that the animal figures are placed at the points of 12, 3, 6, and 9 o'clock. The starlike pattern is in the vicinity of the 11 o'clock point. The cup is in poor condition, so it was hard to identify the sacred animals clearly. Opinions are likely to differ in this case. We think that they are the figures of the Boar (in the 12 o'clock site), the Horse (3 o'-clock), the Lion (6 o'clock), and the Deer (9 o'clock). The creator of that design seems to have symbolized by these animals celestial luminaries or even constellations. It is noteworthy that the animals are shown in motion, producing an impression of their permanent motion along the orbit. Wasn't that an attempt made by the ancient author to convey the perpetual motion of the celestial sphere or succession of the seasons?

The starlike component could serve as a kind of reference point for keeping time—a day, a month, a year. If our considerations prove to be correct, then the artifact can be regarded as the simplest form of a metal calendar. The design on this ancient silver cup seems to have been based on a geocentric (or some other similar) point of view.

Conclusion

We have described here for the first time an astronomical analysis of three objects from the early history of Georgia. Thousands of such objects have been discovered there, but they have not been analyzed from this viewpoint before. We feel that much remains to be done to study the evidence of early astronomical traditions and practices in this part of the world. We hope that others will join us in this study.

Acknowledgments

We would like to thank the Organizing Committee of the Conference and, in particular, Dr. Rolf Sinclair, for their concern and interest in our work which made it possible for us to present this report at such an important meeting as this is.

References

O. Gambashidze, et al. Principal Results of the Expedition to Meskheti and Javakheti. G. Field Archeological Research, 1986, Tbilisi.
O. Lordkipanidze. What Secrets Does Vani Keep? 1984, Kutaisi.
O. Lordkipanidze. Argonautics and Ancient Colchis. 1986, Tbilisi.

Editor's Note

The authors present here one possible (but not necessarily unique) interpretation of these objects, arguing that they symbolize the sun, the moon, and other astronomical bodies. They present no further evidence to corroborate this interpretation. However, this is (to our knowledge) the first description of early Georgian artifacts from an astronomical viewpoint, and as such warrants publication to make them better known.

Similar mirrors have been found in China (but assigned later dates) and seen there in several museums by J. Fountain (private communication 1998). There is the possibility of Hellenistic influence in the design of the spoon and the cup, as R. Poss points out (private communication 1997).

32

The Equinox in Pre-Hispanic Canary Islands

César Esteban and Juan A. Belmonte
Instituto de Astrofísica de Canarias, La Laguna
Tenerife, Spain

María A. Perera Betancor
Unidad de Patrimonio
Cabildo Insular de Lanzarote
Lanzarote, Spain

We report on the discovery of six possible equinoctial markers in some of the most important archaeological sites at four islands of the Canarian archipelago (Gran Canaria, Lanzarote, Fuerteventura, and Tenerife). The large fraction of equinoctial markers over the total number of sites with well defined astronomical alignments found is very high, indicating the high probability that the equinox or a date near it could be important in the calendar or ritual of the ancient islanders. We discuss the most likely astronomical target of this markers (astronomical equinox or half-way point in time between both solstices). Although the problem is not settled definitively, several clues favor the second possibility.

* * *

Introduction

The Canary Islands were populated before the time of Christ by migrations of proto-Berber groups from the north-west of Africa, as is accepted by present-day prehistorians and archaeologists (e.g. Tejera Gaspar & González Antón 1987). The islanders do not seem to have had knowledge of navigation, at least at the time of the Castillian conquest of the archipelago during the 15th century. This fact, and the remarkable ecological differences among the islands, are essential factors to explain the cultural differences among the people of each island. These differences are reflected in many aspects as, for example, the economic activities, the social structure, and the cultural remains. In contrast, the religious world appears to have had strong similarities in the whole archipelago. The historical references written after the conquest by the Europeans indicate the presence of similar kind of cults, costumes and rituals spread all over the archipelago. The presence of celestial gods is almost a constant in all the islands (Serra 1940; Jiménez González 1990) as well as the importance of certain rocks or mountains in some of them. It is interesting to remark that such elements were also common ingredients of the ancient north-African Berber religion (see Camps 1995).

The presence of sacred places in pre-Hispanic Canary Islands is reported by many ancient historians and modern archaeologists. The typology of these sites is variable from one island to another and even inside the same island. However, there are elements in some of these precincts which are common in many of the islands as, for example, the presence of small carved basins—*cazoletas*—with connecting canals—*canalillos*—as well as certain kind of glyphs, fact that suggests a possible common origin for, at least, some basic elements of the religious world.

We are carrying out a long-term project to study the astronomical aspects related to the pre-Hispanic Canarian cultures. This work is based on intense and multidisciplinary archaeoastronomical fieldwork (including astronomers: J. A. Belmonte and C. Esteban and archaeologists: M. A. Perera Betancor and others) in archaeological sites all over the archipelago. The surveyed sites include mostly those classified as sacred and funereal places. Some results of this investigation have been published or are going to be published in the near future (Belmonte et al. 1994, 1997; Esteban et al. 1996, 1997, 1998; Perera Betancor et al. 1996; Jiménez et al. 1997) and clearly point out the importance that the observation of the position of the celestial bodies had for the religion of this ancient people.

The Equinox in the Historical Chronicles

There are numerous but rather vague ethnological references about the existence of a calendar and the use of the position of the Sun and the Moon for time computing in the ancient historical references—the so-called chronicles—from the time of the conquest (Abreu Galindo 1977; Marín de Cubas 1993; Morales Padrón 1993). Of special interest are the references about certain sacred places called *almogarenes*, which are usually located on the tops of certain mountains. In these places, rituals involving the pouring of milk of goats over the soil took place at precise moments of the year.

Among the ethnological references of the pre-Hispanic calendar in use in Gran Canaria is one attributed to Antonio Sedeño (1505?) which especially attracts our attention (Morales Padrón 1993). He tells us:...*they counted the year by 12 months, the month by moons, and the day by suns, and the week by 7 suns. They called the year Achano. They terminated the year at the end of the fourth month, that is, their year started with the spring equinox, and, in the fourth month when they had finished the growing season, which was at the end of June, made great festivals for nine consecutive days....* The importance of this reference is that records the only mention about the possible use and determination of the equinox by the aboriginal culture of the archipelago for calendrical purposes.

In this paper we present some results from our archaeoastronomical fieldwork carried out at some of the most important archaeological sites of the archipelago. All the sites present a remarkable possible equinox marker based on the use of alignments with artificial elements or natural features of the horizon.

Gran Canaria

This island presents the richest and most culturally evolved pre-Hispanic society of the archipelago. The number and spectacularity of the astronomical relations and hierophanies we have discovered indicates the strong archaeoastronomical potential of this island (see Esteban et al. 1998). The three archaeological sites (one necropolis and two shrines) reported in this section were prospected in previous fieldwork campaigns and visited at the equinox sunrise or sunset in 1995 and 1996 to check the predictions and to record the phenomena on photographic or video support. The results on the equinoctial markers discovered in this island have been presented and discussed in more detail in a series of two papers (Esteban et al. 1996, 1997). We recommend looking up those references mainly for further useful graphical information.

Necropolis of Arteara

This important and well known archaeological site (Schlueter 1977–79) is the largest necropolis in the whole archipelago and is located at the small village of Arteara, in the very deep valley of Fataga. It comprises 800 tumular tombs constructed by a pile of drystones, many of them in the form of a truncated cone with oval base. The average dimensions of the tombs are about 2 m long and 1.5 m high. The radio-carbon datings available give epochs as early as the 5th century B.C. for the first inhumations.

There is an old tradition in the village telling that, at the dawn of an uncertain summer day, the first sun rays—appearing inside a natural notch (see Figures 1 and 2 of Esteban et al. 1996)—illuminate one of the tumuli named the *King's Tomb*. The origin of this name for the tumulus is unknown. However, it may have originated because of its relatively larger size and its location on a small promontory close to the easiest access to the necropolis. Taking into account this ethnological clue, we selected the *King's Tomb* as the most suitable observer's position in the necropolis.

The horizon is very near—about 1 km—and appears rather flat and featureless, having an average angular height of about 30 arc min. The only remarkable feature on the eastern part of the horizon is the natural inverted trapezoidal notch, almost 2° deep and about 1° wide in its bottom, where the tradition spoke of the sunrise phenomenon taking place. In this sense, it was remarkable to find that, standing at the *King's Tomb*, the bottom part of the notch corresponded to a declination close to zero. Several months later, in our visit at the autumnal equinox (23 September 1995; declination of +0° 03' for the center of the solar disk at dawn) we corroborate that the first appearance of the solar disk occurred at the southern edge of the bottom of the notch. Later on, the Sun climbed on the southern slope of the

Figure 1. Equinox sunrise illuminating the King's Tomb at the tumular necropolis of Arteara, (Gran Canaria).

notch (see Figure 2 of Esteban et al. 1996). As can be see Figure 1, the illumination of the *King's Tomb* by the sun's rays from the notch at dawn gives an amazing and suggestive impression.

Fortaleza Grande

This archaeological site belongs to the village of Santa Lucía and extends along a sharp, long and narrow rocky mountain rising 85 m over the surrounding plateau. The archaeological context is very rich. It has many natural and artificial caves, a defensive wall along the top with towers, and the remains of four circular drystone constructions called *goros* by the local people (see Figure 3 of Esteban et al. 1996 for a general photograph of the site). There have been numerous findings of richly painted pottery and idols, a fact that suggests the special character of the site.

The horizon is very rich in conspicuous and sharp mountains. To the north, the highest peaks of the island can be seen. On the other hand, the western horizon is dominated by the impressive cliffs of the mountain range of Amurga, about 1100 m high and some 3 km distant. The presence of an outdoor archaeological site in one of the peaks of this mountain range is remarkable. This is the *almogaren* of Amurga (also known as *almogaren* of Fataga), which is considered a sacred place by the archaeologists (see del Arco Aguilar, Jimínez Gómez & Navarro Mederos 1992). It consists of more than 40 *cazoletas* of different sizes carved into the flat rocky ground (see Figure 5 of Esteban et al. 1996).

It was striking to notice that, seen from Fortaleza Grande, the precise location of the *almogaren* of Amurga on the horizon is very close to the geographical west. We visited the site one day after the spring equinox of 1996. At the moment of the local sunset, the center of the solar disk was at a declination of +0° 34', corresponding to the moment of the half-way point between both solstices. We selected the most prominent feature of the archaeological site as the observer's position which is a volcanic tube that passes through the mountain from east to west, and where a number of idols were found during the excavations (a fact that suggests the sacred character of the cave). From this particular position we could observe the sun setting just over where the *almogaren* of Amurga is located, which occupies an angular extension very similar to the solar diameter (see Figure 4 of Esteban et al. 1996).

Roque Bentayga

This archaeological site shows the most impressive and clear astronomical implications among all the places we have studied in the Canary Islands so far. The archaeological context and the striking lunar relations found here have been described extensively in other works (Belmonte et al. 1994, Esteban et al. 1998).

The *almogaren* of Roque Bentayga is located at the eastern side of the huge rocky promontory giving its name to the site. The archaeological context includes a number of artificial caves, carved platforms, glyphs and basins (Grau-Bassas y Mas 1980). The most conspicuous feature of the site is a circular glyph which, in fact, has been considered the centre of the sanctuary (Jiménez González 1990).

In previous work (Belmonte et al. 1994), we predicted that, standing at the circular glyph, an observer could be able to see the sun rising at the equinox behind a notch carved on the top of a nearly 5 m high rocky peak located at the eastern edge of the almogaren. We visited the site at the morning of 1995 September 23 confirming this prediction (see Figure 2). To our surprise, a more striking phenomenon took place on the almogaren which could be noticed only by *in situ* observation. After the moment of sunrise on the horizon, the shadow projected by the notch moves down on the eastern side of the monolith of Roque Bentayga reaching the floor of the carved platform where the circular glyph is. Soon after, the profiles of the notch and of the circular glyph become both coincident and with the same linear size. At this moment, the declination of the center of the solar disk was at +0 03'. A site plan and a photographic sequence of this striking phenomenon can be found in Figure 7 of Esteban et al. (1997).

Lanzarote and Fuerteventura

These are the easternmost islands of the archipelago and, therefore, the closest ones to the coast of Africa, about 100 km apart. They were the first islands to be conquered and settled by the Europeans at the beginning of the XV century. The ar-

Figure 2. Equinox sunrise as seen from the circular central glyph of the *almogaren* of Roque Bentayga (Gran Canaria). The sundisk appears behind a notch located at the top of a nearby rocky promontory. The pole is placed in a deep carved little basin behind the notch.

chaeological and ethnological data recorded indicate the similarities between their pre-Hispanic populations. This fact could be explained by the relative closeness of both islands. In fact, the aboriginal people of Lanzarote and Fuerteventura called themselves with the same term: *Majos*. It is suggestive the similarity of this word with *Amasek* of *Amahek*, that was the name that the ancient people (from the first millennium before Christ) of the north of Africa called themselves. This people were called Lybians or Maxies by the Greeks.

Our group has surveyed these two islands very intensely during the last three years. In the case of Fuerteventura we have found significant astronomical relations, mainly in the petroglyph station of Tindaya and some of the large stone circles spread along the entire island (Perera Betancor et al. 1996). In the case of Lanzarote, we have not found clear astronomical implications except, perhaps, the one reported here: the "Castle" of Zonzamas.

The "Castle" of Zonzamas (Lanzarote)

This site is located in the large pre-Hispanic settlement of Zonzamas, which was possibly the aboriginal capital at the time of the conquest and, in fact, the most important archaeological place of the island. It is thought that the "Castle" was probably the residence of the king or leader of the community of islanders.

From our first visit, we noticed the presence of the remarkable Tahiche volcano just to the east of the archaeological site and dominating this part of the horizon. We visited the place at the dawn of the autumnal equinox of 1995 just to check the occurrence of the sunrise at this precise geographical point. Our predictions were confirmed, and the solar disk appeared just on the middle of the volcano (2.5° wide as seen from the "Castle", see Figure 3), suggesting the presence of a possible equinoctial marker in this archaeological site. The possible astral relations of the "Castle" of Zonzamas would have further support with the reported finding of a large engraved stela with a possible solar motive inside the site (see Cabrera Pérez 1992). In this sense, it is also of special interest the finding of an important ritual deposit with the

Figure 3. Equinox sunrise over Tahiche volcano as seen from the archaeological site of Zonzamas (Lanzarote).

presence of an anthropomorphic idol (with possible religious meaning) in one of the slopes of Tahiche volcano (see Cabrera Pérez 1992).

Tablero de los Majos (Fuerteventura)

This place is located in one of the present-day most isolated areas of Fuerteventura which instead was intensely settled in pre-Hispanic times: the peninsula of Jandía. The Tablero de los Majos consists of two stone circles of 15–20 m diameter constructed with two rows of middle-sized drystones.

The landscape surrounding the site is impressive. The eastern horizon, from north to south, consists of the highest mountains of the island, and is very rich in geographical features. On the other hand, the western half of the horizon corresponds to the sea line. The most interesting fact of the eastern horizon is La Zarza mountain, the highest peak of the island, which is located near the geographical east. Moreover, one of the circles has a well preserved entrance—marked by two vertical stones—just aligned with this mountain if one stands at the center of the circle. Correcting for the altitude of the peak of La Zarza (8.5°), we found that the sunrise when the solar disk has a declination of about +0° 30' (very close to the equinox and coincident with the average declination of the Sun at the mid-point between both solstices) occurs just on the top of that mountain (see Figure 4). This result appears quite suggestive. However, observations of the equinox sunrise from the site are needed to confirm this prediction.

Tenerife

This is the westernmost island of the ones included in this work and witnessed an apparently less developed material culture from the archaeological point of view. The social and political system was distinguished by a division into nine tribal territories at the time of the conquest. The people lived mostly in caves and small huts, and the economy was based on cattle raising and on agriculture. Islanders employed very simple pottery, generally lacking any kind of ornament or decoration.

Our archaeoastronomical work on this island has been centered on the study of sites containing *cazoletas* and *canalillos* which are similar, in basic aspects to

Figure 4. Eastern horizon from the center of one of the stone circles of Tablero de los Majos (Fuerteventura). The arrow indicates the top of La Zarza mountain, the highest peak of Fuerteventura, where the Sun rises at a declination of +0° 30' (about the mid-half point in time between both solstices). It is striking that the entrance of the stone circle (marked by two vertical stones) is pointing to that direction.

the *almogarenes* of Gran Canaria. These outdoor precincts have been interpreted as sacred places (Tejera Gaspar, 1992) in the basis of their archaeological elements and contexts. Usually, they are related with the presence of petroglyphs, some of them with possible astral interpretation (Tejera Gaspar 1992). Some results obtained for this island have been published in several works (Belmonte et al. 1994, Jiménez et al. 1997). In the following we will focus on a recently surveyed site where we have found—for the first time in Tenerife—a possible relation with the equinox.

La Pedrera

This archaeological site was discovered and reported by Perera López (1992). It is located in the coastal rocky promontory of Dos Hermanos at the north of the island. Its difficult access has probably helped to preserve it without significant damage.

The site consists of two different platforms of petroglyphs and carved basins. We have surveyed only the western nucleus of the complex, which is the most extended and interesting, having the largest number and most representative glyphs. Among these, there is one interpreted as a fish with a cross inside and two carved circular basins. There are also two (or three) anthropomorphic glyphs of oval shape with sexual appendices probably representing a woman (one or two) and a man

Figure 5. First sunrays at equinox sunrise at La Pedrera (Tenerife).

(one). Perera López (1992) reported that the fish and the anthropomorphic glyphs were oriented to the east.

When we visited the site we noticed the impressive eastern horizon, which coincides with a very abrupt coastline. It is remarkable that there is not any other part of the horizon seen from the site with a similar richness in geographical features. We visited the place at the autumn equinox of 1996 and we could take photographs of the sunrise (see Figure 5). Figure 6 illustrates the fact that the two best preserved anthropomorphic glyphs (the possible third one is also oriented roughly in the same direction) are oriented in the general direction where the sun rises near the equinox. In any case, the typical rough making of the aboriginal glyphs and the small sizes of the anthropomorphic ones of La Pedrera only permit us to consider their possible astronomical alignment as ritual, without the necessary precision to be considered as markers.

Perera López (1992) interprets this station as a precinct devoted to the worship of the feminine fecundity on the basis of the iconography of the glyphs. The possible astronomical relation with the equinox (the autumnal equinox represents the beginning of the rainfall in the island) could give a further support to this hypothesis. It would be of major interest to follow the investigation on this promising site in a wider point of view in order to shed some light on the religious world of the ancient islanders.

Final Remarks

The percentage of equinoctial alignments over the total number of the sites with well defined astronomical relations found in the Canary Islands is in general quite high and varies from one island to other. In the case of Gran Canaria we have found 3 equinoctial markers versus 2 very probable solstitial markers, one at Cuatro Puertas (summer solstice, Belmonte et al. 1994 and discovered independently by Barrios 1998) and another one at the necropolis of La Guancha (winter solstice, Belmonte et al. 1997). There is a further possible equinox marker reported by Aveni & Cuenca (1994) in Gran Canaria, in the complex of truncated conical drystone structures of Tauro mountain, as well as another solstitial site at Llanos de Gamona. As has been

Figure 6. Orientation of the anthropomorphic glyphs of La Pedrera (Tenerife). The arrow indicates the point where the equinox sunrise takes place.

commented above, the only possible astronomical alignment found in an archaeological site at Lanzarote up to date is the one reported here at Zonzamas, which is a possible equinoctial marker. In the case of Fuerteventura, we have found a very likely solstitial marker at Tindaya mountain (winter solstice) and two other possible sites related with the summer solstice but with lower rate of confidence (Perera Betancor et al. 1996) versus the single equinoctial site at Tablero de los Majos. Finally, Tenerife shows 4 solstitial sites with high degree of confidence (some others with much lower confidence, see Jiménez et al. 1997) versus 1 equinoctial at La Pedrera.

There is an interesting problem concerning to which was the true astronomical target of this markers or alignments: was it the equinox or the position of the sun at the half-way point in time between both solstices? Ruggles (1997) has discussed this problem and pointed out the important (but sometimes forgotten) fact that the concept of the equinox has only a precise meaning within the framework of Greek geometrical astronomy. Although the aboriginal people came from north-west of Africa and, in its origin, these proto-Berber groups could have Punic and Roman influences and, in principle, some relation with Greek astronomical concepts, it is hard to believe how the abstract concept of equinox was important and maintained by an almost isolated people during centuries. The use of a more definite and simple concept, and one relatively easy to determine observationally, is more convincing as it could be the half-way point in time between solstices. This method would involve the counting and manipulation of large number of days—at least 183—and the precision of one day in the determination of the solstices. In this sense, the possibility of the counting of large numbers of days by the aboriginal people has been suggested by Barrios (1997) and the possibility of the precise determination of the solstice is illustrated in the large number of solstitial markers found in all the islands. A strong reason for the utility of the equinox or a date near the equinox is that the determination of that calendrical date using the rising or setting of the sun, can be determined with higher precision at these moments. In fact, the rate of change of the declination of the sun reaches its largest annual value, and the day-to-day displacement of the solar disk at either sunrise or sunset is more noticeable for naked eye observations. In our case it is difficult to discern unambiguously which is the true astronomical target in our Canarian sites. In any case, in addition to the reasons given above, there are some clues from our results that suggests that the astronomical target could be the mid-half point in time between solstices (herafter termed MHPBS):

1. At the necropolis of Arteara, the sunrise at the MHPBS would occur at the center of the notch. At the equinox, the rising occurs at the southern edge of the notch (see Figure 2 of Esteban et al. 1996).

2. At Fortaleza Grande, the astronomical alignment between the *almogaren* of Amurga and the cave (the most relevant feature of the site and where the religious idols where found) is more precise at the MHPBS.

3. In Tablero de los Majos, the top of La Zarza gives a declination more consistent with the MHPBS.

4. Analysing the photographs, we cannot reject a possible phenomenon of tangential sunrise over the abrupt profile of the eastern horizon at the MHPBS at La Pedrera. In this case, we would have a more striking effect that could be used as a marker.

We consider that, although some evidences favour the use of the MHPBS by the ancient islanders, this is still an open question that deserves further careful *in situ* observations to settle it definitively. At any rate, we think that the results presented in this paper, strongly suggest the possibility of the practical determination and use—ritual and/or calendrical—of the equinox or the mid-point in time between both solstices, by the pre-Hispanic people of, at least, several islands of the Canarian archipelago.

Acknowledgements

We would like to thank the enthusiastic support of Margarita Sanz de Lara and Carmen Pinza during and after the hard fieldwork. The collaboration of R. Schlueter, J.J. Jiménez, J.V. Febles, O. González and R. Chinesta was essential in different parts of the work. We acknowledge an anonymous referee for very useful suggestions. This study was possible thanks to the financial support of the Instituto de Astrofísica de Canarias under project P7/93.

References

Abreu Galindo, F. J. 1977, Historia de la conquista de las siete Islas de Canaria (Santa Cruz de Tenerife: Goya Ediciones).

del Arco Aguilar, M. C., Jiménez Gómez, M., & Navarro Mederos, J. F. 1992, La arqueología en Canarias: del mito a la ciencia (Santa Cruz de Tenerife: Interinsular/Ediciones Canarias).

Aveni, A. F., & Cuenca, J. 1994, "Archaeoastronomical fieldwork at the Canary Islands" El Museo Canario XLI, p. 29.

Barrios, J. 1997, "Tara: A study of the Canarian astronomical pictures. Part II", in Proceedings of the IVth S.E.A.C. meeting "Astronomy and culture", Ed. C. Jaschek & F. Atrio Barandela, Salamanca: Universidad de Salamanca, p. 47.

Barrios, J. 1998, "A Berber solstitial marker earlier than the 16th century in Grand Canary Islands", in Proceedings of the Fourth Oxford International Conference on Archaeoastronomy, in press.

Belmonte, J. A., Esteban, C., Aparicio, A., Tejera Gaspar, A., & González, O. 1994, "Canarian astronomy before the conquest: the pre-Hispanic calendar" Revista de la Academia Canaria de Ciencias VI, p. 133.

Belmonte, J. A., Esteban, C., Schlueter, R., Perera Betancor, M. A., Génova, R., & Cruz, M. 1997, "Archaeoastronomy and Sacred Places in Tenerife (Canary Islands)", in Proceedings of the IVth S.E.A.C. meeting "Astronomy and culture", Ed. C. Jaschek & F. Atrio Barandela, Salamanca: Universidad de Salamanca, p. 247.

Cabrera Pérez, J.C. 1992, Lanzarote y los Majos (Santa Cruz de Tenerife: Centro de la Cultura Popular Canaria)

Camps, G. 1995, Les Berbères. Mémorie et Identité (Paris: Editions Errance).

Esteban, C., Belmonte, J. A., Schlueter, R., & González, O. 1998, "Pre-Hispanic Gran Canaria: an archeoastronomical paradigm", in Proceedings of the Fourth S.E.A.C. Conference, in press.

Esteban, C., Schlueter, R., Belmonte, J. A., & González, O. 1996, "Pre-Hispanic equinoctial markers in Gran Canaria, part I" Archaeoastronomy (supplement to Journal for the history of astronomy) 21, p. S73.

Esteban, C., Schlueter, R., Belmonte, J. A., & González, O. 1997, "Pre-Hispanic equinoctial markers in Gran Canaria, part II" Archaeoastronomy (supplement to Journal for the history of astronomy) 22, p. S52.

Grau-Bassas y Mas, V. 1980, Viajes de exploración a diversos sitios y localidades de la Gran Canaria (Las Palmas de Gran Canaria: El Museo Canario).

Jiménez González, J. J. 1990, "Elementos astrales de la arqueología prehistórica de las Islas Canarias" Investigaciones arqueológicas en Canarias II, p. 93.

Jiménez, J. J., Esteban C., Febles, V., Belmonte, J. A. 1997, "Archaeoastronomy and Sacred Places in Tenerife (Canary Islands)", in Proceedings of the IVth S.E.A.C. meeting "Astronomy and culture", Ed. C. Jaschek & F. Atrio Barandela, Salamanca: Universidad de Salamanca, p. 185.

Marín de Cubas, T. A. 1993, Historia de las siete islas de Canaria (La Laguna: Canarias Clásica), p. 204.

Morales Padrón, F. 1993, "La crónica de Sedeño" in Canarias: Crónicas de su conquista, Ed. F. Morales Padrón, Las Palmas de Gran Canaria: Ayuntamiento de Las Palmas. El Museo Canario.

Perera Betancor, M. A., Belmonte, J. A., Esteban, C., & Tejera Gaspar, A. 1996, "Tindaya: Un estudio arqueoastronómico de la sociedad prehispánica de Fuerteventura" Tabona IX, p. 165.

Perera López, J. 1992, "Los grabados de La Pedrera, Tenerife" Eres (Arqueología) 3, p. 33.
Ruggles, C. L. N. 1997, "Whose equinox?" Archaeoastronomy (supplement to Journal for the history of astronomy) 22, p. S45.
Schlueter, R. 1977–79, "Necrópolis de Arteara" El Museo Canario XXXVIII–XL, p. 101.
Serra, E. 1940, " Los mallorquines en Canarias" Revista de Historia VII, p. 281.
Tejera Gaspar, A. 1992, La religión de los Guanches: ritos, mitos y leyendas (La Laguna: Edicolor).
Tejera Gaspar, A., & González Antón, R. 1987, Las Culturas Aborígenes Canarias (Santa Cruz de Tenerife: Interinsular/Ediciones Canarias).

33

Hunting the European Sky Bears: A Proto-European Vision Quest to the End of the Earth

Roslyn Frank
Department of Spanish and Portuguese
University of Iowa
Iowa City, IA

Askotan esan izandu dute gizon
ikastun argiak nola piztiak izan
litezkeen gure erakuslariak.
Txepetxa: Otaño

The two Sky Bears, Great Bear and Little Bear, have been classified as belonging to the most archaic strata of the star figures. It has been assumed that scenes portrayed by certain other star figures were associated with some half-forgotten sky text handed down to us, albeit incompletely, through Greek mythology (Gingerich 1984). Although previously, in Europe, no archaic set of stories connected to the Sky Bears had been clearly identified, extensive field work over the past twenty years in the Basque region of the Pyrenees allowed me to identify a cycle of stories and related ritual performances, found throughout Europe, dealing with the adventures of an archetypal hero, the Bear Son, linked symbolically with both Ursa Major and Ursa Minor, and his encounters with a series of celestially encoded creatures, painted onto non-zodiacal constellations located mainly in the southern sky abutting the "zone of avoidance." There is evidence that the archetypal quest experience embedded in the Pan-European set of sacred stories was externalized collectively, being projected conceptually onto the stellar screen and acted out ritually in song and dance on the ground below.

* * *

Figure 1. The Ancient Constellations South of the Ecliptic. Source: Maunder 1909:155.

At present the data suggest that certain star figures familiar to Europeans, constellations which are non-zodiacal in nature and clustered for the most part in the southern sky, were placed there to illustrate the actions of Little Bear, the Bear Son hero, and the spirit animals that accompanied him on his ritual journey to the End of the Earth (Figure 1). The Pan-European hero is known as Hartzkume in Basque, Juan el Osito in Spanish, Jeun l'Ours in French, Giovanni l'Orso in Italian and Hans Bär in German.[1] In this fashion, the star figures painted onto the sky served to facilitate the collective retelling of the tales. In the latter, the main character, the shaman-hunter initiate, often goes by the name of John Little Bear. In Basque the protagonist is simply called Hartzkume, Little Bear, a term, as in English, with no indication of gender. In fact, I would note that there is also a female variant of this story of initiation. However, we shall concentrate our attention only on the initiation of the Bear Son shaman-healer, his vision quest and efforts to restore the cosmic order.[2] While in Europe the shamanic aspects of the adventure cycle may date back ultimately to the Mesolithic or even the Upper Paleolithic, that is, the story's ecocentric discourse reflects a cosmovision typical of hunter-gatherer societies, the process of repeated projection or mapping of scenes from the stories onto the stellar screen, as we best can recuperate them using data found in star charts of Western astronomy, is probably much more recent, and lasted a millennium or more, from ap-

1. For a discussion of the Pan-European variants as well as Germanic versions of the Bear Son tale, cf. Cosquin 1887; Panzer 1910. For more recent interpretations of the same shamanic thesis, cf. Stitt 1992 and Glosecki 1989, 1988.

2. Over the past decade a remarkable number of studies have been published dealing with the Bear Son tales and their ritual reenactments, reflecting the increased interest by scholars in the enigmatic coding of the European cycle of Bear Son tales (cf. Frank 1996a:131).

Figure 2. Above: Padded Bear Costume with Sheep and Dogskins. Vijanera de Bárcenas de Pie de Concha, 1924. Below: Padded Bear Figure with Waist Chain. Based on the Vijanera de Arenas de Iguna, 1919. Source & Design: Molina González & Vélez Pérez (1986:144).

proximately 4000 to 3000 B.C.[3] Furthermore, it is assumed that the cognitive mechanisms involved in painting these scenes onto the stars were the direct result of the use of these asterisms as celestial supports, mnemonic devices illustrating the action of the story. The latter were employed year after year in the time-factored ceremonies, performance art and rites of initiation intrinsic to a non-anthropocentric cosmovision based on Bear Ceremonialism which had as its center piece the archetypal hero, the Bear Son, the Celestial Bear Ancestor (Figure 2).

Basing our research model, in part, on conclusions drawn earlier by Maunder (1909), Ovenden (1966), Roy (1984); and more recently by Arregi and Frank (1996), particularly in reference to their analysis of the "zone of avoidance," we assert that the Bear Son saga already had achieved its celestial anchoring in a set of asterisms by approximately 3000 B.C., and that therefore this stage, by necessity, must have been preceded by a long process of careful celestial observation lasting

3. For example, the ecocentric and consequently non-anthropocentric nature of the Bear Son saga suggests that it antedates the cognitive frame of reference understood to be characteristic of the Indo-European pantheon with its anthropomorphic high gods. Furthermore, we should assume that in Europe stories centered on a bear ancestor found a variety of celestial projections among hunter and gathering societies. For instance, shamanic practices in Siberia, also based on Bear Ceremonialism, would have painted the shaman's flight onto a different set of asterisms, constellations that were accessible to and imagined by practitioner healers living in the northern latitudes in question. In our case, we are interested only in those projections that took place in Europe along a band from 36% to 41% or possibly as far north as 43%, and in examining the Bear Son's narrative in terms of its application particularly to constellations bordering the zone of avoidance and its immediate vicinity.

perhaps a millennium or more. Eventually these activities would lead the observers to choose certain sets of stars, from among the some three thousand available, upon which they would project different episodes from the sacred tale. Furthermore, such activities appear to be confirmed and reflected in the astronomically encoded social practices of the megalithic cultures of Europe, practices particularly abundant in Western Europe, and perhaps, more concretely in the design and ritual use of the Basque stone octagons (Frank and Patrick 1993; Frank 1996d; Zaldua 1996a, 1996b).

Previous studies dedicated to examining the origins of these non-zodiacal constellations of the southern sky have focused on internal reconstructions of the data, primarily through recourse to the conclusions that can be drawn from the zone of avoidance, namely, that these star figures were already firmly anchored in the sky by approximately 3000 B.C. and that those drawing these pictures lived along a band in Europe between approximately 36% and 41%N. or even as far north as 43% (Arregi and Frank 1996). In other words, those who gave the constellations limiting this zone their final form operated with a horizon coincident with a circle of declination: the space in the south invisible to them was circular, with the south pole of the time at its center (Maunder 1913:330). The outline of this unmapped region, the zone of avoidance, must therefore correspond roughly to the horizon of the place where the constellations were originally designed, or at least be roughly parallel to it, since we may well suppose that stars which rose only two or three degrees above that horizon might have been neglected (Maunder 1909:157). For example, because of precession, today for an observer in north latitude 40% a good deal of the constellation figure of Centaurus killing the Wolf would be hidden from view,[4] while around 3000 B.C. the pole star was in Alpha Draconis, located in Draco, the Serpent Dragon (Figures 3 and 4).

In previous efforts to determine the origin of these non-zodiacal constellations bordering the zone of avoidance, scant attention has been paid to the fact that any attempt to explain the way in which these asterisms came to be fixed on the sky screen must, by necessity, address mechanisms of cultural storage and put forward a model capable of explaining how these cultural artifacts were transmitted over several thousand years, not only from one generation to the next in a specific zone, but also how they came to be diffused throughout a much larger area. For example, at some point in time Greeks astronomers, such as Eudoxus of Cnidus (c. 400 B.C.) and the poet Aratus (c. 250 B.C.), acquired knowledge of such constellations, even though Greek mythology did not offer them any unified celestial narrative that would explain in a convincing manner the placement of these asterisms in the southern sky (Arregi and Frank 1996). Yet the star charts of Hipparchus-Ptolemy show these southern circumpolar constellations, apparently not inherited from Babylonia nor from the Egyptians,[5] as part of the traditional set of classic constellations, that is, of an astronomically encoded cultural assemblage that ended up portraying a horse-like figure killing a wolf in the southern sky (Figure 5).

---

4. Yet as Maunder(1909:150) readily admits: "It is, of course, not possible to fix either time or latitude very closely, since the limits of the unmapped space are a little vague."

5. "From this [study of the zone of avoidance] we learn that the constellations were designed by people living not very far from the 40th parallel of north latitude, not further south than the 36th or 37th. This is important as it shows that they did not originate in ancient Egypt or India, nor even in the city of Babylon, which is in latitude 31 1/2%" (Maunder 1909:157).

Figure 3. Non-Zodiacal Constellations Abutting the Zone of Avoidance. Source: Ovenden 1966:7.

The geographical diffusion of the Bear Son stories and related performance art stretches across all of Europe, from west to east, and extends far beyond the narrow north/south latitudinal band indicated above.[6] In fact, the Bear Son stories represent a Pan-European phenomenon, for they are classified as the most wide spread motif in European folklore, having been transmitted orally from one generation to the next by means of the mechanisms intrinsic to orality (cf. Cosquin 1887; Fabre 1968, 1986; Peillen 1986; Praneuf 1989; Shepard and Sanders 1992; Urbeltz 1994). The cycle of oral tales is present in all Indo-European languages and, more significantly, in European languages classified as being far older, such as Basque (Euskara).[7] In-

6. In fact, the European Bear Son tales were carried to the New World only a few centuries ago, and have been found in Louisiana, New Mexico, and Missouri, as well as in various parts of Mexico and the French, English and Spanish speaking areas of the Caribbean. In addition, they have passed into the folkloric repertoire of indigenous languages of North America for they have been recorded among the Micmac, Shoshoni, Assiniboin, Ojibwa, as well as the Thompson River Indians, the Chilcotin Indians of British Columbia and the Zapotec Indians of Teohuantepec in Mexico (cf. Frank 1996a:119–121 for full bibliographical references).

7. They are also found in Finno-Ugric languages, Finnish, Saami (Lapp), Magyar (Hungarian), as well as Caucasian languages, all linguistic codes that antedate structurally those of the Indo-European group. In the non-Indo-European languages and cultures the tales retain more clearly the earlier shamanic cosmovision which served the teller cognitively allowing him/her to transmit the traditional tales orally from generation to generation; and thereby continue to communicate the celestially encoded message to their audiences (Arratibel 1980; Charachidzé 1986; Dégh 1965; Erdész 1961, 1963; Hartsuaga 1987; Ortutay 1962; Shepard and Sanders 1992).

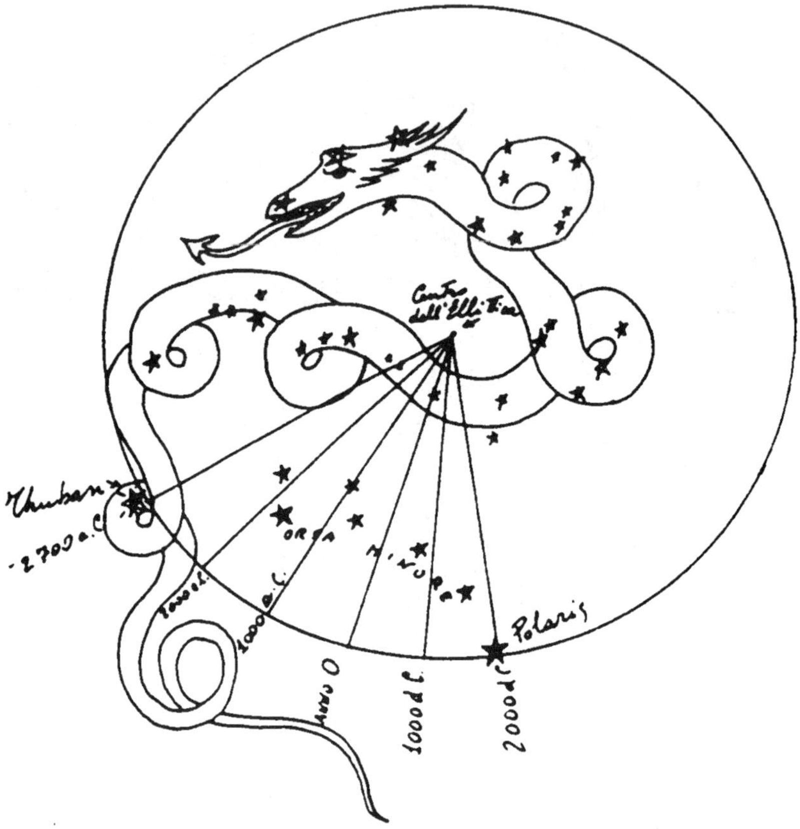

Figure 4. The movement of the North Pole over a period of approximately five thousand years. Source: Sesti 1991:473.

Figure 5. Bayer's Centaurus (1603). Source: Sesti 1991:299.

deed, we allege that these tales and related ritual performances derive from a Pan-European ecocentric cosmology in which earthly bears played a fundamental role as representatives of the Celestial Bear Ancestor.[8] Variants of these stories and similar ritual practices, found throughout Europe, are connected to a much earlier Pan-European cosmovision and to the belief that humans descended from bears. Within this belief system the Bear Son, half-human and half-bear, acts as the intermediary between humans and their bear ancestors, linking humans to the Celestial Bear and Bear Ceremonialism. The stories and related ritual practices form a type of *Bildungsroman* revealing time-factored activities and sky-texts with a strong shamanic flavor. Cross-cultural cognitive parallels from North America indicate the hero's adventures fit a scenario not unlike that of the Native American vision quest with its animal helpers and medicine bundle.[9] Similarly, in Europe, there is evidence that the archetypal quest experience, embedded in the sacred stories, was externalized collectively, being projected conceptually onto the stellar screen and acted out ritually in song and dance on the ground below (Figure 6). The tales and public folk performances narrate episodes from the life of this character as well as his encounters and combats with a remarkable series of celestially encoded creatures. In fact, each year folkloric reenactments of certain scenes from the stories still take place in many parts of Europe, such as the dance of the Grey Mare as well as the battle between the Grey Mare and the Black Wolf. In the Basque performances, the character playing the role of the wolf wears a furry mask and, while moving about, imitates the actions of the animal through his body language (Urbeltz 1994). In summary, these materials hold a key for recuperating, at least in part, that lost celestial narrative and the primal ecocentric cosmovision upon which it was founded.[10]

The present investigation attempts to bring into clearer focus the significance of the Bear Son cosmology by constructing a model capable of integrating and organizing the data collected to this point. The simulation, naturally hypothetical, has

---

8. These oral tales should not be confused with literary folk tales, such as those written down by Perrault, the Brothers Grimm, and Hans Christian Anderson. The latter tales have often undergone radical modifications under the pens of their modern authors and, most recently, under the nefarious influence of Disney and Warner Brothers (cf. Zipes 1983).

9. In reference to the long and involved process that eventually allowed me to recognize the remarkably ecocentric vision of the European Bear Son tales and their intrinsically sacred nature, I am greatly indebted to those Native American Wisdom-Keepers who shared their celestially coded tales with us at the Indian Star Knowledge II Conference held in Boulder, Colorado, November 5–7, 1993, under the sponsorship of AISES, the American Indian Science and Engineering Society. It was the decision on the part of those individuals who, under the inherently difficult circumstances imposed by sharing, often for the first time, such knowledge with outsiders, those not familiar nor at times fully cognizant of the deeply held ecocentric beliefs of Native American peoples, that allowed me finally to perceive the profoundly ecocentric nature of the coding of the Bear Son tales and the celestially oriented meanings behind many of the terms found in these European tales, such as the reference to travelling to "the End of the Earth." I am particularly indebted to Floyd Westerman and Curly Bear Wagner, Cultural Director of the Blackfeet Community College in Browning, Montana, for their astute comments on the European materials.

10. The Franco-Cantabrian region along with that of the Pyrenees has retained this older belief pattern remarkably well in which the bear shaman still dances away the ills of the community and, therefore, guarantees the health of the people, their animals and their crops. For a more general Pan-European overview of the role of the bear and the other celestially encoded characters, cf. Molina González and A. Vélez Pérez 1986; Montesinos González 1984 and Giroux 1984.

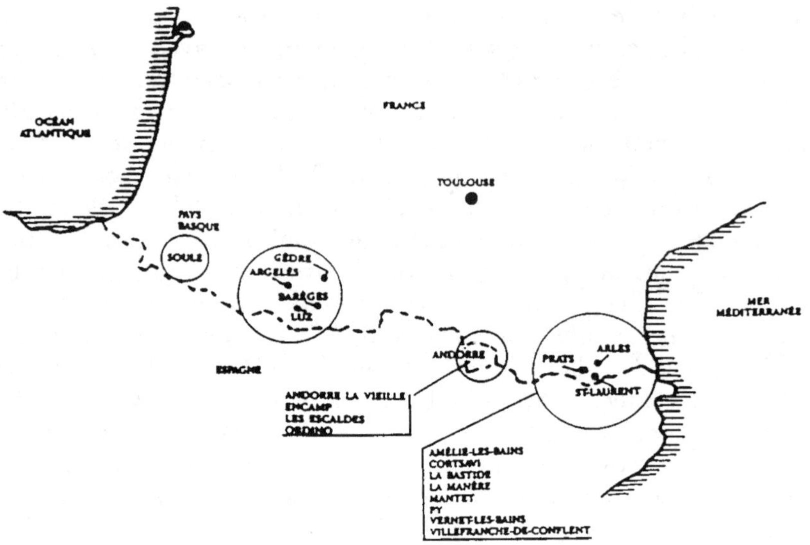

Figure 6. Sites of Candlemas Bear Celebrations in the Pyrenees. Source: Gaston (1987:20).

as its overall goal to chart the cognitive prehistory of astronomical thought in Europe in a more detailed and systematic manner than has been possible until now. Hence, the present study attempts to bring into focus the celestial coding of the Bear Son materials and map them onto these non-zodiacal constellations. The on-going research project has several other related goals. Just as in earlier studies (Maunder, Ovenden and Roy), in examining these non-zodiacal constellations one of the aims is to determine more accurately the moment, the time frame, when the star figures in question came to be projected onto the sky screen and at the same time to determine the latitudinal band along which such a projection took place. In contrast to these earlier investigations, ours focuses far more on exploring the motives that led the constellation makers to choose to paint their narrative onto the heavens. Also, special attention is placed on process, on the complex mechanisms of cultural storage that allowed for the retention and retransmission of the sky text from one generation to the next. In other words, without any evidence of the existence of a central authority and in the absence of written means of communication, the process of retention and transmission of these cultural artifacts must be explained through recourse to those vehicles of cultural storage and diffusion inherent to oral cultures.[11] Lastly, the present investigation has, above all, the goal of setting forth a research model that by integrating the Basque data, will allow us to probe far deeper,

11. Ovenden, for example, assumed that the diffusion of these non-zodiacal constellations took place according to processes inherent to the *elite-dominance* model in which a relatively small group of highly organized people transmit and/or impose their vision of the sky upon indigenous populations. In contrast, in our simulation we argue against the elite-dominance model, by asserting that the constellation figures projected onto the sky screen formed part of a preexisting cultural repertoire. Consequently, our model must attempt to analyze the implications of the zone of avoidance taking into account the mechanisms of cultural storage and transmission that allowed for the diffusion of astronomically encoded information, celestial texts and performances across Europe.

to go beyond the anthropocentric celestial discourse of the Greeks in an attempt to reconstruct the ecocentric cosmovision characteristic of hunters and gatherers from which the Bear Son saga ultimately derives.

For example, in reference to these non-zodiacal constellations, we shall allege that during the fourth millennium B.C. certain stars were selected, from the thousands available, to illustrate selected scenes from the Bear Son master narrative. Quite obviously, over time some of the scenes, constellation figures, placed there originally to illustrate the story may have been lost or misplaced. This appears to have happened in the case of the placement of the Mountain Lion and the Great Porcupine. Furthermore, we argue that this selection process was motivated by two interlocking sets of motives. First, there would have been the desire on the part of these Europeans to paint the value system intrinsic to their cosmovision in a permanent fashion on the largest and most visible screen available to them. Then, we should consider a second factor. Since the master narrative itself is both time-factored and celestially encoded, the second interlocking motivation must have been the desire to exteriorize the astronomical knowledge which they had already acquired, doing so in such a manner that the asterisms themselves would end up being utilized as mechanisms for the cultural storage of astronomical data. In one sense the constellation figures themselves serve mnemonically to illustrate the time-factored actions of the archetypal hero and the spirit animals that accompany him each year on his ritual journey, including his descents and ascents to different cosmic planes. Yet in order for the stars to be chosen and synchronized in such a way that the resulting asterisms would lend direct support to the time-factored narrative, those involved in this selection process had to have acquired prior knowledge of these stars and the way in which, on a yearly basis, they were perceived moving about on the sky screen. In other words, the decisions which these Europeans took in the process of sketching these scenes onto the stars, that is, in creating constellation figures capable of illustrating their text, require us to imagine a previous stage in the overall process: the period in time during which the tellers must have been debating over how to best project their story, a period that, consequently, must have been characterized by an intense interest in astronomical phenomena and one accompanied by extensive cultural and informational exchanges, activities that must have involved detailed observations of the night skies.

In summary, the conceptual projection onto the circumpolar stars of the figure of a Great Bear along with the placement of a Little Bear nearby appears to have resulted from this ecocentric cosmovision in which the veneration of earthly bears played a major role in the symbolic order of humans.[12] Little Bear, however, was also conceptualized as a archetypal shaman-initiate whose journeys were charted through the stars, while other initiates acted them out on the ground (Figure 7). Quite remarkably these celestially coded oral traditions encapsulated in the Bear Son tales have survived in Europe on the margins of elite discourse. Without passing fully into the literate world, over a period of several millennia they have functioned orally to preserve and transmit the Bear cosmogony, the earlier Pan-European myth of origins, from generation to generation.[13] The set of stories, performance

12. In contrast, a strikingly anthropocentric version of the origins of the Sky Bears is found among the Greeks (cf. Krupp 1991:232–234).

13. It should be noted that it was not until the latter part of the 19th century that folklorists attempted to collect these oral tales and commit them to writing. Yet already by the beginning

Figure 7. Bear Leader and Musicians. Source: Engraving from Olaüs Magnus, *Historia de gentibus septentrialibus*. Rome, 1555. Reproduced in Michel and Clébert (1968:329).

pieces and related beliefs, particularly abundant in the Pyrenean region, can be found in one form or another across all of Europe, particularly in the Caucasus region (Charachidzé 1986). In short, it was this archaic European ecocentric framework that ended up painting earthly bears onto the center of the canvas of heaven. And, these bears were originally accompanied by a number of other celestially encoded animals: an ant; a mountain lion and a porcupine; a dog and a hare; an eagle and a dove; several snakes; as well as a grey mare and a black wolf.

At first glance only a few of these animals would strike one as being readily recognizable among the constellation figures seen by European peoples in the night sky. Yet as we shall see, first impressions can be deceiving. In fact, in the course of our analysis it will be asserted that stars found in the following asterisms or adjacent to them, once formed an integral part of the sequence of scenes that were chosen from the stories and subsequently projected onto the celestial screen: Eridanus, Draco, Orion, the Hunter, Lepus and Canis Major, Columba and Puppis, then crossing over the Milky Way, Procyon in Canis Minor and Hydra, Centaurus killing Lupus, the Wolf whose lair is at the mouth of the Underworld, Ophiuchus triumphantly clutching Serpens, and, finally, Aquila flying back up along the Milky Way. Also, other inferential evidence suggests the prior existence of several smaller constellation figures which were projected onto the space inside or in the vicinity of the very large ancient constellation of Argo, the Navis. Finally, evidence derived from Pre-Greek oral traditions implicates Andromeda, Perseus, and Cetus in the latter part of tale (Gingerich 1984). In reference to the manipulation of these nonzodiacal constellations, we need to note that the story-line is sequential, requiring that the scenes appear one after another. Furthermore, over the last five thousand years there has been little change with respect to the these southernmost constellations, primarily a slippage of approximately one month. For example, around 3500

of the 20th century Panzer (1910), for example, had documented 221 European variants of the 301 story type, the descent of the Bear Son into the Under World.

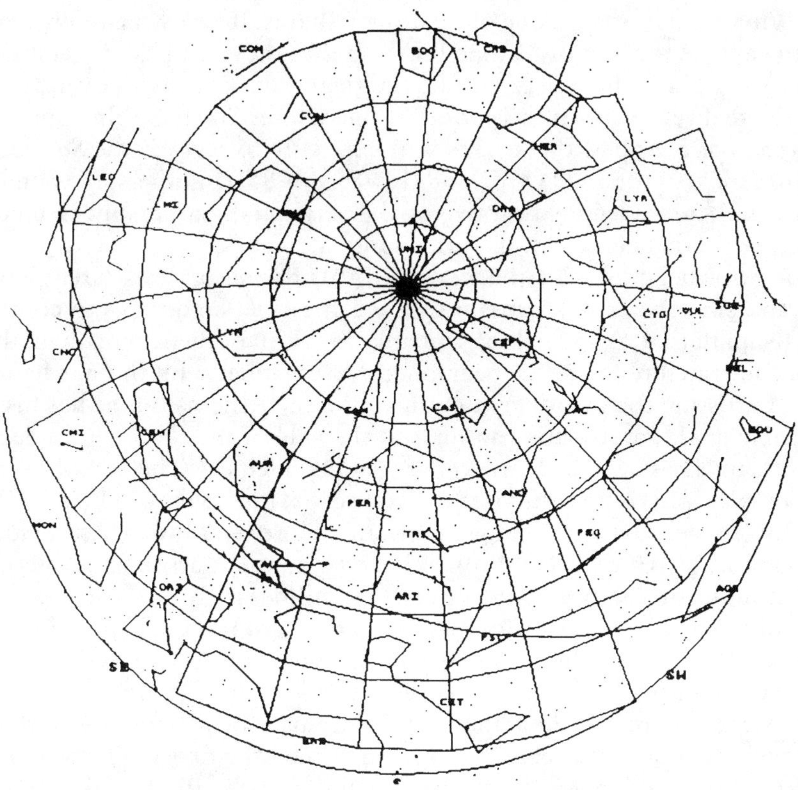

Figure 8. 12:00 AM, Oct 01 3500 BC, Barcelona. 41% 25' 2% 10' E, Dir 180% = South, Elev 89%.

B.C. Eridanus appeared passing the meridian at midnight at the beginning of October (Figure 8), while today it does so around the first of November (Arregi and Frank 1996).

The first chapter of this study (Frank 1995, 1996a) addressed data relating primarily to the supernatural birth and childhood of Hartzkume, the Bear Son, i.e., how he was born in the Sky Cave to a human female and a Great Bear, and how, when the hero was seven years old, he left the cave and went out into the world, a shamanically coded act coinciding with the descent of the intermediary being from the Upper World to Middle Earth. Because of his unique birthright the Bear Son is different from other humans. Sometimes he has "bear-ears." And he is almost always hairier than the rest. Being quite burly and exceedingly cunning, he manages to stave off the taunts of others, although occasionally punishing their impudence roundly. While he is still a youth, he sets out on a journey, a vision quest, that will take him to the End of the Earth. And since Middle Earth where we live is flat like a disk and the firmament that we view, set down on its rim, as a tent when it is placed, the Bear Son ends up at the horizon, at the End of the Earth, where the firmament is "so low that the swallow has to drink water kneeling on the black cottonwood (Erdész 1963, 1961)". Also, we need to keep in mind that there is preexisting problem, often not mentioned explicitly in the stories themselves: the hero must rescue the Sun from the clutches of the Under World Serpents (Erdész 1963,

1961). While the latter have one of their own, that is, their own Sun, they persist in trying to capture the Sun of Middle Earth so that they can bask in its rays all day and all night. The light source will be theirs and darkness will overtake Middle Earth, a coded reference to the period of the Winter Solstice.[14] In some versions, such as the Basque ones, the hero is sent explicitly to recover the *Su-Haziak*, the "Seeds of Fire" (Cerquand [1875–1882] 1985:84–85). In others, the Sun becomes conflated with a maiden held captive by the Serpent(s) and, finally, in feudally encoded variants, this figure is transformed into a princess.

Soon after beginning his journey skywards the hero sees a Great River (Eridanus) and since he cannot cross it to the other side, he decides to continue following the path that he spies along its bank. In our hypothetical reading, this scene would be narrated today at the beginning of November with the star figure of the Sky River crossing the meridian at midnight. Shortly thereafter, he has his first adventure. He comes across four animals at the side of the road with a dead ass in front of them. These will be Hartzkume's Animal Helpers. The first of the four, the Mountain-Cat, calls out to the youth: "Excuse me, but could you help us? We have a real problem because we're hungry, but can't figure out how to best divide up this beast's carcass" (Arratibel 1980). Walking over to the foursome, our hero doesn't hesitate a moment and begins to cut up the dead beast. To each of the animals he offers a different part: to the Mountain-Cat, the hank; to the Dog, a smaller piece of meat; to the Eagle, the liver and intestines—the latter has no teeth—and finally, the Ant gets the ass's hide and bones, the latter so that after Ant eats the marrow, they can serve the creature as a home when it rains. The Mountain-Cat thanks Little Bear, telling him that because he has performed the division so well, each animal is going to give him a talisman as a sign of appreciation, and if he is ever in need of their skills, all he has to do is call out their names and he will have the power to shape-shift into their form.

Apparently the tokens, for example, the tuft of hair of the Mountain-Cat, the Eagle feather and the Ant's leg—for some reason the tales never mention what the Dog gives him—are intended to form part of the young Bear-Shaman apprentice's Medicine Bundle. By manipulating the token of each Animal Guardian, the hunter-apprentice is allowed to shape-shift, taking on the identity of the creature in question; acquiring from it the skills necessary to become a Great Hunter: strong of body like a mountain cat; fleet of foot, quick, like a fox; keen of eye like an eagle; and silent like the ant, who moving about in nature can make herself nearly invisible. In turn, the encounter with the four animals in itself must be read as a parable. It serves to test the hero's powers of perception, concretely, his understanding of the profound interrelationship of all beings: the important role played by each in nature's scheme. The Parable of the Dead Ass speaks to us of the web of nature. Large predators bring down the prey. Smaller predators, wild dogs, jackals and foxes, approach to feast on it, while carrion birds soon gather about gorging themselves on the soft parts, the innards, and then, in the end it is the tiny ant who draws near to chew on the hide and finally pick clean the animal's bones, leaving only the skeleton behind.

---

14. In the folktales the Bear's adversary is traditionally a Wolf (Díaz 1994). In Euskara the month of February is called *Otsaila*, the month/moon of the wolves; while January is sometimes called *Ilbeltza*, the month/moon of darkness. Also, it was in this period that bears disappeared from the earth, into hibernation, leaving the land to the wolves. A similar coding exists in Bulgaria (Koleva 1995).

Shortly after this encounter, the protagonist puts one of his talismans to work, the Ant's. With no difficulty he silently climbs up the Apple Tree (the Sky-High-Tree) and, hidden among its branches, overhears the Apple Tree Serpent (Draco) explaining the Magic Formula that will bring about the latter's death. Little Bear discovers that the Apple-Tree Serpent has a (Twin) Brother, also a serpent and a shape-shifter, a shaman, just like the Apple-Tree Serpent himself. The Serpent Twin (Hydra) is often referred to in Euskara as the Herensuge, the Great Ancestor Serpent. The hero overhears the old Apple-Tree Serpent saying that the only way he will die is if the following formula is utilized: the challenger will see a Giant Porcupine (the other Serpent Brother in disguise) and, changing into a Mountain-Cat, he will fight the former. Then when the Giant Porcupine is killed—a scene we assume was once located in Orion—a Hare will appear; changing into a Dog (Canis Major) the protagonist will chase the Hare (Lepus); having killed the Hare, a Dove (Columba) will appear, and changing into an Eagle (Puppis perhaps), the shape-shifted hero will swoop down snatching the Dove in mid-flight; and having killed it, an Egg will appear. At this point the challenger must take the Egg (Procyon) and hit the Dragon, the Apple Tree Serpent's Brother (Hydra), on its forehead, causing its death. When this is done, the egg inside the head of the Apple Tree Serpent will burst, and he, too, will die (Apalategi 1987; Arratibel 1980; Carrière 1937; Cosquin 1887).

In the Apple Tree Serpent's discussion of the ritual formula there are clear hints of shamanic combat in which two shamans, each capable of shape-shifting, engage in battle on behalf of their respective constituents.[15] As Gloseski (1989:180) has noted, characters found in shamanic narratives are quite often shape-shifters, which holds true for the shaman himself, for his enemies, as well as for his allies. Furthermore, we see that the combats themselves are those reflecting the natural order of things, the interrelationship between prey and predator: between a porcupine and a wild cat, a hare and a fox, a dove and an eagle. In this ecocentric drama, Little Bear plays the role of the predator, while his adversary, the Serpent shaman, plays the role of his prey, his next meal. Taking wilderness as its standard, the ritual landscape acts as a repository for the interlinking ecological relations that permit and actually require the coexistence of predator and prey.[16]

---

15. In reference to the combat between two shamans and the role of their Animal Helpers, Kalweit (1992: 198–199) offers the following comments which are pertinent to our discussion: "For the Lapps, the underworld is a place of danger. Not only are the beings that live there capable of harming shamans, but foreign shamans with hostile intentions can make them lose their way so that they remain there forever. It is said that many shamans have fallen by the wayside in this manner. As the Yakut shamans stage wars between animal spirits every year, the Lapp shamans combat each other by confronting each other and setting their helping spirits on each other in the form of reindeer bulls or spirit birds. Among the Scandinavian Lapps, the helping spirit is called the *sueje*, or shadow. There are shadow fish, shadow reindeer, and shadow snakes, and they accompany shamans on journeys to the underworld to bring back dead or kidnapped souls. If such a shadow animal is injured in a fight, the injury is transferred to the possessor of the animal, and if the animal is killed, the shaman must die as well."

16. Rather than pitting an anthropomorphic hero against a wild beast and, hence, acting out the dualism typical of the tenets of modern Western thought, the Bear Son narrative reaffirms interrelationships, that eating is a sacrament. In this manner the text highlights Gary Synder's notion of "re-inhabitation" (Synder 1974, 1990, 1995)—closely related to Polyani's concept of "embedded" cultures (cf. Rogers 1994)—and further underlines the importance of Shepard's assertion concerning "native" cosmologies, that they demonstrate the presence of a

Once having acquired the Magic Formula, our hero has no problem whatsoever in carrying it out because he has the talismans necessary to do so. With the demise of the Many-Headed Serpent whose stars or heads Little Bear cuts off night after night, this sequence terminates, one that has taken the hero across the path of the Milky Way down into the Under World. We assume that our hero hits the Serpent Twin on its noggin shortly after the Winter Solstice. And as the months pass, Hydra slowly disappears, decapitated, into the western sky, while at the same time the Sun is slowly rising, for it is being gradually released from the clutches of the Under World Serpents. Then, just below the tail end of the Serpent Hydra, the next scene begins to come into view: that of the Grey Mare and the Black Wolf.

But before continuing with the narrative, we need to introduce this furry character who appears when the hero first enters the Dark Woods. It is the female Black Wolf (Lupus), also referred to as a Black Beast and in some versions as a Black Dog with a White Star on its forehead. The latter can be killed only if the White Star on its forehead is pierced. Nearly from the moment the hero enters the Dark Woods, this creature starts harassing him, teasing and playing tricks on him, much like the Trickster character of North American Indian tales, often portrayed as a sly coyote.[17] But in the Hartzkume narrative the Trickster is not a coyote but rather a crafty black wolf. Finally, when the hero who we must recall is a bear shaman, is about to catch his nemesis, the Black Beast dives down a hole with the hero in hot pursuit behind him. The hole, sometimes compared to a mine-shaft or a chimney-like aperture, is the one that leads the protagonist into the Under World where he puts into practice the Magic Formula and finally kills the Dragon. This aperture is known as the Hole of the World which, in shamanic fashion, links the two realms, allowing the shaman to move easily between them (Erdész 1961:332).

Still in the Underworld, a Grey Mare now appears and offers to help Little Bear slay the Black Wolf. In what appear to be somewhat more modern versions of the tale, the Grey Mare is referred to as a Centaur, a man-horse. Furthermore, in some Spanish language versions of the tale the horse-figure intervenes killing not a Black Beast, but rather a Black Moor. In contemporary reenactments of the cosmic battle between the forces of light and darkness, those playing the role of Grey Mares often appear as dancing Hobby-Horses who, in turn, attack their enemies, figures dressed in black. It is a popular folk performance found throughout much of Europe. In parts of western Europe these reenactments are known as the "Morris Dances," while in Poland, the Black Wolves are reinterpreted as more contemporary historicized enemies, Black Tartars (cf. Giroux 1984). In Spain this performance goes by the name of "Moors and Christians." Thus, these ritual replacements along with substantial other evidence, make us suspect that in Spain during the early Middle Ages under the influence of Christianity, the traditional Bear Son portrait in the sky, the scene showing the Grey Mare killing the Black Wolf, was retouched. On top of this shamanically coded sky picture was sketched a slightly different scene, one more acceptable in terms of emerging political-religious norms of the times: the portrait of a Christian knight astride a white mare killing Black Moors. In fact, in Spain this

---

cautious willingness to consume and be consumed, both literally and in a mythic sense (Shepard 1995).

17. For the role of the trickster fox in Dogon Cosmology and its relationship to star knowledge, cf. Griaule and Dieterlen 1986. The classic study, although not informed by ethnoastronomy, is that of Radin 1956.

medieval Christian hero is known by the epithet "Santiago Matamoros" which means "St. James Killer of Moors" (Sánchez Dragó 1992). Hence, we see one set of cultural icons being substituted for another.

Finally, having killed the Black Wolf, the protagonist finds himself trapped in the Under World, often inside Ice Mountain, with no way to escape until a female Eagle appears to him and after a series of negotiations, offers to fly him back to the Upper World. In some versions, Little Bear shape-shifts into an Eagle, but in many others, he is obliged to feed the bird pieces of meat which he has placed in a sack just for that purpose, to sustain the Eagle on her long journey. When the meat gives out, he begins to cut off other pieces of flesh, this time from his own thigh, so that the Eagle will have enough strength to complete her long journey back to the Upper World, along the Milky Way. Apparently this chapter formed part of the repertoire of oral tales that were inherited by the Greeks and later passed on by them, but in a fragmented form. For instance, the Greeks appear to have reframed this chapter of the story, by means of ritual replacement, substituting the figure of Prometheus for that of Little Bear (cf. Charachidzé 1986).[18] Although the annual round of time-coded adventures do not conclude here—the Perseus, Andromeda and Cetus segment lasting today from approximately July through October is still missing—nonetheless, for the moment the previous summary of the main points of the narrative sequence will suffice to give a general idea of the complexity and repetitive nature of the Bear Son saga itself: the way in which year after year selected asterisms were employed in an annual round of ritual practices.[19]

Certain aspects of the tales, for example, their ecocentric focus and, consequently, non-anthropocentic nature, suggest that they should be traced back to the cognitive practices of hunter-gatherer peoples (cf. Oelschlaeger 1991; Shepard 1995). It was the oral repertoire of these peoples that was transmitted directly to the Basques and other European populations still living within orality, while it was passed on, although in a more fragmented and confused form, by the Greeks in their writings (elite discourse). Consequently, we have two different sources for recuperating the same Pan-European materials: elite discourse (written transmission) represented by the celestial stories recorded by the Greeks and popular discourse (oral transmission) which has retained the more archaic versions of the Bear Son narratives.[20] In

18. According to the Greek version, the reason humans were deprived of the celestial fire is because Zeus was angered by the manner in which Prometheus performed the division of the dead ox, an episode that corresponds structurally to the "Parable of the Dead Ass" in the Bear Son saga (cf. Charachidzé 1986; Frank 1995; Vernant 1982:154–169).

19. It should be noted that folk performances still form an integral part of popular theater in the Pyrenean region. This oral legacy motivates the public reenactment of scenes from the Bear Son narrative (cf. Alford 1928, 1930, 1931; 1937; 1969; Dentaletche 1981, 1986; Fabre 1968, 1986; Praneuf 1989). These, in turn, are linked to residual practices of Bear Ceremonialism, e.g., the ritual bear hunt still conducted in this region. In the Basque zone the cast of characters includes a number of the Animal Guardians: the *Katuzain*, the Cat-Keeper (the Mountain Lion); *Zamalzain*, the Horse Keeper (the Grey Mare) as well as the opposing set of shamanically encoded characters: *Pitxou*, the Black Wolf and the *Herensuge* (the Serpent Dragon). In villages of Catalunya, we find yet another of the Animal Guardians, an individual playing the role of the Eagle as well as even the *Tarasca* (Cetus). However, the Ant Animal Guardian seems to have disappeared entirely from such performances.

20. Perhaps the most remarkable example of the efficacy of oral mechanisms of cultural storage and transmission can be identified with respect to the details of Bear Son's encounter with the female Eagle (Aquila) and his ascent from the Under World. Basing his work, in part,

judging the archaic nature of the data retained in the oral transmission of the tales, of special interest to us has been the encounter between the Grey Mare and the Black Wolf. The texts themselves strongly suggest that over time both these creatures underwent a process of ritual substitution. The evidence points to a situation in which the character playing the role of the enemy in the Bear Son tale was transformed and ultimately historicized. In the case of Spain we can see the Black Wolf becoming a Black Beast and, finally, a Black Moor. In turn, as we have noted, the Bear Son hero riding on his Grey Mare came to be identified in the stories themselves with a Centaur, the latter also being the portrayal of the constellation figure encountered on Greek sky charts. In Spain, during the early Middle Ages, the transformation process went a step further for the celestial Bear Son hero appears to have become historicized in the popular imagination and, hence, integrated into the on-going strife between Christians and Moors. There his role was assumed by a newcomer to the sky stage, the celestial hero St. James who, according to pious legend, was seen by Charlemagne in a vision flying across the heavens and killing Black Moors (Peake 1919).

Furthermore, in the tales the expression the End of the Earth is not understood simply as a reference to some vague location to which the hero travels on his journey, but rather also as a reference to a specific geographical location on the European continent, concretely, to the End of the Earth which was found at the westernmost extreme of continental Europe where land mets the sea (Figure 9). There, not far from Finisterre, we discover the port of Padrón in Galicia, renowned as the final destination of pilgrims charting the route of St. James. Yet in popular belief this same location is connected to a confrontation between a Black Wolf and a Grey Mare, a duo known to us on Greek sky charts as Lupus and Centaurus. Thus, it would appear that over time the constellation figure of Centaurus has been rescripted because of the influence of Christianity, as St. James, who is portrayed as riding not some fierce stallion, but rather a white mare. It is this Christian hero who was said to kill the archetypal enemy, the Black Moor (Frank 1995, 1996b, 1996c). Additionally, for many years scholars have recognized that the pilgrimage route itself as well as the veneration of the large dolmen incorporated into the church of St. James

on the oral repertoire of tales related to the Caucasian legendary hero Amirami, Charachidzé (1986) demonstrates their clear filiation with those linked to the Greek Prometheus. However, in contrast to Prometheus's adventures, those associated with Amirami contain the standard motifs found in the more archaic versions of the European Bear Son saga. Furthermore, the episode in which Amirami ascends triumphantly to the Upper World with the aid of the female Eagle replicates with astounding fidelity details found only in the Pyrenean versions of this sequence, including the fact that the female Eagle agrees to help the hero precisely because he has killed the Serpent that had been preying on her eaglets. The significance of this level of detail in both versions, the Caucasian (Charachidzé 1986) and Pyrenean ones (Carrière 1937), at the opposite extremes of Europe, is highlighted by the fact that it does not appear in other versions of this episode documented throughout the rest of Europe. Stated differently, the replication of this episode in nearly identical terms in the western and eastern variants of the tale suggests that we are dealing with elements from the deeper Pan-European celestially coded cognitive stratum which were later retransmitted to the West, although in a fragmented fashion, by Greek mythology, specifically by means of the episodes linked to Prometheus and his unfortunate encounter with an eagle who is portrayed not as a willing and grateful ally, as his Animal Guardian, but as his enemy. In fact, in the more archaic ecocentric versions after the hero has mutilated himself in order to feed the female Eagle, she, in turn, using her own shamanic powers, heals him, making him whole again.

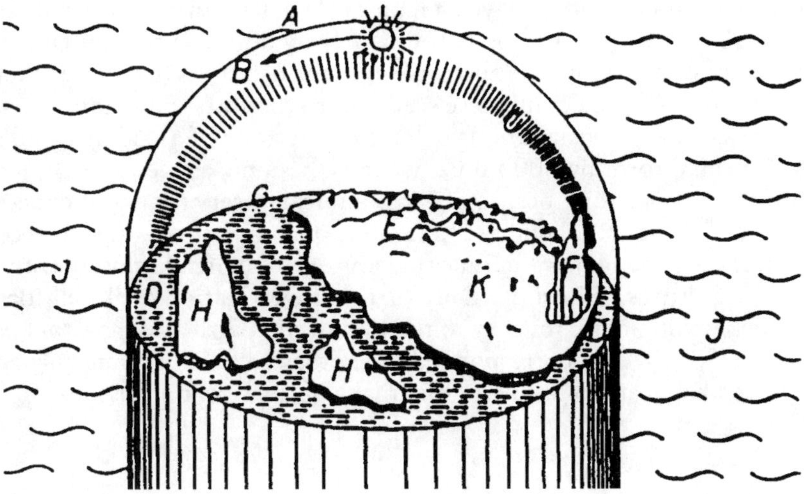

Figure 9. The Middle World after L. Ami. A: the Firmament; B: the orbit of the Sun; C: the thick layer of air; D: the places where the Sun rises and sets; E: the Ice Mountains; F: the Glass Mountains; G: the End of the World; H: the Continents; I: the Sea; J: the Sea within the Universe; K: the Hole of the World. Source: Erdész 1961:331.

Patronensis at Padrón, the last stop on the pilgrimage route to Santiago de Compostella, belong to a much earlier pre-Christian pattern of belief which connected the journey itself in some fashion to the Milky Way (Peake 1919). Similarly, in our reading of the celestial coding of the data, the archetypal quest of the Bear Son hero requires him to traverse almost the entire path of the Milky Way. In short, the ritual landscape associated with the pilgrimage route of Little Bear, a ceremonial pathway traditionally attached to the Milky Way, may serve to illuminate astronomical practices once found among the megalithic peoples of Europe (Sánchez-Dragó 1992:334–428).

In conclusion, our conceptual model for simulating, in a very hypothetical manner, one aspect of the cognitive prehistory of astronomy in Europe has given particular attention to processual questions: storage and transmission of cultural artifacts. For instance, in this simulation the individual and collective ritual journeys modeled on the archetypal cosmic pilgrimage of Little Bear to the End of the Earth would have acted, over several millennia, as central mechanisms in the elaboration, storage and retransmission of these astronomically coded practices. The model argues that the figures were placed in the sky in order to facilitate the collective retelling of time-factored tales and related interactive ceremonies in a public arena, all of which contained a fundamentally educational, transformative dimension. In fact, we are dealing with a ritual landscape in which the adventures of the hero, the keeper of the cosmic order, may have been mapped onto the sky in a direct attempt to take advantage of the path, not of the sun, but rather of the Milky Way: a celestial pathway of initiation for the earthbound in which the individual and collective journey follows the tracks of the ancestors. In this sense the archetypal Bear Shaman is similar in many ways to the remarkable visionary ancestor hero Scar Face of the Blackfoot (Cajete 1994:64–69). The present model argues that the specific set of constellations under consideration emerged from a similar ecocentric cognitive framework, a vitalist one in which shamanic and animistic practices related to Bear Ceremonialism of hunters and gatherers played a major role in structuring the symbolic order of humans. Yet this assertion should not in anyway diminish the fundamental role and importance of astronomical observation in the decisions that were taken in the past in the process of projecting these scenes onto the stars. For example, if we accept that the southern constellations delimiting the zone of avoidance had already received their final form by 3000 B.C., we must assume, as we have alleged previously, that there was a long period of astronomical observation and experimentation that preceded the actual point in time when they were finally able to accomplish the projections desired, a point in time that appears to coincide closely with the rise of megalithic cultures in Europe. Only further investigations will tell for certain whether the Bear Shaman's journey to the End of the Earth along a ritual pathway marked by the Milky Way, corresponded to another, a collective Pan-European one which allowed believers to follow in the tracks of their bear ancestors.

References

Alford, V. 1928 The Basque Masquerade. *Folklore* XXXIX (March): 67–90.
———. 1930 The Springtime Bear in the Pyrenees. *Folklore* XLI (Sept.): 266–279.
———. 1931 The Candlemas Bear. *National and English Review* (Feb.): 235–244.
———. 1937 *Pyrenean Festivals: Calendar Customs, Music and Magic, Drama and Dance*. London: Chatto and Windus.
Apalategi, J. 1987 *Introducción a la historia oral: Kontuzaharrak (Cuentos viejos)*. Barcelona: Editorial Anthropos.
Arratibel, J. 1980 *Kontu zaarrak*. Bilbao: Editorial La Gran Enciclopedia Vasca.
Arregi, J. and R. M. Frank. 1996 An Etnological Approach to the Origin of the Non-Zodiacal Constellations. V Congreso Internacional (SEAC). Astronomía en la Cultura. Salamanca, Spain, 3 September.
Cajete, G. 1994 *Look to the Mountain: An Ecology of Indigenous Education*. Durango, CO: Kivikí Press.
Carrière, J. M. 1937 *Tales from French Folk-lore of Missouri*. Evanston and Chicago: Northwestern University Press, Nos. 10–11.
Cerquand, J. F. [1875–1882] 1985 *Ipar Euskal Herriko Legenda eta Ipuinak. Basajauna, Laminak, Tartaroa*. Anuntxi Aranaren Transkripzioa. San Sebastián: Editorial Txertoa.
Charachidzé, G. 1986 *Prométée ou le Caucase: Essai de Mythologie contrastive*. Paris: Flammarion.
Cosquin, E. 1887 Jean l'Ours. *Les Contes populaires de Lorraine*. Paris: F. Vieweg, Libraire-Editeur. pp. 1–27.
Dégh, L., ed. 1965 *Folktales of Hungary*. Chicago: University of Chicago.
Dentaletche, C., ed. 1986 *L'Ours brun: Pyrénées, Agruzzes, Mts. Cantabriques, Alpes du Trentin*. Pau: Acta Biologica Montana.
Dentaletche, C. and Le Groupe Ours-Pyrénées. 1981 *L'Ours*. Pau: F.L.E.P.
Díaz, J. 1994 Personal Commnication. 14 July.
Erdész, S. 1963 The Cosmological Conceptions of Lajos Ami Storyteller. *Acta Ethnographica* XII: 57–64.
———. 1961 The World Conception of Lajos Ami Storyteller. *Acta Ethnographica* X: 327–344.
Fabre, D. 1986 Le Carnaval de l'Ours á Saint-Laurent de Cerdans. *L'Ours brun: Pyrénées, Agruzzes, Mts. Cantabriques, Alpes du Trentin*, ed. C. Dentaletche. Pau: Acta Biologica Montana. pp. 135–148.

Fabre, D. 1968 *Jean de l'Ours: Analyse formelle et thématique d'un Conte populaire*. Carcassone: Editions de la Revue 'Folklore'.

Frank, R. M. 1996a Hunting the European Sky Bears: When Bears Ruled the Earth and Guarded the Gate of Heaven, eds. V. Koleva and D. Kolev. *Astronomical Traditions in Past Cultures*. Sofia: Institute of Astronomy, Bulgarian Academy of Sciences, National Astronomical Observatory Rozhen. pp. 116–142.

———. 1996b Hunting the European Sky Bears: Ritual Landscape and Ceremonial Pathway of Little Bear Hartzkume. *V Congreso Internacional. Astronomía en la Cultura*. Salamanca, Spain, 3 September.

———. 1996c An Essay in European Ethnomathematics and Orality: Celestial Traditions of John Little Bear in Europe and Euskal Herria (the Basque Country). Fifth International Conference on Mathematics Education. Working Group 21: Mathematics and Culture. ICME-8. Seville, Spain, 15 July.

———. 1996d Epilogue: The Urnieta Project and Methodological Considerations. *Sarobeak Urnietan-Seles en Urnieta-Stone Octagons in Urnieta*, Luix Mari Zaldua. Urnieta, Spain: Kulturnieta. pp. 90–100.

———. 1995 Zeruko bi hartzen aztarnak Europa zahar eta berrian: Etnoastronomia Europako Ahozko Kulturan. Udako Eusko Unibertsitatea. Iruña-Pamplona, Spain, 20 July.

Frank, R. M. and J. D. Patrick. 1993 The Geometry of Pastoral Stone Octagons: The Basque Sarobe. *Archaeoastronomy in the 1990's*, ed. C.L.N. Ruggles. Loughborough, England: Group D Publications. pp. 77–91.

Gingerich, O. 1984 Astronomical Scrapbook. The Origins of the Zodiac. *Sky and Telescope* 67 (March): 218–220.

Giroux, G. 1984 *Carnavals et Fêtes d'Hiver* (Paris: Centre Georges Pompidou, e.d.l.e.c.

Glosecki, S. O. 1988 The Wolf of the Bees: Germanic Shamanism and the Bear Hero. *Journal of Ritual Studies* 2.1 (Winter): 31–53.

Glosecki, S. O. 1989 *Shamanism and Old English Poetry*. New York and London: Garland Publishing, Inc.

Griaule, M. and G. Dieterlen. 1986 *The Pale Fox*. Trans. Stephen Infantino. Chono Valley, AZ: The Continuum Foundation.

Hartsuaga, J. I. 1987 *Euskal Mitologia Konparatua*. Donostia: Kriselu, S.A.

Kalweit, H. 1992 *Shamans, Healers and Medicine Men*. Boston: Shambhala Publications, Inc.

Koleva, V. 1995 Letter to the author. 20 February.

Krupp, E. 1991 *Beyond the Blue Horizon*. New York: HarperCollins Publishers.

Maunder, E. W. 1913 The Origins of the Constellations. *Observatory* 36: 329–334.

Maunder, E. W. 1909 *The Astronomy of the Bible*. London: Hodder and Stoughton.

Molina González, A. and A. Vélez Pérez. 1986 L'Ours dans les Fêtes et Carnavals d'Hiver: La Vijanera en Vallée d'Iguna. *L'Ours brun: Pyrénées, Agruzzes, Mts. Cantabriques, Alpes du Trentin*, ed. C. Dentaletche. Pau: Acta Biologica Montana. pp. 135–146.

Montesinos González, A. 1984 *Fiestas populares de Cantabria 2: Carnavals rurals*. Tantín: Santander.

Oelschlaeger, M. 1991 *The Idea of Wilderness*. New Haven & London: Yale University Press.

Ortutay, G. 1962 *Hungarian Folk Tales*. Corvina, Budapest: Kossuth Publising House.

Ovenden, M. W. 1966 The Origin of the Constellations. *The Philosopical Journal* 3.1: 1–18.

Panzer, F. 1919 *Studien zur germanischen Sagengeschichte, I, Beowulf*. Munich.

Peake, H. 1919 Santiago: The Evolution of a Patron Saint. *Folklore* XXX: 208–226.

Peillen, T. 1986 Le Culte de l'Ours chez les anciens Basques. *L'Ours brun: Pyrénées, Agruzzes, Mts. Cantabriques, Alpes du Trentin*, ed. C. Dentaletche. Pau: Acta Biologica Montana. pp. 171–173.

Praneuf, M. 1989 *L'Ours et les Hommes dans les Traditions européennes*. Paris: Editions Imago.

Radin, P. 1956 *The Trickster: A Study in American Indian Mythology*. London: Routledge and Kegan Paul.

Rogers, R. 1984 *Nature and the Crisis of Modernity*. Montreal: Black Rose Books.

Roy, A.E. 1984 The Origins of the Constellations. *Vistas in Astronomy* 27: 171–197.

Sánchez Dragó, F. 1992 *Gárgoris y Habidis*. Madrid: Planeta.

Shepard, P., and B. Sanders. 1992 *The Sacred Paw: The Bear in Nature, Myth and Literature*. New York: Arkana Press.

Shepard, P. 1995 *The Others: How Animals made us Human*. Washington, D.C./Cavelo, CA.: Island Press/Shearwater Books.

Stitt, J. M. 1992 *Beowulf and the Bear's Son: Epic, Saga and Fairytale in Northern Germanic Tradition*. New York and London: Garland Publishing, Inc.

Synder, G. 1995 Re-Inhabitation. *The Deep Ecology Movement*. Eds. A. Drengson and Y. Inoue. Berkeley, CA: North Atlantic Books. pp. 67–73.

Synder, G. 1990 *The Practice of the Wild*. San Francisco: North Point Press.

Synder, G. 1974 *Turtle Island*. New York: New Directions.

Stitt, J. M. 1992 *Beowulf and the Bear's Son: Epic, Saga and Fairytale in Northern Germanic Tradition*. New York and London: Garland Publishing, Inc.

Urbeltz, J. A. 1994 *Bailar el caos: la danza del la osa y el soldado cojo*. Iruña-Pamplona: Pamiela.

Vernant, J-P. 1982 *Mito y sociedad en la Grecia antigua*. Madrid: Siglo XXI.

Zaldua, J. M. 1996a Configuración y orientación del *Saroe* y su *Artamugarri* in Euskal Herria. V Congreso Internacional: Astronomía en la Cultura. Salamanca, Spain, 3 September.

———. 1996b *Saroeak Urnietan-Seles en Urnieta-Stone Octagons in Urnieta*. Urnieta, Spain: Kulturnieta.

Zipes, J. 1983 *Fairy Tales and the Art of Subversion*. New York: Wildman Press.

34

Solar Eclipses and Encke's Comet on Swedish Rock Carvings

Göran Henriksson
Astronomical Observatory
Uppsala, Sweden

The prehistoric rock carvings in Sweden have been studied for more than a century without general consensus about their meaning. Using a program developed by the author between 1975–1985, the first identifications were made of total solar eclipses depicted on Swedish rock carvings from the Bronze Age, 1800–500 BC. The accuracy of the program has been checked by comparison with written records of total solar eclipses.

In 1991 the author found that important phenomena in the sky such as solar eclipses would have been dated in a calendar with six ships that carried the sun along the ecliptic. The ships have individual shapes and can be identified as constellations. A total solar eclipse was simply dated by a correct position of the symbol for the sun in the actual calendar ship. If planets were visible during the eclipse they were reproduced as cup-marks in correct position in relation to the sun.

A comet was visible during the two total solar eclipses in 1596 BC and 858 BC. The author has been able to identify this as the periodic comet Encke and its orbit has been checked by comparison with the Chinese chronicles. The earliest identification of Encke on the Swedish rock carvings goes back to 1758 BC.

* * *

The Rotation of the Earth

The rotation of the earth is decelerated due to the braking effect of the tidal forces from the moon and the sun. However, the theoretical calculations are useless in this case and we must rely on empirical determinations based on ancient historical records of, preferably, total solar eclipses. The problem is that we can easily enter a situation with circular arguments because the historical chronology is often based on solar eclipses mentioned in the ancient texts and the time corrections that must be used in the computations of the dates for the eclipses are based on the same historical records.

It is therefore necessary to use a more or less unbroken series of well-recorded solar eclipses from different parts of the world and avoid dubious records that cause confusion. The German astronomer Carl Schoch cooperated closely with archaeologists, historians, and linguists and could base his identifications of solar eclipses on the best available ancient texts. His formulas have been improved but his determination of the so called lunar orbital acceleration –24.4 arc seconds/(century)^2 has been used in my computer program (Schoch 1931).

The computer program was constructed between 1975–1985 and has been carefully checked with all known relevant sources. These include Copernicus' personal copy of Calendarium Romanum Magnum published by Johann Stöffler in 1518, now in the library at the Astronomical Observatory in Uppsala, with his notations of the time for the main phases of five solar eclipses, medieval monastery chronicles, Classical Greek texts, Hittite annals and cuneiform tablets from Babylon (Schoch 1931), and Chinese oracle bone texts and chronicles (Pang 1988).

The most important of the cuneiform texts tells us about a total solar eclipse in Babylon in 136 BC April 11 (April 15 Julian Calendar), with the time given for the different phases of the eclipse. This eclipse is described on two separate tablets in the British Museum collection. The following translation of one of the texts is published in Historical Eclipses by F. R. Stephenson (1982) and is made by Abraham J. Sachs of Brown University (1974): "Twenty-four *us* after sunrise a solar eclipse [obscured] on the southwest side when it began.... Venus, Mercury and the normal stars [meaning those that were then above the horizon] were visible; Jupiter and Mars, which were in their period of disappearance [meaning the interval between setting with the sun and rising with the sun], were visible in that eclipse.... [The shadow] moved from southwest to northeast." The totality was reached after a further 18 *us* and the other text tells us that the duration of the entire eclipse was 35 *us*, in good agreement with the time for totality given by the first text. This is the best documented solar eclipse before the 18th century when the astronomers used telescopes.

The difference between the time recorded in the cuneiform text and my computed time is 0.2 minutes. This cuneiform text is an independent check of Schoch's determination of the lunar orbital acceleration because it was completely unknown to him. The published computer programs that I have tested have an error of about 16 minutes, which makes the result of the computations useful only under the most favourable circumstances. The errors depend on the square of the time interval before 1900 AD which rapidly makes the results completely useless for exact identifications for more distant eras.

During the spring 1996 it became clear that no published computer program was even able to reproduce the total solar eclipse at sunrise in Athens 484 AD. However, my program shows that the eclipse was total in Athens and at the moment of maximum eclipse the lower limb of the sun was 0.3° above the horizon. A general review of the problems in the computations of ancient eclipses is given by Stephenson and L. V. Morrison (1984). The value –26.0 arc seconds/(century)^2 has been used as the lunar orbital acceleration in the "Atlas of historical eclipse maps, East Asia 1550 BC–AD 1900" (Stephenson & Houlden 1986). However, this choice of lunar secular acceleration has unfortunately not been successful.

It is very satisfying that my computer program can correctly reproduce the so called double dawn solar eclipse in Zheng, 899 BC April 11, described in the Chinese texts, because it overlaps in time with the youngest of the so far identified total solar eclipses, 858 BC July 5, on Swedish rock carvings from the Bronze Age. Because of the demonstrated great problems for the established authorities to compute ancient solar eclipses, I present the detailed results from my computations of this solar eclipse before sunrise in Zheng, see Appendix.

The Old Testament mentions three miraculous situations that can easily be explained as the partial solar eclipse in 700 BC, 2 Kings 20:9–11, the total solar eclipse in 1131 BC, Josh. 10:13, and the annular solar eclipse in 1207 BC, Josh. 10:12. Another important eclipse occurred in the Hittite capitol Boghazkoi, east of Ankara in Turkey, in 1335 BC, during the 10th year of the reign of king Mursilis II, Bo. 4802, Kol. IV, 25 (Bilabel 1927). These independent written sources give an overlap of more than 500 years with the Swedish rock carvings from the Bronze Age, and are necessary to avoid circular arguments.

The Solar Eclipses

In 1991 it was possible to identify some engravings on rocks in Sweden as observations of the total solar eclipses in 1666, 1596, 1460, 1411, 1366, 1230, 1185, 1169 and 858 BC (Henriksson 1991ab, 1992). The first identifications were made from reproductions of the rock carvings at Ekenberg in Norrköping by Arthur Nordén (1926). The eclipse in 1596 BC was spectacular and has been reproduced on rocks on several places along the zone of totality from Traprain Law, Midlothian, in southern Scotland (Bailey, Clube & Napier 1990) to Lake Onega in northwestern Russia (Hallström 1960), Figure 1. During this eclipse Venus, Mercury, and Saturn were visible to the left of the sun and Jupiter and the comet Encke to the right. The eclipsed sun was reproduced as a series of concentric circles and the planets as cup-marks. The comet was reproduced symbolically as a sword in some cases, but mostly as two close rays perpendicular to the outermost circle. One of the different types of comets in the classification by Pliny the Elder (77) is described as a sword.

The total solar eclipse in 1230 BC July 7, is presented in Figure 2 (Henriksson 1998).

The Calendar Ships

The author has found systematic combinations of the symbol for a total solar eclipse in unique positions in relation to specific types of ships in a series of six different ships. The conclusion was that the ecliptic has been reproduced as a series of six ships: five corresponding to 60 days and the biggest one to 65–66 days beginning with the summer solstice. The six ships can be interpreted as constellations, have individual shapes, and can therefore be recognised in the sky. They may have served in a calendar with six double months corresponding to two of the zodiacal signs. The position above the ship for the sun or the moon marked in which part of the double months something important happened, Figures 1–2. In the case of the moon the phase can even give the correct day. A more detailed presentation of this hypothesis is given in Henriksson (1998).

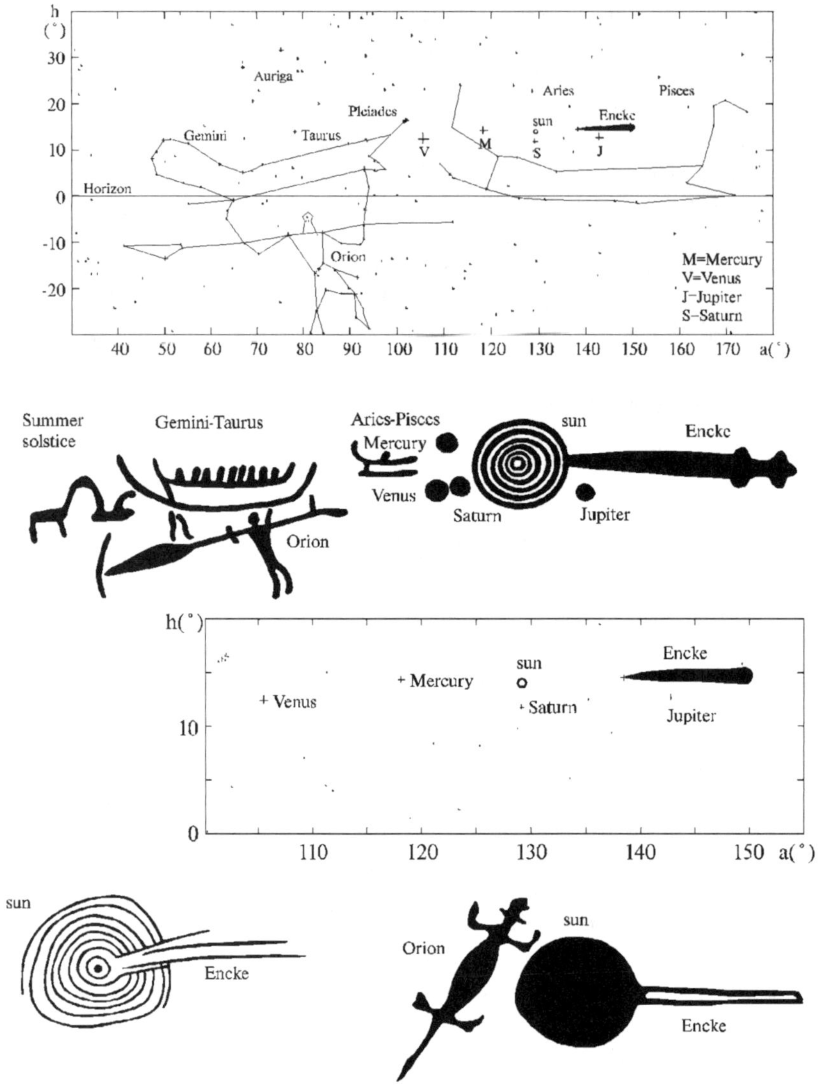

Figure 1. (a) The total solar eclipse on 1596 BC March 3, from the horizon of Ekenberg in Norrköping. The central phase occurred at 08.57.10 local mean solar time, and the duration of the total phase was 2min 46sec. Four bright planets dominated the sky in the vicinity of the sun. But the most spectacular object was comet Encke, visible so close to the sun because of the total solar eclipse. The length of the tail of the comet corresponds to 1.0 AU. (b) The totally eclipsed sun with four planets and Encke in Aries-Pisces. Orion lifts the Gemini-Taurus ship. Rock carving at Ekenberg after Arthur Nordén 1926. (c) The area around the sun during the total eclipse 1596 BC March 3, at 08.57.10 am local mean solar time at Ekenberg. A unique combination of four bright planets and comet Encke must have made a strong impression on the people within the zone of totality. The length of the tail of the comet corresponds to 1.0 AU.(d) Fragment of a rock carving from the prehistoric cult place at Traprain Law, Midlothian, Scotland. This place was also situated within the zone of totality of the solar eclipse on 1596 BC March 3. After Bailey, Clube, and Napier 1990. (e) Rock carving at Lake Onega in eastern Russia, close to the southern limit of the total solar eclipse on 1596 BC March 3. After Gustaf Hallström 1960.

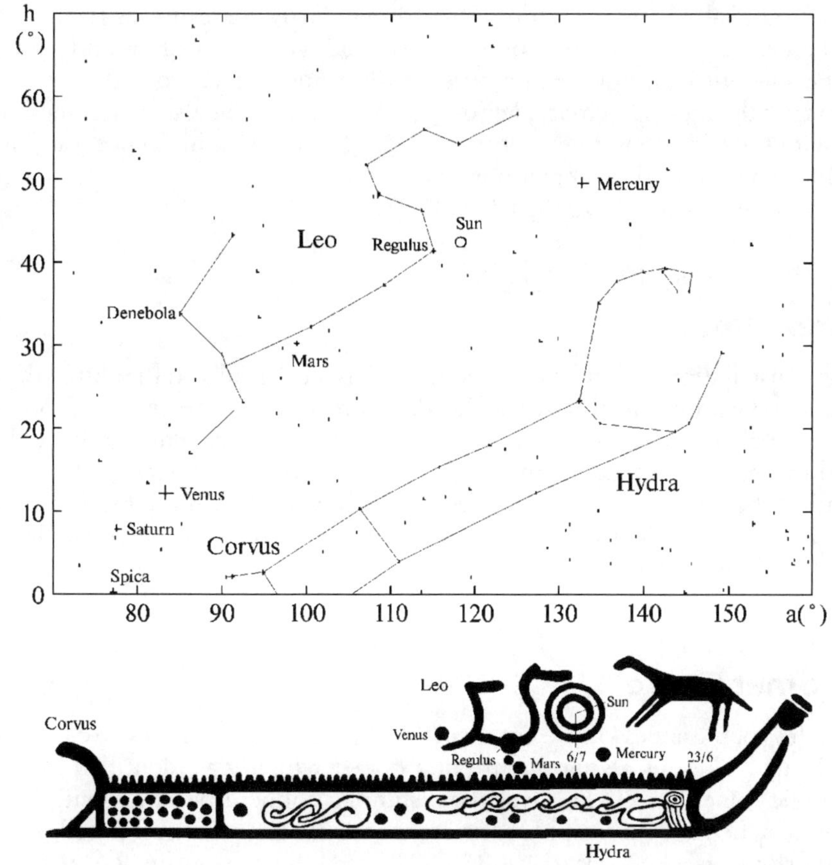

Figure 2. (a) The total solar eclipse on 1230 BC July 7, at 08.59.00 local time at Ekenberg in Norrköping. The sky was dark close to the sun, and it was possible to see planets and bright stars. The stars in Hydra correspond to the largest of the six calendar ships beginning with the summer solstice. The projection makes the upper part of the star map too wide. (The stars are brighter than magnitude 5.0.) (b) In 1869 the King's Custodian of Antiquities, Bror Emil Hildebrand, made the first investigation of this excellent example of Swedish rock carving at Ekenberg. He dated it to the older Swedish Bronze Age, 1800–1000 BC. The ship is for stylistic reasons dated to ca 1500 BC. The length is 2.2 meters. The author identified, in 1991, the scene above the ship as a record of the total solar eclipse on 1230 BC July 6, with the sun in front of Leo and with the planets visible reproduced as cup-marks. The sun was placed above the 14th rail-mark. If this corresponds to July 6, the first one could have been June 23. This may be interpreted as a dating of the eclipse as the 14th day after the summer solstice which took place on June 23 at 04:48 local time. After Hildebrand (1869).

The Moon

While I was looking for a symbol for the moon I realised that the very common footprints often were depicted in similar positions in relation to the ships as the solar symbols. They are never arranged like the footsteps of a walking man, but appear mostly isolated as a left or right foot or as a pair of feet. My conclusion was that the full moon was reproduced as a pair of feet, the first quarter moon as a left foot and the third quarter moon as a right foot, always with the toes pointing downwards. A left hand with the thumb pointing upwards corresponds to the new moon and a left hand with the thumb pointing downwards corresponds to the waning phase, Figure 3. Sometimes the foot was depicted with all five toes and sometimes as a shoe sole but I cannot see any different meaning. There exist also naturalistic depictions of the moon as circles, left or right semicircles or crescents, sometimes together with the corresponding footprint or hand symbols. This hypothesis has successfully been used to date depictions of the comet Encke and different phases of a supernova (Henriksson 1993ab, 1994, 1995).

Constellations

The constellation Orion has been identified as a celestial god that lifted the ecliptical sunship Gemini-Taurus over his head, Figure 1. This is a very common motif.

Leo can also be identified as a specific animal-like constellation above the calendar-ship I have called Leo-Cancer. This ship consists of the stars in Hydra, with the head of the Hydra as the decoration of the bow and with the brightest stars in Corvus as the stern, Figures 2–3. A more complete discussion of constellations is presented in Henriksson (1998).

The Comet Encke

On the rock images of the total solar eclipses on 1596 BC March 3, and on 858 BC July 5 there is also a picture of a comet which was identified in 1994 as the periodic comet Encke. This was the result of a cooperative study with Dr. Mats Lindgren, at the Astronomical Observatory in Uppsala, who computed the gravitational orbit of Encke back to 2000 BC. We used the subroutine RADAU as integrator and took into account the gravitational field from the sun, and all the planets including the centre of mass of the earth-moon system (Everhart 1985). The computed orbit has a very stable evolution with no close encounters with Jupiter. From this preliminary orbit I was able to compute empirical non-gravitational time-corrections and to determine the positions of the comet accurately enough for its identification with about ten comet observations in the old Chinese chronicles (Yoke 1962). The most useful Chinese observations were made in 47 BC and AD 222, 699, 1019, and 1376.

In December 1355 BC and January 1354 BC, the supernova mentioned above and the comet Encke were visible in the same part of the night sky. This dramatic celestial event was depicted on several Swedish rock carvings and can be used as an important double-check for both the correctness of the dating of the supernova and of the determination of the orbit of Encke (Henriksson 1995).

The oldest observation of comet Encke before this investigation was made in 1786 AD. It has now been possible to follow comet Encke back to 1758 BC. A series of four bright passages close to the earth and the sun 1662, 1652, 1646, and 1609 BC have been reproduced around the calendar ship Leo-Cancer which originally was drawn to commemorate the total solar eclipse on the day of the summer solstice, 1666 BC June 22. All these complicated situations eliminate the risk of circular arguments and fit very well with the positions given by the computer program.

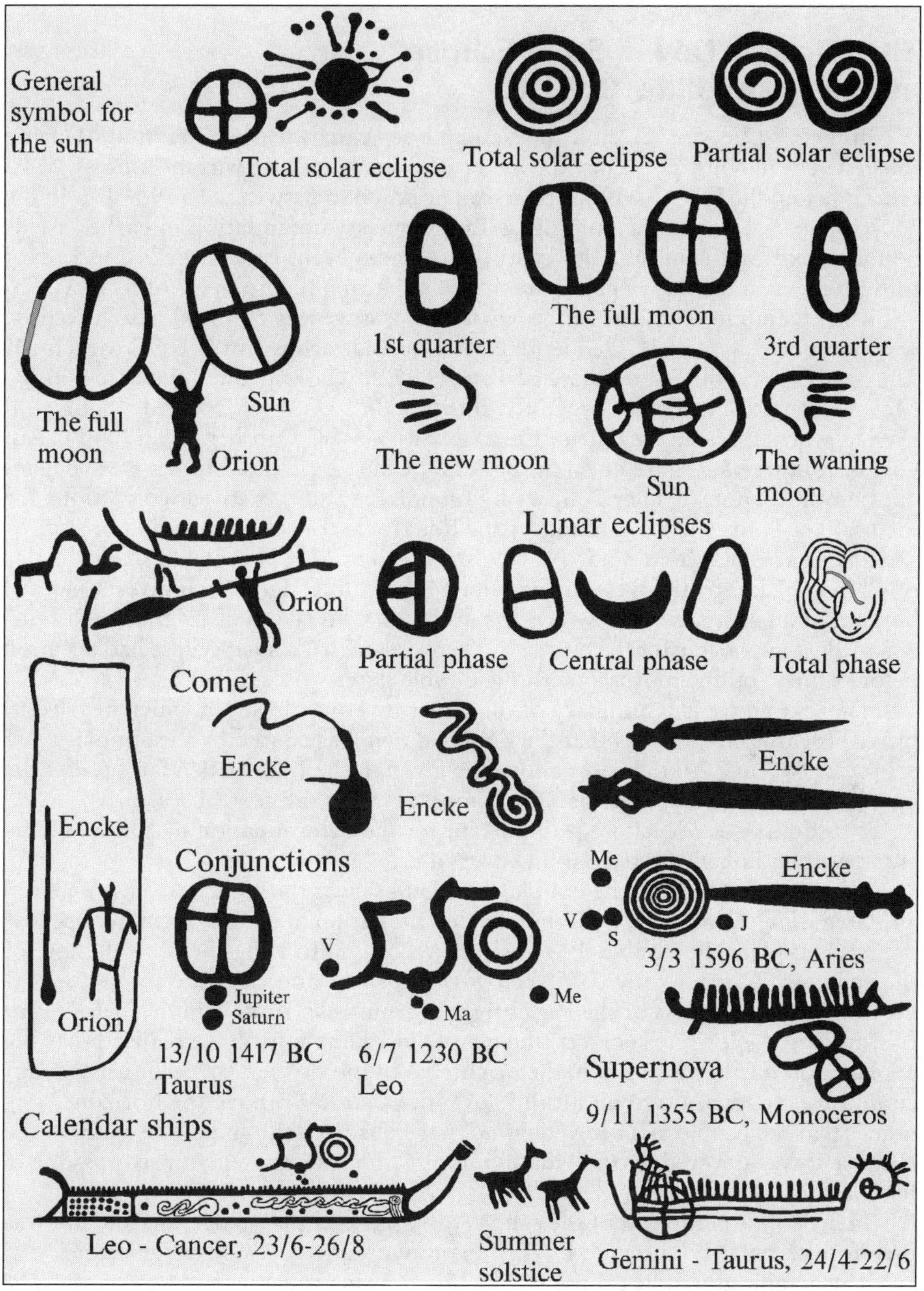

Figure 3. Symbols and astronomical phenomena on Swedish rock carvings.

Appendix

The "Double Dawn" Solar Eclipse at Zheng, in China, 899 BC

In the Old Version of the Chinese Bamboo Annals it is written that "During the first year of King Yi the day dawned twice at Zheng". Yi was the king of Western Zhou and the first year of his reign can be dated to between 895–966 BC. In the encyclopaedia Kai yuan Zhan jing, a Tang Dynasty compilation of earlier astronomical texts, made in the 10th century, the same event is described: "According to Ji jung ji nien the day began twice at Zheng during the first year of King Yi."

The first modern scholar to recognise this passage as a record of a solar eclipse was probably Liu (1944). He identified it as a solar eclipse on 926 BC March 21. This study was followed by those of Tung (1952), who suggested the date 966 BC May 12, Pang (1987), who suggested the date 899 BC April 20, and Zhang et al (1982), who suggested the date of the eclipse as 919 BC October 26 or 903 BC August 3. All these dates are given in the Julian Calendar. These studies were dependant on the Canon of solar Eclipses by Oppolzer (1887), without corrections for the braking of the earth's rotation by the tidal forces of the Moon and the Sun.

Zheng was placed at 34.5° N, 109.8° E, by Tan (1982) in the Historical Atlas of China, 27 km to the west of the Hua Mountain Range which gives a natural horizon of an altitude of a few degrees in the east. As we shall see, this high horizon made it impossible for the people in Zheng to see that a solar eclipse had occurred before sunrise on the morning with the double dawn.

The text above is a summary of the paper by Pang (1988) in which he tried to prove his earlier hypothesis that the double dawn was caused by the annular solar eclipse of 899 BC. All the dates above are given in the Julian calendar. I prefer the Gregorian calendar because then the dates will fall in the correct season.

This double dawn eclipse is important for the determination of ΔT, the difference between Ephemeris time and Universal time.

I agree with Pang that the double dawn eclipse was the annular eclipse on April 11, Gregorian calendar, 899 BC, but his determination of ΔT is not correct because it cannot explain the double dawn effect! According to Pang the central phase of the eclipse took place only 0.75° below the horizon and caused only a short decrease of the brightness of the very bright morning sky some minutes before sunrise. If Pang's solution is correct, the people in Zheng must have realised that the explanation for the variation in the brightness of the sky was the solar eclipse they could observe later, up to an altitude of more than 10° above the horizon. Pang argues that the people in Zheng could not have observed the end of the solar eclipse because it was covered by the Hua mountains, but this was obviously possible in his solution.

In my solution the end of the eclipse took place 2° above the horizon and was therefore completely obscured by the Hua mountains.

The magnitude of the eclipse was 0.95 according to Pang, and he writes: "The ΔT required for optimum solution is 20,900 sec or 5 hr 48 min, with an uncertainty of 500 sec. These results have been independently verified by Fred Espenak of the NASA Goddard Space Flight Centre, who used a modified version of the U. S. Naval Observatory eclipse computing program; and by Rickey W. Parker of Parsec Software, Inc., who used a commercially available program called the 'Visible Universe' (Parker, personal communication, 1987)."

This is considerably smaller than 0.967 according to my computations. It is doubtful if an eclipse of magnitude 0.95 could cause anything but a short constant brightness of the sky, not the striking and unexplained decrease in brightness that caused them to write that "the day began twice". One of my referees has pointed out that this criticism of Pang's interpretation is strongly supported by Stephenson (1992).

With magnitude 0.967 and the Eddington approximation of the limb-darkening of the solar disc, the computed brightness of the sky decreased to the same level that morning as during the nautical twilight with the sun 12° below the horizon. In fact the light level during the central phase corresponded to the light level with the sun 13° below the horizon during a normal morning, Figure 4.

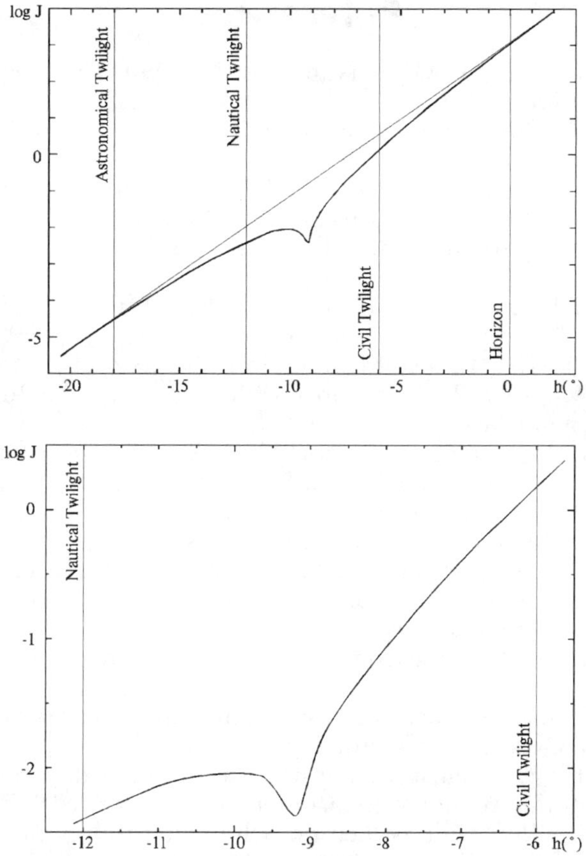

Figure 4. (a) The Old Version of the Bamboo Annals mentions that "During the first year of King Yi the day dawned twice at Zheng". Yi was the King of Western Zhou. The first year of King Yi's reign can be dated to between 895–966 BC. The only possible solution between 800–1000 BC is the annular solar eclipse on 899 BC April 11, at 04:49 local mean solar time. The figure shows the logarithm of the horizontal component of the sky brightness during this solar eclipse as a function of the altitude of the sun. The straight line is the visually observed horizontal illumination of a cloudless sky according to A. Ljunghall (1949). (b) During the central phase of the annular solar eclipse on 899 BC April 11 the brightness of the morning sky in Zheng decreased to the same level as at the nautical twilight. The limb-darkening of the solar surface has been computed in the Eddington approximation.

Acknowledgments

This project has been financed by The Bank of Sweden Tercentenary Foundation between 1990–1996. I am very thankful to professor Bo Gräslund and his colleagues at the Department of Nordic Archaeology in Uppsala for very useful discussions on possible conflicts between my astronomical interpretations and the archaeological interpretations. I am also thankful to Professor Bengt Gustafsson at the Astronomical Observatory in Uppsala for his support and interest in archaeoastronomy. Finally I will express my gratitude to Mary Blomberg who has improved my English and to the two unknown referees who have given constructive criticism.

References

Bailey, M. E., & Clube, S. V. M., & Napier, W. N. 1990, The Origin of Comets, p. 13, Pergamon Press, Oxford.
Bilabel, F. 1927, Geschichte Vorderasiens und Ägyptens vom 16.-11. Jahrh. v. Chr., Heidelberg.
Everhart, E. 1985, "Dynamics of Comets: Their Origin and Evolution", in Proceedings of IAU Coll. 83, eds. A. Carusi, G. B. Valsecchi, Reidel, Dordrecht, p. 185.
Hallström, G. 1960, Monumental Art of Northern Sweden from the Stone Age, Stockholm.
Henriksson, G. 1991a, "Observationer av totala solförmörkelser under åldre brons&ldern?", in Uppsala Observatoriets Nyhetstjänst, 1991-06-17.
Henriksson, G. 1991b, "Kan solförmörkelser lösa forntidsg&tor?", in Röster i Radio och TV, Nr 43, 25-31 oktober 1991. Summary of a 20-minutes lecture in the radio channel P1.
Henriksson, G. 1992, "Archaeoastronomy", in Annual report for 1991, Uppsala Astronomical Observatory, editor B. Edvardsson.
Henriksson, G. 1993a, "Archaeoastronomy", in Annual report for 1992, Uppsala Astronomical Observatory, editor B. Edvardsson.
Henriksson, G. 1993b, "Archaeoastronomy in Sweden", in Proceedings of the Fourth Oxford International Conference on Archaeoastronomy, Stara Zagora, August 1993, submitted.
Henriksson, G. 1994, "Arkeoastronomi i Sverige ", Astronomiska observatoriet i Uppsala.
Henriksson, G. 1995, "Archaeoastronomy", in Annual report for 1994, Uppsala Astronomical Observatory, editor B. Edvardsson.
Henriksson, G. 1998, "Prehistoric constellations on Swedish rock-carvings", in Proceedings of the Annual SEAC Conference in Gdansk 1997, in press.
Hildebrand, B. E. 1869, "Till hvilken tid och hvilket folk böra de Svenska Häll-ristningarna hänföras?", in Antikvarisk Tidskrift II.
Liu, C. 1944, Zhong guo wen hua yen chiu hui kan 4, No. 2, 30.
Ljunghall, A. 1949, "The intensity of Twilight and its connection with the density of the Atmosphere", in Meddelande från Lunds Astronomiska Observatorium, ser II, Nr 125, Hälsingborg.
Nordén, A. 1926, Östergötlands Bronsålder, Linköping.
Oppolzer, T. R. von 1887, Canon der Finsternisse, Wien.
Pang, K. D. 1987, J. Hydrol.
Pang, K. D., & Yau K., & Chou H., & Wolff R. 1988, "Computer analysis of some ancient Chinese sunrise eclipse records to determine the Earth's rotation rate", in Vistas in Astronomy, Vol. 31, pp. 833–847.

Parker, R. W. 1987, personal communication.
Pliny the Elder, 77, Naturalis Historia.
Sachs, A. J. 1974, in Phil. Trans. R. Soc. Lond. A 276, 43–50.
Schoch, C. 1931, "Die säkulare Accelaration des Mondes und der Sonne", in Astronomische Abhandlungen, Ergänzungshefte zu den Astronomischen Nachrichten, Band 8, B2, Kiel.
Stephenson, F. R. 1982, Historical Eclipses, Scientific American, October 1982, p 154–164.
Stephenson, F. R., & Morrison, L. V. 1984, "Long-term changes in the rotation of the Earth: 700 B.C. to A.D. 1980", in Phil. Trans. Roy. Soc. A, 313, 47–70.
Stephenson, F. R., & Houlden M. A. 1986, Atlas of historical eclipse maps, East Asia 1550 BC–AD 1900, Cambridge University Press, Cambridge.
Stephenson, F. R. 1992, "A Re-Investigation of the Double Dawn Event Record in the Bamboo Annals", in QJR Astr. Soc., 33, 91–98.
Stöffler, J. 1518, Calendarium Romanum Magnum.
Tan, Q. 1982, The Historical Atlas of China, Vol. 1. Cartographic Publishing House, Shanghai.
Tung, T. 1952, Studia Serica 4, 585.
Yoke, H. P. 1962, "Ancient and Mediaeval Observations of Comets and Novae in Chinese Sources", in Vistas in Astronomy 5, 127–225.
Zhang, P. et al. 1982, J. Nanjing Univ. [Natural Sci. Ed.] No. 2, 371.

35

Simon Ushakov's Icon "The Archangel Michael Trampling the Devil Underfoot" (1676) and Archangel's Fiery Sword in the Russian Iconographic Tradition in the Context of Bright Comets Appearances

Petko Nedialkov
Department of Astronomy
Sofia University
Sofia, Bulgaria

In the Christian tradition mention of St. Michael the Archangel is often connected with his sword—a symbol of retribution and fight against evil. Gnarling was the first to suggest that comets are the prototypes of the Archangel's fiery sword. Recently, his hypothesis was confirmed on the basis of Bulgarian iconographic material. In the present work we are trying to enlarge the number of possible logical arguments in accordance with this opinion. There are basically two ways of representing St. Michael the Archangel: with an ordinary sword (spear), or with a fiery sword. The correlation between the appearance of the brightest comets and paintings with fiery swords is investigated on the basis of our catalogue of icons of St. Michael the Archangel. The principal differences between the Russian and Bulgarian icons are discussed. Special attention is focused on Simon Ushakov's icon "The Archangel St. Michael trampling the devil underfoot."

* * *

Shining Nothing or Shining Substance?

The coincidence in time in Bulgarian icons between the appearance of bright comets and paintings of St. Michael the Archangel with a fiery sword led us to the assumption of a celestial origin for this symbol (Nedialkov, 1993).

According to the literature available to us, G. Norling was the first scientist who offered the hypothesis that comets are the prototype of St. Michael the Archangel's fiery sword (Kovshun, 1990). He explained the occurrence of "the angel with a fiery sword" to King David as a recurrent return of the comet of Halley in 1010 BC.

We shall put forward the possible logical arguments confirming this thesis. In the first place, we have to note that all sky objects were deified in the remote past. In those times the nature of the comets, meteors, meteorites and bolides was not distinguished clearly. The trails left by bolides are easily warped by the winds in the higher atmosphere and take the shape of twisted dragons. In the folklore of many nations the flight of a meteor is identified with the flight of a fiery dragon on the skies, a snake, etc. For example, according to Hindu mythology comets are the tail of the dragon Dontga-Rahu slaughtered by Vishnu who prevented the theft of the sacred liquid amrit. Very often bright meteors or bolides were described as "fiery dragons with smoky tails" (Ipatiev's chronicle, 1215). According to M. Bekert, the first iron objects used by humans were made from meteoritic iron. The words "iron" and "heavens" are similar in some ancient languages. The worship of sacred stones, most of which were meteorites, was widespread. Like the comets whose appearances had been always taken as omens, some of these stones were used to forecast the future.

In the second place, one can give a lot of examples of divine weapons whose prototypes are easily recognized, not only among terrestrial phenomena (storms, lightning), but also among astronomical phenomena (meteors, meteorites, bolides, comets). In Hindu epics Indra throws fiery spears at the dragon Vritra. The Scandinavian god Thor crashes the stone ball, thrown by the giant Runge to obscure the Sun, and as a consequence a stone-rain spreads over the whole Scandinavian peninsula (Kovshun, 1990).

St. Michael the Archangel is mentioned four times in the Holy Scripture: twice in Daniel (10:13 ff. and 12:1), where the Archangel is represented as the helper of the chosen people (Israel); once in Jude (v. 9), disputing with the devil about the body of Moses, and once in Rev. 12:7–9, fighting with the dragon. In this connection the great heavenly captain St. Michael the Archangel was regarded in the early Church as the helper of Christian armies against the heathen, and as a protector of every believer against the devil, especially at the hour of death, when he conducts the souls to God. The nature of the angels is usually described by analogy with the most thin, filmy and most mobile things in the material world: fire, wind and especially light. The intrinsic nature of angels was thought to be fire. Taking into account the biblical texts, Pseudo-Dionysius Areopagite (late 5th to early 6th century) noted their relationship with the fire of lightning and the purgative fire of sacrifice (Dionysius, 1845). The angels are the angels of light, their "bodies" and "dresses" seem to consist of light—they possess nimbleness, rapidity and brightness. The word "light" itself is a part of the name of one of the Archangels—Uriil (Meletinskij,

1990). We should mention that the non-material and spiritual nature of the angels was not fully recognized by the church in the firsts centuries A.D. while some doctrines were still being worked out. Origenes attributed to angels an eternal body—an opinion eventually shared by St. Augustine. Origenes held the good angels to be no less capable of falling than demons of being saved. This teaching was rejected by most Orthodox church fathers.

The angels were also associated with "fiery" celestial objects. "Heavenly armies" is a common term for the angels in Christianity and the astral gods in Semitic pagan cults. According to mystical legends of Judaism, every Archangel is associated with one of the planets. The Sun is considered to be a cloister of the saints. The existence of certain links between comets and the Sun could be demonstrated in some Orthodox mural paintings (for example in Rila Monastery, Bulgaria) where the angel contending both solar (solar nimbus) and cometic symbolism (fiery sword) is depicted. Plutarch (1st and 2nd century A.D.) in his "Biography of Lysandros" adduced the arguments of some philosophers who considered shooting stars to be aroused from celestial bodies associated with the Sun (an old notion of Anaximander). Byzantine mystics deciphered the word Adam as an abbreviation (Averincev, 1977) in which every letter corresponds to the initial Greek letters of the names of the four cardinal points: "Anathole" (East), "Dysis" (West), "Arctos" (North) and "Mesembria" (South). This idea is repeated also in a Slavic apocrypha. According to this text, "Michael the Archangel appeared as a star in the East and its name was Anatholy" (Porfirev, 1872). As Melamed (1990) wrote, it is "dangerous" to work in the field of Christian symbols, because many of the symbolic images directly related to cosmogonical ideas did not originate from Christianity itself, but were brought in from other religious systems. In this sense we are investigating the origin of the symbol to a greater extent than simply dealing with its meaning in Christianity.

St. Michael is a person who is widely presented in the icons. In the mythologies the sword and/or the spear are weapons of the Cosmic or Solar Hero, subjugator of the Dragons and demonic forces. Though comparatively rare, his sword is depicted as if made of fire, like a flame. There are several similarities between comets and this fiery sword. The first one is the coincidence of shape and color. The flames of the swords resemble the tail of a comet to a remarkable extent. It is also known from the history of astronomy that comets themselves were associated in people's minds with bloody swords (Weber, 1990). For example, the French surgeon Peret wrote about a "terrible" comet in 1528 that frightened the whole world: "... At the top a hand was seen holding a sword tightly and ready to strike. At the edge three stars were seen. From both sides of the beams coming out from the tail of that comet were seen many axes, knives and swords, stained with blood, and among them were noticed terrified men's faces with disheveled beards and with hair standing on end." It is not accidental that the fiery sword is generally depicted as red in color. Their functions are also similar. The sword of St. Michael the Archangel is a symbol of retribution and death, just as comets forecast destruction and death. The apparent fiery nature of comets conforms with the myths according to which the universe is condemned to perish in the flames of cosmic fire (eschatology myths). But even after Tycho Brahe proved in 1577 that comets are cosmic objects, Aristotle's hypothesis of so-called fiery phenomena (comets, aurorae, shooting stars and the Milky Way), which could not be a part of the "perfect" heavens, survived until the beginning of the 19th century. During the same historical period many naive, popular beliefs can be found. According to Afanasiev (1982) during the same period of time meteors were considered to be angels exiled from heavens. Moreover shooting stars were thought to be fallen angels, punished by the seven Archangels in the Book of Enoh, where the angel lore is highly developed (Kovshun, 1990).

Orthodox Icon Painters

A short description of the inner development of the Orthodox icon painters follows. All of them were people who believed in God. They were drawing not only following the rules of the Erminees but also following Scripture. They knew that skies are not only created to mark days, months and years, but they also serve to prompt the omens. Since God takes care of everybody and everything, He has charge of the omens, too. The angels ("messengers," "executors of God's will") descend from the heavens and comets appear in the skies to let the people know about the future. Furthermore, according to Kozma Indicopleystes (6th century) whose opinion was authoritative in the Middle Ages, "not only the Sun and the Moon but all sky objects are moved by the action of God's will. The same angel's power causes the rain, brings the clouds together, rules the winds, the snow, the heat and the cold" (Flamarion, 1872). In his last years Rafaelo Santi (16th century) drew the personifications of the planets surrounded by images of beautiful angels in the Palazzo Kidgi (Bozhkov, 1988). In the 7th century John of Damascus wrote that comets do not belong to the stars but they are called into existence in their own time by the will of God, and they are dispersed later. In the 12th century Honorius Otenskij in his writing "Lucidarius" (very popular in Europe until the end of 17th century) insisted: "Comets are ignited in the air by God in order to call people to repent from their sins" and added "not to give us light, but to coerce changes of kingdoms" (Kovshun, 1990). It's obvious that these points of view were known by the painters, especially by those who acquired their craft in a monastery. At the same time the painters knew that the texts of the Scripture have more than one meaning. They knew that the Scripture contains secrets beyond the limits of human understanding and that the imperfection of human nature is the reason why angels appear to us like a visual substance. Hence it is not surprising that the star of Bethlehem has been considered to be either a mythical, or a miraculous object beyond the bounds of scientific explanation (an angel, according to John Chrysostom (arch. Nikiphor, 1891)), or a real astronomical phenomenon (Humphreys, 1991). As the Pseudo-Dionisius Areopagite wrote "No one should doubt that visibility was used to indicate invisibility." The Orthodox painters possessed a deep knowledge both about the sky and the heavens. They had to contemplate the sky with their material (external) eyes and the heavens with their non-material (inner) eyes. So their creative works connected the imperfect and the perfect heavens (Trubetskoy, 1916

Finally we should not forget that although the Second Commandment forbids the worship of the Sun, the Moon and the stars (Deut. 4:19; 17:3; cf. Ex. 20:) the images of celestial objects can be frequently found in the Orthodox icons. Scripture contributed to this because, according to some scientists, there are plenty of descriptions of observed comets, star rains, bolides and falls of meteorites. Kamenskij speculated that the experience of Abraham: *"a smoking fire pot with a blazing torch"* (Gen. 15:17), was an appearance of a comet, as did Svjatskij concerning the vision of prophet Jeremiah: *"a boiling pot, tilting away from the north"* (Jer. 1:13).

Figure 1. Simon Ushakov's "The Archangel Michael trampling the devil underfoot" (1676), 23×20.5, from Tolmachevskaya-Sosnovskaya collection, Tretyakov Gallery, Moscow.

Ignatiev checked the chronology in the Bible with the appearances of the comet Halley (Kovshun, 1990).

One Unique Icon

We have aimed to check the correlations in time between the appearances of the bright comets and paintings of St. Michael the Archangel with a fiery sword. We compared the icon dating with the times of bright comet appearances, using the "Catalogue of Ancient and Naked-Eye Comets" (Hasegava, 1980) and eleven albums presenting Russian icons from the 12th until the end of the 17th century. Twenty-one images were found, but only one icon with a fiery sword, (Figure 1) the Simon Ushakov's painting "The Archangel Michael trampling the devil underfoot," dated to 1676 (Alpatov, 1984). We would like to add for completeness that a verbal description of an Archangel holding a cross in his right hand and a fiery sword in the left one exists in one of the books by the Russian writer N. Leskov (Ostrovskij, 1988).

On the basis of this sample we made a catalogue which contains the most important characteristics of the icons. The information for each icon includes the title, the date, the author, the origin, place of conservation, right (left) hand of the archangel, the presence of a sword, the orientation of the sword, geometry, the hand of the archangel, the face foreshortening, and the source reference.

Although we do not have a large number of icons at our disposal we are able to point out some principal differences between Bulgarian and Russian icons. The first one is the numerical asymmetry in depicting the Archangel with sword and with spear. We found that the ratio of swords to spears is 5:1 in Bulgarian icons while in Russian icons ones it is 1:5. Both sword and spear are attributes of St. Michael the Archangel, but the first sword is superior to the second (Cooper, 1993). From another perspective, the sword is a weapon to be used in a close combat while the spear is a weapon for striking at a distance. We think that this difference originates from differences in the national spirit of Russians and Bulgarians. The Russians always waged war in vast spaces while the Bulgarians were forced to struggle in distinctly limited areas. According to Gachev (1988) the Russian cosmos is more spacious, not worn out. The vertical line is the main axis of the Bulgarian cosmos. Opposite to it, the horizontal line is the main axis of the Russian cosmos. The direct (vertical) relations with God were never constitutive in Russia, where heresies and Protestantism never sank deep roots. In the Russian cosmos, God, the infinity and the truth are broadly reached but not focused in a point (a definite place). On the other hand, the Bulgarians coalesce with the universe which is an extension of their cosmic body. Probably this was the reason why Ushakov drew a fiery sword, figures of saints and an unknown small figure in the horizontal plane. The last is not an exception. In the beginning of the 1670s figures of the boyars, the Tsar and their wives gradually disappear from iconographic schemes (Omchinnikova, 1965).

It is interesting to mention that in the Russian icons there are several cases of Archangels holding swords and/or spears in their left hands. This transposition could come about for different reasons. The first is to preserve the symmetry of the icon (the case of St. Gabriel). The second is to demonstrate the advantage of the spear over the sword. The third is to encipher some additional information.

The other main difference between Russian and Bulgarian icons is found in coincidences in time between the appearances of bright comets and drawings of St. Michael the Archangel with a fiery sword. We have found a high level of such coincidence in Bulgarian icons (4 of 6). The dating of Ushakov's icon does not coincide with any bright comet event. This fact could be explained by the small number of icons considered, but it seems more probable that the icon was drawn with one of the last passages of a bright comet in mind, the same way as Giotto did. It is possible to conclude that Bulgarian painters are relatively more impressionable than the Russian (Gachev [1988] wrote about the typical Russian style of repulsion from the corporal to the spirituality), but we should not forget that those Bulgarian dated icons with fiery swords were drawn one to two centuries later.

There are several indications which suggest that Ushakov had drawn a comet as a fiery sword. In the first place, the Russians were interested in the remarkable events in the skies from ancient times. A lot of evidence exists about the observations of comets in the Middle Ages. Manuscripts containing astronomical and astrological knowledge had become very popular, mainly as translations and compilations of foreign works. From them the reader could learn about the geocentric system of Ptolemy. In the middle of 17th century and in the 1670s there appeared different works of western European authors in which the heliocentric system of Copernicus was presented (Buganov, 1989). In the second place, a passing glance is enough to recognize that appearances of very bright comets (brighter than 2nd magnitude) in this period were frequent (1658, 1663, 1664, 1665, 1668 or 1672).

Simon Ushakov (1626–1686) was the first of all the painters from medieval Russia whose works became popular in the 19th century. But after discovering the icons of Rublev and Dionysis his glory grew dim (Alpatov, 1955). Although he followed tradition, he tried to convey a sense of reality to the mind of his contemporaries. He wrote a treatise in which he stated that painting is the supreme art because it could reflect like a mirror all natural phenomena (Kaganovich, 1961). Ushakov himself created many icons containing the image of the Savior. He was one of the first who contemplated a new manual on painting where the images of a man, his face, eyes, hands, and feet should substitute for irretrievable old iconographic schemes in Ermenees (Ovchinnikova, 1965). At the same time he was a very educated and gifted man. He knew many western European engravings and drew geographical charts in his workshop. He was interested in astronomy, too. In the Hall of Mirrors (in the Kremlin) his mural paintings (now lost) represented the Sun, the Moon in the night sky and the Zodiac with four figures of women (the seasons) placed around them. Finally it is not accidental that Ushakov drew his St. Michael the Archangel on the globe of the Earth. In this way the painting looks more like a cosmic scheme than the usual icon. The juxtaposition of icons with different dating (Retkovskaja, 1961) shows that until the middle of the 17th century all of them reproduced the cosmogony of Koznia Indicoplevstes. The different areas of, the Earth, heaven and the underworld were depicted as rectangles placed one above another. The first mural painting "according to Ptolemy" with the heavenly spheres on which the planets are moved by the angels appeared in the 12th century and for the first time on an icon in 1658. Until the first stellar map was published in Russia (1699) by Kopievskij, Russian iconography remained a vaulted mirror of the ideas of the structure of the universe (Scheglov, 1986).

The icon of Ushakov refers to texts from Revelation where St. Michael the Archangel is fighting the dragon. The cross in Christianity is a symbol of the weapon of God through which He defeats the devil—and during the time of the Last Judgment, after terrific omens, the crucified Jesus is to appear as glorified and victorious King (Rev. 19:11–16). We would like to underline the connections between the sword as symbol with the symbols of Cross, the Sun and phallus. It is easy to decipher these links in the icon of Ushakov. The solar symbolism is connected mainly with the figure of the Archangel (his nimbus and especially the geometrical center of the icon which occurs not far from the area of solar plexus and phallus) as well as with the sphere of Emanuel (the name of the newborn messiah proclaimed in the Old Testament, cf. Is. 7:14) and the sword. This encouraged us to insist on the hypothesis that the images of a comet, the Sun and the Earth are enciphered on this icon. The trampling of the devil and other fallen angels on the surface of the Earth emphasizes their rough sensual nature.

An alternative hypothesis could be advanced if we try to see this icon in the context of historical development. The simplest meaning could be a symbolic depiction of the war against Turkey and Crimea for Ukraine which began in 1676. It is known from history that the idea of a Third Rome was well received in Moscow. Since those times, a permanent goal of the Russian tsars became the expulsion of the pagan Turks from Constantinople. But in addition to external opponents, there existed many internal enemies that the autocracy of the Tsars and Orthodox church had to contend with. During the fife of Ushakov, Russia was shaken by many uprisings (Balashov's movement 1632–36; Salt revolt in some towns 1648–49; revolts in Pskov and Novgorod in 1650; Cooper revolt in 1662 and the second Peasant war of 1667–71). The great rebel Stepan Razin was captured in 1671 and in the end of same year Astrakhan fell—the last castle of the Peasant war (Buganov, 1989). From 1654 the patriarch Nikon began reforms in the Orthodox church which met strenuous opposition among so-called "razkolniki" (schismatics/dissenters). In 1676 the revolt at Solovki monastery, which lasted more than 12 years, was brutally suppressed (Buganov & Bogdanov, 1991). Thus the year of 1676 marked the end of many wars that the Russian government had begun or won.

Relations between Tsar Aleksej Michailovich, who governed Russia from 1645 until 1676, and Patriarch Nikon were very curious. It seems that the spiritual leader of Russia gazed very often on the heavenly omens. For example he liked to use celestial objects to express his thought, in a paraphrase of Pseudo-Dionysius: "It is not accidental that God created two luminaries in the skies—the Sun and the Moon. The Sun represents the power of prelates, it is shining in the daytime just like the way prelates shine to the souls; the smaller luminary is shining in the night time, just the way temporal power shines to bodies"(Buganov & Bogdanov, 1991). In Europe from the 12th century on many royal courts sheltered famous astrologers, and many important decisions of military, political or intimate character were taken only after consultation with them (Bozhkov, 1988). In the second part of the 17th century images of astrological signs with a Zodiac appeared in the rooms of Aleksej Michailovich, his daughter Sofia and the Princess Golitsina (Bozhkov, 1988). The Orthodox church fought mercilessly against astrology, taking the position that nobody could forecast God's providence. The astrologers invited to Tsar's court were called "the Tsars of astrological fetches." According to Syjatskij (1962) a great quarrel happened between the Tsar and the Patriarch concerning the comet which appeared in 1664. But even before that, in 1658, when Nikon declared he would leave the patriarch stall, the spiritual leader of Russians said to the Tsar's messengers who tried to stop him: "You would be swept by the broom roaming in the heavens," pointing to a comet that had previously appeared. We know little about the relations of Ushakov with Nikon. According to Alpatov (1955) the painter was in good graces, but there is some evidence that Nikon did not approve of the western manner of iconographic depiction at all (Buganov & Bogdanov, 1991). In an Armory palate which Ushakov worked with foreign painters (a Pole and two Dutchmen) from Western Europe in the 1660s (Ovchinnikova, 1965). In 1665 Ushakov was sent for some offense to a monastery, but soon he returned to again draw his icons. In 1677 he was ordered for unknown reasons to stop his work on a portrait of the dead Tsar. In his later years he went again through certain sufferings—his land was taken according to some court judgment. His manner of conveying real sensitivity in the icons met the cruel criticism of the schismatics. The most prominent of them, Avvakum wrote: "Oh, oh! Poor Russia! Why do you want German approaches and customs..." (Ostrovskij, 1989). In Ushakov's icon one can discern some similarity to the composition of Guido Reni "Archangel Michael overcomes the devil" (1636) which was considered to be the height of artistic development of the 17th century and a true symbol of the theory of "Belle ideale" (Bozhkov, 1986).

Our last hypothesis links Ushakov's icon with the death of Tsar Aleksej Michailovich in 1676, the same year that the icon was painted. Ekaterina II called

Tsar Aleksej "a dull, heartless and a callous autocrat." He died in great pain begging the forgiveness of Nikon and of those Solovki monks who remained alive. But he was not given any forgiveness (Buganov & Bogdanov, 1991). Although we have several portraits of the Tsar drawn by Ushakov, we don't know what Ushakov himself looked like.

In our opinion it is realistic to think that all the reasons mentioned above played a role. So it would be more surprising if we lacked such an icon than to have it. We think that the most remarkable element in this icon is the fiery sword of the Archangel. This icon shows the great power of Ushakov's creative imagination.

Acknowledgments

This project was supported by the grant 876/02.07.96 of the Open Society Fund Sofia, Bulgaria, and partially by Mr. Dimitar Dimitrov from Air "Misia" and architect Nikolay Kassabov from "Kokas."

I am grateful also to my colleague Todor Veltchev for his help which improved this paper very much.

References

Afanasiev, A. N., 1982, The Tree of Life, Moskow (in Russian).
Alpatov M. B., 1955, General History of the Art, Iskusstvo, Moscow (in Russian).
Alpatov M. B., 1984, Ancient Russian painting, Iskusstvo, Moscow (in Russian).
Averincev, S. S., 1977, The Poetic of early Byzantine literature, Moscow (in Russian).
Bozhkov, A., 1986, The Bulgarian icon, Bulgarski hudozhnik, Sofia (in Bulgarian).
Bozhkov, A., 1988, The Man and the Art, Bulgarski hudozhnik, Sofia (in Bulgarian).
Buganov, V. 1., 1988, The world of history (Russia in XVII century), Molodaja gvardija, Moskow (in Russian).
Buganov, V. I., Bogdanov A. P., 1991, Rebels and true seekers in Russian orthodox church, Moskow (in Russian).
Cooper J. K., 1993, Illustrated encyclopedia of the traditional symbols, Peter Beron, Sofia (in Bulgarian).
Pseudo-Dionisius Areopagite, 1845, About the Heavenly hierarchy, Synodical typography, Moscow (in Russian).
Evsjukov, V. V., 1988, Myths about the universe, Nauka, Novosibirsk (in Russian).
Flamarion, S., 1872, The history of the sky (in Russ.), SBP m (in Russian).
Gachev, G., 1988, National images of the world, Moscow (in Russian).
Hasegava, 1., 1990, Catalogue of ancient and naked-eye comets, Vistas in Astronomy, v.24, p.52.
Humphreys, C., 1991, The star of Bethlehem—a Comet in 5 BC—and the date of the birth of Christ, 1991, Q. J. R. Astr. Soc., v.32, p.389.
Kaganovich M. L., 1961, History of the Russian art, Academy of Art, Moscow (in Russian).
Kovshun, 1. N., 1990, "And the stars were seized from heavens..." (in Russ.), Naukova durnka, Kiev (in Russian).
Meletinskij, E. M., 1990, Dictionary of mythology, Moscow (in Russian).
Melamed A., 1990, in Multidisciplinary investigations, v. 17, p. 181 (in Bulgarian)

Nedjalkov, P. L., 1993, Comets as prototype of St. Michael the Archangel's fiery sword in Bulgarian iconographic tradition, IV Annual European conference of Archeoastronorny, Smolyan.
Nikiphor, arch., 1891, The Encyclopedia of Bible, Ostozhenka, Moscow (in Russian).
Ovchinnikova, E. S., 1965, The portrait in Russian Art (XVII century), Moscow (in Russian).
Ostrovskij G, 1988, A story about the Russian painting, Izobrazitelnoe izkustvo, Moskow (in Russian).
Porfirev, 1, 1872, Aphocriphical legends about persons and events in Old Testament, Kazan (in Russian).
Retkovskaja, L. S., 1961, In contributions of National Historical museum, v. 33 (in Russian).
Svjatskij D. 0., 1962, Historical and Astronomical contributions, v.8, p. 53.
Scheglov P. V., 1986, Myths of the Earth, reflected on the skies, Nauka, Moscow (in Russian).
Trubetskoy, E., 199 1, Two worlds in ancient Russian iconography, Infoart, Moscow (in Russian).
Weber, M., 1895, The end of the world, Warsaw.

Section 5

Far East and Australia

36

Characteristics of Field Allocation (*fenye*) Astrology in Early China

David W. Pankenier
Department of Modern Foreign Languages and Literatures
Lehigh University
Bethlehem, PA

This article explores aspects of the Field Allocation system of judicial astrology in Zhou Dynasty China (1046–256 BCE) in an effort to trace its origins, characteristics, and development. In so doing it will shed light on the history of astral-terrestrial correspondences in China and their cosmological basis. Within the late Zhou conventional correlations between celestial spaces and terrestrial polities it is possible to discern traces of earlier cosmological conceptions. Analysis of this layer of original correspondences indicates that the earliest principle underlying this astrological system was a cosmological parallelism between the Yellow River and its celestial analogue, the Milky Way. The evidence points to a 2nd millennium BCE date of inception for these astral-terrestrial correlations, consistent with contemporaneous evidence of observational astronomy. The present study is intended to lay the groundwork for subsequent detailed study of the application of field allocation astrology in connection with political and military events of epoch-making significance in Zhou China.

* * *

"The 28 lunar mansions govern the twelve provinces and the Dipper's handle seconds them; the source [of this scheme] is lost in antiquity." Sima Qian (fl. ca. 100 BCE).[1]

1. See Shiji (27:1346). The allusion to the Northern Dipper's role here refers either to the astrological correlation of the Dipper's seven stars with specific terrestrial locations or, more likely since the historian Sima Qian refers specifically to the Dipper's handle, to the ancient calendrical function of the Dipper as a sort of seasonal indicator. For Han and later discussion of the astrological system associated with the Dipper see Kalinowski (1991, 197–98 and 338–340). For a detailed discussion of the keying of meridian passages of the Dipper's stars to those beneath the horizon see Needham (1959, 232–233). Detailed discussion of the Dipper's role, be-

This article explores aspects of the Field Allocation (*fenye*) system of judicial astrology[2] in Zhou Dynasty China (1046–256 BCE) in an effort to delineate its origins, some essential characteristics, and historical development. In so doing it will shed light on the history of astral-terrestrial correspondences in China and their cosmological basis. Such study is intended to lay the groundwork for subsequent detailed analysis of the application of these conceptions in connection with specific political and military events of epoch-making significance in ancient China.[3]

General Considerations Regarding Ancient Chinese Astrology

In early China, as in the West until the 16th century, the connection between celestial and terrestrial bodies, and consequently between celestial and terrestrial events, was considered self-evident. The two traditions part company, however, when it comes to the belief in the reciprocal nature of the correspondence; that is, the fundamental Chinese notion arising from their organismic view of the cosmos that the celestial and terrestrial realms mutually influence one another. In classical China it was thought that various terrestrial conditions were capable of giving rise to both large and small scale effects in the omnipresent essential medium of qi.[4] Such effects could arise from local climatic, topographic, or hydrologic conditions, which in turn were thought to influence human behavior and even the quality of temporal rulership. Propagating through qi, these influences directly affected what transpired in the sky. In the West, by contrast, astrological influence was generally a one-way street proceeding from the empyrean to here below without parallel reflexivity of cause and effect, though allowance was made for divine retribution in response to mortal hubris.

The notion of celestial causality presumably emerged as an explanation for the observed seasonal cycles of generation and corruption in nature and their obvious

---

ginning in the Neolithic, as an object of religious veneration together with its role as a celestial timepiece may be found in Yi Shitong (1996, 22–31).

2. Judicial astrology refers to prognostication about the future course of events and human affairs as an aid to decision making at crucial junctures based on the appearance, movements, etc. of heavenly bodies and transient celestial events. Horoscopic astrology, or prognostication about an individual's destiny based on the locations of heavenly bodies at a particular time, especially the hour of birth, played no role in classical China.

3. For which see Pankenier (in press).

4. See, for example, Major (1993, 173). With regard to the connection between qi and stars, the Jin Dynasty (265–316) scholar Yang Quan had the following to say in his Wulilun (Discourse on the Inherent Principles of Things): "The stars are the blossoming of primordial qi and the essence of [elemental] Water. Qi issues forth and rises, its rarefied essence floats up, winding and flowing at will, to be called the Heavenly River. Another name for it is the River of Clouds; the myriad stars emanate from it"; quoted in Taiping yulan (vol. 1, 8:11a). The conventional view which essentially prevailed from Han times on is succinctly summed up by the Ming dynasty scholar Zhang Huang (1527–1608) as follows: "Now, the heavens are qi, but form patterns; earth is form, and has internal structure. Forms cannot but dissipate and become qi; qi cannot but condense into forms. The heavenly bodies are the rarefied qi of earth which rises up and shows itself in the heavens"; Gu jin tushu jicheng (57:1343).

correlation with the sun and moon, which in turn explains why in early China astronomy and the calendar were indistinguishable.[5] In its earliest form the calendar codified knowledge handed down from antiquity about correlations between celestial cycles and phenomena in nature. This is why primitive calendrical indications such as those preserved in the "Lesser Annuary of Xia" Xia xiao zheng take the form of sequential wuhou "harbingers of the seasons," punctuated by references to two or three of the most prominent, archaic, and seasonally significant constellations — Orion, Scorpio and the Big Dipper.[6]

In addition to conceptions about a generalized celestial influence on terrestrial events, individual celestial bodies like the sun and moon were directly linked to specific meteorological, physical, and political phenomena. Here the evidence ranges from the cardinal orientation of Neolithic burials and sacrifices, to direct observation and representations of the sun and moon in the Neolithic (albeit comparatively rare), to written records of uncommon celestial phenomena such as eclipses, comets, meteor showers, planetary events, and the like, which appear from the Shang period (1554–1047 BCE) on. Not long after the appearance of the heuristic binary complementarity of Yin and Yang and the emergence of correlative cosmology in late Zhou (1046–256 BCE) texts one also finds explicit references to specific solar and lunar effects on the natural world.[7]

If the causal action of celestial bodies could make itself felt on all inanimate and animate beings in this way and be expressed materially through the medium of qi, this explains how such reflexive action could be thought of as an aspect of universal Nature. Moreover, if celestial bodies were considered to exhibit different qualities, exert different influences, and be correlated with (and influenced by) different institutions, terrestrial regions, peoples, etc., all essential elements of late

5. A revival of interest in catastrophism, and research into the climatic and cultural consequences of devastating impacts or near-earth approaches by cometary bodies on time scales measured in centuries or millennia now appears to suggest that such unpredictable disasters and apparitions may also have played a significant cultural role in the comparatively recent past. See, for example, Kobres (1995, 395–405), as well as Peiser's contributions to this volume.

6. Conventional accounts of the history of calendrical astronomy in China which focus primarily on the textually better documented practices of the late 1st millennium BCE tend to insist on a causal relationship between the primacy afforded the heavens and the cosmologically-based imperial ideology then in process of expansive elaboration. See, for example, Cullen (1996, 14): "...we should not miss the ideological significance of the decisive shift towards the primacy of the heavens over mere earthly phenomena. There is certainly a link here with the cosmological underpinnings of the unified empire in which the emperor was to his subjects as heaven was to earth." As Cullen himself points out (1996. 4), however, there is compelling textual evidence from a much earlier epoch of the central importance attached to calendrical astronomy by the early Chinese state. Although there is merit to the cosmological analogy heaven : earth :: emperor: subjects, there is also an abundance of evidence indicating that this fundamental conception, however inchoate, was already operative at least a millennium earlier. In fact, it underpinned the early Zhou claim to a Mandate of Heaven; see Pankenier (1995).

7. Cf. e.g., Huainanzi: "The furred and feathered are the kinds which fly and run, and therefore belong to the Yang; the shelled and scaly are the kinds which hibernate and hide, and therefore belong to the Yin. The sun is ruler of the Yang, and for this reason in spring and summer the herd animals shed hair, and at the solstice the deer shed their horns; the moon is ancestor of the Yin, which is why when the moon wanes the brains of fishes diminish, and when the moon dies the swollen oyster shrinks"; tr. Graham (1989, 333). See also the first twelve chapters ("Twelve Intervals") of L^{n}shi chunqiu (Spring and Autumn Annals of Master Lü) where the seasonal ordinances are described in detail.

Zhou astrology, it follows that by virtue of the intrinsic periodicities of certain celestial bodies, knowledge of past events and their associated astrological conditions logically renders the astrologer not only capable of predicting future celestial events, but also of forecasting in detail the terrestrial events likely to be produced by them.[8] From conceptions such as these, therefore, there flow a number of concrete conclusions about the implications of phenomena, and ultimately about the possibility of foreknowledge of events, both prerequisites for the formulation of a chronosophy such as, for example, that attributed to Zou Yan (fl. 4th century BCE).[9] Of course, once it was understood in late Zhou that eclipses of the sun and moon were in fact periodic phenomena, the same astrological motives lent increased impetus to efforts to improve the accuracy of prediction of those ominous celestial events, efforts which continued under imperial auspices throughout Chinese history.

Preoccupation with the correlation of celestial and terrestrial phenomena preceded by centuries the elaboration of systematized cosmo-political theories, gradually establishing the conceptual parameters within which such theories would develop. Such archaic predilections certainly contributed to the formation of the influential Huang-Lao thought of the 2nd century BCE, a dominant principle of which was that knowledge of the natural world translates directly into political power. Thus, in his discussion of the cosmological chapters of the Huainanzi, John S. Major shows how this assumption underlies the world-view that "cosmology, cosmography, astronomy, calendrical astrology, and other forms of cosmology form a seamless web,

8. A famous remark by Mencius (4B/26) shows how near at hand such thoughts were at the time: "Consider the heavens so high and the stars so distant. If we have investigated their phenomena we may, while yet sitting in the same place, go back to the solstice of a thousand years ago"; tr. Needham (1959, 196). In contrast, D. C. Lau's translation takes Mencius' comment to refer to the predictability of future phenomena: "if one seeks out former instances, one can calculate the solstices of a thousand years hence without stirring from one's seat"; see Lau (1970, 133). That the highly influential Confucian Xunzi (3rd century BCE) found it necessary to argue against the notion that celestial events directly influence terrestrial affairs clearly shows how pervasive such thinking was at the time. See especially his "Discussion of Heaven" in Watson (1967, 81 ff).

9. A chronosophy integrates past, present, and future into a coherent scheme by means of some procedure supposed to afford the possibility of predicting the future. According to David Pingree, the earliest form assumed by chronosophy in the ancient world was generally a natural theology of history which provided history "with intelligibility and/or meaning by looking for sufficient reason outside history itself and outside the world human beings are living in." For an orientation in the theory and practice of astrology informed largely by the experience of Mesopotamia and the Hellenistic West; see Encyclopedia Britannica, Vol. 2, s.v. "Astrology," 219 ff. As Pingree points out, as a natural theology of history, astrology is logically incompatible with a theocentric view. Thus, given the presumed immutability of the sequential alternation of the Five Virtues *wude*, for example, Zou Yan's mutual conquest theory addresses itself only to the intelligibility of the historical process, he makes no claim about historically transcendent meaning. Karl Löwith alluded to this contrast in connection with the fundamental role of correlative thinking in ancient China: "[H]ow can the 'elemental rhythm' of yin and yang and the cycle of growth and decay be adjusted to the belief in a meaningful goal and a 'progressive revelation' of divine truth in history?"; see Löwith (1949, 16). The tension between intelligibility and the expression of divine intentionality in history does, however, emerge in the perennial problem latent in Chinese dynastic ideology; that is, the tension arising from ancient Chinese cosmo-political views about legitimacy attributable to manifest heavenly sanction (i.e., through direct revelation of Heaven's Mandate) as opposed to that conferred by heredity. For the sources bearing on Zou Yan's doctrines and the astral/terrestrial mapping at their core see the conclusions of Brooks (1995, 4–9).

the principles of which a ruler would ignore only at his peril."[10] Similarly, Mark Edward Lewis has described the popular Warring States period (5th to 3rd centuries BCE) mythos in which a non-heroic figure with no combat skills gains victory over a great warrior through the possession of a divinely or magically revealed military treatise, showing that "military theorists thought of their doctrines as an esoteric wisdom that expressed divine patterns inherent in the cosmos."[11] At its simplest, the most basic principle of the so-called "calendrical model of warfare" held that cosmically consonant killing was to be carried out according to the seasons of the year. Other accounts that describe the content of revealed texts such as the so-called River Diagram and the Luo Writing also point to their connection with astral portentology and political ascendancy. More to the point, perhaps, the *Huainanzi*, a book compiled in mid-2nd century BCE, ranks astrological factors first among those to be taken into tactical consideration, in this way pointing directly to the agency by means of which patterns of cosmic order having military application were revealed.[12] Such conceptions are reinforced by still earlier textual accounts in *Zuozhuan* (Zuo Commentary) and *Guoyu* concerning the correlation of political and military actions with celestial events, most notably Jupiter's motion.[13] By the early imperial period, manifestation of heavenly endorsement in the form of a massing of the five planets constituted a definitive sign of the Heavenly Mandate's bestowal on a new dynasty.

Parameters of Field Allocation (*fenye*) Astrology

The regular field allocation scheme whose application is described in the *Zhouli* (Rites of Zhou) (see Table 1) dates from late Warring States or early Han times (3rd to 2nd centuries BCE).[14] The general principles and motives underpinning this astrological scheme are succinctly stated in the job description of the Astrologer Royal *Bao zhang shi*:

---

10. John S. Major (1993, 14) also argues that "the credibility of the Huang-Lao School in the early Han may have rested in part on the degree to which it was grounded in widely shared assumptions that went back to the foundations of Chinese civilization".

11. (Lewis 1990, 98).

12. See *Huainanzi* (15:5a): "Clearly understanding the motions of the planets, stars, sun, and moon; the knack of [using] punishment, moral suasion, and the unexpected; the advantageousness of facing to the front or rear, or going left or right; these are helpful in battle."

13. When one reads in Guoyu (10:11a), "the space Shi Chen (Orion), this is the dwelling place of the Jin people," this statement should not be taken as metaphorical, but rather as indicative of the functional identity at a certain level between the celestial and terrestrial realms. By the same token, as indicated above, a basic axiom of astrological prognostication was that celestial bodies were susceptible to the particular emanations (qi) of terrestrial regions as they traversed their corresponding celestial spaces, and so it is perfectly natural for Sima Qian to say of Jupiter's location, "[The state] wherein Jupiter is located may not be attacked, but it may attack others"; see Shiji, 27:1312, where this axiom is attributed by Sima Qian to the Warring States astrologer Shi Shen (fl. 4th c. BCE).

14. There is, however, strong evidence to suggest that the nucleus of the system, a correlation of the four celestial quadrants with the cardinal directions and perhaps even with the sequence of seasonal powers and their associated colors, may already have been in place by the 2nd millennium BCE; see Pankenier (1995).

"[The Astrologer Royal] concerns himself with the stars in the heavens, keeping a record of the changes and movements of the stars and planets, sun and moon, in order to examine the movements of the terrestrial world, with the object of distinguishing (prognosticating) good and bad fortune. He divides the territories of the nine regions of the empire in accordance with their dependence on particular celestial bodies. All the fiefs and principalities are connected with distinct stars, and from this their prosperity or misfortune can be ascertained. He makes prognostications, according to the twelve years [of the Jupiter cycle], of good and evil in the terrestrial world...."[15]

Table 1. The Field Allocation System of Astrological Correlations[16]

Jupiter station	*Chronogram*	*Field (State)*	*Province*	*Mansion*	*Direction (Zodiacal sign)*
Xingji	chou	Wu-Yue	Yang	Dou, Niu (Sgr-Cap)	NNE
Xuanxiao	zi	Qi	Qing	Nü, Xu (Aqr)	N
Zouzi	hai	Wèi	Bing	Wei, Shi, Bi (Aqr-Peg)	NNW
Jianglou	xu	Lu	Xu	Gui, Lou (And-Ari)	WNW
Daliang	you	Zhao	Ji	Wei, Mao (Ari-Tau)	W
Shi chen	shen	Wei (Jin)	Yi	Bi, Zui, Shen (Tau-Ori)	WSW
Chunshou	wei	Qin	Yong	Jing, Gui (Gem-Can)	SSW
Chunhuo	wu	Zhou	San He	Liu, Xing, Zhang(Hya)	S
Chunwei	si	Chu	Jing	Yi, Zhen (Cra-Cor)	SSE
Shouxing	chen	Zheng	Yan	Jiao, Kang (Vir)	ESE
Dahuo	mao	Song	Yu	Di, Fang, Xin (Lib-Sco)	E
Ximu	yin	Yan	You	Wei, Ji (Sco-Sgr)	ENE

15. *Zhouli zhushu* (The Zhou Li with Notes and Commentaries) in Shisanjing zhushu, 26:181. Tr. Needham (1959, 190). Cf. also Biot (1975, 113ff). The ambiguity of the passage with regard to how the astral and terrestrial correspondences were established has led to varying interpretations, all based on fragmentary and inconclusive evidence from other Zhou and Han texts. The most frequently cited operative principles include: (i) relating the Nine Provinces to the seven stars of the Big Dipper (see n. 16); (ii) tying the five planets and their associated terrestrial regions to stellar locations; (iii) the regular system of allocating the 28 Lunar Mansions among the terrestrial polities; (iv) defining the astral correlate of an ancient feudatory as the celestial location of Jupiter at the time of enfoeffment; (v) identifying the celestial correlate of a locality as the asterism to which ancient inhabitants of that place principally offered sacrifice. Of these, Zheng Qiao (1104–1162) accepts only the last as having a sound historical basis; for his measured criticism of the various theories, see *Gu jin tushu jicheng*, ch. 57, "Stars and Asterisms: General Discussion," 1341.

16. The scheme presented in Table 1 is found in Zheng Xuan's (CE 127–200) commentary to Zhouli and is essentially identical to that attributed to Shi Shen's (fl. 4th c. BCE) *Xing jing* (Canon of Stars) which is reproduced in the astrological monograph in *Jinshu* (11:307ff) and also adduced in a commentary to *Shiji* (27:1346). According to Zheng Xuan, an earlier lost work had elaborated the astral correlates of the Nine Provinces but all that could be reconstructed in his time were the terrestrial correlates of the 12 Jupiter stations (*ci*, lit. "stage of an army's march") in the more familiar field allocation system. In his commentary, *Zhouli shu (Shisanjing zhushu*, 26:181), Jia Gongyan (fl. ca. 650) takes *fenxing* (allocated asterisms) to refer to the stars of the Dipper and reproduces the scheme of correlations shown below in Table 2 from the influential Han dynasty apocryphal work *Chunqiuwei wen yao gou*. Little else is known about the principles of this system except that its prognostications presumably relied on perceived changes in the brilliance, color, or visibility of the stars of Ursa Major. Prognostication using this astrological system involving the Dipper cannot be documented prior to the Han dynasty. Cf. also Major (1993, 135).

Table 1 reproduces the standard scheme of correspondences among the twelve Jupiter stations, solar chronograms, 28 lunar mansions (with corresponding Western zodiacal signs), and terrestrial regions. A similar tabulation found in chapter three of the *Huainanzi* differs in minor detail, on the one hand by omitting the state of Jin and reversing (possibly erroneously) the lunar mansion correlates of Wei and Zhao, and on the other by separating Wu and Yue so that they correspond to Serving Maid (Nü = Aqr) and Southern Dipper/Ox Leader (Dou, Niu = Sgr/Cap) respectively.[17] The scheme of apportionments shown here is obviously the product of a process of systematization that necessarily dates from some time after the *de jure* partitioning of the state of Jin by Zhao, Wei, and Han in 403 BCE. In it an older scheme of Nine Provinces *jiuzhou* has now been expanded to take account of the political and military realities of the early Warring States period. Three additional "provinces" have been carved out of the older scheme of nine—Bing, San He, You—and the territories to which each has been made to correspond (Wèi, Zhou, Yan) have been reassigned accordingly. Table 2, based on the mid-3rd century BCE text Lüshi chunqiu (Spring & Autumn Annals of Master Lü), shows the Nine Fields of Heaven *jiuye* which correspond to the Nine Provinces and can serve to illustrate the contrast.[18]

Table 2. The Nine Fields of Heaven and their Correlations

Direction	*Heavenly Field*	*Province*	*Lunar Mansions*	*States*
Center	Revolving	[Yan]	Jiao, Kang, Di;	[Han, Zheng]
E	Azure	[Yu]	Fang, Xin; Wei,	[Song; Yan ?]
NE	Changing	[Yang]	Ji; Dou, Niu,	[Yan; Wu, Yue ?]
N	Dark	[Qing]	Nü; Xu, Wei; Shi,	[Yue; Qi; Wèi ?]
NW	Somber	[Xu]	Bi; Gui, Lou,	[Wèi; Lu ?]
W	Bright	[Ji]	Wei; Mao, Bi;	[Lu; Zhao;]
SW	Vermilion	[Liang]	Zui, Shen; Dongjing,	[Jin; Qin ?]
S	Fiery	[Yong]	Yugui; Liu, Xing,	[Qin; Zhou ?]
SE	Uplifting	[Jing]	Zhang; Yi, Zhen	[Zhou; Chu]

Here each of the nine fields corresponds to three lunar mansions, except Dark Heaven which has four. As may be seen from the punctuation and arrows in the

17. This curious variation produces an idiosyncratic list of 13 distinct units, rather than the standard 12. John Major speculates that this greater discrimination may be a reflection of the *Huainanzi's* "southern orientation"; see Major (1993, 128). The preeminence of Jupiter's role in astrological prognostication based on this system is evident in the lines immediately following in the *Huainanzi* passage (3:15a–b) on which Major comments: "As shown by the prognostication rules given in lines 20–24, all of the attention given here to the description and apportionment of the lunar lodges is for the purposes of the 'Yin-yang Militarist' School, to indicate good or bad military fortune for the several states." Despite this, and unlike the standard listing show here which begins with the first Jupiter station Xingji, the *Huainanzi* begins the list with the first lunar mansion Horn (Virgo), presumably because that is the sequence the author follows in the passage immediately preceding giving the dimensions of the mansions in degrees. As one might expect, the military value of calendrical astrology is made even more explicit in the *Huainanzi* treatise on military affairs; see note 12 above.

18. *Lüshi chunqiu* (13:1b). Although the passage does not supply the names of the nine provinces, its tabulation of the nine fields of Heaven giving the cardinal and intercardinal directions of each, together with their associated lunar mansions, makes the correspondences unambiguous. The terrestrial correlations are those assigned by Gao You (168–212) in his commentary. These conform with those specified in the parallel passage in the *Huainanzi*, chapter 3, "Treatise on the Patterns of Heaven," 3:2b; see Major (1993, 69).

fourth and fifth columns, this apportionment results in a straddling of boundaries by some lunar mansions which conventionally refer to a single terrestrial location, as for example in the case of the state of Yan in column 5, which ordinarily corresponds to province You and to mansions Wei (Sco) and Ji (Sgr) (see Table 1), but here those two lunar mansions are allotted separately to provinces Yu and Yang (column 3), respectively, and province You does not appear. This is a clear indication that the augmentation of the divisions proceeded sequentially, beginning with the cardinal four, then increasing to nine, and finally to twelve, the second enhancement occurring before, and the third after, the distribution of the 28 lunar mansions among the 12 chronograms (Jupiter stations) was conventionalized. In transforming the scheme of nine divisions into twelve (Table 1), one new province, Bing, was interposed between Qing and Xu and assigned to the state of W-i; another, San He, was placed between Yong and Jing and assigned to Zhou; and a third, You, was placed between Yu and Yang and given to Yan. In addition, the name of the WSW province was changed from Liang to Yi and made to correspond to the territory of the state of Wei, one of the three which succeeded Jin.[19]

On the whole, the two systems' apportionments are in basic agreement, so that it is clear that the expansion by three terms did not alter *preexisting astrological relationships* between celestial and terrestrial locations, but strove rather to accommodate ancient tradition to new political realities by making terminological adjustments "on the ground," as it were. Chief among the historical changes demanding accommodation, besides the break-up of Jin, was the eastward removal and subsequent decline of Zhou after 771 BCE coupled with the rise of Qin in present-day Shaanxi. Taken together these adaptations point to a late Western Zhou or Spring and Autumn period date (8th to 7th centuries BCE) for the Nine Heavenly Fields precursor to the Field Allocation scheme of twelve shown in Table 1.[20]

---

19. The implicit correlation between province Liang and the state of Jin in the Nine Fields system strongly suggests a different identity for Liang than given in later texts. In the "Yu gong" (The Tribute of Yu) chapter of *Shangshu* (Book of Documents), a Zhou text, Liang is identified with southern Shaanxi province, specifically the region between Mt. Hua and a tributary of the Han River known as the Black River; see Nienhauser (1994, 27). But there is no doubt that the state of Jin and its immediate predecessor, Tang, were located near the Fen River in western Shanxi, northeast of the Yellow River's abrupt northward bend near its confluence with the Wei River. Just west of the confluence of the Fen River with the Yellow River is Mt. Liang, one of the peaks which demarcate the watersheds regulated by the legendary Xia Dynasty founder Yu the Great in the border region between provinces Ji and Yong; see Nienhauser (1994, 22). It is probable, therefore, that province Liang was originally located in this area in pre-Zhou times and only later relocated and redefined to include southwestern Shaanxi and Sichuan after the extermination of the principality of Liang by the state of Qin in mid-Zhou Dynasty.

20. Elsewhere I have argued for a 2nd millennium BCE date for the conceptual basis of the correlation between celestial and terrestrial events, locations, etc., based in part on the actual locations and early textual accounts of portents connected with three impressive massings of the five planets. There is considerable evidence to suggest that the correlation between Xia, Shang, and Zhou and the Northern, Eastern, and Southern quadrants of the sky and their cardinal constellations arose very early in Chinese cosmo-political thinking; see Pankenier (1995).

Derivation of Astrological Correspondences

The antecedents of Field Allocation astrology are to be found in certain conceptions already influential in the 2nd millennium BCE. Apart from the rare planetary massings already verified, several passages in pre-Qin works preserve the remnants of etiological myths and traditions which establish the existence of definite connections between certain celestial locations and terrestrial polities or peoples. One of the most famous is the legend of E Bo and Shi Chen which is preserved in the *Zuo Commentary* (Duke Zhao, 1st year):[21]

> Formerly, Gao Xin had two sons, the eldest was named E Bo and the younger one Shi Chen. They lived in Kuanglin but could not get along, daily taking up shield and lance against one another. In the end, Gao Xin could no longer condone it and removed E Bo to Shangqiu to have charge of [the asterism] Chen (Great Fire in Scorpio); the ancestors of the Shang people followed him, therefore Great Fire is the Shang asterism. [Gao Xin] moved Shi Chen to Da Xia to have charge of Shen (Orion), so the people of Tang followed him, and there served the houses of Xia and Shang. The last of their line was Tang Shuyu. When [Zhou Dynasty] King Wu's wife Yi Jiang was pregnant with Tai Shu (i.e., Tang Shuyu), she dreamed that the Lord on High told her: 'I have named your son Yu and will give Tang to him, make Tang belong to Shen (Orion), and cause his descendants to flourish.' When the child was born he had the character Yu on his hand, and so his name was called Yu. When [Zhou] King Cheng extinguished the old house of Tang he enfoeffed Tai Shu there; hence, Orion is the star of Jin. From this we can see that Shi Chen is the spirit of Shen (Orion).

and again (Duke Xiang, 9th year):[22]

> The ancient Regulator of Fire was offered sacrificial nourishment either in the asterism Heart (Sco) or in the asterism Beak (Hya) in order [for the people] to take out and bring in their fires. For this reason Beak is [called] Quail Fire and Heart is [called] Great Fire. Tao Tang's (i.e., Emperor Yao's) Regulator of Fire, E Bo, dwelt at Shangqiu and sacrificed to Great Fire (Sco), using Fire to mark the seasons there. Xiang Tu (grandson of Xie and father of the Shang people) continued in like manner, and so the Shang principally focus on Great Fire. They observed that the incipient signs of their calamities and defeats invariably began with Fire.

The same legend is alluded to in an equaly famous passage in *Guoyu*, "Jinyu" (Discourses of the State of Jin):[23]

---

21. *Combined Concordances to Ch'un-ch'iu, Kung-yang, Ku-liang and Tso-chuan*, Vol. 1, 344.
22. *Combined Concordances to Ch'un-ch'iu, Kung-yang, Ku-liang and Tso-chuan*, Vol. 1, 266.
23. *Guoyu* (10:3a).

I have heard that when Jin was first enfoeffed, Jupiter was in Great Fire, which is the star of E Bo; in truth, it marked the periods of the Shang.

This famous nexus of astral lore involving E Bo and Shi Chen, which has abundant echoes in later literary composition, weaves together various elements of cultural significance in characteristic fashion. At bottom, it is a classic example of the kind of etiological myth dating from the pre-literate period which served both to explain and to memorialize vital astronomical and calendrical knowledge. In so doing it also laid the foundation for the subsequent elaboration of an astrological theory based on celestial and terrestrial correspondences. In this pithy story we can discern an euhemerized tale about the human origins of the deities associated with the principal constellations of spring and autumn, Scorpio and Orion, which are diametrically opposed or "at odds" in the heavens and hence never appear in the sky simultaneously. Just as if unable to abide each other's presence, E Bo invariably ducks beneath the western horizon just before Shi Chen rises in the east. These personified asterisms are then linked to the cardinal directions (East and West), to seasonal activity for which they anciently served as harbingers (carrying out and in of the hearth fires[24] marking the beginning and end of the agricultural season), as well as to the dominant political entities of the archaic period (Shang and Xia), their descendants, and the hereditary lines of astrologers and calendrical specialists who served them. In this and other examples of such astral lore (notably texts linking the above-mentioned bird asterism "Beak" to the fortunes of the Zhou people) it is possible to discern the historical nucleus of the system of astrological correlations which over time was to become amplified and ultimately standardized in the configuration shown above in Table 1.

The prominence and appropriateness of Orion and Scorpio as seasonal harbingers has frequently been remarked upon. More tellingly, however, Scorpio's celestial location and ancient calendrical function may link this constellation with the flood mythos and the cosmogonic exploits of Yu the Great and his father Gun.[25] What principally concerns us here, however, is the fact that both Scorpio and Orion are similarly located adjacent to the Heavenly River, though as we have already noted, at opposite ends of the sky. If we consider the relationship between the asterisms and their terrestrial correlates in the Heavenly Fields scheme in Table 2 it will be noted that Great Fire corresponds to east and province Yu, identified with Song, the successor state of Shang in east-central Henan, while Shen or Orion corresponds to the southwest, province Liang, and the state of Jin, consistent with the traditions relating the feud between Shi Chen and E Bo. Although the connection between

24. Interestingly, as red supergiants, both Alphard (Delta Hydrae) and Antares (Alpha Scorpii) are noticeably orange to reddish in hue.

25. Scorpio's location on the western edge of the Milky Way or "Heavenly River" figures prominently in a recent analysis of the flood mythos and the evidence for the early empirical recognition that what had been thought to be changeless features of heaven were in fact impermanent, as seasonal harbingers in the sky were rendered obsolete by their precessional dislocation relative to the tropical year; see Porter (1996, 27 ff). The recent discovery of a Neolithic tomb at Puyang, Xishuipo, in Henan Province containing a dragon and a tiger mosaic assembled from clam shells and positioned with cosmological accuracy to the east and west of the high status occupant of the tomb indicates that the seasonal role of those cardinal constellations may already have been recognized by the 3rd millennium BCE. For a speculative interpretation of the tomb as a simulacrum of the heavens, see Feng (1990, 52–60).

Shang and the east is well established, how are we to account for the placement of Jin in the southwest? Or, for that matter, for such glaring geographical anomalies as the association of ancient states like Zhou in the west and Qi in the Shandong peninsula on the east coast with cardinal south and north, respectively?

If, however, we plot the provinces and their celestial counterparts on a schematic map of the North China plain the incongruities begin to assume the character of deliberate constructions arising from a preeminent concern with celestial rather than terrestrial topography. Viewed in this way, the scheme may be seen to preserve at its core distinct traces of the original cosmographic organizing principle that was later obscured.

The Heavenly River as Cosmographic Divide

The Milky Way arches across the dome of the heavens (Figure 1) from Sagittarius in the northeast (juncture of the northern and eastern quadrants of the sky) to Gemini in the southwest (juncture of the southern and western quadrants), with Scorpio and Orion strategically located at those opposite extremes. If we imagine that ancient Chinese astrologers practicing their art in the great river's flood plain envisioned the Milky Way as the celestial analogue of the Yellow River, an explanation for the apparently incongruous astrological correlations immediately suggests itself. When the generally southwest to northeast course of the Yellow River (Figure 2) is compared to the SW to NE orientation of the Heavenly River (Figure 1), most of the celestial correlations of the Nine Provinces in the Spring and Autumn period make sense. The "fields" of Qi (Province Qing) in the "north" and Qin/Zhou (Province Yong) in the "south" determine the generally "north to south" orientation of the major cardinal axis, while the area above the Yellow River becomes "west" and below becomes "east." Jin's location (Province Liang) in the astrological "southwest" can now be seen as a logical outgrowth of this paradigmatic celestial frame of reference, as does Chu's (Province Jing) location in the "southeast." Further examination of the figures confirms that the scheme's essential fidelity to celestial topography rather than to terrestrial directions probably arises from an archaic conception in which the Yellow River and the Milky Way were functional analogues in astrological terms.[25]

The heavenly topography, thus, is primary. The complete Heavenly Fields scheme presents a few minor inconsistencies with this portrayal of the system's origins, but it is not difficult to imagine how these could have arisen over time as the growing complexity of the geo-political situation and its astrological ramifications tended to obscure the original cosmographic concept. Interestingly, the identification of latecomers on the political horizon, Wu and Yue (province of Yang), with astrological "northeast," though presumably quite late, is still consistent with the original concept. The one major incongruity, the identification of the states of Wèi and Lu with north-northwest, makes no sense either geographically or astrologically, though the association of the state of Yan with east-northeast, while not astrologically correct, is at least geographically accurate. By the time the expanded Warring States system portrayed in Table 1 was fully elaborated, the original cosmographic

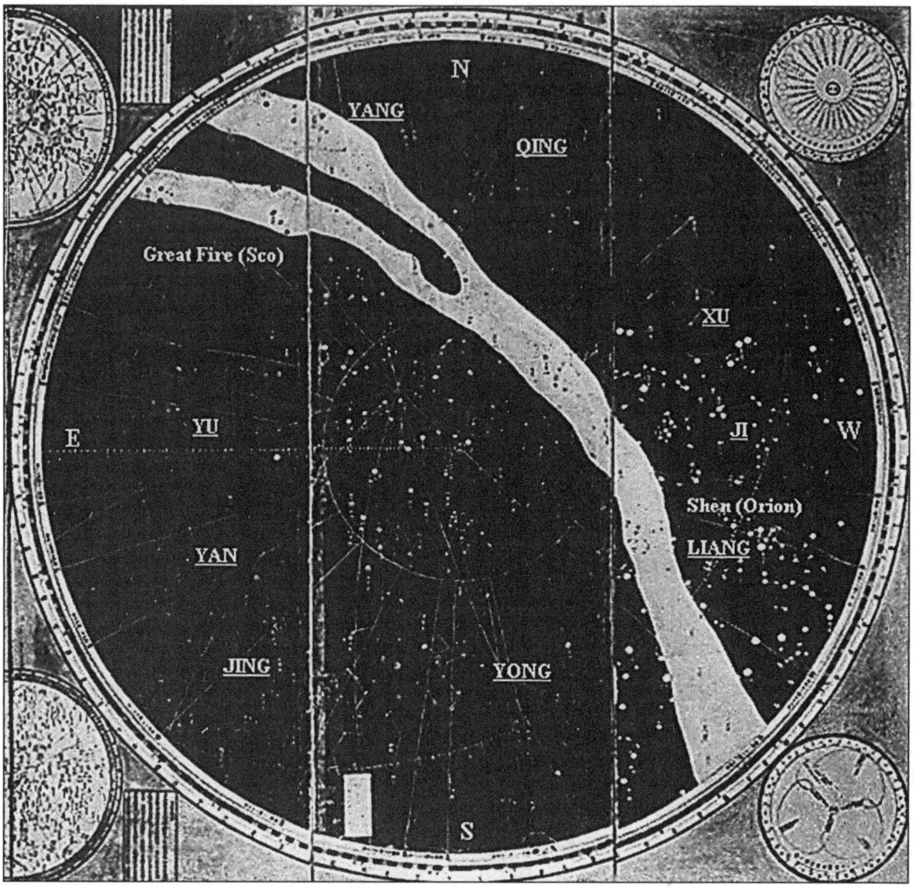

Figure 1. Locations of the Nine Provinces around the Milky Way. After Zhongguo gudai tianwen wenwu tuji (Collected Illustrations of Cultural Artifacts of Ancient Chinese Astronomy) (Beijing: Wenwu chubanshe, 1980, p. 101).

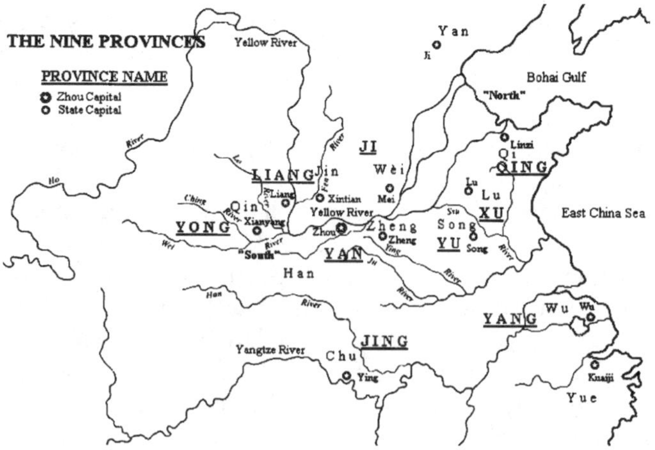

Figure 2. Locations of the Nine Provinces in relation to the Yellow River.

frame of reference appears to have been largely obscured by the accretions. Although the analogy between the Yellow River and the Milky Way was conventionally repeated in Yin Yang and Five Elements (Powers) speculations from this time on, actual references in the late Zhou and Han astrological literature to the Milky Way as the Yellow River's celestial analogue are surprisingly rare.[26]

One noteworthy, but quite late, exception is the brilliant Tang dynasty (618–907) monk-astronomer Yi Xing (fl. 725), who is recorded as having seen this archetypal astral-terrestrial parallelism as fundamental in Field Allocation astrology. By Yi Xing's time the expansion of Hua-Xia civilization had long since brought south China within the mainstream of Chinese culture, so that both of China's major river systems now had to be taken into account, somewhat obscuring the archaic one-to-one correspondence between the Yellow River and the Milky Way. Nevertheless, the fundamental conception is still discernible in his explanation. According to the Song dynasty encyclopaedist Zheng Qiao (1104–1162):

> How excellent was Yi Xing's discussion of the twelve stations in the Tang dynasty! From beginning to end he relied on the River of Clouds (i.e., Milky Way) to talk about them. The River of Clouds is the *qi* of the Yellow and Yangtze Rivers. He discerned the Two Divisions (of China) by the mountain ranges and river systems and recognized the rising and setting of the River of Clouds at the Four Nodes (NE, SE, SW, NW). On comparing below to the ancient Han dynasty commanderies and kingdoms, differentiating and locating the field allocations (*fenye*) was (as easy as) pointing to one's palm. Evidently, the stars are like *qi*, nothing more; the River of Clouds, the Northern Dipper, the Five Planets, there is nothing which is not *qi*. How profound was Yi Xing's learning![27]

Conclusion

As Sima Qian strongly intimated, by his time astrological prognostication based on celestial and terrestrial correspondences (and by implication, an antecedent paradigmatic cosmography), was already very ancient. Archaic astrological traditions preserved in Zhou texts, the history of astronomical observation, and the historical development of field allocation astrology all support Sima Qian's assertion. Even though adaptations to the field allocation scheme in the late Zhou period could apparently be made with little regard to the archaic geo-political precedents, at an ear-

---

26. Like most of his contemporaries, the famous Han Dynasty astronomer and polymath Zhang Heng (78–139) subscribed to the conventional view that the "essence of [elemental] Water becomes the Heavenly Han (River)"; see the quote from his Ling Xian in Taiping yulan, vol. 1, 8:10a. In a similar vein, the Han apocryphal text Hetu kuodi xiang (River Diagram Images Encompassing the Earth) makes the astral-terrestrial correspondence plain: "The essence of the Yellow River rises to become the Heavenly Han"; Taiping yulan, vol. 1, 8:11a. How, precisely, this generally stated correspondence plays out in terms of specific astrological prognostications is not made explicit, however.

27. Zheng Qiao's approving comments conclude his discussion "Distinguishing Field Allocations" in *Gujin tushu jicheng*, chapter 57, "Stars and Asterisms: General Discussion," 1341. For the geographical details of Yi Xing's analytical scheme, see the "Monograph on Astrology" in *Xin Tang Shu* (New History of the Tang Dynasty), 21:817.

lier stage when the new province of "Three Rivers" was incorporated into the system and the state of Zhou was summarily detached from its erstwhile homeland in Shaanxi Province's Wei River Valley in favor of Qin, care was obviously still taken to preserve the pre-existing correspondence between Zhou, cosmographic "south," province Yong, and the Vermilion Bird constellation (Hydra, Corvus, Crater). Evidently, even gross manipulation in geographical terms was unproblematical so long as Zhou's primordial astral correlates and cosmographic cardinality remained intact. If this portrayal of the historical nucleus at the heart of field allocation astrology is correct, further research in this area holds the promise of adding an important new dimension to the study of the political and military history in the early period. Already the evidence points to a long process of evolutionary development in astrological cosmography reaching well back into the formative Three Dynasties period of Chinese civilization. By mid-11th century BCE judicial astrology figured importantly in the epoch-making Zhou conquest campaign that overthrew the Shang Dynasty, and the evidence that continues to emerge indicates that astral/terrestrial correspondences routinely played a similar role throughout the Zhou dynasty.

References

Biot, E. tr. 1975, Le Tcheou-li ou Rites des Tcheou. 3 vols. (Paris: Imprimerie nationale, 1851). Reprint. (Taipei: Ch'eng-wen Publishing Co.).

Brooks, E. B. 1995, "Warring States Working Group: Query 47" 4 July 1995. Photocopy.

Clube, S. V. M. 1995, "The Nature of Punctuational Crises and the Spenglerian Model of Civilization" Vistas in Astronomy 39.4.

Combined Concordances to Ch'un-ch'iu, Kung-yang, Ku-liang and Tso-chuan. Harvard-Yenching Institute Sinological Index Series, Supplement No. 11. Vol. 1. Reprint. (Taipei: Cheng Wen Publishing Co., 1966).

Cullen, C. 1996, *Astronomy and mathematics in ancient China: the Zhou bi suan jing* (Cambridge: Cambridge University Press).

Encyclopedia Britannica, 15th ed. (Chicago: Encyclopedia Britannica, 1984).

Feng Shi. 1990, "Henan Puyang Xishuipo 45 hao mu de tianwenxue yanjiu (Research on the astronomy of tomb no 45 at Xishuipo, Puyang in Henan Province)" Wenwu 3, p. 52.

Graham, A. C. 1989, *Disputers of the Tao* (LaSalle, Illinois: Open Court).

Gu jin tushu jicheng (Comprehensive Collection of Illustrations and Documents Ancient and Modern) Reprint. (Taipei: Wen-hsing shu-tien, 1964).

Guoyu (Discourses of the States) Sibu beiyao edition. (Shanghai: Zhonghua shuju, 1927–35). Reprint. (Taipei: T'ai-wan Chung-hua shu-chü, 1966).

Huainanzi (Master Huainan) Sibu beiyao edition. (Shanghai: Zhonghua shuju, 1927–35). Reprint. (Taipei: T'ai-wan Chung-hua shu-chü, 1966).

Jinshu (History of the Jin Dynasty). Reprint. (Beijing: Zhonghua shuju, 1974).

Lewis, M. E. 1990, *Sanctioned Violence in Early China* (Albany: State University of New York Press).

Löwith, K. 1949, *Meaning in History* (Chicago: University of Chicago Press).

Lüshi chunqiu (Spring & Autumn Annals of Master L^{n}). Sibu beiyao edition. (Shanghai: Zhonghua shuju, 1927–35) Reprint. (Taipei: Chung-hua shu-chü, 1966).

Major, J. S. 1993, *Heaven and Earth in Early Han Thought: Chapters Three, Four, and Five of the Huainanzi* (Albany: State University of New York Press).

Lau, D. C. tr. 1970, *Mencius* (Harmondsworth, Middlesex: Penguin Books).

Needham, J., et al, 1959, *Science and Civilisation in China*. Vol. 3. *Mathematics and the Sciences of Heaven and the Earth* (Cambridge: Cambridge University Press).

Nienhauser, W. H., ed. 1994, *The Grand Scribe's Records*. Vol. 1. The Basic Annals of Pre-Han China. (Bloomington: Indiana University Press).

Pankenier, D. W. 1995, "The Cosmo-Political Background of Heaven's Mandate," *Early China* 20, p. 121.

Pankenier, D. W. in press, "Applied Field Allocation Astrology in Zhou China: Duke Wen of Jin and the Battle of Chengpu (632 BCE)" Journal of the American Oriental Society.

Peiser, B. J. 1998, "Cultural Aspects of Neo-Catastrophism: What are the Implications for Archaeoastronomy?" This Book (Durham: Carolina Academic Press).

Porter, D. L. 1996, *From Deluge to Discourse: Myth, History, and the Generation of Chinese Fiction* (Albany: State University of New York Press).

Karlgren, B. tr. 1950, *Shangshu* (Book of Documents) (Stockholm: Museum of Far Eastern Antiquities).

Shisanjing zhushu (The Thirteen Classics with Notes and Commentaries) 2 vols. Reprint. (Taipei: Wen-hua t'u-shu kung-ssu, 1970).

Sima, Qian. *Shiji* (The Grand Scribe's Records) Reprint. (Beijing: Zhonghua shuju, 1959).

Sun Xiaochun & Kistemaker, J. 1997, *The Chinese Sky During the Han: Constellating Stars & Society* (Leiden: E. J. Brill).

Taiping yulan, 7 Vols. Reprint. (Taipei: P'ing p'ing ch'u-pan-she, 1975).

Watson, B. 1967, *Basic Writings of Mo Tzu, Hsün Tzu, and Han Fei Tzu* (New York: Columbia University Press).

Xin Tang shu (New History of the Tang Dynasty) Reprint. (Beijing: Zhonghua shuju, 1976).

Yi Shitong. 1996, "Beidou ji—dui Puyang Xishuipo 45 hao mu beisu tianwen tu de zai sikao (Sacrifices to the Northern Dipper—further thoughts on the astronomical chart made of shells in tomb #45 at Puyang, Xishuipo)" Zhongyuan wenwu 2.

37

Canton at the Tropic of Cancer

Bella C. Chiu
Cambridge, MA

Archaeoastronomy has been a field of study for only about twenty years, a great part of it on Mesoamerica. The proximity of Canton in China to the Tropic of Cancer has been noticed some years ago, mentioned for example, in Needham, 1959. But not much written history is known about this region. What we know is mostly archaeology. This seems to be the time then, to work on the problem, as a problem in archaeoastronomy.

Prehistoric maps show that Canton is close to the Tropic of Cancer for its value today, and not too far for the time that Canton was built. It was in the third century BC. In other words, Canton was built remarkably close to the location of the limit at which zenith passages of the sun can be observed.

Historical evidences are consistent, but no direct evidence of observing the zenith passages has yet been found.

I have made calculations and graphs to explore the situation and they are consistent with my view, that in the early days of Canton, the watching of the zenith passages of the sun was observed, and this may have influenced the location of the city.

Also, I have made estimates of errors the ancient Cantonese could have made while looking for the limit at which zenith passages occurred, and they are consistent.

Finally, there are some Chinese sources, written during the Ch'ing dynasty (1644–1911), about the beginnings of Canton, and about the 2nd century BC ruler whose tomb is probably the one designated by a cross on A. Herrmann's prehistoric map.

* * *

Introduction

The proximity of Canton in China to the Tropic of Cancer was noticed some years ago. It was mentioned, for example, in Needham, 1959. But not much history is known about this region. What we do know is mostly archaeology. Perhaps then, this is the time to work on the problem, as a problem in archaeoastronomy.

This paper is written to increase interest on the problem. More information is needed to give a proof—proof of the indigenous interest in zenith passages, or of the connection with the history, which is partly legend, on the beginnings of Canton.

Canton is close to the Tropic of Cancer for its value today, and not far for the time that Canton was built, in the 3rd century BC. In other words, Canton was built remarkably close to the location of the limit at which zenith passages of the sun can be observed. I have made calculations and graphs to explore the situation in more detail, and they are consistent with my view that in the early days of Canton, the zenith passages of the sun were observed, and this may have determined the location of the city.

The two prehistoric maps from A. Herrmann's book (1935, 2nd ed. 1966) show that few cities were known to have-been built in southern China during the Iron Ages, or the 3rd century BC. And yet two are found very close to the Tropic of Cancer. More work in archaeology is now being done here which may clarify the situation.

Historical evidence is also consistent, but no direct evidence of watching the zenith passages has yet been found. One story recorded of later history (Joseph Needham, 1959) was that of the military leader, Kuan Sui, who marched southward and noticed that the sun reached zenith and beyond—the shadow of a gnomon appeared south of it when he stood beyond a certain latitude. This happened in 349 AD, in the 5th month of the year.

On the state of mathematics, I have made a simple calculation of distance from the Tropic of Cancer, once given the shadow length of the gnomon. This was probably not done in early Chinese history.

And when such calculations were made later, the forms of the math may have been different However, the ancient astronomers of the 2nd century BC did use similar triangles for math, and they did observe the disappearing of the shadow of the gnomon, although they did not consider the curvature of the earth, and they reached a much smaller value for what they were after, the height of the sun (or heaven).

A few archaeological sites are found west and south of Canton. The migration, according to our prehistoric map, is toward the northeast. I have made estimates of errors the ancient Cantonese could have made while looking for the limit at which zenith passages occurred. They agree with the fact that some sites are about 20 miles south of the ancient value for the Tropic of Cancer. Finally, there are some Chinese sources written during the Ching dynasty (1644–1911 AD), about the beginnings of Canton, and about the 2nd century BC rulers whose tombs, one of which is probably the one designated by a cross on A. Herrmann's prehistoric map. The newer archaeological material, and more interesting maps, tell where the first three kings of Nanyue were buried. Some are still speculations by the archaeologist in charge, Mai Yinghao.

Iron Age Tombs near Canton

On the Herrmann map of prehistoric sites in China, the tombs are 0.2' north of the Tropic of Cancer (taken as 23.5° N), with error of map measurement ~0.1'. According to this map, we have 17 sites in China, during the Iron Ages That would be about 500–200 BC.

At the time of the First Emperor (255–206 BC), which is also during the Iron-Age the city of Hsiang was just 0.05' north of the Tropic of Cancer, taken as 23.5° N, and Nanhai (ancient name for Canton) is 0.24' south. Here we have two cities close to the Tropic of Cancer.

Maps are from Herrmann's book, first published in 1935 by Harvard University Press, and then revised in 1966. They were important maps and much used for commercial and political purposes. But they came from only Western sources. We know that many more archaeological sites have now been found.

Finding Your Location from the Shadow of a Gnomon: The State of Astronomy in Ancient China

An interesting historical event is a good reference point from which to discuss the state of astronomy in China. In particular, this one is about the shadow of a gnomon. It is about the observation of Kuan Sui, who moved south during a military campaign (Joseph Needham, 1959). His latitude is given by a modem calculation, using the length of shadow of a gnomon which he had recorded. A gnomon of height p stands at a distance D from the Tropic of Cancer. The noontime sun casts a shadow of length d during summer solstice. We set d = 9.1 inches, as observed by Kuan Sui, and assume a pole length of p = 10 feet, which is a value used before in Chinese historical records, then del = D/R, where R = radius of earth = 3963 miles. Either del or D measures the distance of our gnomon from the Tropic of Cancer.

We have $d/p = \tan T$.
and $T = A + \text{del}$.

Making the very good approximation, that A is small,

then $\tan(\text{del}) = d/p$
$= 9.\ 1/10$ ft. $= 9.1/120$ in, and del $= 0.0757$ radians

Then $D = R \times \text{del} = 3963$ mi. $\times\ 0.07 = 300$ miles.

This corresponds to about 300' = 5°. In 349 AD, the Tropic of Cancer was at 23° 39'N, so our calculation for the latitude is 18.6° N.

The gnomon height was unlikely less than 8 feet or more than 12 feet, so the latitude might have been from 17.3° to 19.4° N. which is very close to the values given in Needham's book. Notice that in our diagram, without the use of the curvature of the earth, we would have gotten a very wrong answer. The ancient Chinese did not appreciate the curvature of the earth at this time (about 2nd or 3rd century BC). But in the 2nd century BC, they did notice the shortening of the shadow, to find the height of Heaven (or the sun). Without the curvature of the earth, it was only 100,000 li, or 1/3 of that in miles.

They used similar triangles in math. Records of the use of trigonometry came later in the 11th century. It was mathematics similar to trigonometry, and it is possible that another form was used even earlier to arrive at results close to ours.

Discussion of Errors

If our ancient sites were built to be at the Tropic of Cancer, then errors in latitude would probably be toward the south. This is because earlier migration was from the south and small errors in recognizing the zero shadow result in a larger error in finding the limit—in deciding that the zero point location is not further north.

In the case of the Han dynasty tomb northwest of Canton, if the distance from the Tropic of Cancer (at 23.7° N for 1 AD) is about 20 miles, then the shadow of a gnomon 10 feet tall a can be calculated; it is about half an inch. This error of half an inch might have been accepted, not only because of the penumbra fuzziness of image, but because of an additional problem: the noontime sun of the next day reaches zenith at a few miles away. Exactly at the Tropic of Cancer, the distances are small. But 20 miles south, the next one is a few miles away. The point is, they would not be able to compare the unclear half-inch shadow with the appearance of an exact solar zenith with no shadow.

History tells us that three centuries later. an excavation was made, but only the tomb of the third king was found. This was probably the one depicted by a cross on the map. The obliquity of the ecliptic was at 23° 43.3'N in 137 BC, when Chao Tuo was buried. The third king was buried not much before 1 AD. The latitude at this time was 23° 42'N.

The Five Kings of Nanyue

Nanyue is the ancient name for Canton (Guangzhou). South China had a culture of its own, an indigenous people whose main occupation was agriculture. But by the Han dynasty (202 BC–220 AD), it was ruled by the Han people from the north. Kings were sent from the northern capital to rule the South. But still, there was fighting between the kings sent to Nanyue and the emperor/empress in the north.

Now there are questions about the tombs of these kings. Six hundred tombs from the Han dynasty, of kings and lesser officials, were uncovered by archaeologists during the years, 1953–78. Then in 1983, two special tombs were found: one was surely the tomb of Zhao Mei, the second king of Nanyue. The other one, speculates Mr. Mai, who led the expedition, could be the tomb of Zhao Ying Qi, the third king "in view of the size of the chamber and the exquisite quality of the few art objects left by the plunderers."

"It is known that during the period of the Three Kingdoms [3rd and 4th centuries AD], the military leader, Sun Quan, sent troops to look for the tombs of the kings of Nanyue, but they found only the tomb of Zhao Ying Qi." I began my problem of archaeoastronomy because of a prehistoric map in an atlas by A. Herrmann in 1935, newer edition 1966. A tomb site of the Han period was designated as being just Northwest of Canton. That would make it very close to the Tropic of Cancer.

So now I believe that the cross on the map indicates the location of the tomb of Zhao, Ying Qui, because it has been described in history books. And it is Northwest of Canton although not as far north as I would have wished, to agree with the latitude of the Tropic of Cancer in ancient times.

Mr. Mai Yinghao wrote a book as well as an article in the Encyclopedia of China: Archaeology (1986) about the excavations in and near Guangzhou. Professor K.C. Chang of Harvard University kindly translated the article and a letter from Mr. Mai. The quotations are their words.

The tomb of the first king, Zhao Tuo, is still not found. Mr. Mai speculates again, that it is on a hill, overlooking the City, called Yuexiushan. This is interesting. But I would say that the results are inconclusive that one of the tombs, or another such historical marker in the area of Canton, was placed there intentionally because of the Tropic of Cancer. The ancient determination of the limit to zenith passages could have been 32' off in latitude, but it is too far off to be a confirmation. We still need direct evidence.

For example, there is found in Henan Province in the north, a gnomon dated to be from the Zhou dynasty (earlier than 256 BC). It is not just a vertical pole, but quite permanent with a solid base. We would want to find such a gnomon in South China.

Sightings from a hilltop would be interesting too. If the tomb of the first king, Zhao Tuo is indeed on the top of Yuexiushan it would be worthwhile to check from this vantage point the directions of sunrise and sunset on summer solstice or zenith passage. One would check also on this date how long is the shadow of a vertical gnomon. However, the location could have been chosen simply so that the first king's tomb would be overlooking the city,

Conclusion

In conclusion, I would say that we have so far only circumstantial evidences:

1. The math and astronomy of these times are sufficient to note and record solar zenith passages (2nd century BC);

2. South China was agricultural, and had reason to observe the zenith passages of the Sun.

3. The Han kings who ruled here were sophisticated enough to seek the limit at which zenith passages occurred. If the indigenous people observed zenith passages, the kings ruled long enough to have been interested in marking it.

The tombs of the first three kings are nearly found, and almost identified. The history of this region is consistent with our views on what happened here.

And so we propose, if and when archaeological identifications are made:

1. That at the Han dynasty tombs now found, north of the modem city of Canton, scientists should look for benchmarks. The tombs may have been used. If gnomons are used, they should be large and permanent. (A Zhou dynasty gnomon with heavy base was found in the north, in Henan Province, dated ca. 300 BC. Such a gnomon in South China would be evidence of the importance of shadow measurements.)

2. That they should search the horizon from these tombs for mountain peaks or troughs which would give sunrise or sunset positions. But horizon observation is not necessary for zenith observation.

The Kings of Nanyue: A Historical Note

Nanyue, ancient name for Canton or Guangzhou, was built in the 3rd century BC. They were an agricultural people, their legend telling of goats who brought five different colored strands of rice, each with one in its mouth (five strains of rice?).

By the beginning of the 2nd century BC, their government had been taken over by the main, Han nation of the north. Officials from the north were sent to rule as kings of Nanyue and the surrounding areas of South China.

This was during the Iron Age of China. Many necessities including the plow were made of iron. So when the South tried to rebel under Zhao Tuo, it was put down by a boycott of iron products, by the edict of the Empress Lu.

It is no surprise then, that Zhao Tuo late in life was afraid his tomb would be desecrated, and so he had four coffins made. After his funeral, they were sent out in four directions, and today still, have not been found.

History tells us that the military leader, Sun Quan, in the Three Kingdoms period (3rd to 4th century AD) tried to find his tomb, but succeeded in finding only that of the third king, Zhao Ying Qi.

The second king's (Zhao Mei's) tomb was found and identified in 1983, and today it is within a museum at the north-central part of the city.

It is the speculation of Mr. Mai, the archaeologist, that a grand tomb also found in 1983, is that of the third king Zhao Ying Qi, now rediscovered. It is located Northwest of the city, in a region called Xicun.

He further speculates that the first king Zhao Tuo had his tomb located on a hilltop overlooking the city, also looking from the north.

Acknowledgments

I wish to thank Professor Philip Morrison and other colleagues for encouraging my work in this field. Professor Rulan C. Pian encouraged my work on this particular paper, and Lensey Namioka had encouraged me to develop writing skills.

I am grateful to Professor Liu Chun-Jo for her help in Chinese sources, and Mr. Timothy Connor for finding important maps.

Finally, I wish to thank Mr. Mai Yinghao for his informative letter, and Professor K.C. Chang for translating it.

References

A. Herrmann, *Historical and Commercial Atlas of China*, pp. 8,18,19, Harvard-Yenching Inst. Monograph Series, Vol.1, (Harvard Univ. Press, 1935, 1966).

J. Needham, *Science and Civilization in China*, Vol.3, pp. 200, 286–293, (Cambridge Univ. Press, 1959). Includes table of obliquity, discusses state of astronomy in China during Han dynasty.

N. Sivin, "Chinese Archaeoastronomy: between Two Worlds," *World Archaeoastronomy*, ed. by A. F. Aveni, (Cambridge Univ. Press, 1989).
Wang Zhongshu, *Han Civilization*, p. 122. Translated by K.C. Chang et al., (Yale Univ. Press, 1982).
Atlas of Guangdong Province (Cartographic Pub. Hse. of G.P., 1989) pp. 6–7 Yuexiushan & museum from 1983 site of Zhao Mei.
Newest Canton City Roadmap (translation) vol. 1. Good topographical map of hill (1924).
Postal Atlas of China Preface dated 1933). Has Xicun Ch'ing dynasty sources: Shui Jin Zhu, (in Chinese), A note in the Naval History.
China, a Provincial Atlas by Hsieh (1995), shows latitudes.
Chinese Collected Local Records (in Chinese, 1879), based on Ch'ing dynasty sources. Refers to gnomons, legend of goats, etc.
History of Astronomy Workshop in Notre Dame, IN (June 1995). Gave a short talk on this article. Interest was shown on calculations and on origin of Canton.
Encyclopedia of China: Archaeology (1986). Wang Zhongshu kindly wrote me about this source now available in America. Article on sites found in Guangzhou region, written by Mai Yinghao.

38

Discoveries in Aboriginal Sky-Mapping (Australia)

Hugh C. Cairns
Darling Point, N.S.W., Australia

At Oxford 3, a set of working models found that Aboriginal people mapped particular star groups and the moon cycle by means of cup-dots and line-grooves at archaeological sites and on artifacts (Cairns 1993). Then, a large range of the night sky was found (Cairns at Oxford 4) to be symbolized within a particular prehistoric painting at a ceremonial area related to a whole cosmos of indigenous belief. Now, a 1995 discovery produces a further model which works in application to famous enigmatic engravings near Sydney. The Dark Spaces of the Scorpius-Crux section of the Milky Way appear within the artistic polymorph figures inscribed in sandstone. As a new insight, this discovery—that figurative engravings by prehistoric Australians match artistic perceptions of dust cloud nebulae—may be of importance to religio-spirituality as well as to the history of pre-science. By analogy with the Daramulan emu of Aboriginal initiation lore in Scorpius-Crux, the Tiamat chaos-monster of the Middle East, killed by Marduk, now identified as the planet Jupiter, at last makes sense if it is the Scorpius-Cygnus dust-cloud nebula. Such major traditional myths of observant and oral recording pre-settled aboriginal peoples bring into focus astronomical anomalies considered by the natural philosophers and scientists of Babylonia, Egypt and of course Greece, who looked at them anew with mathematical and other tools.

* * *

Introduction and Ethnography

Historic sources acknowledge recorded major star-lore, moon-story and planetary interest in many areas of Australia and in many Aboriginal tribes. This took many forms, and sometimes related to ceremony. The Milky Way, its movements and complexity, was envisioned by metaphor in stories, was used as time-keeper, season-marker, territory-divider, law-reminder and as a major sign-post relating to deep cultural structures.

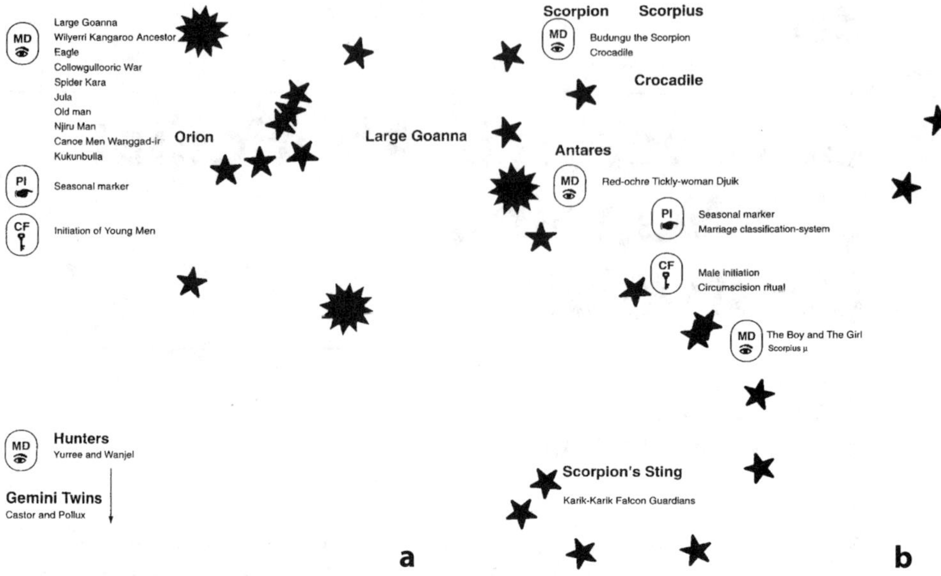

Figure 1. Star maps made by the writer from published (historic) ethnographic material. (a) Stars associated with Orion. (b) Stars associated with Scorpius.

This cognitive mapping was achieved, for instance, by storied relationships between particular stars. Put together, such data can produce star-maps for observed night-sky light-sources such as Scorpio and Orion (Figure 1). This simplification represents, first, with a selection of Metaphorical Descriptions (MD) from acknowledged sources, using material Australia-wide but mainly the southeast. Such a star-map also includes aspects of particular Ceremonial Focus (CF) noted from some tribal source, but not thereby to be securely generalized. Third, certain examples of Practical Information (PI) carried within the stories are noted, like seasons or ceremonial timing which may related to empirically observed night-sky realities.

Crucially for this present paper, Figure 2 shows the published depiction of the Scorpius-Crux 'Emu' (Basedow 1925). This ancestral figure was of ceremonial importance (Matthews 1904, Howitt 1904), and is presently (Cairns: Turin 1995) the basis of discoveries in archaeology near Sydney. These impact by analogy the arena of theories concerning major ancient religious myths and basic concepts. The astronomical models and their discoveries may even throw light onto precursors of science itself, observed within other art and cultures but not yet in written form.

Working Models Testing for Astronomical Content Used with Australian Aboriginal Prehistoric Material

1. Patterns of cups or dots sometimes represent star-groups; when they do, they are figurative art representing specific forms perceived by careful observation of the night sky.

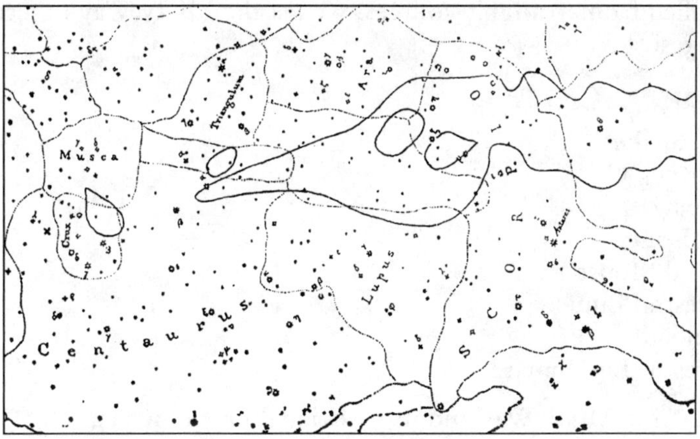

Figure 2. Basedow's 1925 depiction of the Aboriginal 'Emu' constellation.

2. Where star patterns are found in figures, they may be stars.
3. The presence within ancestral figures of star patterns can reasonably suggest astronomical possibilities of spiritual dimension, and possibly of ceremonial significance or relation.
4. Sometimes a single natural or cultural pit or cup or hole may signify an individual star or other astronomical object.
5. Details and whole figures in Australian Aboriginal engravings and paintings may be of celestial relation, and of immense cultural and spiritual significance and cultural importance.

Note that the relationships of stars and planets perceived through their genealogical relationships, and within Aboriginal stories meet within cultural perceptions of Whole Cosmos and Deep Structure. Aboriginal philosophical, experiential and structural wholenesses are always paramount, including the 24-hour human experience put into ancestral perspective.

"Everything in nature . . . (is) divided between the classes." Stewart's description here of indigenous Mount Gambier peoples' feeling/understanding of cosmic relationships corroborates Bridgman's researches (below). The relationship of everything to other levels of empirical reality makes for a personal universe which makes sense, and indeed 'saves the phenomena.' A selection of classified regularly experienced realities is set out in Table 1.

Table 1. Reality Sets in Nature

Black Cockatoo	Black Crestless Cockatoo	Crow	Tea-tree
Moon	Summer	Hail	Owls
Stars	She-Oaks	Thunder	Wallabies
—	Kangaroo	Winter	Ducks
—	Autumn	Rain	Crayfish
—	Wind	Clouds	—
—	Sun	Lightning	—

(Bridgeman in Fison & Howitt 1880)

Here, culled from available sources, we list the Milky Way in terms of **ancestral personages:**

1. Very Long Great Snake
2. Two Snakes
3. Big Rock Cod
4. A Being
5. The Being
6. Fabled Monster
7. Ancestral Emu
8. Ngalbad
9. Ancestor Linjilbmara

In more detail, the Milky Way and its **associated phenomena** are the:

1. River of fish and water-lilies for star people to gather
2. Creek bed
3. Large Native Plum Tree fruiting late December
4. Great Stream of lofty trees and never-failing water
5. Tree Platform relating to the hero Wananganda and the two Brothers, and to the 'Albatross' mark created by the Rainbow Serpent
6. Two Brothers
7. Two Snakes possibly in the Tree Platform of the hero Wananganda
8. Two Snakes in sky as part of Galaxy (Milky Way)
9. Track of the (harmless) Carpet Snake
10. Crocodile
11. Wandjina-type figure with Rainbow Serpent
12. Lights of many campfires
13. Nurrumbungguttia Ancestors and their fires: Warring and two enormous Mindii snakes who made the River Millee (Murray)
14. Orderly Stars responding to the Tribal Hero Priepriggie who needs songs and the stars dancing to him
15. Stars made by our part-human Father Aknittu, who threw the knife to earth, and has stars and sun parading past him
16. Ancestral Hero Unjilbmara, whose hair streams across the sky
17. Big Rock Cod (the fish-shape at base of the Southern Cross) with the two Brothers (the two thin strips of blackness lower down in the Milky Way)
18. Karia—a Great, very Long Snake, 'a dark wavy pattern' almost like a crocodile, and whose son is the Bunyip
19. Bunjip, a "fabulous monster" an animal 'something like a horse' 'in this deep water-hole' 'in the river.' When stars (the blackfellows) chase the game (the smaller ones) into it, spearing them, they find the deep hole in the river is Bunyip's Waterhole.

Sources for the above terms for the Milky Way are: Stanbridge 1857; Dawson 1881; Basedow 1925; Groote Eylandt Community 1940s; Massola 1968 Wells 1937; Strehlow 1947, 1971; NSW Museum paintings; Elkin 1977; Mountford 1976.

Stars related for instance as father, mother, son, daughter, husband, wife, mother-in-law, cousin; or as in Aboriginal stories. These relationships are identified in Table 2. Further research and new sources will doubtless add to these findings.

Table 2. Genealogical star relationships.

Aquila A,B	Cygnus B,D			
Aquila	Lyra	Capricornus		
Aquila	Lyra (Vega)	Hercules E,P		
Aquila	Lyra	Hercules	Ophiuchus	
Altair	Vega	Corona Borealis		
Altair	Delphinus	Vega		
Altair	Achernar	Vega	Arcturus	
Altair	Delphinus	Vega	Canopus	
Altair	Vega	Capricornus	Corona Australis	Formalhaut
Canopus	Achernar	Pictor		
Canopus	Vega	Delphinus	Altair	
Canopus	Magellanic Clouds	Sirius	Gemini	
Capella	Gemini	Mirage		
Gemini	Pleiades	Hyades	Orion	Eridanus

Sources: Stanbridge 1857, MacPherson 1881, Maegrith 1932, Mountford 1976, Tindale 1974, Allen 1975, Isaacs 1980, Haynes 1989. Sky relationships are found in continuing research by the writer, and by others to do with the Communities eg. Rai, Morieson & Pearce at the Museum of Victoria will no doubt enlarge this Table.

The Ethnographical Data

Both Orion and Scorpius may be the focus of the star-stone central to the lower part of Elvina Track site, since both line up with correct angles at the appropriate time of year (see below Figure 7). This would make the site a year-round ceremonial area.

The set of star groups in Figure 3 incorporates a wide range of selected the ethnographic data. The data are nevertheless sparse within the ethnography. However what remains after the ravages of time and external influence links empirical astronomical data, observations with spiritual dimensions, and much else. It reveals personal, ceremonial, seasonal, navigational, educational, and initiation usages Australia-wide, and generalizing within working models does not distort the picture within the ethnography. It continues useful at new sites and within new data. Conveyed into simplifying field models by analogy, and used to suggest possible meanings, lived-in and thought-of orders and personal models, and areas of motivation in art and artistic cultures, it has ongoing fruitfulness.

This seems now true outside of Australia. Applied to field data in Africa, Europe and America, both present and prehistoric, it is not without success at rock art sites (eg. Cairns 1991). Now (see final section) it is most suggestive in providing major interpretation of ancient Middle-Eastern texts and tablets. The predictive ability of the sky-mapping theory means that it is a reasonable tool for use when questions of possible meaning arise, and answers can influence selection of sites and data. Perceptions of cultural and financial value, and of community and scientific goals, may be enhanced if relations to early science and the history of astronomy can be established by research.

Engraved Patterns	Sydney Analogies	Australian Stories	Aboriginal Ceremony	Art	Stellar Interpretation
	Maroota ancestor Polygons	Yes	Yes	Boomerang A42240	Orion's Belt
	Mount Ku-ring-gai Polygons	Yes	Yes	Boomerang 1843 M-Stick X873	Scorpius (Antares)
	Maroota ancestor Elvina (site 2)	Yes		M-Stick X873	Australian Triangle
	Elvina polymorph Polygons	Yes	Yes	M-Stick X873	Crux Southern Cross
	Maroota ancestor Polygons	Yes	Yes	Boomerang A42240	Pleiades (Sagittarius)
	Polygons	Yes	Yes		Orion's Belt (Scorpius sting)
	Lizard Site Polygons	Yes	Yes		Canis Major Hercules
	Polygons		Yes	M-Stick X873	Lupus
	Polygons	Yes	Yes	M-Stick X873 Boomerang A42240	Southern Cross
	Patonga Rd West Polygons	Yes			Corona Borealis Australis
	Elvina ancestor Polygons	Yes	Yes		Southern Cross Pointers
	Elvina Site 1 Mount Ku-ring-gai Woman ancestor	Yes	Yes	Boomerang A42240	Hyades Taurus
	Polygons	Yes	Yes		Aquila
	(Elvina Site 2)	Yes			Milky Way

Figure 3. Interpretations of dot patterns as specific star groups.

Major Night Sky Phenomena within Aboriginal Cultural Material: Experimental Findings

Historic bark paintings of various aspects and stories of the moon, Pleiades, Hyades, Orion, Scorpius, the sun, the Milky Way, and so on, show these celestial entities expressed in Aboriginal art. Some ancient rock art sites, today described by informants as eg. Morning Star or Moon Dreaming areas, and as ancestral creation, show this is ancient tradition. With the night sky there in prehistoric and present rock art, the working models can be applied to both *shapes* in the art, and to groups, especially patterns of engraved cups (cupules) and painted dots. At least some enigmas within Australian rock art yield secrets to this reasonable theory of astronomical content. They also provide an alternative to current theories of entoptic, shamanic processes (Lewis-Williams, 1990) being responsible for most groupings, rows and patterns of cupules/dots.

A summary of certain ethnographic data according to sky mapping interpretation is found in Figure 3. The examples of message-stick X873 (Figure 4), the rock pavement at Elvina Track No. 1 (Figure 5), and the Bunjil painting near Stanford, Victoria (Figure 6) show three different types of material, from three different geographical areas, which emerge from experiment with patterns matching night-sky phenomena.

X873 turns out to be mainly to do with the timing of ceremonies. The Bunjil portrait shows his empirico-cosmic reality as an immediate presence within the thought-of order of belief systems, which include major ceremony revealed by the star-groups painted. At Elvina Track (Figure 7) major observation and recording of star-groups, moon cycle and the Black Space dust nebula Initiation-Monster (Basedow 1925: Figure 2), are recorded within the ceremonial framework.

When the experimental models are applied, some of the most enigmatic archaeological material makes sense. Further analogy from the ethnography may then reinforce the finding once the paradigm is perceived and new data become available. This has happened most recently in August 1996 with Tiamat (see below); but

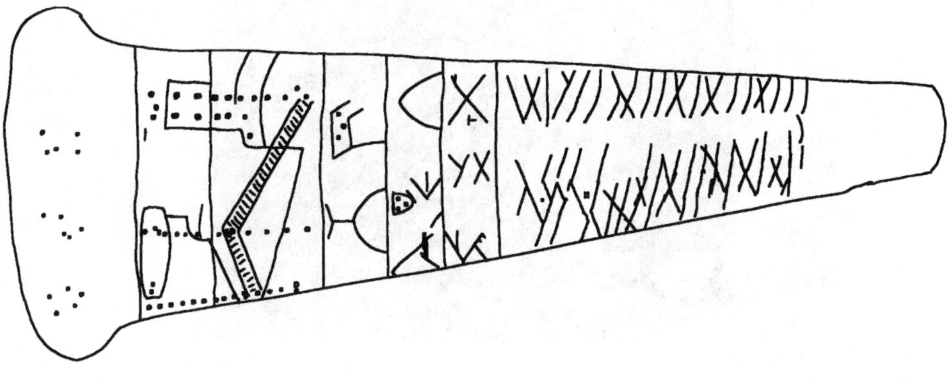

Figure 4. Author's drawing of message stick X873 from Western Australia. Marks gouged in wood. Length 81.8 cm.

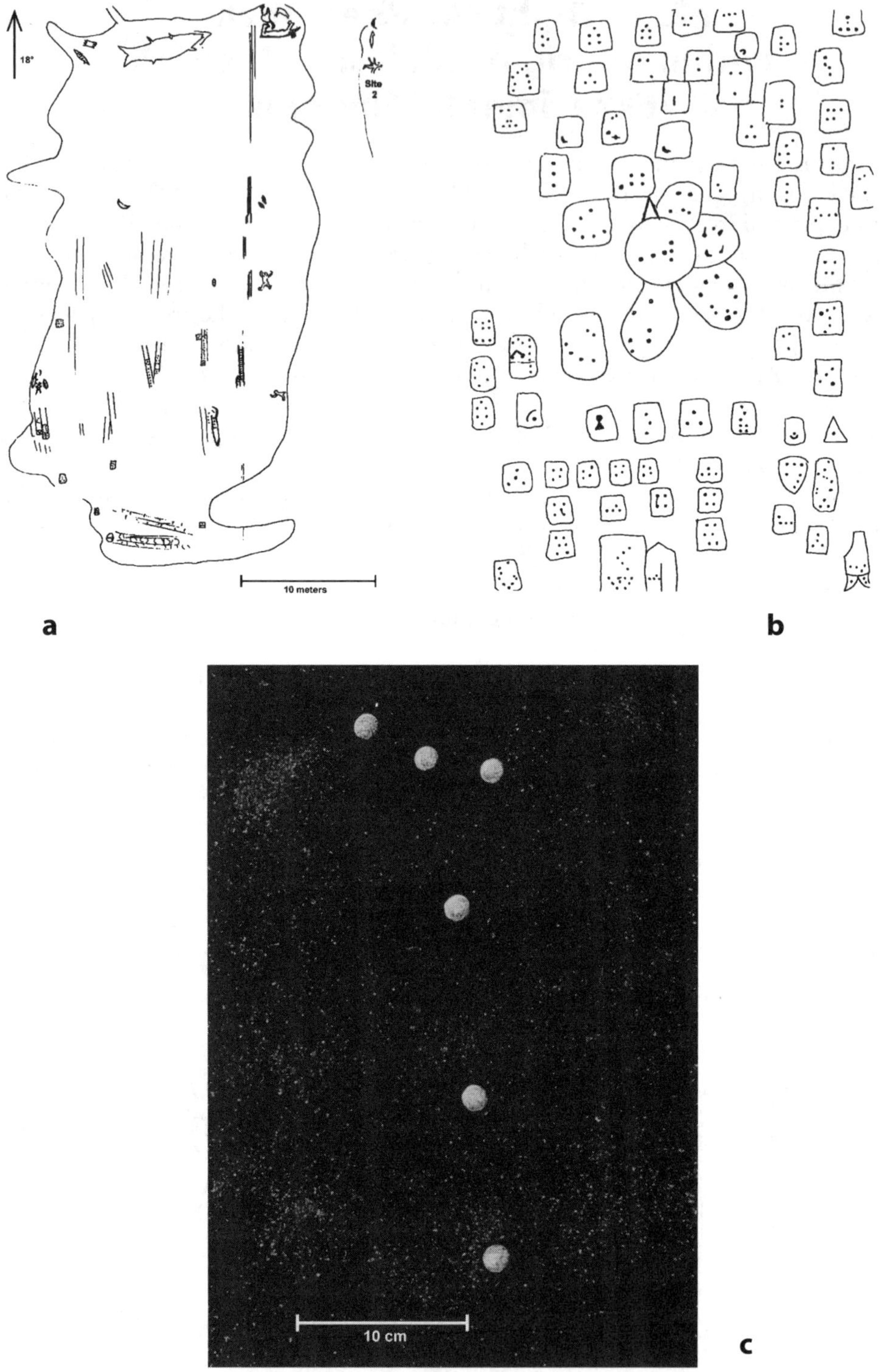

Figure 5. Drawings by author. (a) Park, Sydney. (b) Elvina Track No. 1 (c) An angled pattern fitting Orion's belt and sword and directed accurately at this constellation at its point of rising.

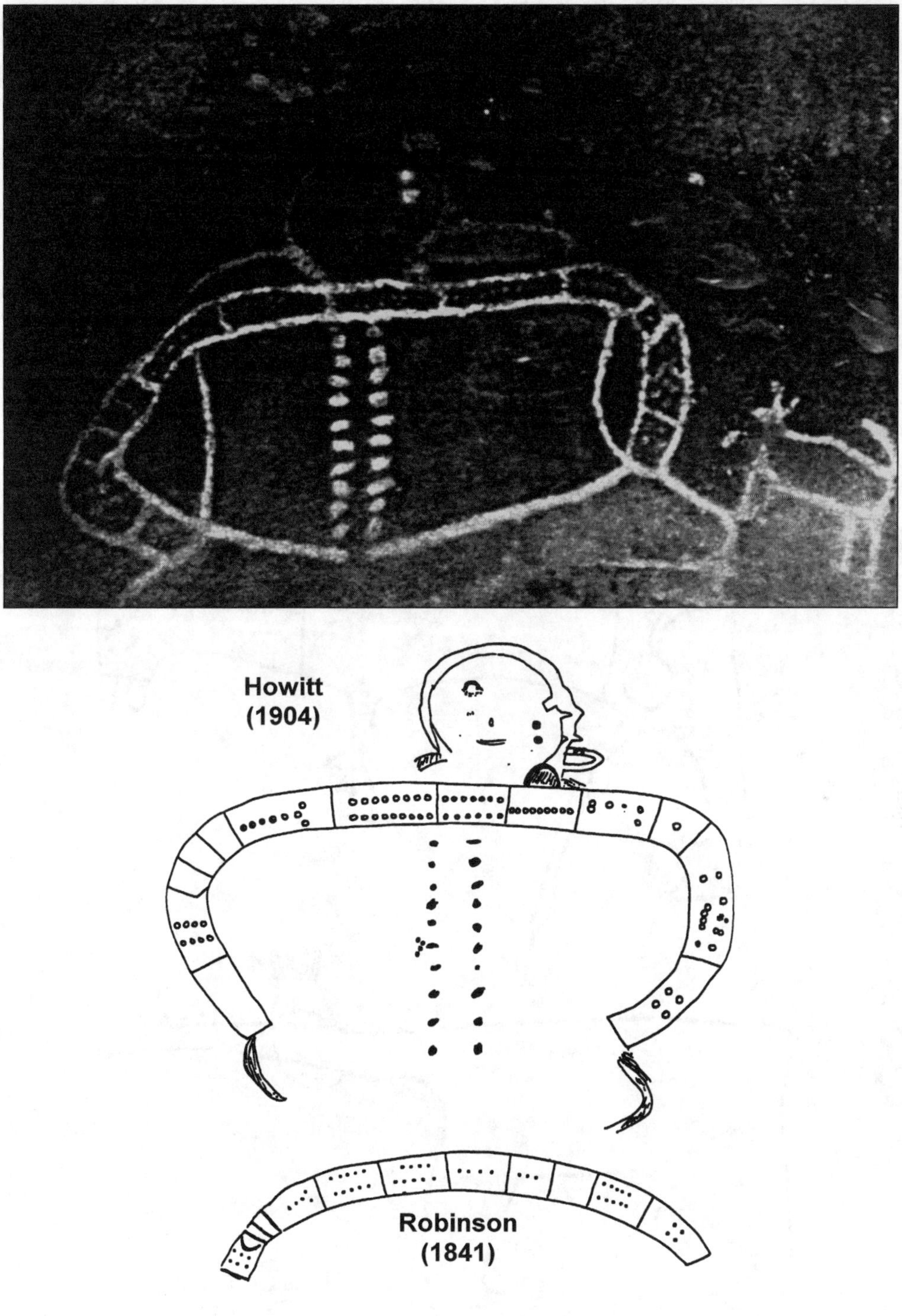

Figure 6. Bunjil rock painting (a) photograph from Grampian, Victoria ceremonial area. (b) Sketch diagrams. In spite of apparent reworking of the painting, dot patterns and crescents retain astronomical form.

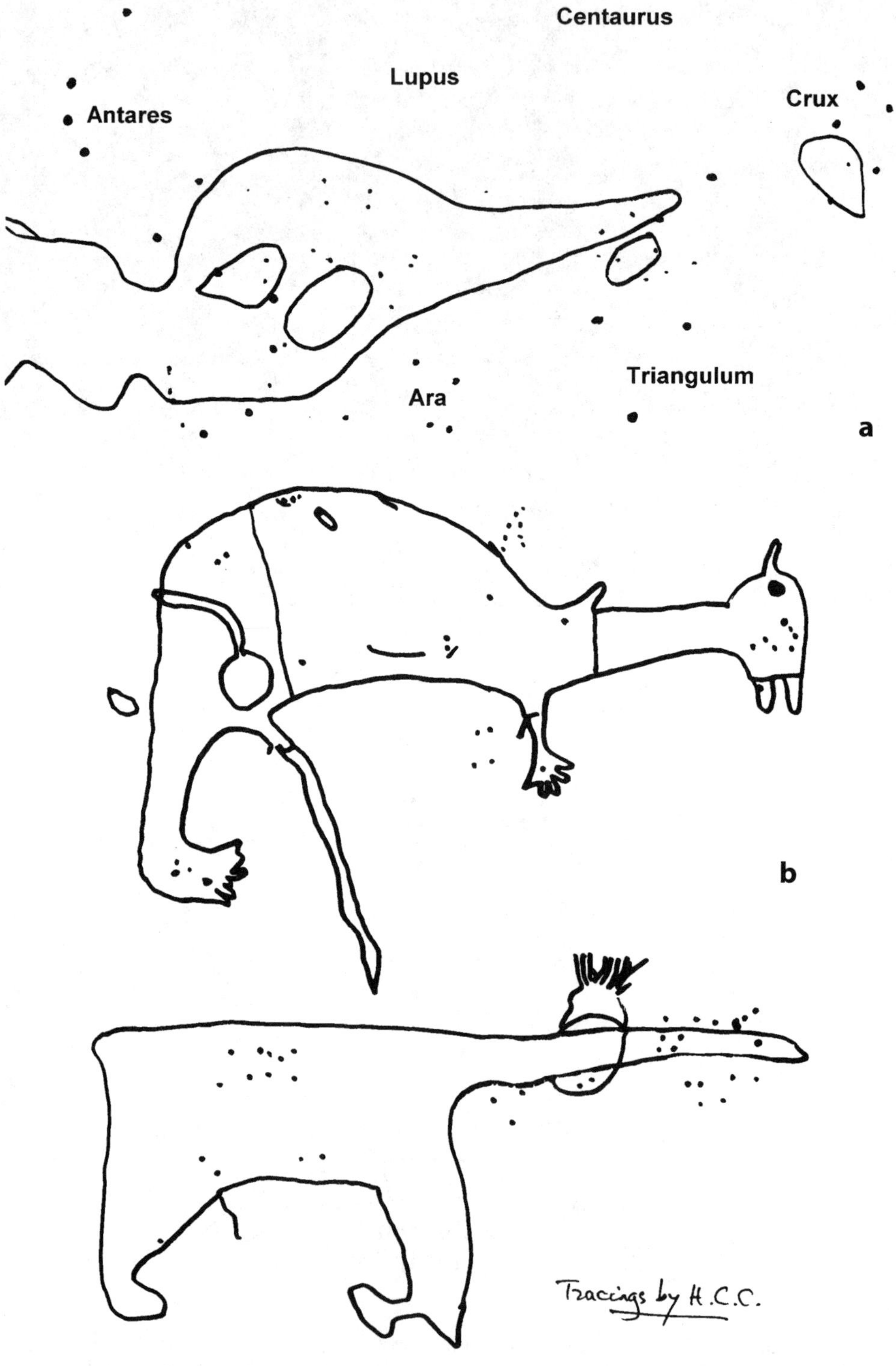

Figure 7. Polymorphs and their associated star fields. (a) Basedow's "Emu" (1925) (b) Elvina Track and America's Track polymorphs (1995). Tracings by the author.

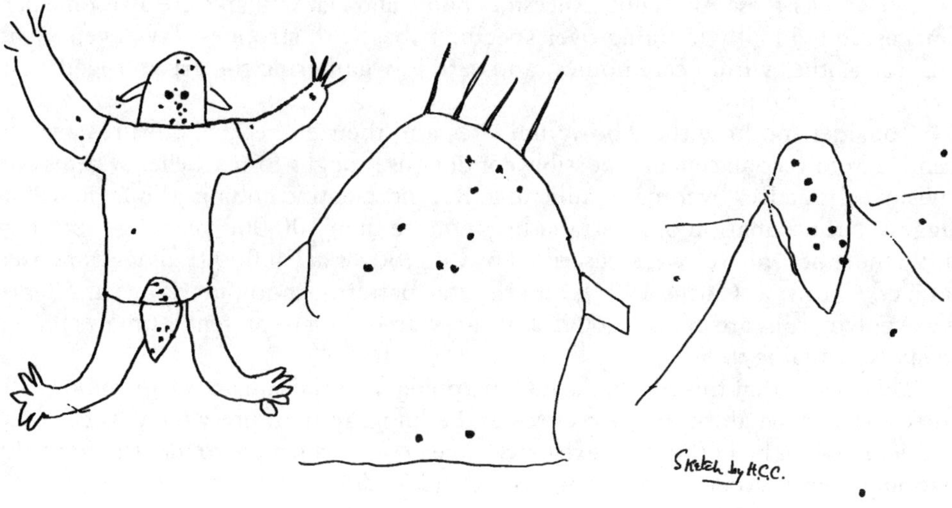

Figure 8. The Maroota ancestral figure: sketch by author of whole figure, head, and lingam-penis.

the process began in 1991, in the Sydney area, after the findings of Cairns (1993). Five new sites were found (Branagan & Cairns 1993 a,b) with identical and some repeating patterns which matched earlier finds and the astronomical phenomena. Regular spacing and type and pattern of cups attests to human cultural tradition in observing and marking night phenomena.

When the patterned figure of Bunjil—the Great Ancestor (Figure 6) who was identified in story as Altair and Fomalhaut with specific relations with Crux, Hercules, Corona Borealis and Orion was found to match in his ornamentation (the painted dots and marks) both the stories' relationships and the observed phenomena of the night sky in the Southern Hemisphere, the chase was on. The famous site at Maroota near Sydney suddenly had star-group patterns within and outside his torso (Figure 8). Then the Elvina Track polymorphs (Figure 7. cf. Cairns 1995) were found to have cups in specific patterns and places as well as in what seem random places which within a few months were found (Figure 8) to more-or-less match up with those in or surrounding the Scorpius-Crux dust-cloud dark spaces. One of these (the Coal Sack) was already known to be significant in Aboriginal lore and ritual. But the overall shapes of the Sydney polymorphs now were seen to match the vast black shape of Scorpius-Crux, and the shape of Daramulan the initiation Ancestor (Figure 2) was even found to be recorded ethnographically as doing what the dark spaces appear to do: change shape.

Clearly, the archaeological enigmas of Bunjil and the Daramulan polymorphs are elucidated by the night sky phenomena, as well as figures, cup-or-dot patterns, and many groups of lines and grooves. Both the ceremonial Ogre of Initiation and the cosmic Presence of (what missionaries inaccurately called the Aboriginal) 'All-Father', are shown in their nightly empirical power at their own sites, in the form of Presence-in-the-Heavens, and in their engraved or painted shapes out of which (or in relation to which) the cultic reality of the ceremonial actor could jump to thralling music, and loom over the initiates for real—mask, body-paint, ceremonial objects and all.

No wonder these Aboriginal Ancestors Bunjil and Daramulan were awesome personages, individually looming over specific tribes with strongest law, even vying with each other within ceremonies, and certainly gathering the group together in social ceremonial power.

Consider too how the Aboriginal sites and their embedded cultures are descended from very ancient and possibly continuous thought-forms, belief-systems and envisioned practices over many millennia. Recent scientific findings (Fullagar 1996) suggest that occupation of Australia began more than 100,000 years ago, but certainly the moon-phases were described by 28 grooves at 30,000 (Fullager's revised date: cf. Cairns: in Cairns 1993), and the star patterns, moon markings and Dark Space polymorphs are in the 4–6000 B.P. range according to present estimates (based on stylistic analogy).

This means that this Australian sky mapping material can act as an analogy for other areas of the globe where records and ethnography of pre-village, pre-settled peoples may not be in (known) existence. The area most simple to identify for a sky mapping theory experiment was the Near and Middle East.

Northern Hemisphere, Mid-East, Dark Space Form: Tiamat

The week after Oxford 5, under Namibian desert skies, the writer's hunch about the Dark Spaces, noted in the Oxford 5 poster exhibition and in conversation, but enlarged at the South African Rock Art Research Association (SARARA) conference at Windhoek received major confirmation.

The Milky Way was further West than when the Daramulan Form was first made out by the author (Jan.–Feb. 1995). I found myself staring up at the Scorpius-Cygnus range of Dark Spaces. The dust-clouds were the other way up (ie. Emu was upside down); and of course I suddenly realized that Crux is not seen from Mediterranean lands, so their Black Spaces would be different. Since Australia sees Crux to Cygnus very clearly, it is now not surprising that *Two Serpents* in the Milky Way (as noted above) are features in Aboriginal lore, and outstanding in the art.

Immediately I saw an almost exact match with the clay tablet depictions of Tehom/Tiamat (Figure 9) which I had used as a slide the previous week at SARARA. The black form was highlighted by the 'flares' (actually, of course, the brightnesses of the centre area of our galaxy) erupting from its back and belly, and its head (I realized) was the same dust-cloud nebula which formed the back leg of the Elvina Track and other polymorphs.

A moment later, the Tiamat Dragon-Leviathan monster of the Babylonian, Canaanite and Biblical myths, I recognized in the Black Spaces of the Milky Way, was even the Water-Chaos-Monster killed by Marduk, for Marduk was Jupiter, and there was the planet moving across the Tiamat dark-space dust-clouds, metaphorically speaking, while I looked.

The Scorpius-Cygnus area reveals the Tiamat figure as per the Ugaritic tablet, and the Crux-Scorpius dust-clouds match the Sydney Basin polymorphs and what little ethnography there is. The working models and the Australian material with sky-mapping theory had received solid confirmation by observation. Like all speculation

Figure 9. Clay tablet of the slaying of Tiamat.

about the past, even yesterday, certainty is impossible. But the observable night skies match this archaeological and classical material. Whether this understanding of Tiamat-Leviathan (Lotan/Crocodile/Serpent/Rahab in its many guises in different tribal traditions) will elucidate Genesis 1:2 where the Water-Chaos Tehom. appears, remains to be seen (cf. Jack Miles 1995, and the El, Ea, Shaddai, Bea, Bel, Baal, Yahu-Elohim story).

In the Aboriginal setting in Australia, a further aspect is that some new sites such as Pearl Beach have star patterns without mythological figures nearby. This might fit with possible observations of secular usage of stars for navigation (MacPherson 1881); or with possibilities of mathematical imagination as in Figure 10 where both regularities of number and of shape may be being observed in Sydney engravings. There might be further analogy to the Middle East: their pre-urban, pre-settled, tribal aboriginal astronomical interests may have resulted in number, shape and pattern on the Sydney Basin model, with artistic expression and story, the necessary fore-runners for perusal and critical thought in the eastern Mediterranean schools.

Conclusion Further Enigma-Mysteries Elucidated by the Sky-Mapping Theory

The work in Ku-ring-gai Chase and other sites in the Sydney Basin has shown major astronomical underpinnings to be present in art and ceremonial sites (as the massive star-lore of Aboriginal culture might have suggested along time ago: anal-

Figure 10. A recurring triangle oriented to Aquila falling in the west.

ogy from other indigenous cultures round the globe, likewise). Ancestors of mighty spiritual and ceremonial importance were precisely located in the night skies, looming over the people and present in the art, music, song and rituals, with the story also written in the. stone of Sydney sites.

By analogy—and seen specifically through displacement of the working models used on rock art sites for observation and detail these Presences in their global aboriginal forms are now seen to inform the Western myth-traditions of the Middle-East (as well as South American traditions: Urton 1982; Aveni 1981). The Tiamat Dragon is now seen as night-sky looming Presence, ever-present indeed and of threatening power, located in the Milky Way and only defeated by the planet observed to pass through her unscathed. Observational astronomy is a very ancient human activity, and intertwined with important religion, spirituality and theology. The Australian evidence is now important interpolation into this world arena.

In Australia, the astronomies embedded in the indigenous cultures will await further elucidation from the communities themselves. The working models need to continue to be applied to art, ancient and modern, to enigmatic markings not presently accepted as either art or carrying meaning at all, and to present-day ceremonial and storied relationships of the experienced and observed phenomena of Aboriginal individual and communal life.

Acknowledgments

Special acknowledgment for data-theory background is due to:

A. Aveni; R. Bednarik; D.F. Branagan; J. Clegg; J. Flood, Goobalathaidin; L. Haynes; E. Krupp; D.Lewis-Williams; E. Moore; A. Marshack; J. Morieson; W. B. Murray; C. Ruggles; P. Trezise; D. H. Turner; P. Vinnicombe; G. Urton. Major field work was and is due only to D. F. Branagan and J.Clegg.

References

Allen, L.A. 1975. Time before morning. Melbourne: T. Y. Crowell.

Angas, G.F. 1847. Savage Life and Scenes in Australia and New Zealand. London: Smith Elder.

Aveni, A. 1981. 'Tropical Archaeoastronomy' Science, Vol. 213, pp. 161–171.

Basedow, H. 1925. The Australian Aboriginal. Adelaide: Preece & Sons.

Branagan, D.F. and Cairns, H.C. 1993a. 'Tessellated pavements in the Sydney region, New South Wales'. Journal and Proceedings, Royal Society of New South Wales, Vol.126, pp. 63–72.

Branagan, D.F. and Cairns, H.C. 1993b. 'Marks on sandstone surfaces: Sydney Region, Australia: cultural origins and meanings?' Journal and Proceedings, Royal Society of New South Wales, Vol. 126, pp. 125–133.

Cairns, H.C. 1991. 'Is ancient sky-mapping expressed in prehistoric artistic cultural material?' In Rock Art—The Way Ahead. (ed) Pager, S-A. Parkhurst, South Africa: SA.R.A.RA.

Cairns, H.C. 1992. 'Information within Aboriginal Rock Art Material'. In Conservation, Preservation and Presentation of Rock Art. (ed). Mandt, G., Michelsen, K., Riisoen, K.H. pp. 131–139. Bergen: Arkeologiske Skrifter, University of Bergen.

Cairns, H.C. 1993. 'Aboriginal sky-mapping?' In Archaeoastronomy in the 1990s. (ed) Ruggles, C.I.N. Loughborough, UK: Group D Publications (University of Leicester).

Cairns, H.C. 1995. International Rock Art Congress, Turin, Italy.

Dawson, J. 1881. Australian Aborigines. Melbourne: Robertson.

Elkin, A.P. 1967. 'Religion and Philosophy of the Australian Aborigines'. In Essays in Honour of G.W.Thatcher. (ed.) E. C. B. MacLaurin. Sydney: University of Sydney Press.

Elkin, C.H. 1977. Aboriginal Men of High Degree. Brisbane: University of Queensland Press.

Fison, L. and Howitt, R.A. 1880. Kamilaroi and Kurnai. Group-marriage and relationship, and marriage by elopement. Melbourne: G. Robertson.

R.L.K. Fullagar, D.M. Price & L.M. 1996. On Luminescence Dating, early art and the early human occupation of northern Australia. Antiquity 70:751–73.

Howitt, R.A. 1904. Native Tribes of South-east Australia. 1895–1904. Sydney: Howitt Collection Vol. 5, No. 2, pp. 108–146.

Isaacs, J. 1980. Australia's Living Heritage. Sydney: Landsdowne.

MacPherson, P. 1881. 'Astronomy of the Australian Aborigines'. Journal and Proceedings of the Royal Society of New South Wales Vol. XV, pp. 71–80.

Maegraith, B.G. 1932. 'The astronomy of the Aranda and Luritja tribes.' Trans. Roy. Soc. South Australia.

Matthews, R.H. 1904. 'Ethnological Notes on the Aboriginal tribes of New South Wales and Victoria'. Journal and Proceedings of the Royal Society of New South Wales. Vol. 38, pp. 203–381.

Miles, J. 1995. God : a biography. New York: Alfred A. Knopf.

Mountford, C.P.1976. Nomads of the Australian Desert. Sydney: Rigby.

Stanbridge, W.E. 1857. 'On the Astronomy and Mythology of the Aborigines of Victoria'. Proceedings of the Philosophical Institute of Victoria, Vol. 2, pp. 137–40.

Strehlow, T.G.H. 1947. Aranda Traditions. Melbourne: University of Melbourne Press.

Urton, G. 1982. At the Cross-roads of Earth and Sky. Austin: University of Texas.

Wells, A.E. 1937. Skies of Arnhem Land. Sydney: Angus & Robertson.

Section 6

Education and Archaeoastronomy

39

Archaeoastronomy in the Core Curriculum

Sarah M. Nelson
Department of Anthropology
University of Denver
Denver, CO

Robert E. Stencel
Department of Physics and Astronomy
University of Denver
Denver, CO

Archaeoastronomy is an appropriate topic for an interdisciplinary course. When such courses are mandated, as in the University of Denver's Core Curriculum, various strategies are possible. Here we describe the course we created, and have since offered twice. It focuses on a single site and challenges the students to see whether they can discover evidence that the site is in any way a reflection of ancient astronomies. Most students rose to the challenge, and student responses were quite creative.

Introduction

The opportunity to create a new course in Archaeoastronomy that would fit into the concept of an upper division Core course was a challenge that neither of the authors of this paper could resist, especially in view of an interesting site in China to work with, which is, as yet, unexplored in terms of astronomical interpretations. We will begin discussing our plan for this course with a description of the site, Niuheliang in northeastern China (See Figure 1), a ceremonial center of the Hongshan culture, on the border between the grasslands of Inner Mongolia and the forests of Manchuria. This is followed by a discussion of the expectations and opportunities provided by the new concept of the Core Curriculum at the University of Denver, which provides us with advanced-standing students who have the skills and maturity to learn quickly. Finally, we consider the kinds of skills that would be needed if we were training the class to be research assistants in the field at Niuheliang, and explain how we hope to impart the required knowledge and skills to the students, as well as what we hope the students will be able to accomplish by the end of the class.

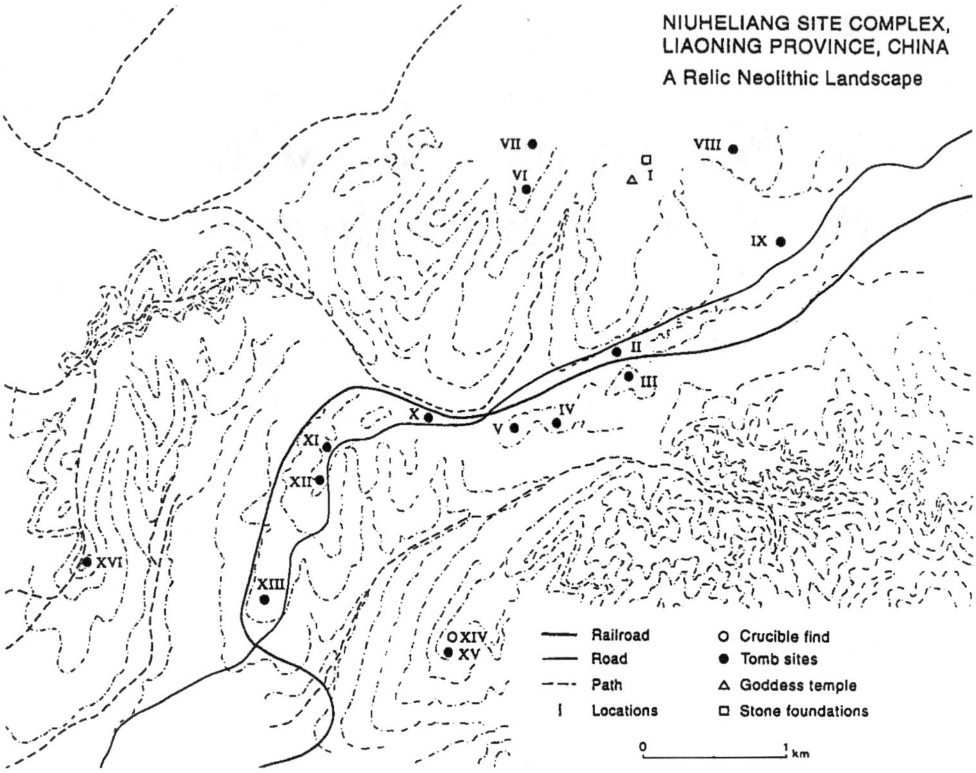

Figure 1. Map of the Niuheliang Complex.

The Hongshan Culture and Niuheliang

The Hongshan Culture was first named in the early part of this century, from the Hongshan ["red mountain"] site in the city of Chifeng, now in Inner Mongolia, which contains a rather ordinary Neolithic assemblage (Torii & Torii 1904, Nelson 1990). Both ground stone and chipped stone tools were found at the site, as well as bone tools, and both brown surface-roughened and painted black-on-red pottery. The painted pottery was unusual for a site so far north of the Yellow River, but otherwise the site was unremarkable. It was assumed that the notion of painting pottery had diffused from the south (Hamada & Mizuno 1938). Over the years many such sites were found (e.g. Andersson 1923, Fang & Liu 1984, Sun 1984, Li 1986)—we will refer to them as Hongshan villages, although in fact too little is known about the structure of these sites, and many are known only from surface collections. A few carbon-14 dates place the culture in the range of 4000–2500 BC [calibrated dates] (Guo 1995).

The discovery of ceremonial sites with "altars" and a "temple," large and small sculptures, finely crafted jades, enormous stone-mounded tombs, and other features brought new attention to the Hongshan Culture, especially the site of Niuheliang [NHL], referred to informally as the "Goddess Temple" site (Fang & Wei 1986a, b; Guo 1989; LPARI 1986; Nelson 1991, 1994; Guo 1995). At first it hardly seemed credible to the Chinese archaeologists that these large sites, each covering several square kilometers, could be related to the unremarkable Neolithic villages. The presence of the same kinds of painted pottery, appropriate dates from NHL, and a stone-mounded tomb in one of the village sites, convinced them that Dongshanzui (Guo & Zhang 1984) and NHL were indeed part of the Hongshan culture complex.

A number of features make the NHL site outstanding. The building called the Goddess Temple has a ground plan which ranges from 2 to 9 meters wide and 27 meters long. Three circular projections make up the widest part, while a long narrow section leads to an off-center rectangular room. Footings have the remnants of painted geometric designs in three colors.

Inside the building fragments of large statues made of unbaked clay were discovered. The most compelling of these is a face with inset green jade eyes, slightly larger than life-sized. This is believed to be female because other statue fragments include a breast and shoulder. It is identified as a representation of a deity because of the ceremonial nature of the building in which it was found. The unbaked clay was molded over armatures of wood or bone, with finer clay on the outside and paint applied to the surface. Other statues include the jaw and trotters of a pig, and a claw that may have belonged to a bird or a dragon. Black-on-red painted cylinder sherds and a nearly whole incense burner were also found.

Another oval building lies next to the Goddess Temple but not connected to it. Behind the Goddess Temple is a large stone-edged platform, of unknown use. A few refuse pits have been excavated nearby. From the Goddess Temple a mountain that resembles a pig's head is visible, with peaks representing ears.

Several areas of stone-mounded tombs have been identified. These are variously round or square in plan, and most of those excavated have both primary and several secondary burials. Black-on-red cylinders stand in rows around the edges of the mounds. These amount to an estimated 10,000 ceramic vessels in NHL as a whole (Guo 1995: 42). Grave goods placed with the bodies include well-crafted jades, in specific shapes, including turtles, birds, clouds, a comb-shaped object, and the *zhulong*, or "pig-dragon." A "hoof-shaped" jade is often found under or near the head, and is thought to be a holder for the hair. One clay figurine of a nude female, similar to those found at another ceremonial site [Dongshanzui], was found in a burial. These are clearly high-status burials of elite members of the population (Nelson 1996b). Non-elite burials have not been discovered, although a stone-mounded tomb was found on a hill above a village site.

An enigmatic artificial hill which was not a burial has also been investigated. It contained concentric rows of stone at three levels, and near the top a layer of crude, thick potsherds. The sherds have been interpreted as crucible fragments. Evidence of metal working, however, is rare. The only metal excavated at the site is a piece of copper wire which, with a chunk of jade, formed a decorative object, presumably an earring since it was found beside the skull (Han 1993). Another possible indication of metallurgy is a clay mold was reported from NHL, but it was out of context, so it is hard to know whether it is associated with the Hongshan culture or a later culture also present in the region (*Renmin Ribao* 1988).

The site of NHL is clearly a ceremonial precinct, with its large building and groups of tombs. The nature of the ceremonies is unknown, although the iconography seems to center on pigs and women (Nelson 1996a). The written record in China, although it is the longest continuous such record in the world, is of no direct help in this case. Writing first appears in China in the late Shang dynasty [around

1400 BC], with characters clearly related to those presently in use, although it takes a specialist to read them. The writing has lasted because it was carved into bones used for divination, in the form of requests to the spirits/ancestors for prognostication of the future. These oracle bones, as they are called, reveal much about the Shang culture (Keightley 1978), but there is little written on them that suggests any connection with the cultures to the north, especially the realistic statuary and porcine iconography.

Documents recorded at a much later time refer to a span of time approximately equal to that of the Hongshan culture as being that of a series of legendary emperors. One of these, of uncertain date, is Paoxi, who is memorialized in the I Ching: "When Paoxi ruled the world, he looked up, observing the phenomena of the heavens..." (Wu 1987: 53). This was *before* the creation of agriculture by Shennong, who is said to have taught people how to cultivate grains, implying that astronomy in China is older than plant and animal domestication. The dynasty of Shennong became known as the Flame Emperors, who were conquered by the Yellow Emperor. The final battle is said to have taken place "in the wild plains on the borders of modern China proper and Inner Mongolia" (Wu 1987:52). Perhaps the Yellow Emperor battled the group that succeeded the Hongshan, identified archaeologically as Lower Xiajiadian. Fortifications, weapons, highly stratified burials, and so forth indicate a war-like people in this region perhaps as early as 2000 BC.

Thus, the Hongshan culture seems to be just beyond the reach of both history and legend. But although no direct connection with the Hongshan culture can be established, there are ways in which Hongshan participates in the general characteristics of Chinese Neolithic. The most prominent of these is the use of jade, for Chinese culture has always valued jade more highly than precious metals. Square and round earth altars have also been claimed as a demonstration of continuity of "Chineseness" from Hongshan to later periods, and the *zhulong*, the "pig-dragons," are said to be China's earliest dragons (Sun & Guo 1984).

Folklore may also provide tenuous clues. The fact that the stars do not rise and set directly overhead was explained by the notion that the sky was once on four pillars, but one of them collapsed in a giant earthquake. The sun is featured in various forms of iconography, often related to a bird. By Han times (206 B.C. to A.D. 220), a three-legged crow represents the sun. The moon was as important as the sun for calendrical reckoning, and the lunar calendar is still in use for establishing the dates of traditional ceremonies, especially in Korea, where many Chinese traits have been preserved. Both Korean and Chinese groups have used the constellation Ursa Major for divination and for worship. A divination board with the *Chilsong*, the "Seven Stars" [of Ursa Major] was found in a tomb in North Korea as early as the first century B.C. Another avenue to explore is the relationship of pigs, used for divination as well as for sacrifice in woman-led shamanistic ceremonies in Korea, with the Chilsong, an important group of deities in the mudangs' pantheon. The cardinal directions are associated with colors and animals, painted on the walls of Han dynasty tombs.

Written records in China demonstrate a long continuity of interest in the night sky, and the behavior of celestial bodies. Chinese astronomers are known to have been able to predict eclipses at least as early as the Zhou dynasty, beginning about 1100 BC, and a "guest star" appears to be recorded on a Shang dynasty oracle bone (Breuer 1970:5,6), although Keightley (1978:65, n.38) points out a possible alternative reading. Both the writing system and the astronomical knowledge must have roots deep into Neolithic times. Thus it may be possible to learn more about the meaning of the site of NHL by trying to understand its relationship to astronomical phenomena.

The landscape around NHL is hilly. It would be possible to sight astronomical phenomena rising in the east through various notches from the constructed hill, as well as from the summit of any of the tombs, or from the hills themselves. To test this possibility, we have constructed the class as an extended exercise. The students explore the relationship of the site to the sky, and acquire both astronomical and archaeological skills in the process. To explain why we will structure the class this way requires a digression into the concept of the Core Curriculum at the University of Denver, especially the move to Upper Division courses that are not merely interdisciplinary, they are interdivisional. Our course cuts across the sciences and the social sciences.

The Core Curriculum

The University of Denver recently recast its undergraduate curricular distribution requirements under a new strategy called the "Core." As part of Core requirements, students take not only the usual arts and humanities, social science, and natural science distribution courses, but during Junior/Senior years, tackle a series of interdisciplinary courses that combine pairs of these usually separate disciplines. Our new Archaeoastronomy course is such an example, being taught jointly by an anthropologist and an astronomer. Goals of the Core include cultivating student abilities to write clearly and speak articulately, grasp historical achievements and comprehend contemporary challenges using tools of various disciplines and cultures, and to approach problems with awareness of moral dimensions and ethical consequences.

Astronomical Skills to Be Acquired

Practical astronomy needed for interpretation of any archaeological site can be summarized as including surveying fundamentals, the general appearance of the sky, and the tools for understanding the specific motions of stars, the sun, planets and the moon. Surveying fundamentals will include lecture and lab sessions covering altitude above horizon, azimuth or compass directions, basic surveying methods, navigation, latitude and longitude and GPS exercises. Recognizing features of the sky will involve learning common constellations, asterisms and bright stars (and their cross-cultural equivalents), stellar magnitude system, the daily, seasonal and annual motions of the stars and factors controlling the rising and setting of the sun and stars. Observing sessions and computer simulations will be employed for these studies. Further, the motions of planets and the moon will be studied with an eye toward their predictability and role in calendric systems. Transient phenomena such as comets will be examined as well. We were fortunate to have a bright comet (Hale-Bopp) apparent during our first course offering in spring of 1997, to serve as a

Table 1. Practical Astronomy for Archaeological Fieldwork

1. **General appearance of the sky**
   - altitude, azimuth, latitude and longitude, GPS
   - constellations by culture (star wheel)
   - daily, seasonal and annual variations, rising and setting of the sun
   - astronomical software and web pages
2. **Planetary motions**
   - apparent (synodic vs actual (sidereal) orbital periods
   - conjunctions, quadratures, elongations, oppositions
   - retrograde motion: apparent orbital motion toward the west
   - special case: Venus & the Mayans
3. **Lunar motions**
   - phases, tides, librations (59% visible) features visible
   - types of "Moonths"—moon advances 13.176 deg/day
   - orbital elements: axis, period
   - precession of the orbit & multiyear cycles
   - humans watch for repeat patterns: 52 year lunar cycle (Stonehenge)
4. **Comets**
   - physical nature, recorded in history and interpretation
5. **Other astrophysics**
   - solar-terrestrial interactions; stellar distances, motions
   - stellar masses, surface temperatures, evolution; nebulae

timely example. Modern astrophysics concerning gravitation theory, the nature of the planets, comets and nebulae, were interspersed in order to provide the students with a solid foundation for comparison of modern theory with more ancient cosmologies that become embodied in archaeological structures.

Archaeological Skills Needed to Understand the Site

To interpret an archaeological site, students need to understand how a site is excavated and the chain of inferences by which conclusions are reached about the nature of the site. How do we know this is Neolithic? What can be said about the subsistence base and the economy? How about the social and political structure? Can we say anything about gender? What do those female statues mean, anyway?

But for the purposes of archaeoastronomy, the primary skill that students need is map-reading. They will have to learn to interpret a topo-map of the site, and be able to imagine themselves into the landscape. If "Pig Mountain" is significant, might it have lent additional significance to whatever planets or constellations rose above it? Was the artificial hill a sighting platform, or a backsight? Do the groups of tombs have geographic orientations that point them to true north? Especially the square tomb—do their sides align with north, or with each other within a group, or between groups? Is there any formation that might reflect Ursa Major, or any other constellation?

Table 2. Archaeological Skills Needed

1. **Site excavation**
   - procedures: datum point, site grid, excavation plan, sampling
   - moving and processing soil and artifacts
   - interpreting strata
2. **Laboratory procedures**
   - artifacts: lithics, ceramics, bone tools, perishables
   - ecofacts: soil, flora, fauna
3. **Site interpretations**
   - age and stage
   - subsistence base and economy
   - political and social structure
4. **Survey skills and landscape interpretation**
   - map reading, orientation
   - purposes of survey
   - techniques of survey

The Structure of the Class

The class will be given the incomplete map of NHL on the first day, and told that its interpretation is their final exam. The students will be divided into small teams, to focus on cultural aspects, surveying, astronomical calculations, etc., with the option for rotation to other teams during the term. We hope that problem can be made so engrossing as to give them "nightmares", as they become increasingly engaged with its interpretation and consider how to apply the knowledge they are acquiring about astronomy and archaeology to the "map problem".

Our lectures will focus on providing the essential skills of astronomy and archaeology necessary to enable these students to be sent to do fieldwork in a relatively short time. We will make use of area sites for field practicum and the campus observatory for reinforcing geometric skills.

The last third of the course will draw on student skills to being a careful, guided discussion of results being derived in the cultural and astronomical aspects of the site analysis. Because we have no preconceived notion as to whether the site is inherently astronomically oriented, we hope to discover along with the students what answers may be possible given available information. This way, the students have the maximum chance to become engaged in the disciplines and excited by emerging results or setting of direction for follow-on studies. In the best case, the course will not only help develop conclusions about this site, but also inspire students to appreciate modern and historic cultures with new insight.

References

Andersson, J. G. 1923, "The cave deposit at Shaguotun in Jingxi, Fengtian" Palaeontologia Sinica, ser. D, 1, Vol. 1, p. 1.

Anonymous 1988, "Another Important Discovery of the Chinese Northern Cultural Relics—Neolithic Casting Mold Unearthed in Inner Mongolia" Renmin Ribao Jan. 10, 1988, p. 3.

Breuer, H. 1970, Columbus Was Chinese (New York: McGraw-Hill Book Company).

Fang D.C. & Liu B. H. 1984, "Discovery of the Hongshan Culture Jade Tombs at Hutougou of the Buxing County in Liaoning" Wenwu 6, p. 1.

Fang D. C. & Wei F. 1986, "Excavating a Lost Culture" China Reconstructs, December p. 33.

______. 1986, "Brief Report on the Excavation of Goddess Temple and Stone Graves of the Hongshan Culture at Niuheliang in Liaoning Province" Liaohai Wenwu Xuegan 8, p. 1.

Gao M. X. 1989, "On Hongshan Burials" Beifang Wenwu 1, p. 25.

Guo D. S. 1995, "Hongshan and Related Cultures," in The Archaeology of Northeast China, Ed. S. M. Nelson, London: Routledge, p. 21.

Guo D. S. & Zhang K. J. 1984, "Brief Report on the Excavation of Construction Sites of Hongshan Culture at Dongshanzui in Kezuo County, Liaoning Province" Wenwu 11, p. 1.

Hamada K. & Mizuno S. 1938, Prehistoric Sites at Hung-shan, Chihfeng, in the Province of Jehol, Manchukuo. Archaeologia Orientalis, Series A, Vol. VI. 1938.

Han R. B. 1993, "Recent Archeological Metallurgical Achievements at the University of Science and Technology Beijing." The International Conference: Chinese Archaeology Enters the 21st Century. May 29.

Keightley, D. 1978, Sources of Shang History, The Oracle-Bone Inscriptions of Bronze Age China (Los Angeles and Berkeley: University of California Press).

Li G. D. 1986, "Liaoning Province, Lingyuan, Sanguandianzi, Chengzishan Excavation Report." Kaogu 10, p. 497.

Liaoning Province Archaeology Research Institute (LPARI). 1986, "Brief Report on the Excavation of the "Goddess Temple" and the Stone Graves of the Hongshan Culture at Niuheliang in Liaoning Province" Wenwu 8 p. 1.

Nelson, S. M. 1990, "Neolithic Sites in Northeastern China and Korea" Antiquity 64, p. 234.

______ 1991, "The Goddess Temple and the Status of Women at Niuheliang, China" The Archaeology of Gender. Eds. D. Walde and N. Willows. Proceedings of the 22nd Annual Chacmool Conference. Calgary, Alberta, Canada, p. 302.

______ 1994, "The Development of Complexity in Prehistoric North China" Sino-Platonic Papers 63, p. 1.

______ 1995, The Archaeology of Northeast China: Beyond the Great Wall (London: Routledge).

______ 1996, "Ideology and the Formation of the State in Northeast China," in Ideology and the Early State, Eds. H.J.M. Claessen and J.G. Oosten. Leiden: University of Leiden, p. 115.

______ 1996, "Ritualized Pigs and the Origins of Complex Society: Hypotheses Regarding the Hongshan Culture" Early China 20, p. 1.

Sun S. D. 1984, "On the Hongshan Culture Jade Dragon at Sanxingtala." Wenwu 6, p. 7.

Sun S. D. & Guo D. S. 1984, "On the Primitive Civilization of the Liao River Basin and the Origin of Dragons." Wenwu 6 p. 11.

______ 1986a, "Discovery and Study of the ' Goddess Head Sculpture' of the Hongshan Culture at Niuheliang" Wenwu 8, p. 18.

______ 1986b, "Hongshan: A Lost Culture" China Pictorial 8, p. 2.

Torii, R. & K. Torii. 1904, "Etudes Archeologique et Ethnologique—Populations Primitive de la Mongolie Orientale" Journal of the College of Science, Imperial University of Tokyo, Vol. 36, article 4.

Wu, K. C. 1982, The Chinese Heritage (New York: Crown Publishers, Inc.).

40

Ancient Cosmology Resource Center

Bryan Penprase
Department of Physics and Astronomy
Pomona College
Claremont, CA

We describe our internet based resource center in ancient astronomy (http://arcturus.pomona.edu) which has been created to serve as a tool for both research and education. Our ancient cosmology resource center includes a variety of student research papers, original essays and images related to ancient astronomy, and a compilation of links to other ancient astronomy sites around the world. Our resource center includes an interactive world atlas of ancient astronomy, which serves to organize materials on the astronomy practiced by various ancient cultures around the world. We also offer on-line digitized images for the Dresden and Borgian codices, and are expanding to include higher resolution images of a variety of ancient texts. We are able to serve as a clearinghouse for materials related to student work, research and teaching of ancient astronomy, and welcome submissions of curricular or research materials for posting on the world-wide web. During the Oxford V conference we also developed a strategy for creating a working group of educators in this field, and details of this project are briefly described.

* * *

Introduction

The study of the heavens is one which people from around the world used to define themselves, and to unify their cultures. We have developed a set of resources with which we hope to extend research and teaching in ancient astronomy, by serving as a clearinghouse of materials resulting from student and faculty research in ancient astronomy, curricular materials, and original essays and images related to the subject. This project has been developed with the help of students from Pomona College and other Claremont Colleges, who have participated in our seminar in "Multicultural Cosmology", "Ancient Astronomical Cosmology", and our new course entitled "Archeoastronomy and World Cosmology." Our resource center includes a research library of over 130 volumes on ancient astronomy, and has made extensive use of the holdings of the Special Collections department of the Claremont College library, in which are found rare facsimile editions of the complete set of Edward King's color reproductions of Mesoamerican codices and field notes from the early 19th century.

Structure of the Ancient Cosmology Resource Center

We have created a web site which is one part of a resource created for the use of both students and educators in ancient astronomy. The web page is a clearinghouse of information to help initiate student research in the subject, and to offer an overview of the great ancient cultures of the world specifically addressing the astronomical sites and practices of these people. This user interface allows visitors to instantly access any part of the world with sky lore, background materials and other information on line.

The current available resources can be divided into four parts, which are outlined below:

1. An interactive Atlas of the Ancient World, which shows the locations of major astronomical sites on all the continents of the earth, and related links and original essays on relevant aspects of Ancient astronomy.
2. A glossary and timeline of Ancient Astronomy terms and events.
3. A set of course materials for educators and students which includes useful classroom activities and exercises for the study of ancient astronomy
4. On-line versions of ancient Maya and Aztec codices, currently including the Dresden and Borgian codices.

The web page has the current URL of http://arcturus.pomona.edu, and can be linked to from the Pomona College astronomy web page http://shanti.pomona.edu. In addition to these resources, we also are developing a lending library for use by area educators and are expanding into multimedia materials in teaching ancient astronomy, including videos, CDROMs and audio tapes.

Contents of the Resource Center Web Page

In the following sections we detail the contents of our on-line resource, and describe the user interface and contents of the resource currently available.

User Interface: Interactive Atlas of World Astronomy

The Ancient Astronomy Resource Center provides an interactive world atlas of astronomy to organize the collection of materials. The opening page, presented in Figure 1, offers an interactive map of the world in which to explore the world of ancient cosmology and astronomy. Developed by Pomona College students Leigh

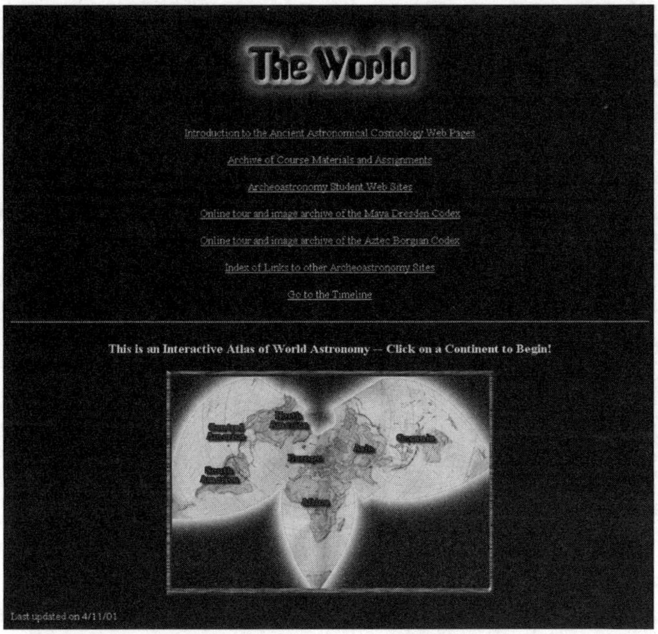

Figure 1. Sample of ancient cosmology resource center, with an on-line interactive atlas of world astronomy. The internet is used to allow collections of essays and images to be available to educators and researchers across the U.S. and the world.

Ann Sabicer, David Jensen, and recently updated by students in our Pomona College archeoastronomy class, the web site currently contains extensive original essays on Stonehenge, the Anasazi, ancient Norse cosmology, Inuit civilization, the Pythagoreans, and a class expedition to a petroglyph site near Havasu City, Arizona, and other diverse topics related to ancient astronomy. The contents are continuously updated and expanded, as each new year we offer students class projects which provide new research and images which are useful both for teaching ancient astronomy and for the research in ancient astronomy.

On-line Gallery of Images from Mayan and Aztec Codices

In addition to the world atlas of astronomy, we offer a gallery of images from both the Dresden and Borgian codices, compiled from the Claremont College special collections version of the 1830 facsimile copy which was the life's work of Edward King, sometimes referred to as the Viscount Kingsborough. The Claremont College special collections holdings include full color facsimiles of the Dresden, Nutall, Paris, Madrid, Grolier, Nutall, Borgian, Vindobenesis, and Mendoza Codices, which are in the process of being captured into digital images for use in our resource center.

The images currently on the web were captured from a video camera suspended above the rare book, and we are experimenting with more advanced techniques of capturing pages from rare books, primarily using high resolution digital cameras with accessory lenses for acquiring higher quality digital images. These higher res-

Figure 2. Another sample of our web page, which shows how we archive images for display. Miniature frames of the images allow for rapid access to larger images. In this case, we show the web page for our on-line collection of images scanned from the Aztec Borgian Codex.

olution images will be placed in the resource center as they are acquired. Each page of the codices currently available contain thumbnail images in a separate frame of the web page, which when clicked allows the user to view a full sized image of the codex page. The type of user interface used is shown in Figure 2, which displays the result of an on-line viewing of the Borgian Codex.

In addition to the gallery of images from the Borgian and Dresden codex, we have an interactive version of the Venus tables in the Dresden Codex, in which the user may click on any of the subsections of the codex to view an enlarged image with a translation and identification of the depicted figures. This type of user interface is a valuable learning tool, and could be used for other pages of the codices in which translations are being worked on by other investigators.

Curricular Materials in Ancient Astronomy

In addition to a diverse array of holdings related to research in archeoastronomy, some curricular materials are being gathered on the web site for use in teaching archeoastronomy. We welcome submission from other educators to help develop a database of lab exercises, class projects, and writing assignments in ancient astronomy education. Several writing assignments are provided on the web site, which include traditional research projects in ancient cultures and astronomy, and less traditional projects which involve developing an imaginary culture (often non-human) with a complete calendric and astronomical tradition.

Several of the student research projects on the web site resulting from these writing assignments have been very original and demonstrate the potential contributions education in ancient astronomy can provide to further research in the field. One of our students, Matt Green, has written the article entitled "The Sacred Sky of the Navajo and Pueblo Cultures", which was recently published in the Griffith

Figure 3. Mural painted onto our campus wall, prepared by students in the Pomona College Ancient Astronomical Cosmology seminar from Fall semester 1995, led by Sue Lin Hilbert. The mural depicts an Aboriginal Creation myth, in which the Ancestors walking across the newly made earth, singing and leaving a trail across Australia linking the Dreamings together. Above is the blue river in the sky, which the Aborigines identified with the Milky Way, and fish in the river indicate bright stars important in the early Australian sky lore. (Myth from Chatwin, p. 58).

Observer and was ranked in the top 10 in their Hughes Science Writing competition. In addition to writing, some students have chosen to develop a non-verbal method of communicating astronomical knowledge. This included several groups who painted large murals on our campus to communicate astronomical themes from Inuit, Aboriginal Australian, and Navajo cultures to their fellow students. One example of such a mural is shown below in Figure 3, which was painted on the Pomona College campus during Fall Semester 1995, and represents the efforts of modern college students to artistically convey the cosmology of the ancient Aboriginal people. One of our student research projects documents field work at a petroglyph site in the Mopah Mountain range, home of the Chemihueve Indians, and investigates the possible astronomical use of a desert oasis skywatching site in the desert of Eastern California. This work is being prepared for publication, and also illustrates the useful interplay between teaching and research in this field.

Working Group in Ancient Astronomical Education

We would like very much to expand this web site to be a more general clearinghouse of materials for teaching and research in archeoastronomy and world cosmology. Any contributed URL's, images, laboratory or classroom activities, or essays are being collected and will be posted for other educators to use, with full attribution to the author(s). In this way it should be possible for teachers and professors of world astronomy to avoid duplicating efforts in creating a comprehensive resource enhance all of our courses. Using the web it should be possible to create a 'virtual department' of world astronomy. This is especially important for this field, since only by the combined efforts of individuals in a variety of disciplines and locations is it possible to present this fascinating subject to our students.

During the Oxford V conference we discussed creating a "Working Group for Educational Aspects of Cultural astronomy". The coordination of the group was discussed, and the author offered to help develop such a group using the web based resource center to share course materials among the group. We intend to cross-link all the web pages currently offered by the Oxford V participants, and to add libraries of course materials, and archeoastronomy images to aid professional educators in bringing this fascinating subject to their students. We are also interested in testing the materials here with K-12 teachers who would like to add an ancient astronomy component to their science classes. A number of additional resources including software and videotapes are available for examination by local educators interested in teaching ancient astronomy. For more information, please contact Dr. Bryan Penprase at Pomona College, Claremont CA, or by email to bpenprase@pomona.edu.

Acknowledgments

The efforts of dozens of Pomona College students has made this project possible. In particular the efforts of Leigh Ann Sabicer, David Jensen, Julie Rivera and Eirik Harris have enabled the creation of the web based resource. The imagination and creativity of these and other Pomona College students have shown how enjoyable and intellectually valuable teaching ancient astronomy can be. This work is sponsored by a NASA grant ED-90019.01-94A from the IDEA program administered at the Space Telescope Science Institute. Some of the computer and video equipment used in the course was purchased with funds from the Mellon Foundation grant to the Claremont Colleges, and with support from the Pomona College Physics Department.

References

Chatwin, B. 1987, "The Songlines", (London:Pan Books, Inc).

41

Archaeoastronomical Research and Public Education in a Digistar Computer-Generated Planetarium

Carolyn Sumners and Alan Knox
Houston Museum of Natural Science
Houston, TX

The Burke Baker Planetarium of the Houston Museum of Natural Science is engaged in a program designed to create awareness of archaeoastronomical topics in the general public and in targeted school populations. This paper addresses the collaborations needed to create accurate and meaningful archaeoastronomical experiences in a planetarium environment and to produce curricula that can be readily integrated into a standard school curriculum. Also noted are the benefits of this public outreach in terms of increased public and student awareness of the knowledge and skills possessed by the world's first astronomers. This paper also describes the capabilities of the Digistar computerized star field projection technology to display both the precessed sky and digitized wire-frame images of an archaeological site. This technology allows juxtaposition of a particular architectural perspective with an aligned and precessed star field. The simulation of diurnal motion allows the audience (whether public or professional) to experience observation of celestial phenomena within the context of the archaeological site. This paper discusses the process of creating an archaeological simulation in the planetarium and then producing the educational support program for school and public audiences. Monte Alban near Oaxaca in Central Mexico is the archaeological site selected to demonstrate the technology and research potential of this effort.

Keywords: archaeoastronomy, planetarium, Monte Alban, K-12, astronomy, curriculum

Figure 1. Monte Alban main plaza looking north from the South Platform. (The arrow-shaped Building J is at the south end of the plaza's central spine.)

* * *

Project Overview

The Museum's interest in the ethnography of native American cultures and the ethnic diversity of its 2.3 million annual visitors has lead to the development of this project. A total solar eclipse crossing central Mexico in 1991 provided the setting for collection of the image base that has yielded research and public programming in the Museum's Burke Baker Planetarium.

The Monte Alban site, near Oaxaca in Central Mexico, was chosen because of its antiquity and its compact form which would translate well into a three-dimensional wire-frame model. Existing survey work also provided a baseline for collection of new survey information.

Procedure

Five steps are required to render a site in the Digistar computerized planetarium projection system and to create the sky environment at the time of the site's construction. The resulting display allows an archaeologist without astronomy training to "visit" an archaeological site at any relevant time in history and look for significant architectural alignments with constellations or solar system objects. Celestial objects appear over ruins as they would have at the time specified. By allowing the archaeologist direct access to the star field, this procedure minimizes the pattern bias toward familiar constellations which an observational astronomer can bring to the analysis. Similarly, this process also gives teachers, students, and the general public the opportunity to share in constructing archaeoastronomical hypotheses and in conducting research in the simulated planetarium night sky environment. Each step combines action items and decision points.

Step 1: Creation of a CAD file of the site.

Three dimensional wire-frames must be created for each significant structure at a site. These must be rendered at the same scale with a common reference point so they can be integrated into a uniform site plan. Topographic information and horizon markers can also be included in this transfer.

Once data points are integrated into a wire frame of the ruin, the archaeologist can add roofs to the virtual structures based upon practices at the time of construction. Reconstructed walls and platforms from the ruin can also be adjusted in height, form, or location. The opportunity to modify the virtual ruin allows testing of many different reconstruction options. In this test project no structural modifications were made, but this decision could be revisited.

In like manner, topographical markers can be adjusted based on changes that occurred since construction. In this project, only distant natural horizon markers (such as mountains or valleys) were included. The possibility of change in 2,000 years is minimal.

Step 2: Projection of the site in front of the star field.

The 3-D wire frame rendering of the site is imported and centered on the Digistar coordinate grid. Files representing individual structures are combined to create the complete site plan.

Step 3: Synchronization of star field and site

The star field is oriented for site latitude and star positions are precessed backward to the time of construction. Once rendering and projection are completed, the star field can be positioned correctly around the site. It should be noted that the precession algorithms of Digistar address the first order axis precession only. More precise algorithms exist and can be considered independently. These second order changed can be entered manually into the Digistar's definition of the orientation of the rotation axis. In this test project, these higher-order adjustments were not made. A discussion of this decision is provided below in the "Project Evaluation".

Step 4: Selection of observing location in the ruin

Alignment with site lines from one location to another can be done during a real time fly-through of the site. The projection system can support architectural and astronomical image bases simultaneously.

Step 5: Programming diurnal and annual motions

The Digistar star field can advance through a night, a year or a millennium with the architectural site projected as a constant along the horizon. Alignments with star patterns or the ecliptic can be observed. The researcher can also use a joystick or dials to travel through the ruin, change elevation, peek out from behind structures, or observe along a line created by two or more architectural features.

Summary of Research

Proposed astronomical alignments at Monte Alban date back to the work of Alfonso Caso who suggested that Building J might have been an observatory (Caso 1938). Caso was of the opinion that a doorway of a tunnel that cut through the point of Building J opened southwest to allow for the passage of certain stars. This interpretation has been shown to be incorrect (Krupp 1983).

In 1972, Aveni and Linsley re-examined Building J's astronomical orientation and determined a loose orientation toward the setting positions of Alpha and Beta

Figure 2. Looking to the southwest over the stairs of Building J.

Centauri and Crux in 275 B.C. (their proposed time of construction). (Aveni and Linsley 1972). Aveni and Linsley also measured the direction of a given line perpendicular to the base of Building J's stairway. This line passes through what was once an opening in a building on top of Structure P across the main plaza. Aveni postulated that this line would point to the rising point of Capella around 275 B.C. (Aveni and Linsley 1972). The heliacal rising of Capella could have been associated with the sun's northward zenith passage. In Aveni's work, this alignment formed the basis for the geometric tie-in with Building P and the zenith tube it contains.

Data Collection

The Monte Alban project was designed after consultation with Marcus Winter of the Oaxaca Regional Center of the Instituto Nacional de Antropologia e Historia (INAH) and his assistant, Damon Peeler. Building J was identified as the primary target with the ancillary goal of a complete site survey for the purpose of creating a planetarium display.

A Houston-based data collection team of four local science teachers, two geophysicists with survey training and an architect from Rice University accompanied the Museum staff to Monte Alban. Before data collection began, two existing surveys of Monte Alban were translated into AutoCAD by the Advanced Visualization Lab (RAVL) at Rice University. (This lab was also responsible for the creation of AutoCAD files for all survey data collected by the project.) The primary survey plot by A. Caso (as provided in Marquina 1964) was complemented by the addition of detailed building drawings also provided by Caso (Marquina 1964). This survey was then overlaid with building survey data by H. Hartung (Aveni 1980). The resulting composite survey showed inconsistent orientations for Building J as well as data fit problems with surrounding buildings. The survey team decided to maximize the data points collected to be able to remove as many ambiguities as possible in the existing data base. Laser transit survey points were prioritized based on the buildings whose orientations were most in question.

For the laser transit, a TBM (transit benchmark) was established at the northwest corner of the South Platform. Most plaza structures can be seen from this location. A backsight was made from the TBM to a geodetic benchmark on the North Platform. The latitude and longitude of that benchmark were determined from a geodetic benchmark at the train station in Oaxaca. These data were refined to an accuracy of 15 meters based on the Magellan NavPro 1000 satellite positioning system.

At night three transit bearings and time measurements were recorded for the sighting of Polaris with the average used to provide an absolute azimuth reference for the relational data base of laser survey points. Additional latitude and longitude measurements were taken at seven points on the site. An average was determined from 20 separate readings provided by four satellites for each field point. The mean scatter of the satellite data taken by the Magellan system produced an averaged accuracy of 15 meters for these positions as well.

Three survey rods carried by different team members allowed the laser transit to collect data quickly and efficiently. The transit surveyor shot each site and yelled an identification number to the distant teammember holding the rod. The team member recorded his exact location on the composite survey map while the transit assistant recorded (x,y,z) values for the location as well as its distance. The survey produced 175 raw data points of which 168 were ultimately useful. Errors came when team members did not record exactly where they were standing during a sighting. It was decided to remove any points that were obviously inconsistent with transit data.

Data Reduction

Rendering of Site

Data were entered into an AutoCAD program and a modified plan view was generated. Significant differences in building orientations were documented. AutoCAD data were then used to create three-dimensional structures to be translated into the Digistar projection system in the planetarium. During the field survey, all structures were photographed with a reference height from different directions. These photographs were used to fill in details that the laser transit could not capture. DXF files of wire-frame buildings were transferred to the Digistar and recombined to give a three-dimensional environment through which the planetarium audience can travel in real time.

Rendering of the Star Field

The Digistar software allows the user to specify the longitude, latitude, and epoch for a star field. Once the date and location are specified, Digistar adjusts and precesses the entire star field to any determined time. The visual effect was so dra-

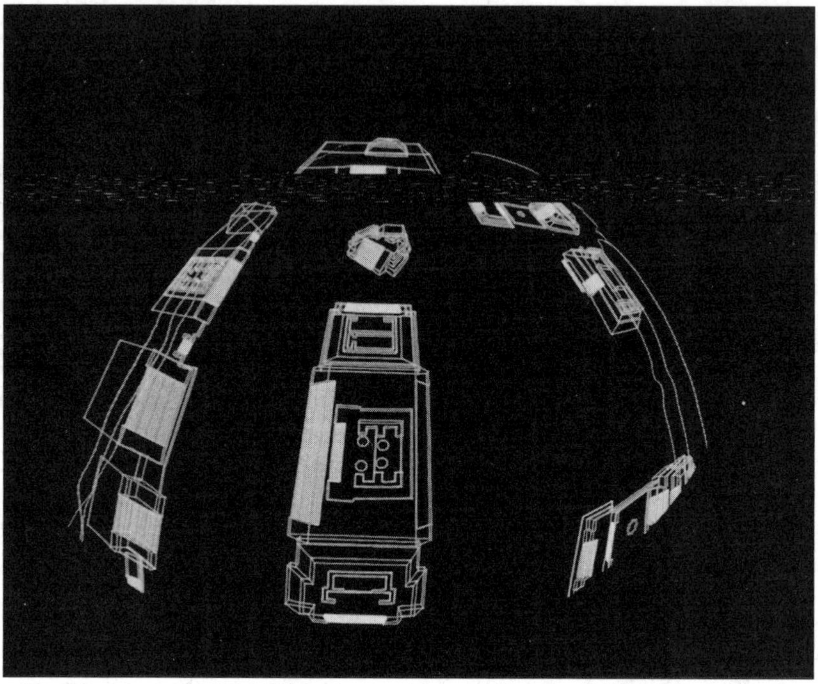

Figure 3. Monte Alban main plaza projected on planetarium dome (facing south).

matic over the wire frame of the North Platform that the planetarium staff decided to use it in the public planetarium program. This precession demonstration includes simultaneous movement of the pole and the month markers along the ecliptic. It is

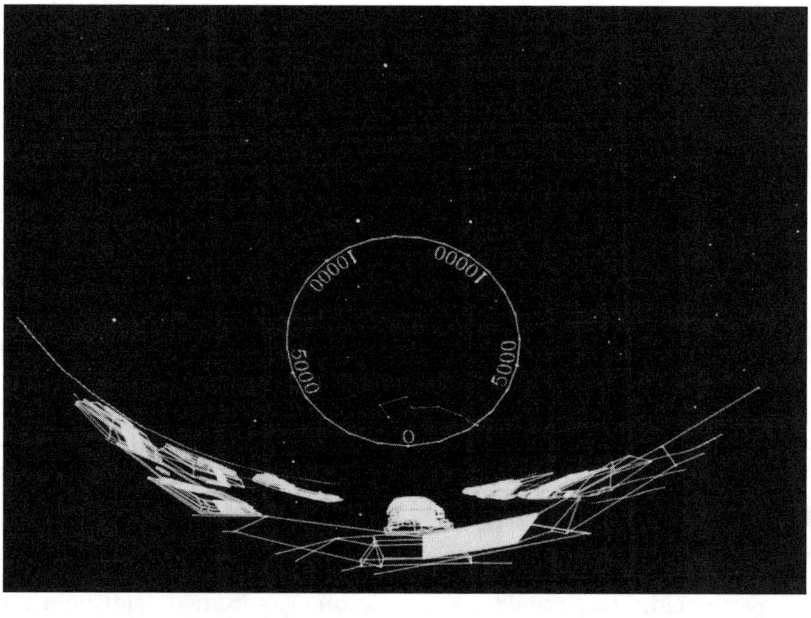

Figure 4. Precession of North Celestial Pole

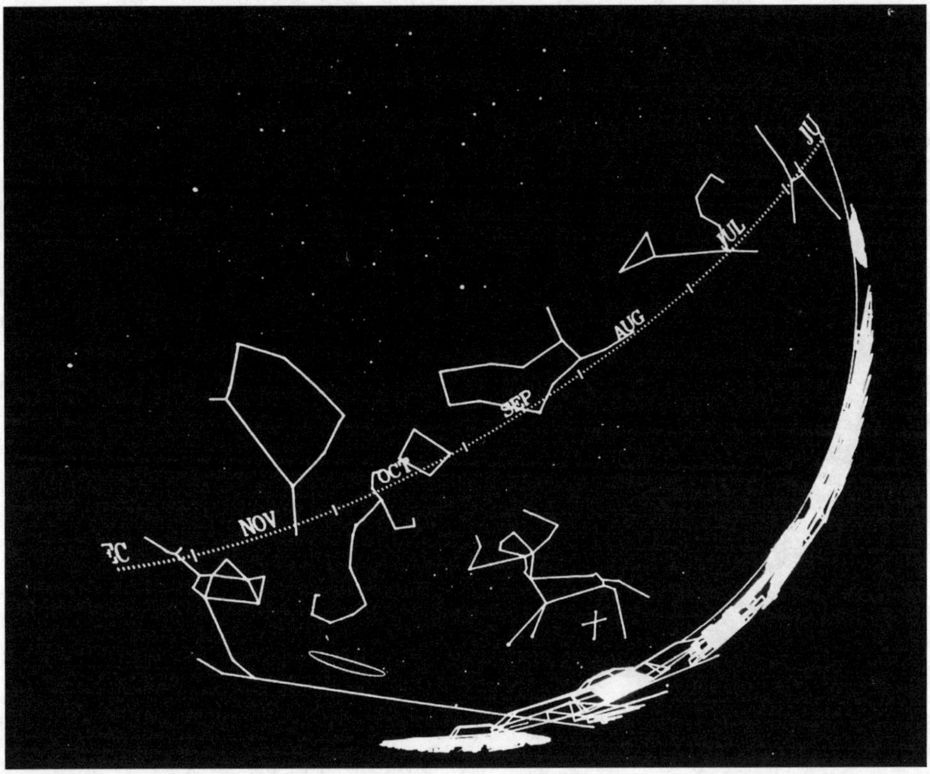

Figure 5. Centaurus and Crux over Building J.

difficult to observe these linked effects of precession on a limited flat window provided by a computer screen.

Integration of the Architectural Model and the Star Field

The advantage of the computer-generated planetarium star field is the flexibility with which the observer can modify the epoch and test alternative sight lines with just minor programming adjustments. More significantly this planetarium environment, complete with site wire frames, is the closest simulation of the real archaeological site at the time when it was constructed. Its 360 horizon provides more realism than any simulation displayed on a computer monitor.

Data Analysis

Evaluating Published Alignments

Inside the planetarium, the research team focused on viewing to the northeast and southwest over Building J. A sight line with the zenith tube on Building P was

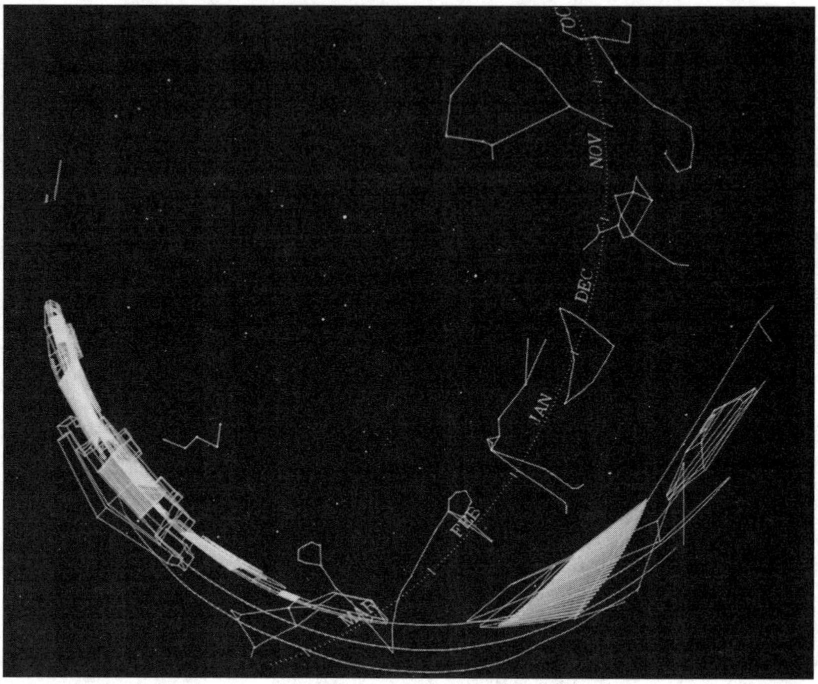

Figure 6. Cassiopeia rising over Building P from Building J as seen in the Planetarium with epoch set at 275 B.C.

used to extend the observational base line. As observed from Building P, Alpha and Beta Centauri and Crux set over the top of the Building J ruin and the loose alignment postulated by Aveni and Linsley was confirmed.

Observing from the center top of Building J over Building P led to a compelling alignment with Cassiopeia and a questionable alignment with Capella (Aveni 1984). This Cassiopeia alignment has been proposed independently by Robbins (Robbins 1991).

Adjustment of the date of construction and redefinition of the viewing angle increase the believability of the Capella alignment and make the Alpha and Beta Centauri and Crux alignments less compelling. Peeler and Winter (1995) recommend these modifications to improve the Capella alignment.

Identifying New Alignments

During data collection at Monte Alban, Sumners became fascinated with the orientation of Building H in the Main Plaza's central spine. There seemed to be a western-facing window in the west wall of this structure. From the proposed sighting tube on Building P, an observer would have a clear view through this window toward the western horizon. There is no structure along the western Main Plaza to obstruct this view. The sun would set through this window in late March and again in late September (as the window is currently reconstructed). The rainy season on Monte Alban occurs when the sun sets to the right (north) of this window. The ex-

Figure 7. Looking toward the proposed zenith tube opening in Building P.

istence of an equinox or near-equinox alignment at Monte Alban has been proposed independently by Robbins for the Danzantes building which lies along the western side of the Main Plaza (Robbins 1991).

Project Evaluation

The most effective role of the Digistar planetarium is in discovering an alignment and evaluating its likeliness based on observations of the annual and diurnal motions of the sky from the reconstructed archaeological site. Once a compelling alignment is observed, very accurate positional calculations can refine and confirm planetarium observations. Seeing the alignment happen in the planetarium places the significance of the event in context of the entire sky and the entire ruin.

The planetarium environment is also critical for participation of science teachers, students, and the general public in archaeoastronomical research. Complex positional calculations and an incomplete understanding of star field motions (especially precession) place archaeoastronomical research beyond the reach of most lay audiences. Yet the constellation patterns and archaeological sites are easily identified and alignments can be readily observed and evaluated in the planetarium.

In this project, two proposed alignments were supported and two new alignments were proposed. Both of the proposed alignments have been supported by independent research. As an additional benefit, the data base used in development of the planetarium display can be used to generate cardboard models of the ruin and

Figure 8. Looking through the "window" on the top of Building H from the vantage point of Building P.

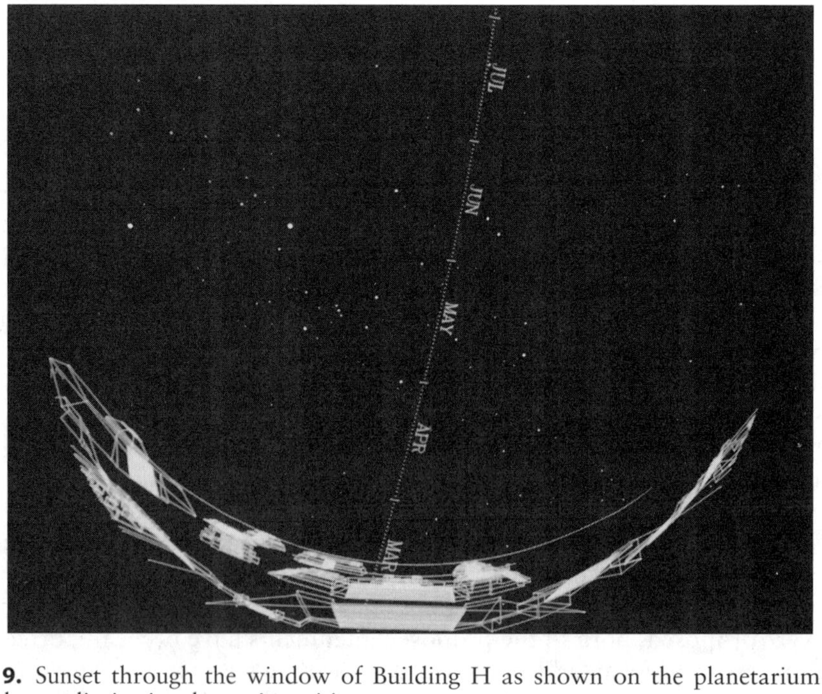

Figure 9. Sunset through the window of Building H as shown on the planetarium dome. Dates along ecliptic give the sun's position.

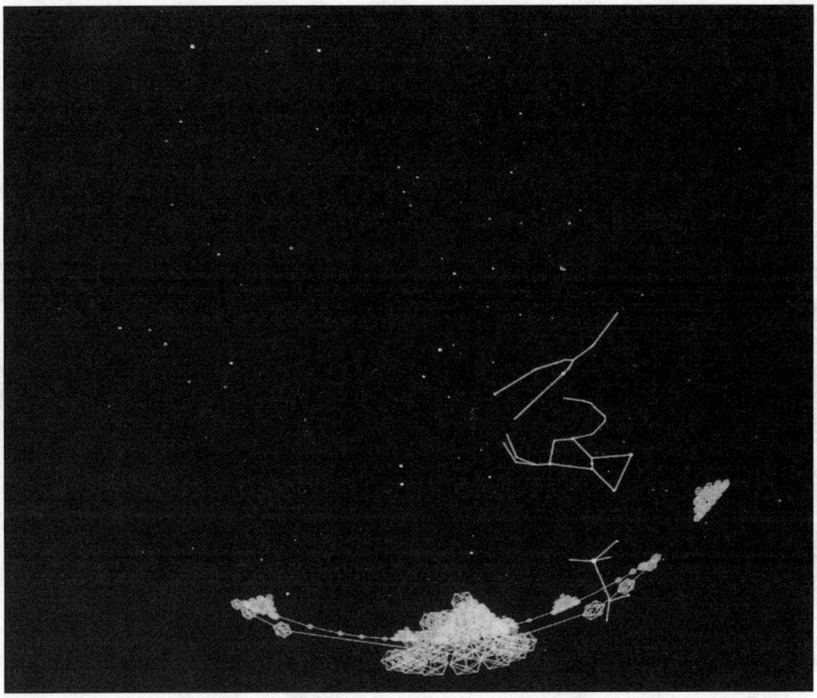

Figure 10. Orion region rising above the cairns of the Big Horn Medicine Wheel.

to aid in the research efforts of archaeoastronomers. The Building J positional data base has been sited in a recent publication (Peeler and Winter 1995).

Development of a Planetarium Program

The Monte Alban research produced an image base for a planetarium public program on archaeoastronomy. The resulting program (Astro Safari: In Search of the First Astronomers) uses Houston evening news anchor teams on simulated assignment in MesoAmerica, Africa, China, and Europe. Local radio broadcasters bring reports from other locations: South Pacific, Peru, Senegal, and Wyoming. AutoCAD files were created from existing image bases for the Big Horn Medicine Wheel, Teotihuacan (Sun and Moon Pyramids and the Street of the Dead), Chichen Itza (El Caracol, El Castillo, and the ball court), and the Great Pyramids at Giza. Other ruins were illustrated with videos and stills.

Audiences journey to each site at the time of its construction and make the astronomical observations that might once have regulated the secular and religious calendars of an ancient society. Up to 218 visitors can share these archaeoastronomy adventures in each planetarium program.

For live lectures, archaeoastronomical events can be simulated on command. During these presentations, a button box allows the planetarium lecturer to play written Digistar programs. A large monitor shows what is projected on the dome.

Figure 11. Planetarium operator selecting files from a button box. The projection is on the screen behind the operator and the lens of the Digistar projector is in the center of the theater, to the center right in the photograph.

The projector is under the fish-eye lens in the center of the planetarium chamber (seen just above the console in the photograph to the left).

This Astro Safari program has been used for 3.5 years as a public venue and as a school program. An accompanying curriculum carries the experience into the classroom. Admission covers all delivery costs and the program is self-funding.

Establishment of a Connection between Archaeoastronomy and Public Education

Each year the Houston Museum of Natural Science provides an expanded version of the Astro Safari program with accompanying curricula for over 10,000 seventh grade students in the Houston Independent School District. Several non-site-specific factors have contributed to this program's success:

- Research in astronomy concept development has shown that sun-moon-earth relationships are difficult to grasp with graduating college seniors having difficulty explaining lunar phases, eclipses, and seasons.
- A constructivist archaeoastronomical curriculum provides students with many different opportunities to construct and apply their understanding of the sun-moon-earth system as they interpret ancient architectural sites.

- Educators are searching for curricula that link science to the achievements of diverse cultures—especially those represented in the school population.
- Multi-cultural linkages allow team-teaching of science with social studies, mathematics, language arts, and studio art classes. Each of these factors contributes to a teacher's eagerness to use this program in the classroom.

Selection of a Target Student Population

The effective integration of archaeoastronomical curricula depends on a collaboration with the local educational community. Local school districts use national standards and benchmarks to address specific needs of their student population. A local scope and sequence for science and social studies will assist in identification of the most appropriate target grade level for inclusion of archaeoastronomy. For instance, in the United States, the sun-moon-earth system first appears in the third or fourth grade science curriculum and is revisited at sixth or seventh grade. High school earth science and physics courses may also discuss the topic in a more quantitative manner. For the reasons which follow, we decided that the most effective curriculum intervention would be in conjunction with the science program at the middle school level.

1. Student cognitive level: Archaeoastronomical investigations require students to change their reference frame as they consider how an astronomical event might have appeared in an archaeological context. This same conceptual ability is required to translate an observational event (like the changing azimuth of sunrise) to the orientation of the earth's rotation axis as the planet revolves around the sun. The ability to change reference frames is associated with formal operational thought as defined by Piaget. This cognitive process develops as students reach middle school.

2. Student experience base: Research in science education emphasizes the importance of spending time assisting students in the development of a few concepts rather than covering many concepts superficially. Students need time to learn how to translate the rotation and revolution of earth into changing orientations of objects in the sky. Unlike their rural grandparents, modern urban students do not observe changes in the position of the sun, moon or stars during the night or year. Rarely do they identify a constellation or a planet. A school project is often required to force students to leave their rich indoor environment long enough to observe changing moon phases, sunrise angles, or planet motions. The observational data base for interpreting motions of the sun-moon-earth system does not exist for most urban students. Archaeoastronomy can give students rich and exciting simulated experiences on which they can create and then test their understanding of the sun-moon-earth system.

3. Multi-cultural, multi-disciplinary focus: National standards and local objectives support the inclusion of science contributions from non-European civilizations to balance and enrich the science curriculum. The study of archaeoastronomy emphasizes the establishment of calendars and chronicling of changing sky positions by civilizations around the world. Table A below lists the different archaeological ruins used in our curriculum product, their location, and the type of student

construction activity associated with each. The presentation of linkages between science and other disciplines is also encouraged. In this project the links with world history and geometry are obvious. The middle school environment allows for more multi-disciplinary activity than at the high school level. Middle school science is also offered to all students and this program is designed to appeal to a diverse student population.

The Curriculum Product

The curriculum focuses on the construction of ruins with materials ranging from paper patterns to packing peanuts and aquarium gravel. The goal is to give students an awareness of how archaeological ruins actually looked. A decision was made to focus on observation of similar events at a variety of settings with a global distribution. Whenever possible, students would construct models of these sites using readily available materials. Table A lists the activities in the curriculum packet accompanying the Astro Safari.

Table A: Activities in Archaeoastronomy

Site or Concept Site/Culture	Curriculum Activity
Medicine Wheel Big Horn Mtn.	construction with gravel
Monte Alban Oaxaca, Mexico	paper construction of Building J
Ancient Cities Mexico	paper & pencil activity
Great Pyramids Giza, Egypt	paper construction of pyramids and sphinx
Egyptian temple Dendera	night observation of Sirius
Stonehenge England	model constructed with packing peanuts
Megalithic circles Senegal	measuring angles activity
Pacific Ocean Hawaii	observing Polaris and Arcturus
Zodiac Chaldea	determining sun signs
Zodiac China	determining birth year sign
Lunar calendars 3 cultures	matching calendars

Copies of this curriculum supplement and the Astro Safari planetarium script are available at the Museum's web site: http://hmns.org.

Acknowledgments

We would like to thank Ernesto Gonzalez Licon, Director, and Maria de los Angeles Romero, then interim Director of the Oaxaca Center of the Instituto Nacional de Antropologia e Historia (INAH) for their support of our data collection efforts. Marcus Winter of Centro INAH Oaxaca and Damon Peeler were critical in designing and implementing the project. Royce Nelson provided the laser transit and operated it expertly. Bill Bavinger, then of the RAVL at Rice University, assisted in identifying the critical laser transit measurements. Shisha VanHorn reduced all raw data into AutoCAD files and Ryan Wyatt transferred these images to the Digistar projection system. Science teachers Dan Kutso, Jane Tomsula, and Loretta Durst recorded all data and produced their own research results from the project. Gary Young supervised all photographic efforts. Jon Sumners and Royce Nelson Jr. carried transit rods all over the Monte Alban site. Patricia Reiff, Chairman of the Space Physics and Astronomy Department of Rice University, directed the National Science Foundation grant which funded teacher participation and publication of the curriculum materials. Barbara Foots, Director of Science for the Houston Independent School District, has provided unfailing support for implementation of this project in HISD.

References

Aveni, A. F. 1980, Skywatchers of Ancient Mexico (Austin: University of Texas Press).

Aveni, A. F. and Linsley, R. M. 1972, "Mound J, Monte Alban: Possible Astronomical Orientation" American Antiquity 37, No. 4.

Caso, A. 1938, "Exploraciones en Oaxaca" Quinta y Sexta Temporadas 1936–37. Publicacion No. 34. Instituto Paramericano de Geografia e Historia.

Marquina, I. 1964, Arquitectura Prehispanica (Mexico: Instituto Nacional de Antropologia e Historia).

Peeler, D. E. and Winter, M. 1995, "Building J At Monte Alban: A Correction and Reassessment of the Astronomical Hypothesis" Latin American Antiquity 6 No. 4, p.362–369).

Robbins, R. R. as reported by Tindol R. 1992, "Astronomer proves Mexican ruin was ancient solar observatory" On Campus, University of Texas.

Sumners C. 1992, "Astro Safari: Search for the First Astronomers" Houston Museum of Natural Science Planetarium Script.

Winter, M. 1989, Oaxaca, The Archaeological Record (Mexico: Minutiae Mexicana).

Appendix

Oxford V Conference
St. John's College
Santa Fe, New Mexico, August, 1996

Final Program

Saturday, Aug. 3

3-8 PM Registration

6:00 Buffet Dinner

8:30 Welcome (Peter Pesic, St. John's College)

8:45 Welcome (Oxford V Organizing Committee—the "Gang of Ten")

Poster group #1: Saturday-Tuesday

Ambruster, C. W. and T. Hull; Chamberlain, V. D., W. R. Latady, and T. Prince; Eddy, F. W., J. McK. Malville, and D. R. Parker; Freeman, K., R. Bliss, and J. Thompson; Hoskinson, T.; Lekson, S. H.; Malville, J. McK. and J. Jacobs; Mooney, R. M.; Munson, G. E. and J. McK. Malville; Olowin, R. P.; Robbins, R. R.; Roussel-Dupre, D.; Rudolph, J. H.; Schaefer, B. E.; Stekel, F. D.; Sullivan, W.; Teames, S.; Walton, J. and J. McK. Malville; Watson, R. P., J. McK. Malville, and GB Cornicopia; Westerman, J.

Sunday, Aug. 4
General Topics

Ray White, Presiding

8:15 AM Sinclair, R. M. Introduction

8:45 Gumerman, G. J. Future Planning

9:30 Schaafsma, P. If a Rose is a Rose is a Rose, Then is it True that a Star is a Star is a Star?

9:50 Gumerman, G. J. The Universe in a Cultural Context

10:30 Coffee

11:00 McCluskey, S. Different Astronomies, Different Cultures, and the Question of Cultural Relativism

11:20 Schaefer, B. E. Astronomy that Changed the World

11:50 Discussion

12:15 PM Lunch

General Topics (Continued)

Suzanne Chippindale, Presiding

1:30 PM Ulansey, D. The Cosmic Mysteries of Mithras

2:00 Peiser, B. J. Cultural Aspects of Astronomical Neo-Catastrophism

2:30 Masse, W. B. Sky as Environment: Archaeology and Temporary Celestial Events in the Ancient Southwest

3:00 Coffee

3:30 Discussion

4:00 Lekson, S. A History of the Southwest

5:00 Discussion

5:30 Dinner

6:00 Bar

Monday, Aug. 5
Mainly US Southwest

Mike Zeilik, Presiding

8:30 AM Chamberlain, V. D. Tracking Stars in Dinetah

9:10 Poss, R. L. Interpreting Rock Art: European and Anasazi Representations of Spirituality

9:30 Malville, J. M. and N. J. Malville. Pilgrimage and Astronomy at Chaco Canyon

9:50 Bates, B. C. A Cultural Interpretation of Astronomical Alignments: Crack'n Rock Community, Wupatki National Monument

10:10 Discussion

10:30 Coffee

11:00 Bostwick, T. W. The Shaw Butte Hilltop Site: a Prehistoric Hohokam Observatory

11:30 Discussion

12:15 PM Lunch

US; Hawaii; Crete

Von Del Chamberlain, Presiding

1:30 Preston, R. A. and A. L. Preston. Consistent Forms of Solstitial Interaction Between Sunlight Petroglyphs Throughout the Prehistoric American Southwest

1:50 Hollabaugh, M. Celestial Imagery in Lakota Culture

2:10 Lankford, G. E. Walking the Milky Way: the Astronomy of Death

2:30 Warther, F. X. Search for Unity. Kilo Lokahi. Astronomy and the Voyage of Goddess Pele in Ancient Hawai'i

2:50 Henriksson, G. and M. Blomberg. Evidence for Elements of Greek Astronomy and Religion in Minoan Crete

3:10 Discussion

3:30 Coffee

4:00 Poster Session #1.

5:30 Dinner

6:00 Bar

8:00 INVITED TALK:

"Light of the Dawn—Astronomical Legends of the Abenaki People" Gerard Tsonakwa (Tucson AZ)

Tuesday, Aug. 6
Outings:

Comanche Gap Petroglyph Field, Lamy NM

School of American Research, Santa Fe NM

5:00 PM INVITED TALK:

"Stardance"

John David Mooney (Chicago IL)

7:00 BBQ Dinner on Terrace at St. John's

Music by the "Red Hot Chili Peppers"

Poster group #2: Wednesday-Saturday

Achar, N. B. N.; Ammarell, G; Cairns, H. C.; Carlson, J. B.; Chiu, B. C.; Esteban Lopez, C.; Morante Lopez, R. B.; Nedialkov, P.; Penprase, B. E.; Simonia, I. and T. Simonia; Soltysiak, A. and A. Lebeuf; Sumners, C. and R. Wyatt; Sumners, C. and D. Iadevaia.

Wednesday, Aug. 7
World Archaeoastronomy

Peter Pesic, Presiding

8:30 AM Ruggles, C. Astronomy, Cosmology, and Stonehenge

9:10 Henriksson, G. Solar Eclipses, Supernovae, and Comets on Swedish Rock Carvings

9:30 Gardner, S. L. Scratching the Surface of Astronomy in the Land of the Bible

9:50 Krupp, E. C. Bedroom Politics and Celestial Sovereignty: The Cosmo-Political Function of Hittite Rock Art

10:10 Discussion

10:30 Coffee

11:00 Nedialkov, P. The Fiery Sword of St. Michael the Archangel in Russian Icons and Bright Comet Appearances: One Unique Icon

11:20 Gurshtein, A. A. The Evolution of the Zodiac in the Context of Ancient Egyptian History

11:40 Locher, K. The Decans of Ancient Egypt: Timekeepers for Worship, or Worshipped Beyond Time?

12:00 M Discussion

12:20 PM Lunch

Afternoon Free

Dinner: Hunt and Gather in Santa Fe

Thursday, Aug. 8
World Archaeoastronomy (Continued); Mesoamerica

Ed Krupp, Presiding

8:30 AM Nelson, S. M. and R. E. Stencel. Archeoastronomy as Part of a Core Curriculum

8:50 Snedegar, K. The Pleiades over Africa

9:10 Mannikka, E. The Role of Astronomy in the 12th Century Temple of Angkor Wat

9:30 Discussion

10:00 Coffee

10:30 Pankenier, D. W. Applied Astrology in Zhou China: the Battle of Chenpu (632 BCE)

10:50 Carlson, J. B. La Malinche and San Miguel: Sacrifice and Pilgrimage to the Mountain of Sustenance in Ancient Mexico

11:30 Discussion

12 M Lunch

Mesoamerica (Continued)

Kim Malville, Presiding

1:30 PM Fountain, J. W. A Database of Solar Markers

1:50 Macri, M. and D. Beattie. The Lunar Cycle and the Mesoamerican Counts of Twenty, Thirteen, Nine, and Seven

2:10 Brown, R. B. Mound of the Cross, Paquim, Casas Grandes, Chihuahua

2:30 Morante Lopez, R. B. Astronomical Observations at La Ciudadela, Teotihuacan, Mexico

2:50 Coffee

3:20 Milbrath, S. Jupiter in Classic and Postclassic Maya Art

3:40 Discussion

4:20 Poster Session #2

5:30 Dinner

6:00 Bar

8:00 PUBLIC LECTURE: jointly sponsored by the School of American Research and the Oxford V Conference

"Skywatchers, Shamans, and Kings"

E. C. Krupp (Director, Griffith Observatory & Planetarium, Los Angeles CA)

Friday, Aug. 9
South America; Spain

Steve McCluskey, Presiding

8:30 AM MacKie, E. The Chambered Tomb of Maes Howe, Orkney Island, and the Winter Solstice

8:55 Pitluga, P. B. An Analysis of The Nazca Spirals

9:20 Zuidema, R. T. The Astronomical Significance of Processions and Pilgrimages in the Calendrical Organization of Cuzco, Peru

9:45 Dearborn, D. S. P. Where the Sun Returns to Earth

10:05 Discussion

10:25 Coffee

11:00 Frank, R. Hunting the European Sky Bears: a Proto-European Vision Quest to the End of the Earth

11:30 Esteban, C. and J. A. Belmonte Aviles. Archeoastronomical Fieldwork in the Canary Islands: Overview and Implications

12:00 M Thomson, M. F. Megalithic Sites in Malta

12:05 Discussion

12:20 PM Lunch

Conference Summary

Rolf Sinclair, Presiding

1:30 Business

3:00 Coffee

3:30 Rapporteur #1: R. M. Sinclair

4:00 Rapporteur #2: E. C. Krupp

4:30 Conclusion of Oxford V

5:00 Meeting ends

5:30 Dinner

6:30 Bar

Saturday, Aug. 10

7:00 AM Breakfast.

8:00 Leave St. John's. Start of Chaco Canyon Trip

Monday, Aug. 12

Return from Chaco Canyon Trip
6:00 PM Dinner at St. John's

Tuesday, Aug. 13

7:00 AM Breakfast. Leave St. John's
Sayonara!

Oxford V Conference
St. John's College
Santa Fe, New Mexico, August 1996

Attendees

Dr. B.N. Narahari Achar
University of Memphis

Dr. Carol Ambruster
Villanova University

Dr. Gene Ammarell
Ohio University

Mr. Bryan Bates
Coconino Community College

Mr. David Begay
Navajo Community College

Dr. Mary Blomberg
Uppsala University, Sweden

Mr. Todd W. Bostwick
Pueblo Grande Museum

Dr. R. Ben Brown
Chihuahua, Chih., Mexico

Dr. Hugh Cairns
Darling Point, N.S.W., Australia

Dr. John B. Carlson
College Park MD

Dr. Von Del Chamberlain
Salt Lake City UT

Dr. Carlson R. Chambliss
Kutztown University

Mrs. Suzanne P. Chippindale
Santa Fe Community College

Dr. Bella C. Chiu
Cambridge, MA

Mr. GB Cornucopia
Chaco Culture National Historical Park

Srta. Adriana Davila
Pontifica Universidad Catolica del Peru

Dr. David S.P. Dearborn
Lawrence Livermore National Laboratory

Dr. César Esteban Lopez
Instituto de Astrofisica de Canarias, Spain

Mr. Victor B. Fisher
Towson State University

Prof. Maelee Thomson Foster
University of Florida

Mr. John Fountain
Tucson AZ

Prof. Roslyn Frank
University of Iowa

Ms. Katherine Freeman
Farmington NM

Dr. Sara L. Gardner
Tucson AZ

Dr. Alan Gillespie
University of Washington

Dr. George J. Gumerman
Santa Fe Institute

Dr. Alex A. Gurshtein
Mesa State College

Mr. O.W. (Bud) Hampton
College Station TX

Ms. Carol Harper
Bonny Doon CA

Dr. Goran Henriksson
Astronomiska Observatoriet Uppsala, Sweden

Dr. Mark Hollabaugh
St. Paul MN

Mr. Tom Hoskinson
1000 Oaks CA

Mr. Tony Hull
OCA Applied Optics

Mr. John Jacobs
Kirkland WA

Dr. Beth Jewell
Villanova University

Dr. Edwin C. Krupp
Griffith Observatory

Mr. George E. Lankford
Batesville AR

Dr. Steven H. Lekson
University of Colorado at Boulder

Dr. Irene Little-Marenin
Wellesley College

Dr. Steven J. Little
Glen Haven CO

Prof. Kurt Locher
Gruet, Switzerland

Dr. Euan MacKie
The Hunterian Museum, United Kingdom

Dr. Martha J. Macri
University of California—Davis

Dr. J. McKim Malville
University of Colorado at Boulder

Dr. Nancy Malville
University of Colorado at Boulder

Dr. Eleanor Mannikka
University of Michigan

Ms. Nancy Maryboy
Navajo Community College

Dr. W. Bruce Masse
Los Alamos National Laboratory

Professor Steven McCluskey
West Virginia University

Ms. Kim McLean
Chaco Culture National Historical Park

Dr. Susan Milbrath
Florida Museum of Natural History

Mr. John David Mooney
Chicago IL

Mr. Richard M. Mooney
Soddy Daisy TN

Dr. Ruben Morante-Lopez
Orizaba, Ver., Mexico

Ms. Abbie Morris
New York City NY

Mr. Clifford H. Morse, A.I.A.
Kensington CA

Mr. Greg Munson
Mesa Verde CO

Mr. Petko L. Nedialkov
Sofia University, Bulgaria

Dr. Sarah M. Nelson
University of Denver

Ms. Dawn Offenbecker
Stony Brook NY

Dr. Ronald P. Olowin
St. Mary's College of California

Mr. Joseph Orr
Santa Fe NM

Dr. David W. Pankenier
Lehigh University

Dr. Benny J. Peiser
Liverpool John Moores University, United Kingdom

Mr. Bryan Penprase
Pomona College

Dr. Peter Pesic
St. John's College—Santa Fe

Ms. Phyllis B. Pitluga
Adler Planetarium & Astronomy Museum

Mr. Stanley B. Plum
Phoenix AZ

Jon R. Polansky, M.D.
Mill Valley CA

Dr. Richard L. Poss
University of Arizona

Dr. Robert Preston
Jet Propulsion Laboratory

Dr. R. Robert Robbins
University of Texas at Austin

Dr. Diane Roussel-Dupre
Los Alamos National Laboratory

Dr. Marvin Rowe
Texas A&M University

Mr. John H. Rudolph
Bainbridge Island, WA

Dr. Clive L.N. Ruggles
Leicester University, United Kingdom

Dr. Polly Schaafsma
Santa Fe NM

Dr. Bradley E. Schaefer
University of Texas at Austin

Ms. Christine M. Schooley
TREX—Museum of New Mexico

Dr. Rolf M. Sinclair
National Science Foundation

Dr. Keith Snedegar
Utah Valley State College

Mr. Arkadiusz Soltysiak
Warsaw University, Poland

Prof. Frank D. Stekel
University of Wisconsin-Whitewater

Dr. Karen M. Strom
University of Massachusetts

Dr. William F. Sullivan
Orange MA

Dr. Carolyn Sumners
Houston Museum of Natural Science

Ms. Sallie Teames
Hurst TX

Gerard Tsonakwa
Tucson, AZ

Dr. David Ulansey
Berkeley, CA

James Walton
Boulder CO

Dr. Miranda Warburton
Navajo Nation Archaeology Department

Mr. Francis X. Warther
Killauea/Kaua'i HI

Dr. Dorothy Washburn
Maryland Institute, College of Art

Mr. Jim Westerman
Chicago IL

Professor Raymond E. White
University of Arizona

Mrs. Ruby White
Tucson AZ

Professor Michael Zeilik
University of New Mexico

Professor R. Tom Zuidema
Urbana IL